Thermodynamic Data for Biochemistry and Biotechnology

Edited by Hans-Jürgen Hinz

With 55 Figures
and 132 Tables

Springer-Verlag
Berlin Heidelberg New York Tokyo

Professor Dr. HANS-JÜRGEN HINZ
Universität Regensburg
Institut für Biophysik und Physikalische Biochemie
Universitätsstraße 31
8400 Regensburg
FRG

ISBN 3-540-16368-9 Springer-Verlag Berlin Heidelberg New York Tokyo
ISBN 0-387-16368-9 Springer-Verlag New York Heidelberg Berlin Tokyo

Library of Congress Cataloging in Publication Data. Thermodynamic data for
biochemistry and biotechnology. Bibliography: p.   . Includes index. 1. Thermodynamics. 2. Biomolecules. 3. Biological chemistry. 4. Biotechnology. I. Hinz,
Hans-Jürgen, 1942–   . QP517.T48T47 1986   574.19′283   86-1812

Typesetting, printing and bookbinding: Brühlsche Universitätsdruckerei, Giessen
2131/3130-543210

Preface

The strong trend in the Biological Sciences towards a quantitative characterization of processes has promoted an increased use of thermodynamic reasoning. This development arises not only from the well-known power of thermodynamics to predict the direction of chemical change, but also from the realization that knowledge of quantitative thermodynamic parameters provides a deeper understanding of many biochemical problems.

The present treatise is concerned primarily with building up a reliable data base, particularly of biothermodynamic and related quantities, such as partial specific volumes and compressibilities, which will help scientists in basic and applied research to choose correct data in a special field that may not be their own. Most chapters reflect this emphasis on data provision. However, it was also felt that the expert user deserved information on the basic methodology of data acquisition and on the criteria of data selection. Therefore all tables are preceded by a critical evaluation of the techniques as well as a survey of the pertinent studies in the corresponding areas. The surveys are usually self-consistent and provide references to further sources of data that are important but not covered in the present volume.

The reader will realize that in different chapters, different symbols have been used for the same properties. This unfortunate situation is particularly obvious in those chapters where partial specific or molar quantities had to be introduced; however, it also occurs in those contributions concerning phase changes of macromolecules. The editor saw no possibility of unifying the nomenclature in all chapters because there exist no universally accepted rules for classification of biothermodynamic quantities. Furthermore, a unified nomenclature created for the book, if introduced to the unprepared user, would greatly reduce its usefulness, since the reader would find himself confronted with symbols different from those used in the original literature. Such a solution would introduce new problems rather than solve the existing ones. As a compromise, each chapter is therefore preceded by a list of symbols which are usually those used in the corresponding article. These symbols are employed by the majority of authors of the original publications. If there are slight deviations for the sake of consistency, these new symbols have been properly defined. The reader should, however, be aware of the ambiguity that arises where the same quantity is referred to by different symbols in different chapters.

The editor sincerely hopes that in the future scientists will convert to a unified nomenclature to facilitate identification and proper use of data. First steps are being made in a forthcoming suggestion for nomenclature approved by IUB, IUPAB and IUPAC.

It is almost unavoidable that, despite all efforts, some errors have remained undetected. Editor and publisher will appreciate very much if such unfortunate mistakes are brought to their attention.

The Editor

Acknowledgments. There are many people I would like to thank for their help in the realization of this volume.

The original idea of providing a critical selection of chemical thermodynamic data for the biological sciences stemmed from the late GEORGE T. ARMSTRONG, with whom I had the privilege and pleasure of cooperating in the first stages of the book. Further development of the material would have been unthinkable without the many valuable and critical suggestions of the referees and authors. Therefore the present compilation is entirely the result of the gratifying international cooperation with the authors and referees, whose devoted engagement and authoritative expertise rendered the editor's job enjoyable.

I would particularly like to acknowledge the fruitful discussions with R. BILTONEN, H. EISENBERG, F. FRANKS, S.J. GILL, H.-D. LÜDEMANN, N. PRICE, P. PRIVALOV, and I. WADSÖ, who have given me every possible support, encouragement and reassuring confidence. Finally I want to thank my collaborator Mrs. R. WEIGERT for her tireless efforts and almost unlimited patience in accurately typing several revisions of various manuscripts.

Contents

Section III   Interactions in Solution

Contents   XIII

Referees

Chapter 2: W. Doster   *Chapter 10:* S. J. Gill
Chapter 3: H. Eisenberg   *Chapter 11:* J. T. Edsall
Chapter 4: W. Doster   *Chapter 12:* G. Olofsson
Chapter 5: P. L. Privalov   *Chapter 13:* J. Brandts
Chapter 6: F. Franks   *Chapter 14:* H. Klump
Chapter 7: M. R. Eftink   *Chapter 15:* H. Klump
Chapter 8: W. Pfeil   *Chapter 16:* S. Miller
Chapter 9: S. J. Gill   N. Price

Addresses of Authors and Referees

Prof. Dr. GEORGE BARISAS, Department of Chemistry, Colorado State University, Fort Collins, Colorado 80523, USA

Dr. I. V. BERENZIN, Thermochemical Laboratory, Department of Chemistry, Lomonosov State University, Moscow 117234, USSR

Prof. Dr. JOHN BRANDTS, Department of Chemistry, University of Massachusetts, Amherst, Massachusetts, USA

Prof. Dr. KENNETH BRESLAUER, Department of Chemistry, Douglass College, Rutgers University, New Brunswick, New Jersey 08093, USA

Prof. Dr. SERGIO CABANI, Istituto di Chimica Fisica, Università di Pisa, Italy

Prof. Dr. ATTILIO CESÀRO, Istituto di Chimica, Università di Trieste, P. Le Europa, 34127 Trieste, Italy

Dr. WOLFGANG DOSTER, Physik Department E 13, Technische Universität München, 8046 Garching, FRG

Dr. HELMUT DURCHSCHLAG, Institut für Biophysik und Physikalische Biochemie der Universität Regensburg, Universitätsstraße 31, 8400 Regensburg, FRG

Prof. Dr. JOHN T. EDSALL, Harvard University, The Biological Laboratories, 7 Divinity Avenue, Cambridge, MA 02138, USA

Prof. Dr. MAURICE R. EFTINK, Dept. Chemistry, College of Liberal Arts, University of Mississippi, MS 38677, USA

Dr. A. M. EGOROV, Thermochemical Laboratory, Department of Chemistry. Lomonosov State University, Moscow 117234, USSR

Prof. Dr. HENRYK EISENBERG, Polymer Department, Weizman Institute of Science, Rehovot, Israel

Dr. VLADIMIR FILIMONOV, Institute of Protein Research, Academy of Sciences of the USSR, 142292 Poustchino, Moscow Region, USSR

Prof. Dr. FELIX FRANKS, University of Cambridge, Dept. of Botany, Botany School, Downing Street, Cambridge CB2 3ER, England

Dr. G. L. GALCHENKO, Thermochemical Laboratory, Department of Chemistry, Lomonosov State University, Moscow 117234, USSR

Prof. Dr. PAOLO GIANNI, Istituto di Chimica Fisica, Università di Pisa, Italy

Prof. Dr. STANLEY J. GILL, Department of Chemistry, University of Colorado, Boulder, Colorado 80306, USA

Prof. Dr. HANS-JÜRGEN HINZ, Institut für Biophysik und Physikalische Biochemie der Universität Regensburg, Universitätsstraße 31, 8400 Regensburg, FRG

Prof. Dr. HEINZ HOFFMANN, Lehrstuhl für Physikalische Chemie, Universität Bayreuth, Universitätsstraße 30, 8580 Bayreuth, FRG

Prof. Dr. HARALD HØILAND, Kjemisk Institutt, Universitetet i Bergen, 5014 Bergen, Norway

Prof. Dr. HORST KLUMP, Institut für Physikalische Chemie der Universität Freiburg, Hebelstraße 38, 7800 Freiburg, FRG

Dr. MADELEINE LÜSCHER-MATTLI, Institut für Anorganische, Analytische und Physikalische Chemie der Universität Bern, Freiestraße 3, CH-3000 Bern 9, Switzerland

Prof. Dr. STANLEY MILLER, Department of Chemistry, B-017, University of California, San Diego, La Jolla, California 92093, USA

Dr. GEORGE MREVLISHVILI, Dept. of Physics of Biopolymers, Academy of Sciences of the Georgian SSR, Tbilisi 380077, USSR

Dr. GERD OLOFSSON, Thermochemistry Chemical Center, P.O.B. 740, 22007 Lund 7S, Sweden

Dr. WOLFGANG PFEIL, Akademie der Wissenschaften der DDR, Zentralinstitut für Molekularbiologie, Abt. Biokatalyse, Lindenberger Weg 70, 1115 Berlin Buch, DDR

Prof. Dr. N. PRICE, Department of Biological Science, University of Stirling, Stirling FK9 4LA, Scotland, Great Britain

Prof. Dr. PETER L. PRIVALOV, Institute of Protein Research, Academy of Sciences of the USSR, Poustchino, Moscow Region, USSR

Dr. M. V. REKHARSKY, Thermochemical Laboratory, Department of Chemistry, Lomonosov State University, Moscow 117234, USSR

Prof. Dr. PHILIP D. ROSS, Laboratory of Molecular Biology, National Institutes of Health, Bethesda, Maryland 20205, USA

Dr. WERNER ULBRICHT, Lehrstuhl für Physikalische Chemie, Universität Bayreuth, Universitätsstraße 30, 8580 Bayreuth, FRG

Dr. HEINRICH WIESINGER, Institut für Physiol. Chemie, Hoppe-Seyler-Straße 1, 7400 Tübingen, FRG

Section I **Introduction**

Chapter 1 Present and Future Uses and a Bit of History

JOHN T. EDSALL

Workers in biochemistry and biophysics frequently have need for thermodynamic data relating to molecules, both small and large, and to the equilibrium properties of systems involving such molecules. Knowledge of the partial specific or molar volumes, heat capacities, and compressibilities of molecules of biological interest is frequently important, both in the planning of experiments and in understanding molecular interactions. The characterization of reversible conformational transitions, in nucleic acids, proteins, and other molecules, is a central problem today, whether those transitions are associated with ligand binding, with temperature or pressure changes, or with various other conditions. Thermodynamics is often relevant to kentic studies, as in many, perhaps most, enzyme-catalyzed reactions, for which certain intermediate steps in the total process may be essentially at equilibrium at any moment. The special thermodynamic properties of water and aqueous solutions, as they relate to ions, and to polar and nonpolar molecules, are naturally of supreme importance in biology.

A vast amount of information is now available in the scientific literature, relating to these and other thermodynamic properties of systems of biochemical interest. The investigator who needs such data for current work often has difficulty in finding the particular information that is needed, or in judging which of the reported values are the most trustworthy. It is the aim of the present volume to present selected and critically evaluated data for many of the important fields of research in biochemistry and biophysics. Those who have compiled and evaluated the data in this book are all experts in their respective fields, and their labors should save other workers much time and trouble by guiding them to the best available data.

This book represents one important product of the work of the Interunion Commission on Biothermodynamics, supported by the International Union of Pure and Applied Chemistry (IUPAC), the International Union of Biochemistry (IUB) and the International Union of Pure and Applied Biophysics (IUPAB), and wisely guided by its Chairman, Professor Ingemar Wadsö of Lund. This book has been in preparation for a number of years, first under the direction of the late Dr. George T. Armstrong, of the United States National Bureau of Standards. His admirable guidance of the project was ended by his untimely death, and it is to the devoted work of Dr. Hinz and his collaborators that we owe the completion of this project.

1 Some Early History

The relation of thermodynamics to biochemistry and biophysics reaches back to its beginnings. Several of the great pioneers in the study of energy transformations were much concerned with the study of living organisms. Lavoisier was deeply interested in the study of respiration. In his calorimetric work with Laplace (1785), they measured the heat produced by a guinea pig in relation to its carbon dioxide production, comparing it with the heat of formation of the same amount of carbon dioxide produced by burning carbon in an oxygen atmosphere, and found some discrepancies. A few years later, working with Armand Seguin (1789–91), he carried the work further, recognizing that oxygen consumption increased with rising temperature and during digestion and exercise [1].

Later two medical men played a central part in developing the principle of the conservation of energy. The physician Julius Robert Mayer stated the concept clearly, though it was J. P. Joule who independently provided the decisive experimental evidence. Hermann von Helmholtz, who was trained in medicine and had served as an army surgeon, presented a mathematical analysis of the concept of conservation, which was highly influential. Helmholtz, who was both a great physiologist and a great physicist, was probably the greatest biophysicist that the world has yet seen.

The calorimetric study of whole organisms, in its relation to their total metabolism, was assiduously pursued by Carl Voit, Max Rubner, Wilbur O. Atwater, Francis G. Benedict, Graham Lusk, and numerous others throughout the late nineteenth and early twentieth centuries. Their work led, among many other findings, to the firm conclusion that the first law of thermodynamics was valid for living organisms as for the inorganic world; a conclusion now universally accepted.

Here we are concerned, however, not so much with whole organisms as with defined systems of biochemical importance. It was Willard Gibbs, in 1876–78, (in work now more readily acessible in reference [2]), who laid the foundations for the thermodynamic study of multicomponent systems, although he himself made no mention of applications to biochemical systems. The work of Lord Kelvin and Rudolf Clausius had already clearly established the Second Law, and Clausius had defined and named the concept of entropy. Gibbs started from there, formulating the concept of chemical potential for the components of any system, and defining the system in terms of the relative masses and the potentials of its components, among the different phases. The fundamental requirement for equilibrium was that the potential of any component that was free to move between one phase and another must be the same in all phases. The phase rule was only one of the many consequences that flowed from these fundamental assumptions. Gibbs was the first to develop those fundamental functions that we now know as enthalpy, Helmholtz energy, and Gibbs energy, and to show the power of their use in the study of thermodynamic systems. In one brief passage he stated the basic equations for what later became known as the Donnan equilibrium, since F. G. Donnas was to formulate the same relations, independently and in different

---

[1] For a brief account of Lavoisier's work, with further references, see for instance Guerlac [1] and especially the longer work of Holmes [1 a].

form, 35 years later. Gibbs also devoted much of his great memoir to surface phenomena, with results that were fundamental for later work.

J.C. Maxwell and other contemporary physicists recognized the greatness of the work of Gibbs, but scarcely any chemists at that time could assimilate his condensed and abstract presentation and apply his concepts to actual experimental systems. The concept of chemical potential must, in those days, have been difficult to grasp. Recognition of Gibbs by chemists came earlier in Europe than in America. J.H. Van't Hoff, Wilhelm Ostwald, and Svante Arrhenius were to lead in the development of the new physical chemistry, and Gibbs influenced them all.

Van't Hoff's treatment of osmotic pressure was a striking example of the influence of biology on the advancement of physical chemistry [3]. The osmotic studies of the botanist Wilhelm Pfeffer, arising from his work on the flow of sap in plants, provided Van't Hoff with the experimental data that led him to his limiting law of osmotic pressure in dilute solutions. By showing the equivalence between osmotic pressure and freezing point depression or boiling point elevation in solution, he provided a powerful tool for determination of molecular weights, with compelling evidence, for salts, acids and bases, of Arrhenius's concept that these substances existed largely as free ions in aqueous solution.

Gilbert Newton Lewis, more than any other one person, translated the concepts of Gibbs in terms that made them readily applicable to the analysis of experimental results. In particular his use of activities and activity coefficients for components of nonideal solutions – the activity of a component bearing an exponential relation to its chemical potential – proved highly useful. He initiated a far-reaching experimental program for the determination of the thermodynamic properties of a wide variety of systems, and his treatise on *Thermodynamics* with Randall [4] had great influence on the younger generation of chemists and biochemists in the early twentieth century.

2 Acids, Bases, and Oxidation-Reduction Equilibria

Extensive thermodynamic data concerning the strength of acids and bases were being collected by the early twentieth century, especially in the laboratories of Wilhelm Ostwald and his school. A landmark for biochemists was the formulation of the principle of buffer action by L.J. Henderson in 1908 [5], which was soon followed by the great work of S.P.L. Sørensen in formulating the concept of pH and setting forth in detail the procedures for its measurement [6]. This was an essential preliminary to Sørensen's fundamental work on the physical chemistry of proteins, in which he treated them as definite chemical compounds, and defined their solubilities in terms of the phase rule, in systems of specified pH, salt concentration, and temperature, and studied their molecular weights by osmotic pressure measurements [7].

Closely related to the studies of acid-base equilibria was the problem of oxidation-reduction potentials. Work in this area for inorganic systems had been pursued since the late nineteenth century; but the first systematic studies of such potentials in organic compounds began about 1920 in the laboratory of W.M.

Clark, whose admirable book, published 40 years later, gave a lucid survey of the field with a critical evaluation of the experimental studies then available [8]. It was Clark who first formulated the interrelations between oxidation-reduction potentials and acid-base equilibria in the same compounds. He set high standards for care and skill in experimental procedures. His early studies involved synthetic dyes, rather than substances of biochemical origin, but their impact on biochemists was great. Among the earliest studies of biochemical compounds was the work of J. B. Conant on the hemoglobin-methemoglobin system [9, 10], which not only established the potential of the system but made clear the distinction between oxygenation and oxidation of hemoglobin.

Many biological oxidation-reduction systems are sluggish and require the presence of a catalyst, which usually is an enzyme, to attain a stable equilibrium potential. A crucial example in early work was the succinate-fumarate system, which readily established an equilibrium in the presence of succinic dehydrogenase, but not without the enzyme. The work of Lehmann [11] and of Borsook and Schott [12] was particularly important, for it clearly showed that the same redox potential was established under given conditions, for systems catalyzed by any one of a variety of succinic dehydrogenases from different sources. The enzyme, in other words, functioned as a perfect catalyst. Borsook and Schott also calculated the potential of the system from the third law of thermodynamics. They made use of heat capacity and other thermal measurements on succinic and fumaric acids, together with the Gibbs energy changes on solution and ionization of the acids in water, and showed that the resulting calculated potentials agreed with the experimental values within very narrow limits of error. This was only one aspect of a comprehensive program carried on by Borsook, H. M. Huffman, and others to determine enthalpies, entropies, and Gibbs energies of formation for a large series of important biochemical compounds. This work continued for about a decade.

All of Clark's early studies dealt with compounds in which two electrons appeared to be transferred simultaneously between oxidant and reductant. Many workers believed for a time that this was characteristic of such reactions in organic compounds generally. In 1930, B. Elema in Holland and L. Michaelis at the Rockefeller Institute reported independently on systems in which electron transfer took place in two steps, with formation of an intermediate free radical (a semiquinone). During the next decade Michaelis, already eminent for his work in biophysical chemistry, systematically developed the study of semiquinones, gradually overcoming the initially widespread disbelief that such free radicals could really exist [13]. Their importance in biochemical systems has become abundantly clear in subsequent years. More recently, Walz [14] has given a valuable discussion of the thermodynamics of oxidation-reduction reactions in their relation to bioenergetics, which relates the subject closely to modern biophysics and biochemistry.

3 Thermodynamics of Major Metabolic Processes

Otto Meyerhof [15, 16] was a biochemist of unusual intellectual breadth, with a deep interest in physics and philosophy as well as in the biomedical sciences. His early close association with Otto Warburg, and later with A. V. Hill, directed him into the application of thermodynamics to biochemical systems. The long series of researches in his laboratory on glycolysis and the biochemistry of muscle involved the quantitative study of heats of reaction and equilibrium constants of most of the intermediate reactions in these fundamental processes. The pivotal role of ATP in biochemical processes became clearly apparent in the course of these investigations. Fritz Lipmann, who had worked for some years in Meyerhof's laboratory, was later to bring all this and other related work into focus in his immensely influential review on phosphate bond energy in biochemical processes [17]. Although his terminology was somewhat different from that of the physical chemists, the fundamental relations that he presented have greatly influenced all further research in biothermodynamics. At about the same time, another important review by Kalckar [18] illuminated many of the same problems, from a different point of view.

Sixteen years later, H. A. Krebs and H. L. Kornberg [19] gave a comprehensive survey of the energy transformations in living matter, with detailed discussion of the Gibbs energy changes involved in a wide range of key processes of intermediary metabolism. An appendix by K. Burton tabulated the Gibbs energies of formation of a large number of key metabolites, critically evaluated from the best available data, in a series of five comprehensive tables. This work still stands as a major source of reliable data for biothermodynamics. Since that time, of course, more work in this field has been reported, and we may look forward in future to another such critical survey of the whole field.

The most radical change in outlook since the review by Krebs and Kornberg is certainly the introduction by Peter Mitchell of the concept that in active cells and tissues "...chemical reactions, like osmotic reactions, are transport processes when looked at in detail," ([20], p. 47) and that they represent vectorial diffusion reactions in ways earlier unrecognized. Mitchell's concepts have not yet been applied in sufficient detail to supply quantitative data for a book like this one, but they have already had a far-reaching effect in serving to unify the diverse views of biochemists and biophysicists.

4 Thermodynamics of Amino Acids and Proteins: Electrostatic and Hydrophobic Interactions

Edwin J. Cohn, at Harvard Medical School, beginning shortly after 1920, carried further the work on protein solubilities to which he had been exposed in the laboratory of S. P. L. Sørensen. With constant advice from his close friend George Scatchard, Cohn realized early the similarity of the effects of salts on protein solubility to the interactions of simple salts with one another, and with small organic molecules, as described by the electrostatic theories of Debye and Hückel. This led to a long series of researches on amino acids, peptides, and other related small molecules, involving Scatchard, J. G. Kirkwood, T. L. McMeekin, J. Wyman, J. P. Greenstein, myself, and others. They involved the influence of electric charge interactions, and of nonpolar groups, for molecules in water and other polar solvents, with respect to solubilities, partial molar volumes, heat capacities, and compressibilities, and emphasized the importance of what were later to be termed hydrophobic interactions [21]. J. A. V. Butler in England also made major contributions to the study of hydrophobicity, and recognized the unusual thermodynamic properties of aqueous solutions of compounds containing nonpolar groups. Much that then seemed mysterious was later to be importantly, though only partially, clarified by the work of Frank [22] and especially of Kauzmann [23]. The enormous outpouring of studies on hydrophobic interactions in recent years has led to important advances, but the molecular interpretation of the observed thermodynamic properties of such systems is still highly debatable (see for instance [24]).

5 Thermodynamics of Ligand Binding to Macromolecules

The reversible binding of ligands to macromolecules is of major concern to biochemists and biophysicists. The ligands may be protons, for acid-base equilibria; they may be metallic ions, or other ions such as the anions, both inorganic and organic, that bind to serum albumin. They may be very simple molecules, such as oxygen or carbon monoxide, which bind to hemoglobin and myoglobin. For biological macromolecules several features of special importance are that: (1) many ligand molecules of a given kind may be bound by a single macromolecule (2) the binding of successive ligands may be cooperative, or anti-cooperative, or they may bind independently of one another; (3) there is often interaction between the binding of different kinds of ligands, as in the Bohr effect in hemoglobin, where the binding of protons promotes release of oxygen, and vice versa. (4) There may be important conformational transitions in the macromolecule, associated with ligand binding. Here again the conformation change in hemoglobin, associated with binding of oxygen or carbon monoxide, is a prime example. Another example is the effect of proton binding on the reversible unfolding of protein molecules that generally occurs in acid solution.

All these effects give rise to complexities that go far beyond the simpler types of mass action equilibria to which chemists had become accustomed. We may note two major advances in understanding of such phenomena, that occurred some 60 years ago. One was the formulation by Linderstrøm-Lang of the effect of electrostatic forces on ion binding by proteins [25], in which he skilfully applied the interionic attraction theory of Debye and Hückel to a spherical model of a protein. The other was the formulation by Adair [26] of the equation for multiple binding of the four ligands that could attach themselves to the heme iron of hemoglobin. The equation could be immediately generalized for n identical ligands.

Other important advances, in the treatment of multiple ligand binding, followed; the work of George Scatchard calls for particular mention, for cases in which the number of binding sites in the macromolecule, as well as their binding constants, are unknowns that have to be determined [27]. Countless subsequent papers have made use of Scatchard's approach.

The greatest advances in the treatment of multiple ligand binding, however, are those due to Jeffnies Wyman. Wyman's general treatment of linkage phenomena derives directly from the fundamental treatment of Gibbs, but it is developed in subtle and powerful detail, such as no previous author had achieved, both for homotropic interactions, as in the successive binding of oxygen molecules by hemoglobin, and for heterotropic interactions, as in the Bohr effect [28]. Later he introduced the concept of the binding potential, describable usually in terms of a binding polynomial which is in fact a partition function describing the mean distribution of bound ligand molecules among the individual macromolecules in the system. (For an elementary treatment, see for instance [29].) He also extended the treatment to include systems of molecules that can undergo allosteric transitions between different conformations [30, 31]; to polysteric systems, involving reversible association or dissociation of the macromolecule, linked to ligand binding; and to linkage between phase transitions and ligand binding, as in sickle cell hemoglobin ([31], and references there given).

Wyman has also generalized the thermodynamic potentials of Gibbs to cover systems in which some components are held at constant mass, and others at constant potential [32]. For multicomponent systems this greatly expands the range of applicable binding potentials, beyond Wyman's earlier formulation, to include the wide variety of such systems that can be experimentally investigated. I have discussed this at more length in an article about the work of Jeffries Wyman and myself [33].

The linkage between ligand binding in hemoglobin and the reversible dissociation of the hemoglobin tetramer into dimers has been most thoroughly explored by G. K. Ackers and his associates [34]. Their work on mutant and chemically modified human hemoglobins has also revealed important correlations between these linkage relations and the locations of the altered residues in the three-dimensional hemoglobin structure [35], as revealed by the crystallographic work of Perutz and his associates [36]. This, of course, takes us beyond the realm of pure thermodynamics; it involves the use of specific structural models and their correlation with the thermodynamic data. Thermodynamic evidence, of course, cannot prove the validity of a particular model; it may, however, eliminate certain models from consideration. With the increasingly detailed knowledge of the structure and dynamics of biological macromolecules, the interplay between thermodynamics and structural studies is sure to become increasingly important.

6 The Contributions of this Book

This book provides a series of informative critical surveys in its 14 chapters, each of which brings together the best available data in its particular field, in conveniently available form, for use in experimental work, in calculations, and as material for possible theoretical interpretations. The subject just discussed above, that of protein-ligand interactions, is involved in several chapters. The first of these, by H. Wiesinger and H.-J. Hinz (Chap. 7), is general in scope, but with particular emphasis on the use of microcalorimetric studies in the thermodynamics of ligand binding. P. D. Ross (Chap. 8) treats the thermodynamics of association between protein molecules, a subject of great and increasing importance. B. G. Barisas (Chap. 9) evaluates the many-sided evidence for the linked phenomena in the binding of ligands to hemoglobin, and the linkage between ligand binding and subunit dissociation.

H. Durchschlag (Chap. 3) evaluates data on partial specific volumes of macromolecules, and some smaller molecules, in water, while H. Høiland (Chaps. 2, 4) does the same for partial molar compressibilities of organic solutes in water; these reveal important information concerning hydrophobic and electrostatic interactions in water solution, and compressibilities are significant for such problems as the nature of conformational transitions in proteins under high pressure. G. M. Mrevlishvili (Chap. 5) deals with the closely related topic of the heat capacities of macromolecules, especially proteins and DNA, and their constituent monomeric units. A. Cesaro (Chap. 6) evaluates the best available data on aqueous solutions of carbohydrates and polysaccharides. This is an important field that has often received too little attention. S. Cabani and P. Gianni (Chap. 10) evaluate solutility data, for gases and solids, in aqueous nonelectrolyte solutions. These data are not only valuable in themselves, but they can be used as guides in estimating expected values for other substances for which data are not at present available. M. Lüscher-Mattli (Chap. 11) deals with thermodynamic parameters of many biopolymer-water systems. The data cover a wide range of composition, from nearly dry biopolymers up to rather concentrated solutions, and they are of value both for practical purposes and for theoretical understanding of the interactions involved. The data for micelles, evaluated in Chapter 12 by H. Hoffmann and W. Ulbricht, are – like much else in this book – important both to physical chemists and biochemists.

W. Pfeil (Chap. 13) deals with the extensive data on the reversible unfolding of proteins, V. V. Filimonov (Chap. 14) with the conformational transitions of polynucleotides, and K. J. Breslauer (Chap. 15) with similar transitions for oligonucleotides. These data are naturally of central importance for understanding the chemical basis for the biological organization of proteins and nucleic acids. In the final chapter 16 M. V. Rekharsky, G. L. Galchenko, A. M. Egorov and I. V. Berezin consider, in systematic fashion, the thermodynamics of enzymatic reactions for six different classes of enzymes, and the enthalpies involved in the biosynthesis of various classes of bond.

Professor Hinz, as Editor, and all his contributors, deserve thanks and appreciation from all workers in the field. May this book be widely used, to the benefit of all workers in biochemistry, biophysics, and the modern biological sciences in general.

Acknowledgment. In preparing this introduction, I thank the U.S. National Science Foundation (SES 8308892) for support.

References

1. Guerlac H (1975) Antoine-Laurent Lavoisier: chemist and revolutionary. Scribners, New York, pp 171
1 a. Holmes FL (1985) Lavoisier and the chemistry of life: an exploration of scientific creativity. University of Wisconsin Press, Madison pp. xxiv + 565
2. Gibbs JW (1931) The collected works of J Willard Gibbs, vol 1. Longmans Green, New York, pp 55–353
3. Van't Hoff JH (1887) Z Phys Chem 1:481–508 [English translation in Alembic Club Reprints No 19 (1929) The foundations of the theory of dilute solutions. Alembic Club, Edinburgh, pp 5–42]
4. Lewis GN, Randall M (1923) Thermodynamics and the free energy of chemical substances. McGraw-Hill, New York, xxiii + 653 pp (Rev edn by Pitzer KS, Brewer L 1961)
5. Henderson LJ (1908) Am J Physiol 21:173–179
6. Sørensen SPL (1912) Ergeb Physiol Biol Chem Exp Pharmakol 12:393–532
7. Sørensen SPL et al. (1917) Comp Rend Trav Lab Carlsberg 12:68–372
8. Clark WM (1960) Oxidation-reduction potentials of organic systems. Williams and Wilkins, Baltimore, xi + 584 pp
9. Conant JB (1923) J Biol Chem 57:401–414
10. Conant JB, Fieser LF (1925) J Biol Chem 62:595–622
11. Lehmann J (1930) Scand Arch Physiol 58:173–312
12. Borsook H, Schott HF (1931) J Biol Chem 92:535–557
13. Michaelis L, Schubert MP (1938) Chem Rev 22:437–470
14. Walz D (1979) Biochim Biophys Acta 505:279–353
15. Muralt AV (1952) Otto Meyerhof (1884–1951) Ergeb Physiol Biol Chem Exp Pharmakol 47:i-xx
16. Nachmansohn D, Ochoa S, Lipmann FA (1960) Otto Meyerhof (1884–1951) Biog Mem Natl Acad Sci 34:153–182
17. Lipmann FA (1941) Adv Enzymol Relat Areas Mol Biol 1:99–162
18. Kalckar HM (1941) Chem Rev 28:71–178
19. Krebs HA, Kornberg HL (1957) Ergeb Physiol Biol Chem Exp Pharmakol 49:212–298 (Appendix by K Burton is on pp 275–285)
20. Mitchell P (1981) In: Semenza G (ed) Of oxygen, fuels, and living matter, part 1, with reprints of five of Mitchell's papers on pp 57–160. Wiley, Chichester, pp 1–56
21. Cohn EJ, Edsall JT (1943) Proteins, amino acids and peptides as ions and dipolar ions. Reinhold, New York, xviii + 686 pp
22. Frank HS, Evans MW (1945) J Chem Phys 13:507–532
23. Kauzmann W (1959) Adv Protein Chem 14:1–63
24. Edsall JT, McKenzie HA (1978) Adv Biophys 10:137–208; (1983) 16:53–183
25. Linderstrøm-Lang K (1924) Comp Rend Trav Lab Carlsberg 15(7):1–29
26. Adair GS (1925) Proc R Soc Lond A 109:292–300; J Biol Chem 63:529–545
27. Scatchard G (1949) Ann NY Acad Sci 51:660–672
28. Wyman J (1964) Adv Protein Chem 19:223–286
29. Edsall JT, Gutfreund H (1983) Biothermodynamics: the study of biochemical processes at equilibrium. Wiley, Chichester, xiii + 248 pp (see in particular pp 177–209)
30. Wyman J (1967) J Am Chem Soc 89:2202–2218
31. Wyman J (1981) Biophys Chem 14:135–146
32. Wyman J (1975) Proc Natl Acad Sci USA 72:1464–1468
33. Edsall JT (1985) In: Semenza G (ed) Comprehensive biochemistry, vol 36. Selected topics in the history of biochemistry, personal recollections II. Elsevier, Amsterdam 99–195
34. Ackers GK (1980) Biophys J 32:331–346
35. Pettigrew DW, Romeo PH, Tsapis A, Thillet J, Smith ML, Turner BW, Ackers GK (1982) Proc Natl Acad Sci USA 79:1849–1853
36. Perutz MF (1976) British Med Bull 32:195–208

Appendix: Physical Quantities, Units, and Conversion Factors

1. SI Base Quantities and Units

Base Qunatities name	Symbol	Base Units name	Symbol
Length	l	metre	m
Mass	m	kilogram	kg
Time	t	second	s
Electric current	I	ampere	A
Temperature	T	kelvin	K
Amount of substance	n	mole	mol
Luminous intensity	I_v	candela	cd

2. Derived Quantities

Quantity	Symbol	Units (name)	Relation to old Units
Force	$N\ (kg \cdot m \cdot s^{-2})$	Newton	$1\ N = 10^5\ dyn = 0.10197$
Pressure	$Pa\ (N \cdot m^{-2})$	Pascal	$1\ Pa = 10^{-5}\ bar = 0.98692 \cdot 10^{-5} atm$ $= 0.0075\ Torr$
Work	$J\ (N \cdot m = W \cdot s)$	Joule	$1\ J\ = 0.239\ cal$ $= 2.778 \cdot 10^{-7}\ kW \cdot h = 10^7\ erg$
Power	$P\ (J \cdot s^{-1})$	watt	$1\ W = 859.8\ cal \cdot h^{-1}$
Molar mass	$M\ (kg \cdot mol^{-1})$	mol	the old unit is $g \cdot mol^{-1}$ its use can be continued
Concentration	$c\ (kg \cdot mol^{-1} \cdot m^{-3})$	$mol \cdot m^{-3}$	the old unit is $mol \cdot dm^{-3}$ or $mol \cdot l^{-1}$ its use can be continued

3. Physical Constants

Quantity	Symbol	Value	SI units
Gas constant	R	8.3143	$J \cdot K^{-1} \cdot mol^{-1}$
Boltzmann constant	k	1.38062	$J \cdot K^{-1}$
Avogadro number	L	$6.0220943 \cdot 10^{23}$	mol^{-1}
Electronic charge	e	$1.602192 \cdot 10^{-19}$	C
Faraday constant	F	$9.64867 \cdot 10^4$	$C \cdot mol^{-1}$
Velocity of light	c	$2.997925 \cdot 10^8$	$m \cdot s^{-1}$
Planck constant	h	$6.62620 \cdot 10^{-34}$	$J \cdot s$

4. Defined Constants

Quantity	Symbol	Value
Thermochemical calory	cal	$4.1840\ J$
Triple point of water	T_{tp}	$273.16\ K$
Molar standard volume of ideal gas (1 atm, 273.16 K)	V^0	$0.0224136\ m^3$
Standard acceleration of gravity	g	$9.80665\ m \cdot s^{-2}$
Dielectric constant of vacuum	ε_0	$8.854 \cdot 10^{-12}\ C^2 \cdot J^{-1} \cdot m^{-1}$ $\cong 8.854 \cdot 10^{-14}\ As \cdot V^{-1} \cdot cm^{-1}$
Standard atmospheric pressure	atm	$1.013250 \cdot 10^5\ Pa$ $\cong 1.013250 \cdot 10^6\ dyn \cdot cm^{-2}$

5. *Energy Conversion Units*

$1\,\text{J}\cdot\text{mol}^{-1}$	$= 0.2390\,\text{cal}\cdot\text{mol}^{-1}$
	$= 8.3594\cdot10^{-2}\,\text{cm}^{-1}\cdot\text{molecule}^{-1}$
	$= 1.036\cdot10^{-5}\,\text{eV}\cdot\text{molecule}^{-1}$
$1\,\text{cal}\cdot\text{mol}^{-1}$	$= 4.184\,\text{J}\cdot\text{mol}^{-1}$
	$= 0.3498\,\text{cm}^{-1}\cdot\text{molecule}^{-1}$
	$= 4.336\cdot10^{-5}\,\text{eV}\cdot\text{molecule}^{-1}$
$1\,\text{eV}\cdot\text{molecule}^{-1}$	$= 9.649\cdot10^{4}\,\text{J}\cdot\text{mol}^{-1}$
	$= 2.306\cdot10^{4}\,\text{cal}\cdot\text{mol}^{-1}$
	$= 8.066\cdot10^{3}\,\text{cm}^{-1}\cdot\text{molecule}^{-1}$
$1\,\text{cm}^{-1}\cdot\text{molecule}^{-1}$	$= 11.9626\,\text{J}\cdot\text{mol}^{-1}$
	$= 2.859\,\text{cal}\cdot\text{mol}^{-1}$
	$= 1.2398\cdot10^{-4}\,\text{ev}\cdot\text{molecule}^{-1}$
$1\,\text{Joule}$	$= 10^{7}\,\text{erg}$
	$= 0.239\,\text{cal}$
	$= 6.242\cdot10^{18}\,\text{eV}$
	$= 2.778\cdot10^{-7}\,\text{kW}\cdot\text{h}$
	$= 9.8694\cdot\text{l}\cdot\text{atm}$
$1\,\text{erg}$	$= 10^{-7}\,\text{J}$
	$= 2.390\cdot10^{-8}\,\text{cal}$
	$= 6.242\cdot10^{-11}\,\text{eV}$
	$= 2.778\cdot10^{-14}\,\text{kW}\cdot\text{h}$
	$= 9.8694\cdot10^{-10}\,\text{l}\cdot\text{atm}$

$1\,\text{cal}$	$= 4.184\,\text{J}$
	$= 4.184\cdot10^{7}\,\text{erg}$
	$= 2.612\cdot10^{19}\,\text{eV}$
	$= 1.162\cdot10^{-6}\,\text{kW}\cdot\text{h}$
	$= 4.1292\cdot10^{-2}\,\text{l}\cdot\text{atm}$
$1\,\text{eV}$	$= 1.6021\cdot10^{-19}\,\text{J}$
	$= 1.6021\cdot10^{-12}\,\text{erg}$
	$= 3.829\cdot10^{-20}\,\text{cal}$
	$= 4.450\cdot10^{-26}\,\text{kW}\cdot\text{h}$
	$= 1.581\cdot10^{-21}\,\text{l}\cdot\text{atm}$
$1\,\text{kWh}$	$= 3.600\cdot10^{6}\,\text{J}$
	$= 3.600\cdot10^{13}\,\text{erg}$
	$= 8.604\cdot10^{5}\,\text{cal}$
	$= 2.247\cdot10^{25}\,\text{eV}$
	$= 3.552\cdot10^{4}\,\text{l}\cdot\text{atm}$
$1\,\text{l}\cdot\text{atm}$	$= 1.101328\cdot10^{2}\,\text{J}$
	$= 1.101328^{9}\,\text{erg}$
	$= 24.2179\,\text{cal}$
	$= 6.325\cdot10^{20}\,\text{eV}$
	$= 2.815\cdot10^{-5}\,\text{kW}\cdot\text{h}$

Section II Nonreacting Systems

Chapter 2 Partial Molar Volumes of Biochemical Model Compounds in Aqueous Solution

HARALD HØILAND

Symbols

V:	volume of the solution $[dm^3\,mol^{-1}]$
V_1^*:	molar volume of the pure solvent $[cm^3\,mol^{-1}]$
V_2:	partial molar volume of solute $[cm^3\,mol^{-1}]$ in a two-component system
V_x^0:	partial molar volume of solute x at infinite dilution $[cm^3\,mol^{-1}]$
V_{NaA}^0:	partial molar volumes of sodium carboxylates at infinite dilution $[cm^3\,mol^{-1}]$
V_{HA}^0:	partial molar volumes of carboxylic acids $[cm^3\,mol^{-1}]$ at infinite dilution
V_ϕ:	apparent molar volume $[cm^3\,mol^{-1}]$
$V_{\phi x}$:	apparent molar volume of substance x $[cm^3\,mol^{-1}]$
$V_{\phi\,BH\,O}$:	apparent molar volume of a hypothetical neutral aquoamine complex, BH_2O, that can dissociate into the hydrolysis products BH^+ and OH^-
V_w:	van der Waals volume
r_w:	van der Waals radius
ΔV^0:	change in molar volume
ϱ^*:	density of solvent $[g\,cm^{-3}]$
ϱ:	density of solute $[g\,cm^{-3}]$
M_2:	molar mass of solute
m:	molality of solute [mol solute/1000 g solvent]
α:	degree of hydrolysis

1 Introduction

Partial molar volumes have proved very useful in the study of molecular and ionic interactions in solution. The values at infinite dilution provide information about solute-solvent interactions and the concentration dependence will reflect solute-solute interactions. The present review is aimed at volumes of relatively simple organic molecules that may be looked upon as biochemical model compounds. It is believed that the partial molar volumes of these compounds in water represent a basis for the understanding of complex biochemical systems. With this in mind, it is especially interesting to examine homologous series. Systematic trends, such as additivity relations, are of special importance and will be discussed at the end.

2 Definitions

The partial molar volume of a solute is defined as:

$$V_2 = (\partial V/\partial n_2)_{T,p,n_1}. \tag{1}$$

However, in this form it is practically impossible to determine partial molar volumes. Instead, the apparent molar volume is defined:

$$V_\phi = \frac{V - n_1 V_1^*}{n_2},$$ (2)

where n_1 and n_2 are the number of moles of solvent and solute respectively, and V_1^* is the molar volume of pure solvent. Equation (2) can be rearranged to read:

$$V_\phi = \frac{1000(\varrho^* - \varrho)}{m\varrho\varrho^*} + \frac{M_2}{\varrho}.$$ (3)

In Eq. (3), ϱ^* and ϱ are the densities of solvent and solution, m is the solute molality, and M_2 the solute molar mass. The factor 1000 appears when densities are given in g cm^{-3}, which is generally the case, instead of kg m^{-3}.

The relation between apparent and partial molar volume is as follows:

$$V_2 = V_\phi + \frac{m^{1/2}}{2}(\partial V_\phi / \partial m^{1/2}) = V_\phi + m(\partial V_\phi / \partial m).$$ (4)

From Eq. (4) it is seen that the apparent and partial molar volumes become equal at infinite dilution. Otherwise it is possible to equate the two quantities, provided the apparent molar volume is known as a function of molality. Similar equations can be evaluated on the molarity scale.

In order to obtain reliable V_ϕ data, it is necessary to measure densities with great precision. This is easily demonstrated by differentiating Eq. (3) with respect to m at constant ϱ, or with respect to ϱ at constant m. One then obtains:

$$\text{Probable error in } V_\phi: \left(\frac{M_2}{\varrho} - V_\phi\right)\frac{\delta m}{m};$$ (5)

$$\text{Probable error in } V_\phi: \left(\frac{1000}{m\varrho^*} + V_\phi\right)\frac{\delta\varrho}{\varrho}.$$ (6)

Equations (5) and (6) demonstrate that in dilute solutions V_ϕ is not seriously affected by errors in m. However, errors in ϱ may cause large errors in V_ϕ. If, for example, m is 0.01 m, an error in ϱ of magnitude 10^{-5} g cm^{-3} will cause an error of about 1 cm^{-3} mol^{-1} in V_ϕ. It is therefore necessary to measure densities with great accuracy. Today's standard with vibrating tube densitometers is of the order 10^{-6} cm^3 mol^{-1}. This in turn means that the temperature must be controlled very well, approximately ± 0.005 K will do.

3 Extrapolation to Infinite Dilution

The functional relationship between apparent molar volume and molality reflects solute-solute interactions. It is thus not surprising when experiments show distinct differences between ionic and nonionic solutes. Concerning nonionics, experience has shown that a linear functional relationship between apparent molar volume and molality holds to quite high concentrations, i.e.,:

$$V_\phi = V_\phi^0 + b_u m = V_2^0 + b_u m. \tag{7}$$

The slope of Eq. (7), b_u, is usually small.

Apparent molar volumes of ionic solutes were first fitted to the equation [1]:

$$V_\phi = V_\phi^0 + S_v' c^{1/2} = V_2^0 + S_v' c^{1/2}. \tag{8}$$

Like the b_u coefficient of nonionic solutes, S_v' is an empirical slope varying with the charge of the ions and the type of ions.

Redlich and Rosenfeld [2] predicted that a constant limiting slope should be obtained for a given ionic charge. This argument was based on application of the Debye-Hückel limiting law. On this basis, Redlich and Meyer [3] suggested the following extrapolation function:

$$V_\phi = V_2^0 + S_v c^{1/2} + b_v c. \tag{9}$$

Here S_v is the theoretical limiting slope calculated from the Debye-Hückel limiting law and b_v an empirical constant. An extrapolation using Eq. (9) means a plot of $(V_\phi - S_v c^{1/2})$ versus c. Values of S_v are 1.868, 9.706, and 14.944 cm^3 mol$^{-3/2}$ dm$^{1/2}$ for 1:1, 2:1, and 2:2 electrolytes respectively.

4 Tabulation of the Experimental Data

In the following pages, partial molar volumes of homologous series of organic solutes in water will be presented. Within each series a comprehensive table is presented. The values given are at infinite dilution and 298.15 K. All relevant literature has been examined, and the data have been critically evaluated. The recommended values are thus generally mean values and the given error limits are standard deviations. However, in several instances, some of the data are old and do not agree well with more up-to-date investigations. Under such circumstances, these data have been omitted when calculating the recommended value, but a reference to then will still be given. In other instances, only one set of data can be found. In these cases the partial molar volumes have been tabulated as they were reported in the original paper. The error limits in these cases are usually a measure of the reproducibility, and must not be directly compared with standard deviations calculated on the basis of several sets of data.

Since the concentration dependence of the apparent molar volumes may be of considerable interest, b_u values of Eq. (7) have been included in the tables. If the compound in question is an electrolyte, the b_v values of Eq. (9) are presented. Generally it will be safe to apply Eq. (7) and (9) up to at least 0.5 m, in many cases much higher.

Polymers and proteins have not been included. Although they certainly have biochemical significance, it would extend both the scope and the number of data beyond that of a single review paper. Further, the molar mass of proteins and polymers is often not well defined, and the treatment of such solutes is often quite different from those presented here.

5 Alcohols, Polyols, and Phenols

Aqueous solutions of alcohols are probably the most extensively studied systems as far as partial molar volumes are concerned. Comprehensive studies have been carried out by several workers [4–21]. The data have been presented in Table 1.

Since alcohols and other mono- and bifunctional solutes vary greatly in size and shape, a common standard or reference state is important. For want of anything better, the normal practise has been to subtract the molar volume of the pure liquid alcohol from the aqueous partial molar volume, thus obtaining the excess volume, V_2^{0E}. The pure solute is then the reference. The excess volumes of alcohols are always negative, more negative the longer the alkyl chain of the alcohol. Though the pure liquid alcohol is far from a fully satisfactory standard state, the negative excess volumes are still thought to reflect stabilization of the

Table 1. Partial molar volumes of alcohols, polyols, and phenols at infinite dilution in aqueous solution; recommended values at 298.15 K [a]

	$\dfrac{V_2^0}{cm^3\,mol^{-1}}$ (b_u) [b]	Reference
Methanol	38.17 ± 0.10 (-0.2)	[4, 8, 9, 13, 16, 19, 23]
Ethanol	55.08 ± 0.07 (-0.6)	[4, 8, 9, 13, 16, 19, 23]
1-Propanol	70.70 ± 0.10 (-0.9)	[4, 8–10, 13, 16, 19, 23]
2-Propanol	71.90 ± 0.10 (-1.3)	[10, 19, 22]
1-Butanol	86.60 ± 0.10 (-1.1)	[8–10, 13, 15, 16, 19, 23]
2-Butanol	86.62 ± 0.07 (-1.7)	[6, 10, 13, 15, 21]
2-Methyl-1-propanol	86.70 ± 0.10 (-1.4)	[13, 15]
2-Methyl-2-propanol	87.75 ± 0.07 (-1.6)	[13, 15, 23]
1-Pentanol	102.68 ± 0.13 (-1.9)	[8–10, 13, 16]
2-Pentanol	102.55 ± 0.10 (-1.8)	[10]
3-Pentanol	101.21 ± 0.10 (-2.2)	[6, 10, 13]
2,2-Dimethyl-1-propanol	101.87 ± 0.06 (-2.1)	[13]
1-Hexanol	118.65 ± 0.10	[8, 10]
2-Hexanol	118.49 ± 0.10	[10]
3-Hexanol	117.14 ± 0.09 (-3.2)	[6]
1-Heptanol	133.43	[8]
2-Heptanol	134.39 ± 0.10	[10]
3-Heptanol	133.3 ± 0.2	[10]
4-Heptanol	133.3 ± 0.2	[10]
Cyclobutanol	75.6 ± 0.1	[12]
Cyclopentanol	89.0 ± 0.5 (-2.2)	[6, 12]
Cyclohexanol	103.5 ± 0.5 (-2.5)	[6, 12]
Cycloheptanol	116.88 ± 0.1 (-3.0)	[6, 12)
Cyclooctanol	129.7 ± 0.4	[12]
Cyclopropanemethanol	76.0 ± 0.2	[12]
α-Methylcyclopentanemethanol	92.4 ± 0.2	[12]
Cyclopentanemethanol	103.6 ± 0.7	[12]
α-Methanecyclopentanemethanol	118.2 ± 0.7	[12]
Cyclohexanemethanol	118.1 ± 0.7	[12]
1,2-Ethanediol	54.63 ± 0.08 (-0.1)	[9, 12, 13, 16, 19, 20, 27–31]
1,2-Propanediol	71.22 ± 0.10	[10, 20, 27]
1,3-Propanediol	71.90 ± 0.10 (-0.3)	[9, 12, 16, 20, 27, 30
1,3-Butanediol	88.32 ± 0.10 (-0.7)	[10, 20]
1,4-Butanediol	88.33 ± 0.10 (-0.7)	[9, 10, 12, 13, 16, 20, 30]

Table 1 (continued)

	$\dfrac{V_2^0}{cm^3\,mol^{-1}}$ $(b_u)^b$	Reference
2,3-Butanediol	$86.56 \pm 0.10\ (-0.8)$	[10]
1,5-Pentanediol	$104.39 \pm 0.10\ (-1.2)$	[9, 10, 12, 16, 30]
2,4-Pentanediol	$104.64 \pm 0.10\ (-1.3)$	[10]
2,2-Dimethyl-1,3-propanediol	$102.34 \pm 0.01\ (-1.0)$	[13]
1,6-Hexanediol	$120.44 \pm 0.10\ (-1.3)$	[9, 12, 13]
2,5-Hexanediol	120.5	[32]
1,7-Heptanediol	$136.40 \pm 0.10\ (-2.0)$	[10]
1,8-Octanediol	$152.6\ \pm 0.1$	[12]
1,10-Decanediol	$184.4\ \pm 0.6$	[12]
cis-1,2-Cyclohexanediol	$101.3\ \pm 0.7$	[12]
trans-1,2-Cyclohexanediol	$103.0\ \pm 0.7$	[12]
1,4-Cyclohexanediol	105.3	[32]
1-Methyl-trans-1,2-Cyclohexanediol	$116.6\ \pm 0.7$	[12]
trans-1,2-Cycloheptanediol	$117.0\ \pm 0.1$	[12]
cis-1,5-Cyclooctanediol	$132.5\ \pm 0.3$	[12]
1,3-Adamantanediol	$138.2\ \pm 0.3$	[12]
1,4-Adamantanediol	$139.3\ \pm 0.4$	[12]
1,2-Benzenediol	87.07	[33]
1,3-Benzenediol	88.92	[33]
1,4-Benzenediol	88.70	[33]
2-(Hydroxymethyl)-1,3-propanediol	$102.30 \pm 0.01\ (-0.5)$	[13]
2,2-Bis(hydroxymethyl)-1,3-propanediol	$101.81 \pm 0.01\ (-0.2)$	[13]
α-Methylcyclopentanemethanol	$118.2\ \pm 0.7$	[12]
Cyclohexanemethanol	$118.1\ \pm 0.7$	[12]
2-Methoxyethanol	$75.15 \pm 0.05\ (-0.4)$	[35, 36]
2-Ethoxyethanol	$91.15 \pm 0.15\ (-1.0)$	[35, 36]
2-Propoxyethanol	$107.10 \pm 0.05\ (-1.4)$	[35, 36]
2-Butoxyethanol	$122.95 \pm 0.05\ (-1.6)$	[35, 36]
Tetrahydro-2-furanmethanol	93.8	[37]
Tetrahydro-2h-pyran-2-methanol	108.1	[37]
2-Aminoethanol	$59.25 \pm 0.02\ (-0.03)$	[38]
3-Amino-1-propanol	$75.21 \pm 0.04\ (-0.2)$	[38]
2-(methylamino)ethanol	$77.07 \pm 0.02\ (-0.6)$	[38]
2-(Dimethylamino)ethanol	$94.17 \pm 0.03\ (-1.1)$	[38]
2-(Ethylamino)ethanol	$92.91 \pm 0.02\ (-1.1)$	[38]
2-(Diethylamino)ethanol	$123.04 \pm 0.03\ (-1.9)$	[38]
Diethylene glycol	$92.22 \pm 0.05\ (-0.5)$	[36, 41]
Triethylene glycol	$129.27 \pm 0.04\ (-0.8)$	[36, 41]
Tetraethylene glycol	$166.31 \pm 0.08\ (-1.1)$	[36, 41]
Phenol	86.17 ± 0.2	[24–26]
Phenol sodium salt	66.3	[26]
m-Nitrophenol	99.7	[26, 49]
m-Nitrophenol sodium salt	84.9	[26, 49]
p-Nitrophenol	98.2	[26, 49]
p-Nitrophenol sodium salt	85.1	[26, 49]
m-Cyanophenol	97.5	[26]
m-Cyanophenol sodium salt	83.4	[26]
p-Cyanophenol	98.3	[26]
p-Cyanophenol sodium salt	84.1	[26]

[a] Values at other temperatures can be found in references [4, 5, 9, 11, 15, 20].
[b] In units of $cm^3\,kg\,mol^{-2}$.

water structure around the hydrophobic groups of the solute molecule, i.e., hydrophobic hydration.

The concentration dependence of the apparent molar volumes of alcohols, the b_u coefficient of Eq. (7) is negative in all cases. It becomes more negative as the alcohol size increases. The b_u coefficient is generally taken to reflect solute-solute interactions, but the interpretation is far from straightforward. Franks and Smith [15] have questioned the basis for linearity, suggesting that in the absence of solute-solute interactions in the limiting region, zero slope is to be expected. Perron and Desnoyers [23] have adapted a formalism that describe a thermodynamic function, the apparent molar volume in our case, in terms of an ideal contribution plus a series of pair, triplet, and higher order interaction parameters.

$$V_{\phi,x} = V_x^0 + v_{xx}m_x + v_{xxx}m_x^2 + \dots, \tag{10}$$

where V_x^0 is the partial molar volume of solute x at infinite dilution, m_x its molality, and v_{xx} and v_{xxx} the pair and triplet interaction terms respectively. For methanol only pair parameters are evident up to 1.4 mol kg^{-1}. For butanol it appears that triplet interaction terms are significant at low molalities, and above 0.4 mol kg^{-1} even higher interaction parameters are observed. This method of evaluating interaction parameters appears promising, but for most alcohols it is presently not possible to evaluate triplet and higher order parameters because of the uncertainty of experimental results. However, by inspecting the apparent molar volumes of the totally water-soluble alcohols, it seems that higher order parameters must surely become significant as the alcohol content increases. This is reflected in the b_u coefficient. The negative slope only persists to a certain molality. The apparent molar volume reaches a minimum value and then increases rapidly towards the molar volume of the pure alcohol [8, 22, 34]. As the size of the alkyl chain increases, this minimum moves towards lower molalities and also becomes more sharply defined. Beyond this minimum the solution properties become similar to normal nonaqueous systems.

Cabani et al. [7] have measured apparent molar volumes at several temperatures and fitted the data to polynomials of the following type:

$$V_\phi(T) = V_\phi(25\ °C) + \alpha(T-25) + \beta(T-25)^2 + \gamma(T-25)^3. \tag{11}$$

In this form it is possible to differentiate V_ϕ with respect to temperature and obtain partial molar expansibilities. Cabani's data show a wide range of expansibility values at low temperatures, but this becomes narrower at higher temperatures. The expansibilities of monofunctional alcohols increase with temperature, while they decrease for polyhydric alcohols. This result emphasizes the significance of water-hydroxyl group interactions in determining the solution properties of such solutes.

Another interesting feature of the partial molar volumes of alcohols is the additivity. Although a general discussion of additivity follows at the end of this chapter, it seems worth noting here the remarkably constant contribution of the methylene group. In the series of 1- and 2-alcohols, excluding methanol and 2-propanol, the CH_2 group value becomes (16.0 ± 0.2) cm^3 mol^{-1}. The data on cyclic alcohols appear less regular, but the experimental error margins are larger and could account for the irregularities. Still the methylene group contribution can be evaluated to (13.5 ± 0.8) cm^3 mol^{-1}.

6 Ethers, Ketones, and Aldehydes (Table 2)

The partial molar volumes of ethers, ketones, and aldehydes are also generally smaller than the molar volumes of the pure liquids; i.e., negative excess volumes

Table 2. Partial molar volumes of ethers, ketones and aldehydes at infinite dilution in water; recommended values at 298.15 K

	$\dfrac{V_2^0}{cm^3\,mol^{-1}}$ $(b_u)^a$	Reference
1,1′-Oxybisethane (ethyl ether)	90.4 ±0.5	[12]
2,2′-Oxybispropane	115.0 ±0.6	[12]
Oxethane	61.35±0.05 (−0.4)	[12, 38]
Tetrahydrofuran	76.85±0.07 (−1.1)	[12, 37, 39]
2-Methyltetrahydrofuran	94.00±0.1 (−1.7)	[12, 39]
2,5-Dimethyltetrahydrofuran	111.00±0.03 (−1.9)	[12, 39]
Tetrahydropyran	91.74±0.05 (−1.7)	[12, 37, 39, 40]
Oxepane	105.46±0.1 (−2.6)	[12, 38]
Dimethoxymethane	80.42±0.1 (−0.4)	[12, 38]
1,2-Dimethoxyethane	95.86±0.05 (−1.3)	[12, 16, 36, 38]
Tetramethoxymethane	126.2 ±0.4	[12]
Diethoxymethane	113.88±0.03 (−1.3)	[38]
1,2-Diethoxymethane	127.29±0.03 (−2.6)	[38]
1,3-Dioxane	80.82±0.02 (−0.3)	[38]
1,4-Dioxane	80.94±0.05 (−0.2)	[12, 37, 39, 40]
1,3-Dioxolane	65.37±0.1 (0.2)	[12, 39]
1,3-Dioxepane	95.98±0.1 (−0.7)	[12, 38]
1,1′-Oxybis(2-methoxyethane)	132.70±0.05 (−1.9)	[12, 16, 36, 41]
1,1′-Oxybis(2-ethoxyethane)	165.0	[46]
1,3,5-Trioxane	69.51±0.1 (0.2)	[12, 38]
2,4,6-Trimethyl-1,3,5-trioxane	123.71±0.02 (−1.5)	[38]
2,5-Dimethoxytetrahydrofuran	122.34±0.04 (−1.5)	[38]
2,5,8,11-Tetraoxadodecane	169.73±0.1 (−2.5)	[36, 41, 42]
2,5,8,11,14-Pentaoxapentadecane	206.76±0.1 (−3.4)	[36, 41]
2,5-Dimethoxy-2-(dimethoxymethyl)-tetrahydrofuran	177.1 ±0.7	[12]
1,4,7,10-Tetraoxacyclododecane (12-Crown-4)	149.9 ±0.2 (−1.1)	[43]
1,4,7,10,13-Pentaoxacyclopentadecane (15-Crown-5)	186.06±0.1 (−0.3)	[43]
1,4,7,10,13,16-Hexaoxacyclooctadecane (18-Crown-6)	223.35±0.1 (−2.8)	[38, 43]
2-Propanone (acetone)	66.92±0.1 (−0.4)	[12, 44, 45]
2-Butanone	82.52±0.1 (−0.8)	[12, 45, 46]
2-Pentanone	98.0 ±0.1	[12]
3-Methyl-2-butanone	95.0 ±0.3	[12]
Cyclobutanone	70.9 ±0.1	[12]
Cyclopentanone	84.5 ±0.3	[12]
Cyclohexanone	99.7 ±0.4	[12]
Cycloheptanone	113.8 ±0.3	[12]
Cyclooctanone	128.0 ±0.1	[12]
Cyclononanone	142.0 ±0.3	[12]
Acetaldehyde	43.7 ±0.3	[47]
Benzaldehyde	96.08	[48]
3-Hydroxybenzaldehyde	97.87	[49]
3-Hydroxybenzaldehyde sodium salt	83.32	[49]
4-Hydroxybenzaldehyde	96.94	[49]
4-Hydroxybenzaldehyde sodium salt	83.53	[49]

^{a} In units of $cm^3\,kg\,mol^{-2}$.

are observed. The concentration dependence of the apparent molar volumes is also similar to that of alcohols. It is linear and negative in the water-rich region. The apparent molar volume reaches a minimum and increases rapidly towards the molar volume of the pure liquids.

The thermodynamic behavior of the lower homologs of ethers and ketones, as well as of alcohols, shows that they can be classified as structure-makers [102]. Polyethers, on the other hand, appear to be structure-breakers. It has also been suggested from volume and expansibility data that the ether oxygen promotes regions of unbounded and easily expansible water, since this oxygen atom is unable to form cooperative hydrogen bonds with water.

Cabani et al. [39, 50] compared mono- and di-ethers, observing that the entropy loss in the transfer process gas→solute is large for the monoethers and considerably less for the diethers. This is probably because the two polar sites of the diethers prevent cluster formation. However, both types of solute have negative excess volumes. The monoethers can probably penetrate the water structure, only one oxygen disturbs the water structure around the hydrophobic groups. This is the usual interpretation of negative excess volumes. In order to explain that the diethers also have negative excess volumes, Cabani et al. [39] argue that a dense water region is created around the two oxygens of these solutes.

7 Carboxylic Acids, Sodium Carboxylates, and Esters (Table 3)

The ionization (or dissociation) equilibria of the carboxylic acids dissolved in water are important in relation to their partial molar volumes. Indeed the volume (changes) of ionization, ΔV^0, have been the prime target for several investigations concerned with the aqueous solution properties of carboxylic acids. Hamann [59] has reviewed the data up to 1974. Further data have been supplied by Høiland [58, 103, 104].

The volume of ionization can be obtained by two independent methods of measurement. First densities can be measured, and the volume of ionization can be calculated as:

$$\Delta V^0 = V_{NaA}^0 - V_{HA}^0 - (V_{Na+}^0 - V_{H+}^0). \tag{12}$$

Here V_{NaA}^0 and V_{HA}^0 are the partial molar volumes of the sodium carboxylates and the corresponding carboxylic acids respectively, at infinite dilution. The difference $(V_{Na+}^0 - V_{H+}^0)$ is -1.2 cm^3 mol^{-1} [38]. ΔV^0 can also be calculated by measuring the equilibrium constant as a function of pressure:

$$\Delta V^0 = RT (\partial \ln K / \partial P)_T. \tag{13}$$

Values calculated from the two methods of measurement can be seen to agree very well [56, 59, 104].

ΔV^0 is always negative as expected from electrostriction theory [139, 140]. ΔV^0 decreases with increasing size of the carboxylic acid, but a constant value is obtained for the higher homologs of each series. ΔV^0 of the ordinary monocarboxylic acids reach a constant value of (-14.2 ± 0.2) cm^3 mol^{-1} from butanoic acid onwards [104]. For the α-carboxylic acids ΔV^0 become (-13.9 ± 0.2) cm^3

Table 3. Partial molar volumes of carboxylic acids, sodium carboxylates and esters at infinite dilution in water; recommended values at 298.15 K[a]

	$\dfrac{V_2^0(HA)}{cm^3\,mol^{-1}}$	(b_u)[b]	$\dfrac{V_2^0(NaA)}{cm^3\,mol^{-1}}$	(b_v)[c]	$\dfrac{V_2^0(Na_2A)}{cm^3\,mol^{-1}}$	(b_v)[c]	Reference
Formic acid	34.69±0.05	(0)	25.05±0.1	(0.1)			[51–54]
Acetic	51.92±0.05	(−0.1)	39.23±0.1	(0.1)			[51–57]
Propanoic	67.90±0.1	(−0.1)	53.84±0.1	(0.1)			[51, 53–55]
Butanoic	84.61±0.1	(−0.1)	69.19±0.1	(0.1)			[51, 53–55]
Pentanoic	100.50±0.1	(−0.2)	85.20±0.1	(0.1)			[54, 55]
2-Methylbutanoic	100.5						[54]
3-Methylbutanoic	100.5						[54]
Hexanoic	116.55±0.1	(−0.2)	101.10±0.1	(0.2)			[55]
α-Hydroxyacetic	51.75±0.1	(0.6)	38.74±0.1	(1.7)			[58]
α-Hydroxypropanoic	69.38±0.1	(−0.2)	55.14±0.1				[58]
α-Hydroxy-2-methylpropanoic	86.78±0.1	(−0.2)	71.52±0.1	(1.8)			[58]
α-Hydroxybutanoic	85.45±0.1	(−0.4)	70.50±0.1	(1.9)			[58]
α-Hydroxypentanoic	100.47±0.1	(−0.5)	85.60±0.1	(3.6)			[58]
α-Hydroxy-3-methylbutanoic	100.83±0.1	(−0.3)	85.73±0.1	(1.9)			[58]
α-Hydroxyhexanoic	117.26±0.1	(−0.3)	102.22±0.1	(4.1)			[58]
Galacturonic	107.6±0.1	(0.1)	93.1±0.1	(−20)			[106]
Glucuronic	110.2±0.1	(0.2)	95.7±0.1	(4.3)			[106]
Ethanedioic	49.12±0.1		41.21±0.1		28.08±0.1	(−2.2)	[60]
Propanedioic	67.22±0.1	(0.5)	55.96±0.1	(1.3)	36.21±0.1	(−3.5)	[60, 76]
Butanedioic	82.94±0.1	(0.4)	68.88±0.1	(1.3)	54.10±0.1	(−3.5)	[60, 76]
Pentanedioic	99.14±0.1	(0.7)	84.77±0.1	(2.0)	69.98±0.1	(−4.0)	[60, 76]
Hexanedioic	115.66±0.1	(0.8)	100.98±0.1	(2.3)	86.24±0.1	(−5.6)	[60, 76]
Heptanedioic	131.93±0.1	(1.0)	116.59±0.1	(1.7)	101.84±0.1	(−5.2)	[60]
Tartaric	83.99±0.1	(0.5)	70.83±0.1	(1.3)	56.26±0.1	(−7.8)	[58]
Formic acid ethylester	72.8±0.7						[12]
Acetic acid methylester	72.46±0.05	(−0.1)					[46]
Acetic acid ethyl ester	88.90±0.1	(0.1)					[12, 46]
Ethanedioic acid dimethyl ester	91.4±0.5						[12]
Methoxyacetic acid methyl ester	93.3±0.1						[12]
Acetic acid methoxymethyl ester	94.4±0.2						[12]

[a] Values at other temperatures can be found in [58].
[b] In units of $cm^3\,kg\,mol^{-2}$.
[c] In units of $cm^3\,dm^3\,mol^{-2}$.

mol^{-1} from α-hydroxybutanoic acid onwards [58]. ΔV^0 for the first ionization step of dicarboxylic acids exhibits the same lower limiting value as the monocarboxylic acids, but this value is first obtained for heptanedioic acid [103]. The results demonstrate that the negative charge of the carboxylate group only affects the hydration of the nearest methylene groups. Beyond the γ-carbon no volumetric effects are observed.

Due to the ionization equilibria, the partial molar volumes of the acids, as directly calculated from densities, will include a contribution from ionized species. It is, however, possible to correct for this by the following equation derived by King [52]:

$$V_\phi - \alpha \Delta V^0 = V_2^0 + b_u m, \qquad (14)$$

where V_ϕ is the directly calculated apparent molar volume, α is the degree of ionization, and b_u an empirical constant that is comparable to b_u of Eq. (7). V_2^0 is the
interesting quantity, i.e., the partial molar volume of the unionized acid at infinite
dilution. Since ΔV^0 depends on V_2^0, and both are unknown, it is necessary to carry
out an iteration procedure in order to determine both quantities.

For dicarboxylic acids the monosodium salts must also be corrected for further ionization. This can be achieved in much the same way as shown by Høiland
[103].

The data in Table 3 have all been corrected for ionization. The monosodium
salts of monocarboxylic acids and the disodium salts of dicarboxylic acids have
been treated as ordinary strong 1:1 and 2:1 electrolytes.

8 Amines, Amides, and Ureas (Table 4)

Amines dissolved in water hydrolyze. This affects the partial molar volumes in
much the same way as ionization (dissociation) affects the partial molar volumes
of carboxylic acids. It is therefore once again necessary to correct the apparent
molar volumes as directly calculated from measured densities. Cabani et al. [39]
have proposed the following equation

$$V_\phi = (1-\alpha)V_{\phi,BH_2O} + V_{\phi,BH+OH-}, \tag{15}$$

where α is the degree of hydrolysis, V_{ϕ,BH_2O} the apparent molar volume of a hypothetical neutral aquo-amine complex, BH_2O, that dissociates to the hydrolysis
products BH^+ and OH^-. The apparent molar volume of the latter is the last term
of Eq. (15). For the neutral aquo-amine complex Eq. (7) was considered valid, and
Eqs. (9) was used to describe the concentration dependence of the ionic species.
BH^+, OH^-. By combining Eq. (7), (9), and (15) the following expression was obtained:

$$\frac{V_\phi - V_{BH+OH-}^0 - S_v c^{1/2}}{(1-\alpha)} = V_{BH_2O}^0 + b_u c(1-\alpha) + K_h b_v, \tag{16}$$

where K_h is the hydrolysis constant. The left-hand side of Eq. (16) can be plotted
versus $(1-\alpha)$ c, and $V_{BH_2O}^0$ is then obtained. The term $K_h b_v$ can be neglected since
K_h is very small in all cases. Finally the partial molar volume of the neutral amine
was obtained by subtracting the partial molar volume of water:

$$V_B^0 = V_{BH_2O}^0 - V_{H_2O}^0. \tag{17}$$

Kaulgud et al. [64], however, have demonstrated that Eq. (16) does not produce
a straight line in all cases. They modified it by multiplying each side by $(1-\alpha)$ c,
thus obtaining:

$$(V_\phi - V_{BH+OH-}^0 - S_v c^{1/2}) c = (1-\alpha) c (V_{BH_2O}^0 + (1-\alpha) c b_u). \tag{18}$$

The term $(1-\alpha)$ c b_u was neglected, and plots of the left-hand side versus $(1-\alpha)$ c
gave perfectly straight lines through the origin, as predicted. The differences in
V_B^0 between the two methods of extrapolation are generally not much larger than
the experimental error, though in some cases it becomes significant; for instance

Table 4. Partial molar volumes of amines and their chloride salts, diamines and their mono- and dichloride salts, cyclic amines, amides, and ureas at infinite dilution in aqueous solution; recommended values at 298.15 K [a]

	$\frac{V_2^0(B)}{cm^3\,mol^{-1}}$	(b_u) [b]	$\frac{V_2^0(BCl)}{cm^3\,mol^{-1}}$	(b_v) [c]	$\frac{V_2^0(BCl_2)}{cm^3\,mol^{-1}}$	(b_v) [c]	Reference
Ammonia	24.85 ±0.1	(−0.1)	36.1 ±0.2				[6, 53, 60–64]
Methaneamine	41.88 ±0.3	(−0.3)	53.8 ±0.1	(−0.4)			[6, 53, 63–65]
Ethaneamine	58.61 ±0.2	(−0.9)	70.7 ±0.1				[6, 64, 66]
1-Propaneamine	74.15 ±0.05	(−1.2)	87.23 ±0.1				[6, 64, 66]
1-Butaneamine	89.80 ±0.1	(−1.4)	103.3 ±0.1				[6, 64, 66]
1-Pentaneamine	105.7 ±0.1	(−2.0)	119.3 ±0.1				[6, 66]
1-Hexaneamine	121.6 ±0.2	(−3.0)	135.2 ±0.1				[6, 66]
1-Heptaneamine	137.6 ±0.3	(−6.5)	151.0 ±0.1				[6, 66]
Cyclopentaneamine	90.98 ±0.18						[67]
Cyclohexaneamine	105.35 ±0.17						[67]
Cycloheptaneamine	118.50 ±0.2						[67]
Cyclooctaneamine	133.4 ±0.2						[67]
1-Adamantaneamine	138.9 ±0.3						[67]
Benzeneamine (anilin)	89.30 ±0.1						[67]
N-Methylmethaneamine	59.20 ±0.2	(−0.5)	72.47 ±0.2	(−1.3)			[53, 67]
N-Ethylethaneamine	91.68 ±0.1	(−2.1)	106.73				[6, 53, 64, 65, 68]
N-Propyl-1-Propaneamine	123.06 ±0.1	(−3.2)	138.66				[6, 63]
N-Butyl-1-butaneamine	155.4 ±0.4	(−25.0)	170.68				[6, 63]
N,N-Dimethylmethaneamine	79.0 ±0.4	(−1.5)	90.59 ±0.2	(−1.0)			[6, 53, 64, 65]
N-Ethyl-N-Methylethaneamine	106.77 ±0.1	(−2.3)	122.8				[6]
N,N-Diethylethaneamine	120.9 ±0.1	(−2.3)	138.6 ±0.2	(−14.4)			[6, 65]
Aziridine	48.87 ±0.02	(−0.4)	61.99 ±0.14	(−0.5)			[39]
Azetidine	63.71 ±0.01	(−0.9)	76.56 ±0.06	(−0.6)			[39]
Pyrrolidine	77.77 ±0.01	(−1.2)	91.72 ±0.02	(−1.0)			[39]
1-Methylpyrrolidine	97.29 ±0.02	(−3.7)	110.64 ±0.02	(−1.5)			[39, 69]
Piperidine	92.3 ±0.5	(−2.4)	107.75 ±0.1	(−1.9)			[39, 69]
1-Methylpiperidine	110.54 ±0.02	(−3.5)	125.50 ±0.05	(0)			[39, 63, 69]
2,6-Dimethylpiperidine	124.4 ±0.1	(−3.3)	141.82	(−3.5)			[71, 175]
Pyridine	77.4 ±0.2	(−1.4)	90.72	(−1.0)			[53, 71, 72, 175]
2-Methylpyridine	94.2 ±0.2	(−1.3)	108.79	(−1.9)			[53, 71, 72, 175]
3-Methylpyridine	93.7 ±0.1	(−0.5)	108.05	(−1.6)			[71, 72, 175]

Table 4 (continued)

	$\dfrac{V_2^0(B)}{cm^3\,mol^{-1}}$	$(b_u)^b$	$\dfrac{V_2^0(BCl)}{cm^3\,mol^{-1}}$	$(b_v)^c$	$\dfrac{V_2^0(BCl_2)}{cm^3\,mol^{-1}}$	$(b_v)^c$	Reference
4-Methylpyridine	94.0 ±0.3	(−2.0)	108.01	(−1.3)			[71, 72, 175]
2,6-Dimethylpyridine	109.9 ±0.1	(−2.9)	126.1	(−5.4)			[71, 72, 175]
2-Aminopyridine	83.02	(−0.2)					[175]
3-Aminopyridine	83.13	(−0.3)					[175]
4-Aminopyridine	82.74	(0.0)	99.63	(−1.2)			[175]
3-Hydroxypyridine	75.35	(0.2)					
1,2-Ethanediamine	62.70±0.2	(−0.1)	73.9 ±0.2	(0.4)	80.0 ±0.2	(−5.1)	[67, 73−75]
1,3-Propanediamine	78.4 ±0.4	(−0.5)	90.7 ±0.4	(0.2)	98.5 ±0.4	(−5.4)	[67, 74−76]
1,4-Butanediamine	93.6 ±0.1		107.48±0.1	(0.3)	116.30±0.3	(−8.0)	[67, 75]
1,5-Pentanediamine	109.40±0.1		123.86±0.2	(0.1)	133.57±0.3	(−7.9)	[67, 75]
1,6-Hexanediamine	124.8 ±0.3		139.84±0.2	(−0.2)	151.28±0.2	(−9.8)	[67, 75, 76]
1,7-Heptanediamine	141.52±0.1		155.56±0.2	(−0.8)	167.20±0.2	(−8.0)	[67, 75]
1,8-Octanediamine	157.3 ±0.1		171.62±0.2	(−0.4)	183.76±0.3	(−10.6)	[67, 75]
1,9-Nonanediamine	172.85±0.1		187.6 ±0.4		199.7 ±0.5	(−13.0)	[67, 75]
1,10-Decanediamine	188.3 ±0.9		203.3 ±0.3		215.51±0.2	(−12.0)	[67, 75]
trans-1,4-Cyclohexanediamine	108.9 ±0.9						[67]
Piperazine	83.41±0.1	(−1.0)			99.26±0.1	(−4.1)	[40, 70]
1-Methylpiperazine	102.30±0.03	(−2.0)			117.48±0.07	(−3.6)	[70]
1,4-Dimethylpiperazine	121.00±0.02	(−2.9)			135.76±0.09	(−3.6)	[70]
Diethylenetriamine	101.2 ±0.4						[73]
Triethylenetetraamine	137.6 ±0.5						[73]
Tetraethylenepentamine	175.9 ±0.6						[73]
Pyrazole	62.26±0.5	(0.1)					[72]
Imidazole	59.95±0.5	(0.1)					[72]
Pyrazine	70.82±0.5	(−0.1)					[72]
Pyrimidine	70.33±0.5	(0.1)					[72]
Pyridazine	70.42±0.5	(−0.2)					[72]
1,3,5-Triazine	54.79±0.5	(0)					[72]
1H-Benzimidazole	98.53±0.5	(0)					[72]
1H-Benzotriazole	94.49±0.5	(0)					[72]
2,2'-Bipyridine	134.29±0.5	(−5)					[72]
4,4'-Bipyridine	133.21±0.5	(6)					[72]

Compound			Reference
4-Phenylpyrimidine		134.09 ± 0.5 (−28)	[72]
Quinoline		115.53 ± 0.5 (1.4)	[72]
Quinazoline		108.48 ± 0.5 (−3.1)	[72]
Quinoxaline		109.16 ± 0.5 (−2.7)	[72]
1,10-Phenanthroline		142.42 ± 0.5 (−12)	[72]
2-Methoxyethaneamine		79.59 ± 0.03 (−0.2)	[74]
Morpholine	92.99 ± 0.03 (−0.2)	82.56 ± 0.04 (−0.5)	[70]
4-Methylmorpholine	111.71 ± 0.06 (−0.5)	101.28 ± 0.03 (−1.2)	[70]
3-Methoxy-1-propaneamine		95.55 ± 0.04 (−0.6)	[74]
Formamide		38.5 ± 0.1 (0.1)	[19, 29, 77–80]
Acetamide		55.7 ± 0.2 (−0.1)	[19, 29, 44, 77]
Propaneamide		71.5 ± 0.2 (−0.3)	[19, 29, 46, 77]
Butaneamide		87.1 ± 0.1	[19, 78]
Hexaneamide		119.2	[82]
N-Methylformamide		56.83 ± 0.1 (−0.2)	[77, 83]
N-Ethylformamide		74.02	[77, 83]
N-Methylacetamide		74.04 ± 0.1	[77, 83]
N-Propylformamide		87.89	[77]
N-Ethylacetamide		90.72	[77]
N-Methylpropaneamide		89.75	[77]
N-Propylacetamide		105.09	[77]
N-Ethylpropaneamide		105.39	[77]
N-Propylpropaneamide		121.53	[77]
N,N-Dimethylformamide		76.85	[77]
		74.50	[84]
N,N-Dimethylacetamide		89.7 ± 0.8 (−0.9)	[77, 85]
N,N-Dimethylpropaneamide		105.36	[77]
N,N-Diethylformamide		106.7 ± 0.5	[77, 85]
N,N-Diethylacetamide		121.5 ± 0.3	[77, 85]
N,N-Diethylpropaneamide		137.67	[77]
		135.3	[85]
N,N-Dipropylacetamide		154.22	[77]
		152.4	[85]
N,N-Bis(1-Methylethyl)acetamide		152.0	[85]
1-Methyl-2-Pyrrolidinone		90.4	[85]
2-Piperidinone		90.3 ± 0.2	[67]

Table 4 (continued)

	$\dfrac{V_2^0(B)}{\text{cm}^3\,\text{mol}^{-1}}$ (b_u)[b]	$\dfrac{V_2^0(BCl)}{\text{cm}^3\,\text{mol}^{-1}}$ (b_v)[c]	$\dfrac{V_2^0(BCl_2)}{\text{cm}^3\,\text{mol}^{-1}}$ (b_v)[c]	Reference
Hexahydro-2H-azepine-2-one	105.0 ±0.2			[67]
Ethanediamide	55.0 ±0.2			[67]
Propanediamide	72.6 ±0.3			[67]
Butanediamide	88.8 ±0.2			[67]
2-Acetamido-N-methylpropaneamide	126.1			[80]
2-Hydroxyacetamide	56.2			[86, 87]
2-Hydroxypropaneamide	73.3 ±0.2			[87, 88]
4-Hydroxybutaneamide	88.90 ±0.13			[87]
4-Hydroxypentaneamide	105.1 ±0.1			[87]
5-Hydroxypentaneamide	105.4 ±0.3			[87]
6-Hydroxyhexaneamide	121.1 ±0.2			[87]
4-Hydroxyoctaneamide	151.9 ±0.3			[87]
4-Hydroxydecaneamide	183.3 ±0.3			[87]
Hydrazinecarbothioamide	64.0 ±0.5			[67]
2-Thioxo-4-imidazoldinone	76.5 ±0.3			[67]
Urea	44.23 ±0.05 (0.1)			[19, 29, 33, 71] [80–82, 89]
Methylurea	62.1 ±0.1			[19, 29]
Ethylurea	80.2			[19]
Propylurea	94.9			[19]
Butylurea	116.1			[19]
(1-Methylethyl)urea	99.0			[19]
N,N-Dimethylurea	78.88			[112]
N,N'-Dimethylurea	80.03 ±0.1			[81, 112]
N,N-Diethylurea	114.9			[19]
N,N'-Diethylurea	116.1			[19]
Tetramethylurea	115.6 ±0.2 (−1.3)			[81, 112]

[a] Tetraalkyl ammonium salts have been reviewed in [68]
[b] In units of $\text{cm}^3\,\text{kg}\,\text{mol}^{-2}$.
[c] In units of $\text{cm}^3\,\text{dm}^3\,\text{mol}^{-2}$.

for trimethyl-amine the difference is 0.93 cm^3 mol^{-1}. In Table 4, differences due to extrapolation procedures have been considered, though the b_u values given refer to Eq. (16).

The ionization equilibria of amines can be described in two ways:

$$B + H_2O = BH^+ + OH^- \tag{19}$$

$$BH^+ = B + H^+. \tag{20}$$

Many authors [53, 59, 60, 91, 92] have considered process (19). However, in order to correspond with the usual reporting of the thermodynamics of amine ionization the views of Cabani et al. [39] to look at process (20) have been adapted. Examination of the available data [6, 59] show that the volume of ionization of monofunctional amines vary a lot less than for carboxylic acids. Cabani et al. [6] sum up the data by noting that ΔV^0 rapidly reaches a constant value of (4.6 ± 0.4) cm^3 mol^{-1} for primary amines, (2.5 ± 0.3) cm^3 mol^{-1} for secondary, and (1 ± 1) cm^3 mol^{-1} for tertiary amines. Changes in other thermodynamic functions like ΔH^0, ΔS^0, and ΔC_p^0 show similar behavior, reaching constant values very rapidly [93–100].

Concerning diamines, ΔV_1^0 (subscript 1 for the first ionization step) is always less than ΔV_2^0. Shahidi and Farrell [75] observed that ΔV_2^0 became constant after 1,7-heptanediamin. It appears that the charge of the amine group may affect the hydration of the neighboring methylene groups as far as the γ-carbon, just as for the carboxylic acids.

Hepler [101] observed a general tendency for ΔV^0 to increase with increasing ΔS^0. His data was mainly on carboxylic and amino acids, although a few amines were included. Since then new data have been added. It still seems that a linear relationship of increasing ΔV^0 with ΔS^0 holds for carboxylic acids, but Cabani et al. [39] now observe that the opposite holds for mono- and bifunctional amines. Since positively charged amines in this respect behave oppositely to the negatively charged carboxylates, it is tempting to speculate that the charges affect the hydration equally far from its center. However, in terms of structural effects, positive and negative charges affect the hydration sheaths quite differently.

It will not come as a surprise that the apparent molar volumes of neutral amines and amides decrease as the concentration increases, reach a minimum, and then increase towards the molar volumes of the pure liquids [77, 141]. The minima appear to be deeper and occur at lower concentrations as the solute sizes increase. This seems to be a general trend for hydrophobic solutes, and the explanation is the same. The hydrocarbon parts of the solute occupy interstitial spaces in the water structure, thereby stabilizing the water structure; "iceberg formation" or hydrophobic hydration [142].

Giaquinto et al. [77] have proposed another model to account for the concentration behavior of the apparent molar volumes of amides. They considered the mixture of amides and water as a mixture of relatively large spheres (amides) and small spheres (water). Since the amount of void volume is proportional to the size of the spheres, water molecules may at first fill this empty space. In the water-poor region, the apparent molar volume of the amides will decrease as observed. However, in order to explain the observed minima and the following increase in the water-rich region going towards infinite dilution, it becomes necessary to return to the "iceberg" model. Water around alkyl groups is thought to adapt a more "icelike" structure. This will reduce the packing efficiency, and the apparent molar volume of the amides will start increasing. Simple as the model is, it still qualitatively describes the concentration dependence of the apparent molar volume of hydrophobic solutes. The model completely ignores solute-solute interactions. However, since by far the largest part of the apparent molar volume is due to the intrinsic volume of the solute, it may not be surprising that models based on the intrinsic volume agree well with observation.

Unlike the amines and amides described above, the apparent molar volume of urea increase monotonically with concentration. Stokes [143] has made a review of the thermodynamic data of urea. The concentration dependence of thermodynamic functions was explained through stepwise association of urea molecules. This idea proved quite successful in explaining the urea activity and the heat of dilution. Urea-urea interactions seemed to be the main factor in determining the aqueous urea solution properties. Hamilton and Stokes [89] have suggested that this can also explain the concentration dependence of the apparent molar volume. However, spectroscopic data have not provided evidence for long-lived urea dimers or oligomeres in solution [144], and it appears that a model based on urea as an overall structure breaker may be more realistic.

9 Carbohydrates (Table 5)

The apparent molar volumes of pentoses, hexoses, and disaccharides exhibit a small positive concentration dependence. Franks et al. [107, 115] have discussed thermodynamic data of carbohydrates in context with NMR and dielectric relaxation data. They find no solute-solute interactions even at fairly high concentrations. They thus favor a specific hydration model where both the number of potential hydration sites and their relative conformations are important factors. It was impossible to distinguish between the hydration properties of glucose, galac-

Table 5. Partial molar volumes of carbohydrates at infinite dilution in water; recommended values at 298.15 K

	$\dfrac{V_2^0}{cm^3\,mol^{-1}}$	$(b_u)^a$	Reference
Arabinose	93.2 $\pm$0.3	(0.7)	[105, 106]
Ribose	95.2 $\pm$0.1	(0.3)	[81, 106, 107]
Xylose	95.1 $\pm$0.3	(0.7)	[105, 106]
2-Deoxy-D-ribose	93.8 $\pm$0.4		[105]
D-Fructose	110.4 $\pm$0.4		[105]
D-Glucose	110.5 $\pm$0.3	(1.1)	[105–107]
D-Glucose	111.7 $\pm$0.3	(1.3)	[17, 81, 105–109, 162]
D-Mannose	111.5 $\pm$0.3	(1.2)	[105, 106]

Table 5 (continued)

	$\dfrac{V_2^0}{\mathrm{cm^3\,mol^{-1}}}$	$(b_u)^a$	Reference
L-Sorbose	110.6 ±0.4		[105]
2-Deoxy-D-glucose	110.4 ±0.5		[105]
Cellobiose	212.0 ±1		[105, 113, 163]
Lactose	209.1 ±1.0	(1.7)	[105, 106]
Maltose	210.0 ±1.0	(3.1)	[105, 106]
Sucrose	211.3 ±0.3	(1.4)	[81, 105, 106, 109–112]
Melibiose	204.0 ±0.7		[105]
Trehalose	206.9 ±0.5		[105]
Cellotriose	309.2 ±1.0		
Maltotriose	307.0 ±2		[105, 162]
Melezitose	308.0 ±2.0	(7.4)	[105, 106]
D-Raffinose	307.0 ±1.0	(18.6)	[105, 106, 109]
2-O-β-Cellotriosylglycerol	359.5 ±1.0		[105]
4-O-β-Laminaribiosyl-D-glucose	312.2 ±0.9		[105]
Cellotetraose	409.0 ±1.0		[105]
3-O-β-Cellotriosyl-D-glucose	410.5 ±1.0		[105]
4-O-β-Laminaribiosylcellobiose	410.5 ±1.9		[105]
Stachyose	401.0 ±1.8		[105]
Cellopentaose	499.6 ±1.0		[105]
Cellohexaose	601.7 ±2.6		[105]
α-Cyclodextrin	601.5 ±1.0		[105, 113, 164]
β-Cyclodextrin	710.0 ±1.0		[105, 113]
γ-Cyclodextrin	803.0		[164]
α-Methyl-D-xyloside	116.30		[168]
β-Methyl-D-xylose	117.24		[168]
3α-D-Glucose-D-fructose(Turanose)	211.30		[168]
Methyl-α-D-galactopyranoside	132.6 ±0.4		[105]
Methyl-β-D-galactopyranoside	132.9 ±0.2		[105]
Methyl-α-D-glucopyranoside	132.9 ±0.3	(0)	[105, 107, 114]
Methyl-β-D-glucopyranoside	133.6 ±0.5		[105, 107]
3-O-Methyl-D-glucose	134.0 ±0.5		[105]
Methyl-α-D-mannopyranoside	132.9 ±0.4		[105]
Ethyl-α-D-glucopyranoside	149.45	(− 2.7)	[114]
Hexyl-α-D-glucopyranoside	215.6	(− 19)	[114]
Octyl-α-D-glucopyranoside	246.8	(−400)	[114]
Cyclohexyl-α-D-glucopyranoside	197.4	(− 1.8)	[114]
Cyclohexyl-α-D-mannopyranoside	197.7	(− 7.8)	[114]
Phenyl-β-D-glucopyranoside	177.5 ±0.7		[105]
Methyl-β-cellobioside	231.0 ±0.4		[105]
Phenyl-β-cellobioside	283.6 ±1.3		[105]
Phenyl-β-lactoside	283.3 ±0.2		[105]
L-Arabitol	103.1 ±0.2	(0.5)	[27, 106]
D-Arabitol	103.31	(0.2)	[27]
Ribitol	103.2 ±0.2	(0.4)	[27, 106]
D-Xylitol	102.4 ±0.3	(0.2)	[106]
Galactitol	119.3 ±0.3	(0.7)	[106]
D-Mannitol	119.4 ±0.3	(0.6)	[10, 12, 27, 106]
Glucitol (Sorbitol)	119.5 ±0.4	(0.6)	[12, 27, 106]
myo-Inositol	121		[17]
meso-erythritol	87.10	(0.1)	[27]

^{a} In units of $\mathrm{cm^3\,kg\,mol^{-2}}$.

tose, and mannose, suggesting that monosaccharides become effective substituents for water when at least three hydroxyl groups are situated equatorially.

The volumetric data do not render this last argument invalid, although there appear to be significant differences in V_2^0 of glucose, mannose, and galactose. Such differences are even more apparent for partial molar compressibilities (see Chap. 2.4). It thus seems that if the number of equatorial hydroxyl groups are of overall importance, their relative positions in the solute molecule must also be considered, and that the solution properties will to some extent depend on this.

10 Amino Acids and Low Molar Mass Proteins (Table 6)

In aqueous solution the amino acids exist as zwitterions [130]. Charges will generally cause a volume contraction due to electrostriction [139, 140]. Cabani et al. [130] have devoted considerable effort to determining the volume change due to zwitterion formation, ΔV_{zw}. The partial molar volume of the neutral form was estimated first by using group contributions. These were obtained by inspecting data for a large number of monofunctional compounds. The second method utilized van der Waals volumes. The resulting ΔV_{zw} were practically independent of the amino acid as long as it was α-amino acids. The mean value was (-14.7 ± 0.8) cm^3 mol^{-1}. When the charged centers, the amino and carboxyl groups, became separated by methylene groups, ΔV_{zw} became more negative, reaching a limiting value of -18.5 cm^3 mol^{-1} when four or more methylene groups interposed.

Millero et al. [120] plotted partial molar volumes of a number of α-amino acids versus their molar mass. A fairly linear relationship resulted. This suggested that the volume contribution of the NH_3^+-CH-COO$^-$ terminal group was independent of the hydrocarbon side chain. This agrees well with the constant ΔV_{zw} values obtained by Cabani et al. [130].

Table 6. Partial molar volumes of amino acids and low molar mass peptides at infinite dilution in water; recommended values at 298.15 K

	$\dfrac{V_2^0}{\text{cm}^3\,\text{mol}^{-1}}$	$(b_v)^a$	Reference
Glycine	43.26±0.07	(0.9)	[52, 87, 116–123, 165–167]
D-α-Alanine	60.58±0.10	(0.8)	[120, 122]
L-α-Alanine	60.54±0.08	(0.7)	[120, 122, 166]
DL-α-Alanine	60.54±0.08	(0.6)	[116, 120–125, 129, 165]
DL-β-Alanine	58.28±0.10	(0.9)	[124–126, 166]
L-Valine	90.75±0.10	(0.3)	[116–118, 120, 122, 129, 165, 166]
DL-Norvaline	91.80±0.20	(0.5)	[122, 165]
L-Leucine	107.77±0.05	(−1.0)	[116, 120, 122, 124, 165]
DL-Norleucine	107.73±0.10	(−0.9)	[116, 122, 124, 165]
L-Isoleucine	105.80±0.07	(−1.3)	[165]
L-Proline	82.76±0.10	(0.4)	[120, 165, 167]
L-Hydroxyproline	84.49±0.05	(0.7)	[116, 165]
L-Phenylalanine	121.5 ±0.2	(9.5)	[120, 123, 127, 165]
D-Tryptophan	143.82±0.13	(5.0)	[120]

Table 6 (continued)

	$\dfrac{V_2^0}{\text{cm}^3\,\text{mol}^{-1}}$	$(b_v)^a$	Reference
DL-Tryptophan	143.85±0.2	(0.5)	[122, 165]
DL-Methionine	105.35±0.02	(1.1)	[120]
L-Methionine	105.57±0.02	(−0.4)	[165]
L-Serine	60.62±0.05	(1.3)	[87, 121, 128, 165, 166, 167]
L-Threonine	76.90±0.10	(1.0)	[132, 165, 166]
L-Cysteine	73.45±0.03	(1.3)	[120]
L-Asparagine	78.0 ±0.5		[136]
L-Asparagine hydrate	95.63±0.05	(1.1)	[165]
DL-Aspartic acid	73.75±0.08	(9.8)	[120, 129]
L-Aspartic acid	74.8 ±0.2	(−4)	[165]
Glutamic acid	85.88±0.01	(6.7)	[120]
L-Glutamic acid	89.85±0.1	(1.1)	[165]
Lysine	108.5		[116]
L-Arginine	127.42±0.01	(1.4)	[120, 165]
Histidine	98.9 ±0.2	(−0.3)	[120, 125]
L-Histidine	98.3 ±0.1	(10)	[165]
Acetylhistidine	134.0 ±0.5		[136]
1-Acetylglycine	88.39±0.05	(−0.1)	[165]
1-Acetyl-DL-alanine	105.78±0.04	(0.2)	[165]
1-Acetyl-DL-aminobutanoic acid	121.44±0.05	(−0.5)	[165]
1-Acetyl-DL-valine	136.63±0.03	(0.6)	[165]
1-Acetyl-DL-leucine	152.47±0.04	(−0.1)	[165]
2-amino-2-methylpropanoic acid	77.55±0.15	(0.2)	[116, 123, 166]
2-Aminobutanoic acid	75.6 ±0.1	(0.5)	[116, 118, 124]
3-Aminobutanoic acid	76.3 ±0.1	(0.6)	[116, 124, 130]
4-Aminobutanoic acid	73.2 ±0.2	(1.1)	[87, 124, 130]
5-Aminopentanoic acid	87.6 ±0.6	(1.5)	[87, 123, 124, 130]
6-Aminohexanoic acid	104.2 ±0.2	(0.0)	[82, 87, 116, 124, 130, 131]
7-Aminoheptanoic acid	120.0 ±0.1		[87]
8-Aminooctanoic acid	136.1 ±0.1		[87]
9-Aminononanoic acid	151.3 ±0.1		[87]
10-Aminodecanoic acid	167.3 ±0.1		[87]
11-Aminoundecanoic acid	182.6 ±0.1		[87]
Diglycine	76.30±0.1	(2.6)	[118, 121, 129, 133, 134, 165]
Triglycine	112.2 ±0.3	(4.4)	[121, 129, 134, 165]
Tetraglycine	149.7 ±0.1	(24)	[121, 165]
Pentaglycine	187.1 ±0.1	(515)	[121]
Glycyl-L-alanine	92.66±0.06	(1.6)	[133]
Glycyl-DL-alanine	92.37±0.02	(2.9)	[165]
L-Alanylglycine	95.01±0.1	(0.9)	[135]
DL-Alanylglycine	95.22±0.06	(0.9)	[135]
Dialanine	110.6 ±0.3	(1.8)	[121, 133]
Trialanine	163.80±0.1	(4.9)	[121]
Tetraalanine	220.1 ±0.1	(30)	[121]
Glycol-DL-2-aminobutanoic acid	107.81±0.05	(2.6)	[165]
Glycyl-L-valine	121.99±0.02	(1.7)	[165]
Glycyl-L-leucine	139.70±0.07	(0.9)	[165]
L-Leucylglycine	145.34±0.04	(0.6)	[165]
Diserine	111.8 ±0.1	(11)	[121]
Triserine	166.0 ±0.2	(4)	[121]

^{a} In units of $\text{cm}^3\,\text{kg}\,\text{mol}^{-1}$.

It has been shown that the formation of the α-helix structure requires a minimum number of amino acids in the peptide; six appears to be the very least [138]. The peptides reported in Table 6 can thus be looked upon as dipolar ions that may undergo rapid conversions between possible conformational states. It also appears that the partial molar volumes of these oligomers vary in a regular manner. It seems likely that the increment per amino acid unit should be constant, but this occurs first after a minimum of four amino acid units have been added.

The volumetric contribution of the peptide bond ($CO-NH_2$) is another important quantity. Edsall [116] calculated the partial molar volume at infinite dilution of this $CO-NH_2$-group to be 20.0 cm^3 mol^{-1} and 36.3 cm^3 mol^{-1} for the CH_2-$CO-NH_2$ (glycyl) residue. Jolicoeur and Boileau [121] report (38.0 ± 0.5) cm^3 mol^{-1} for the latter. As they point out, the glycyl group volume plus an additivity scheme for the side chain contributions will clearly provide an important basis for predicting V_2^0 values of extended polypeptides and proteins in water.

11 Group Additivity of Partial Molar Volumes (Table 7)

The partial molar volume of a solute at infinite dilution can be regarded as the sum of an intrinsic volume plus volumetric effects of solute-solvent interactions. The latter can be further divided into different contributions. The most commonly used are the void volume, structural or cage effects, and for ionic solutes electrostriction [139, 140, 146, 147].

Concerning organic solutes, the intrinsic volume of each functional group should be constant irrespective of its environment. If the hydration of each functional group is only marginally affected by neighboring groups, the partial molar volumes of such solutes at infinite dilution must be additive. This idea is not new. Indeed as far back as 1899, Traube [148] reviewed the available data and found that partial molar volumes could be obtained by summation of group or atomic increments plus a covolume, or void volume, of 12.4 cm^3 mol^{-1}. Later Terasawa et al. [16] and Edward and Farrell [149] proposed similar models based on van der Waals radii to calculate the intrinsic volume.

However, the most straightforward additivity scheme is the empirical method of calculating the increment per methylene group for homologous series. Cabani et al. [6] used data for 28 solutes of the series primary and secondary alcohols, primary and secondary aliphatic amines, and primary and secondary n-aliphatic protonated amines to obtain $V^0(CH_2)=(15.9\pm0.4)$ cm^3 mol^{-1}. Extending this investigation by using data on alkylaminehydrobromides [66], tetraalkylammonium halides [150], azoniaspiroalkanes [151], hydrocarbons [152], and additional data from Tables 1–6 the result is (15.9 ± 0.3) cm^3 mol^{-1}.

In cyclic compounds the methylene group contribution calculated by Cabani et al. [6] from data on 21 solutes, cyclic alcohols, cyclic amines, and cyclic ethers, is $V^0(CH_2)=(14.2\pm0.4)$ cm^3 mol^{-1}. However, it seems that this relatively well-defined result for cyclic solutes may have been somewhat fortuitous. Kiyohara et al. [40] found a cyclic contribution of the methylene group; $V^0(CH_2)=16.9$ cm^3 mol^{-1}, by comparing data for tetrahydropyran, 1,4-dioxane, morpholine, piperazine, and piperidine. Jolicoeur et al. [153] found that the methylene group

Table 7. Group partial molar volumes at 298.15 K, and the shrinkage σ_i of Eq. (23)

Group	$V_2^0/\mathrm{cm^3\,mol^{-1}}$	$\sigma_i/\mathrm{cm^3\,mol^{-1}}$
$-CH_3$	26.5	
$-CH_2OH$	28.2	
$-CONH_2$	28.8	
$-COOH$	25.8	
$-COO^-$	17.7[a]	
$-CHO$	17.5	
$-NH_2$	15.4	6.5[c]
$-CH_2-$	15.9[b]	
$-CHOH-$	16.4	5.9[c, d]
$-\underset{\overset{\|}{O}}{C}-$	13.1	4.8[c]
$-\underset{\overset{\|}{O}}{CO}-$	19.4	4.8[c]
$-\underset{H}{\overset{\|}{C}}=$	14.5	
$=N-$	5.5	
$-\underset{H}{\overset{\|}{N}}-$	11.6	4.3[c]
$\overset{}{C}=C\overset{}{}$	9.0	

[a] Provided that neighboring CH_2-groups are considered (erroneously) unaffected by the charge of the carboxylate group.
[b] In cyclic solutes the value is $14.2\,\mathrm{cm^3\,mol^{-1}}$.
[c] From [67].
[d] The σ_i value of the OH group.
Millero et al. [120] have presented group values for several more groups of special relevance to amino acids.

contribution of cyclic solutes depended markedly on the ring size. The differences between the extremes were as large as $1.6\,\mathrm{cm^3\,mol^{-1}}$.

Despite some discrepancies for cyclic solutes, the group partial molar volumes are still well suited for predicting partial molar volumes of different solutes in water. Usually partial molar volumes calculated from group values are found to agree very well with experimental values, well within $\pm 1\,\mathrm{cm^3\,mol^{-1}}$. Special care should be exercised with cyclic solutes and charged solutes. In Table 7 group partial molar volumes for a number of relevant functional groups are presented. These have been calculated by comparing data for different homologous series, some have been calculated earlier, a few have been added here. It is also worth mentioning that Millero et al. [120] have presented a table of group values relating to the amino acids. This table is therefore not reproduced here.

In a comprehensive review Cabani et al. [154] have chosen another additivity scheme. The partial molar volume has been represented by the following equation:

$$V_2^0 = A_z + \Sigma n_j\, B_z(j) + \Sigma C_z(Y_1, Y_2, \cdots Y_m). \tag{21}$$

Here $B_z(j)$ represents the contribution of the functional group j. A_z is a constant term, a term similar to the covolume as introduced by Traube [148]. It was found to be $13.4 \, \text{cm}^3 \, \text{mol}^{-1}$ which is reasonably close to Traube's old findings. $C_z(Y_1, Y_2, \cdots Y_m)$ is a correction term necessary for polyfunctional solutes.

The group contributions $B_z(j)$, as well as the correction terms, have been tabulated for numerous groups in Cabani's paper, and we shall not reproduce them here. However, it is interesting to compare some $B_z(j)$ values with the group partial molar values of Table 7. B_z of the methylene group is $15.80 \, \text{cm}^3 \, \text{mol}^{-1}$ which is surprisingly close to the methylene group value of $15.9 \, \text{cm}^3 \, \text{mol}^{-1}$. Generally it seems that for internal groups the agreement between the B_z values of Eq. (21) and the group partial molar volumes of Table 7 is quite good. However, large differences appear for terminal groups. For instance, the B_z value of the methyl group is $19.06 \, \text{cm}^3 \, \text{mol}^{-1}$ compared to $26.4 \, \text{cm}^3 \, \text{mol}^{-1}$. There must, of course, be different group values for the two additivity schemes. In Eq. (21) the void volume is added as a separate term, A_z, while the group values of Table 7 include any void volumes. One must, of course, be careful in interpreting these group values in terms of solute-solvent interactions. However, by comparing the two sets of values, it seems that the significant part of the void volume around these solutes in water can be located to the end groups.

Recently Rao et al. [176] have proposed a method for calculating the partial molar volumes of α-amino acids. The method is based on knowledge of the partial molar volume of glycine and methane and group contributions of the CH_3R species. The model works well for the 11 amino acids tested.

12 Models Based on van der Waals Volumes

As mentioned above, the interpretation of partial molar volumes in terms of solute-solvent interactions is obscured by the size of the intrinsic volume. One possible way to estimate the intrinsic volume is to identify it with the corresponding van der Waals volume. This approach has been used both by Terasawa et al. [16] and Edward and Farrell [149].

Terasawa et al. [16] found that the partial molar volume of homologous series could be represented by the following linear function:

$$V_2^0 = a \, V_w + b. \tag{22}$$

The parameters a and b varied for different homologous series. By analyzing the a and b parameters, they found that the void volume of a solute containing a hydroxyl group is $4.5 \, \text{cm}^3 \, \text{mol}^{-1}$ less than the hydrocarbon with the same van der Waals radius when they are dissolved in water.

In their first paper on the subject, Edward and Farrell [149] suggested the following equation for the partial molar volume:

$$V_2^0 = \tfrac{4}{3} \pi (r_w + \Delta)^3 . \tag{23}$$

Here r_w is the van der Waals radius calculated from van der Waals volumes. Δ is a measure of the void volume. The value of Δ was first calculated to 0.053 nm, later it was recalculated to 0.057 nm. It also turned out that it was necessary to add a term when hydrophilic groups became part of the solute molecule, and the final equation read:

$$V_2^0 = \tfrac{4}{3}\pi(r_w + \Delta)^3 + n_i\sigma_i, \tag{24}$$

where σ_i is the correction necessary for each functional group i. This model does not differ very much from the model of Cabani et al. [154] [Eq. (21)], and it also turns out that σ_i is negative just as the C_z term of Eq. (21) became negative. Shahidi et al. [67] have tabulated σ_i values for several groups. For instance, for the hydroxyl group they find $\sigma_i = -5.9$ cm^3 mol^{-1} which can be compared to -4.5 cm^3 mol^{-1} from Terasawa et al. [16]. Some σ_i values have been included in Table 7. In addition to these values, it is interesting to note that the ether oxygen does not lead to any volume decrease according to this model, i.e., $\sigma_i = 0$.

Finally, the scaled particle theory [155–157] should be mentioned. Here the introduction of a solute is represented by three distinct steps. First, a point particle is introduced in the solvent. The volume will increase due to the kinetic contribution to pressure. Second, the point particle is expanded against the surface forces to make room for the real solute particle. Third, the solute is allowed to interact with the solvent. The reversible work required to create the cavity can be derived from the scaled particle theory [158, 159]. Within this framework it is thus possible to calculate the sum of the intrinsic and void volumes; i.e., the volume of the total cavity. By subtracting this from the measured partial molar volume, one may obtain information about solute-solvent interactions and their volumetric effect. Hirata and Arakawa [160] applied this principle to simple ionic solutions, showing that in water the interaction volume decreased with increasing size of the ions. Høiland and Vikingstad [161] checked the theory on aqueous alcohol solutions. The validity of treating alcohols as spheres within the framework of the scaled particle theory may be questioned. However, the result showed that the interaction volume was negative. In this respect it corresponds to negative σ_i in the Edward Farrell model. The volume of interaction also turned out to be more negative for the diols. It was also seen that the volume of interaction increased (became less negative) as the alkane chain increased. The result seems reasonable, but should not be taken too seriously in view of a rather primitive model.

13 Conclusions

The large amount of partial molar volume data acquired over the years clearly represents an important source of information about solute-solvent interactions. However, since no rigorous statistical thermodynamic theory has yet emerged, the interpretations must rely on models. The models reviewed in this paper are all rather simple, and they do not discriminate well between specific interactions. For instance, the partial molar volume of carbohydrates of the same size may differ by about 1 cm^3 mol^{-1}, probably due to solute-solvent interactions that reflect the stereochemical differences. Yet none of the models will predict or explain such differences. However, a few interesting facts still emerge. First the intrinsic volume is by far the largest contribution to the partial molar volume. This is the drawback, since it may well overshadow contributions from interesting solute-solvent interactions. For pure hydrocarbons it seems that the sum of the intrinsic volume and a void volume is sufficient to describe the solution properties. Hydrophilic groups apparently decrease the partial molar volume, probably because hydrogen bonding disrupts the open water structure around hydrophobic groups. It further seems that several neighboring hydrophilic groups strengthen this volume contraction, though this is obvious only from the model proposed by Cabani et al. [154].

In any case, the models represent a useful rationale in the fact that they can be used to predict the partial molar volume of solutes at infinite dilution, and that they represent a basis for the interpretation of solute-solvent interactions of large and complex molecules in solution.

14 Addendum

Recently two articles have been published that deal with partial molar volumes of nucleobases and nucleosides [169] and n-alkane-nitriles [170]. The results can be found in Tables 8 and 9.

Table 8. Partial molar volumes and the volume change of stacking of some nucleobases and nucleosides at infinite dilution in water; values at 298.15 K. (AK) calculated from attenuated constant model. (SEK) calculated from sequential equal constant model

	$\dfrac{V_2^0}{cm^3\,mol^{-1}}$	$\dfrac{\Delta V_{stack}^0(AK)}{cm^3\,mol^{-1}}$	$\dfrac{\Delta V_{stack}^0(SEK)}{cm^3\,mol^{-1}}$	Reference
Purine	85.40	− 3.3	− 4.7	[169]
Caffeine	145.20	− 5.4	− 6.6	[169, 177]
Cytidine	153.50	17.0	19.0	[169]
Uridine	151.45	4.4	5.3	[169]
Thymidine	167.55	0.0		[169]

Table 9. The partial molar volumes of n-alkanenitriles at infinite dilution in water at 298.15 K

	$\dfrac{V_2^0}{cm^3\,mol^{-1}}$	Reference
Acetonitrile	47.42	[170]
Propanenitrile	64.34	[170]
Butanenitrile	80.59	[170]
Pentanenitril	96.73	[170]
Hexanenitrile	111.91	[170]
Octanenitrile	146.18	[170]

The nucleobases and nucleosides self-associate vertically, which is generally called stacking. The apparent molar volumes are thus no longer linear with concentration. From the volumetric data the changes in partial molar volume as these compounds associate have been calculated. This has also been included in Table 8; i.e.:

$$\Delta V^0_{stack} = V^0_{2, stack} - V^0_{2, mono}. \tag{25}$$

Here $V^0_{2, stack}$ is the partial molar volume at infinite dilution of a nucleobase or a nucleoside in the aggregated state (in the stack), and $V^0_{2, mono}$ the partial molar volume at infinite dilution of the monomeric species.

In order to calculate ΔV^0_{stack}, one has to know the equilibrium constant for stacking. Unfortunately, this is not known for every step as the stack grows, and only an average quantity can be determined. This means that one must rely on models for calculating the equilibrium constant. Two main models have been described in the literature [171–174]. The sequential isodesmic model (SEK) assumes that the enthalpy change is independent of the aggregate size. In the other, the attenuated constant model (AK), the equilibrium constant for the addition of the i-th monomer to the stack is given as $K = K_i/i$. As seen from Table 8, ΔV^0_{stack} varies with the model used, but the general feature remains; ΔV^0_{stack} is positive for the nucleosides and negative for the nucleobases.

References

1. Masson DO (1929) Phil Mag 8:218
2. Redlich O, Rosenfeld P (1931) Z Electrochem 37:705
3. Redlich O, Meyer DM (1964) Chem Rev 64:221
4. Alexander DM (1959) J Chem Eng Data 4:252
5. Friedman ME, Scheraga HA (1965) J Phys Chem 69:3795
6. Cabani S, Conti G, Lepori L (1974) J Phys Chem 78:1030
7. Cabani S, Conti G, Matteoli E (1976) J Solution Chem 5:751
8. Manabe M, Koda M (1975) Bull Chem Soc Jpn 48:2367
9. Nakajima T, Komatsu T, Nakagawa T (1975) Bull Chem Soc Jpn 48:783
10. Høiland H, Vikingstad E (1976) Acta Chem Scand A30:182
11. Høiland H (1980) J Solution Chem 9:857
12. Edward JT, Farrell PG, Shahidi F (1977) J Chem Soc Faraday Trans I 73:705
13. Jolicoeur C, Lacroix G (1976) Can J Chem 54:624
14. Franks F, Smith HT (1968) Trans Faraday Soc 64:2962
15. Franks F, Smith HT (1968) J Chem Eng Data 13:538
16. Terasawa S, Itsuki H, Arakawa S (1975) J Phys Chem 79:2345
17. Neal JL, Goring DAI (1970) J Phys Chem 74:658
18. Kuppers JR (1975) J Phys Chem 79:2105
19. Herskovits TT, Kelly TM (1973) J Phys Chem 77:381
20. Nakanishi K, Kato N, Maruyama MJ (1967) J Phys Chem 71:814
21. Brower KR, Peslak J, Elrod J (1969) J Phys Chem 73:207
22. Bruun SG, Hvidt A (1977) Ber Bunsen-Ges Phys Chem 84:930
23. Perron G, Desnoyers JE (1981) J Chem Thermodynam 13:1105
24. Hamann SD, Linton M (1974) J Chem Soc Faraday Trans I 70:2239
25. Desnoyers JE, Page R, Perron G, Leduc P-A, Platford RF (1973) Can J Chem 51:2129
26. Hopkins HP, Duer WC, Millero FJ (1976) J Solution Chem 5:263
27. Dipaola G, Belleau B (1977) Can J Chem 55:3825
28. Harada S, Nakajima T, Komatsu T, Nakagawa T (1979) J Solution Chem 8:267

29. Longsworth LG (1963) J Phys Chem 67:689
30. Manabe M, Koda M (1976) Mem Niihama Tech Coll 12:68
31. Ray A, Nemethy G (1973) J Chem Eng Data 18:309
32. Cabani S, Lepori L, Matteoli E (1976) Chim Ind (Paris) 58:221
33. Indelli A (1963) Ann Chim 53:605
34. Roux G, Roberts D, Perron G, Desnoyers JE (1980) J Solution Chem 9:629
35. Roux G, Perron G, Desnoyers JE (1978) J Solution Chem 7:639
36. Harada S, Nakjima T, Komatsu T, Nakagawa T (1978) J Solution Chem 7:463
37. Franks F, Quickenden MAJ, Reid DS, Watson B (1970) Trans Faraday Soc 66:582
38. Lepori L, Mollica V (1978) J Chem Eng Data 23:65
39. Cabani S, Conti G, Lepori L (1972) J Phys Chem 76:1338
40. Kiyohara O, Perron G, Desnoyers JE (1975) Can J Chem 53:2591
41. Lepori L, Mollica V (1978) J Polym Sci A 2:1123
42. Wallace WL, Shephard CS, Underwood CJ (1968) Chem Eng Data 13:11
43. Høiland H, Ringseth J, unpublished results
44. Kiyohara O, Perron G, Desnoyers JE (1975) Can J Chem 53:3263
45. Bøje L, Hvidt A (1976) J Chem Thermodynam 8:105
46. Roux G, Perron G, Desnoyers JE (1978) Can J Chem 56:2808
47. Morild E, Tvedt I (1978) Acta Chem Scand B 32:593
48. Desnoyers JE, Ichhaporia M (1969) Can J Chem 47:4639
49. Liotta CL, Abidaud A, Hopkins JP (1972) J Am Chem Soc 96:8624
50. Cabani S, Conti G, Lepori L (1971) Trans Faraday Soc 67:1933
51. Reyher F (1888) Z Phys Chem 2:744
52. King EJ (1969) J Phys Chem 73:1220
53. Hamann SD, Lim SC (1954) Aust J Chem 7:329
54. Palma M, Morel JP (1976) J Chem Phys 73:643
55. Høiland H (1974) Acta Chem Scand A 28:699
56. Redlich O, Nielsen LE (1942) J Am Chem Soc 64:761
57. Watson GA, Felsing WA (1941) J Am Chem Soc 63:410
58. Høiland H, Vikingstad E (1975) J Chem Soc Faraday Trans I 71:2007
59. Hamann SD (1974) In: Conway BE, Bockriss JOM (eds) Modern aspects of electrochemistry, vol 9. Plenum, New York
60. Stokes RH (1975) Aust J Chem 28:2109
61. Jones G, Talley SK (1933) J Am Chem Soc 55:624
62. Longsworth LG (1935) J Am Chem Soc 57:1185
63. Laliberté LH, Conway BE (1970) J Phys Chem 74:4116
64. Kaulgud MV, Bhagde VS, Shirastava A (1982) J Chem Soc Faraday Trans I 78:313
65. Verrall RE, Conway BE (1966) J Phys Chem 70:3961
66. Desnoyers JE, Arel M (1967) Can J Chem 45:359
67. Shahidi F, Farrell PG, Edward JT (1977) J Chem Soc Faraday Trans I 73:715
68. Millero FJ (1971) Chem Rev 71:147
69. Cabani S, Conti G, Lepori L, Leva G (1972) J Phys Chem 76:1343
70. Cabani S, Mollica V, Lepori L, Lobo ST (1977) J Phys Chem 81:982
71. Mollica V, Lepori L (1977) Chim Ind (Paris) 59:877
72. Enea O, Jolicoeur C, Hepler LG (1980) Can J Chem 58:704
73. Lawrence J, Conway BE (1971) J Phys Chem 75:2353
74. Cabani S, Mollica V, Lepori L, Lobo ST (1977) J Phys Chem 81:987
75. Shahidi F, Farrell PG (1978) J Solution Chem 7:549
76. Sakurai M (1973) Bull Chem Soc Jpn 46:1596
77. Giaquinto AR, Lindstrom RE, Swarbrick J, LoSurdo A (1977) J Solution Chem 6:687
78. Dunstan AE, Mussell AG (1910) J Chem Soc 97:1935
79. de Visser C, Pel P, Somsen G (1977) J Solution Chem 6:571
80. Bøje L, Hvidt Å (1971) J Chem Thermodynam 3:663
81. LoSurdo A, Chin C, Millero FJ (1978) J Chem Eng Data 23:197
82. Daniel J, Cohn EJ (1936) J Am Chem Soc 58:415
83. de Visser C, Heuvelsland WJM, Dunn LA, Somsen G (1978) J Chem Soc Faraday Trans I 74:1159

84. de Visser C, Perron G, Desnoyers JE, Heuvelsland WJM, Somsen G (1977) J Chem Eng Data 22:74
85. Assarson P, Eirich FR (1968) J Phys Chem 72:2710
86. Gucker FT, Ford WL, Moser CE (1939) J Phys Chem 43:153
87. Shahidi F, Farrell PG (1978) J Chem Soc Faraday Trans I 74:858
88. Gucker FT, Allen TW (1942) J Am Chem Soc 64:191
89. Hamilton D, Stokes RH (1972) J Solution Chem 1:213
90. Evans AG, Hamann SD (1951) Trans Faraday Soc 47:34
91. Kauzmann W, Bodanzky A, Rasper J (1962) J Am Chem Soc 84:1777
92. King EJ (1965) Acid-base equilibria. Pergamon, Oxford
93. Bates RG, Hetzer HB (1949) J Res Natl Bur Stand 42:419
94. Bates RG, Hetzer HB (1961) J Phys Chem 65:667
95. Cox MC, Everett DH, Landsman DA, Munn RJ (1968) J Chem Soc B 1373
96. Christensen JJ, Izatt RM, Wrathall DP, Hansen LD (1969) J Chem Soc A 1212
97. Everett DJ, Wynne-Jones WFK (1941) Proc R Soc Edinb Sect A 177:499
98. Öjelund G, Wadsö I (1968) Acta Chem Scand 22:2691
99. Cabani S, Conti G, Lepori L (1971) Trans Faraday Soc 67:1943
100. Cabani S, Conti G, Martinelli A, Matteoli E (1973) J Chem Soc Faraday Trans I 69:2112
101. Hepler LG (1965) J Phys Chem 69:965
102. Franks F (1973) In: Franks F (ed) Water, a comprehensive treatise, vol 2. Plenum, New York
103. Høiland H (1975) J Chem Soc Faraday Trans I 71:797
104. Høiland H (1974) J Chem Soc Faraday Trans I 70:1180
105. Shahidi F, Farrell PG, Edward JT (1976) J Solution Chem 5:807
106. Høiland H, Holvik H (1978) J Solution Chem 7:587
107. Franks F, Ravenhill JR, Reid DS (1972) J Solution Chem 1:3
108. Neal JL, Goring DAI (1970) Can J Chem 48:3745
109. Longsworth LG (1955) In: Shedlovsky T (ed) Electrochemistry in biology and medicine. Wiley, New York
110. Garrod JE, Herrington TM (1970) J Phys Chem 74:363
111. Sangster J, Teng TT, Lenzi FJ (1976) J Solution Chem 5:575
112. Philip PR, Perron G, Desnoyers JE (1974) Can J Chem 52:1709
113. Høiland H, Hald LH, Kvammen O (1981) J Solution Chem 10:775
114. Brown GM, Dubreuil P, Ichhaporia FM, Desnoyers JE (1970) Can J Chem 48:2525
115. Tait MS, Sugett A, Franks F, Ablett S, Quickenden PA (1972) J Solution Chem 1:131
116. Cohn EJ, Edsall JT (1943) Proteins, amino acids, and peptides as ions and dipolar ions. Reinhold, New York
117. Gucker FT, Ford WL, Moser CE (1939) J Phys Chem 43:153
118. Ellerton HD, Reinfels G, Mulcahy DE, Dunlop PJ (1964) J Phys Chem 68:398
119. Tyrrell HFV, Hennerby M (1968) J Chem Soc A 2724
120. Millero FJ, LoSurdo A, Shin C (1978) J Phys Chem 82:784
121. Jolicoeur C, Boileau J (1978) Can J Chem 56:2707
122. DiPaola G, Bellau B (1978) Can J Chem 56:1827
123. Kirchnerova J, Farrell PG, Edward JT (1976) J Phys Chem 80:1974
124. Ahluwalia JC, Ostiguy C, Perron G, Desnoyers JE (1977) Can J Chem 55:3364
125. Gucker FT, Allen TW (1942) J Am Chem Soc 64:191
126. Devine W, Lowe BM (1971) J Chem Soc A 2113
127. Greenstein JP, Winitz M (1961) Chemistry of the amino acids, vol 1. Wiley, New York
128. Krescheck GC, Benjamin LJ (1964) J Phys Chem 68:2476
129. Cohn EJ, McMeekin TL, Edsall JT, Blanchard MH (1934) J Am Chem Soc 56:784
130. Cabani S, Conti G, Matteoli E, Tiné MR (1981) J Chem Soc Faraday Trans I 77:2377
131. Bondi A (1964) J Phys Chem 68:441
132. Cabani S, Conti G, Matteoli E, Tiné MR (1981) J Chem Soc 77:2385
133. Dyke SH, Hedwig GR, Watson ID (1981) J Solution Chem 10:321
134. Høiland H unpublished results
135. Kumaran MK, Hedwig GR, Watson ID (1982) J Chem Thermodynam 14:93
136. Greenstein JP, Wyman J (1936) J Am Chem Soc 58:463

138. Hruby VJ (1974) The chemistry and biology of amino acids, peptides, and proteins, vol 3. Dekker, New York
139. Drude P, Nernst W (1894) Z Phys Chem 15:79
140. Desnoyers JE, Conway BE, Verrall RE (1965) J Chem Phys 43:243
141. Kaulgud MV, Patil KJ (1974) J Phys Chem 78:714
142. Frank HS, Evans MW (1945) J Chem Phys 13:507
143. Stokes RH (1967) Aust J Chem 20:2087
144. Finer EG, Franks F, Tait MJ (1972) J Am Chem Soc 94:4424
146. Glueckauf E (1965) Trans Faraday Soc 61:914
147. Millero F (1967) J Phys Chem 71:4567
148. Traube J (1899) Samml Chem Vorträge 4:19
149. Edward JT, Farrell PG (1975) Can J Chem 53:2965
150. Desnoyers JE, Arel M (1969) Can J Chem 47:547
151. Wen W-Y, LoSurdo A, Jolicoeur C, Boileau J (1976) J Phys Chem 80:466
152. Masterton WL (1954) J Chem Phys 22:1830
153. Jolicoeur C, Boileau J, Bazinet S, Picker P (1975) Can J Chem 53:716
154. Cabani S, Gianni P, Mollica V, Lepori L (1981) J Solution Chem 10:563
155. Reiss H, Frisch HL, Lebovitz JL (1969) J Chem Phys 31:369
156. Reiss H, Frisch HL, Helland E, Lebovitz LL (1960) J Chem Phys 32:119
157. Reiss H (1965) Adv Chem Phys 9:1
158. Pierotti RA (1963) J Phys Chem 67:1840
159. Pierotti RA (1965) J Phys Chem 69:281
160. Hirata F, Arakawa K (1973) Bull Chem Soc Jpn 46:3367
161. Høiland H, Vikingstad E (1976) Acta Chem Scand A 30:692
162. Miyajima K, Sawada M, Nakagaki M (1983) Bull Chem Soc Jpn 56:1954
163. Herrington TM, Pethybridge AD, Parkin BA, Roffey MG (1983) J Chem Soc Faraday Trans I 79:845
164. Miyajima K, Sawada M, Nakagaki M (1983) Bull Chem Soc Jpn 56:3556
165. Mishra AK, Ahluwalia JC (1984) J Phys Chem 88:86
166. Ogawa T, Yasuda M, Mizutani K (1984) Bull Chem Soc Jpn 57:662
167. Vliegen J, Yperman J, Mullens J, Francois JP, Van Poucke LC (1984) J Solution Chem 13:245
168. Jasra RV, Ahluwalia JC (1984) J Chem Thermodynamics 16:583
169. Høiland H, Skauge A, Stokkeland I (1984) J Phys Chem 88:6350
170. Janelli L, Pansini M, Jalenti R (1984) J Chem Eng Data 29:266
171. Ts'o POP, Melvin IS, Olson AC (1963) J Am Chem Soc 85:1289
172. Ts'o POP, Chan SI (1964) J Am Chem Soc 86:4176
173. Heyn MD, Bretz R (1975) Biophys Chem 3:35
174. Garland F, Christian SD (1975) J Phys Chem 79:1274
175. Mollica V, Lepori L (1984) Z Phys Chem Neue Folge 135:11
176. Rao MVR, Atregi M, Rajeswari MR (1984) J Chem Soc Faraday Trans I 80:2027
177. Cesaro A, Russo R, Crescenzi V (1976) J Phys Chem 80:335

Chapter 3 Specific Volumes of Biological Macromolecules and Some Other Molecules of Biological Interest

HELMUT DURCHSCHLAG

Symbols

B: binding of a component to a macromolecule ($g\ g^{-1}$)
E: electrostatic (Donnan) exclusion of a component from a macromolecule ($g\ g^{-1}$)
M: molecular weight
N: number of components
P: pressure
T: temperature (K)
V: total volume (cm^3)
$\bar{V}$:   partial molar volume of a component ($cm^3\ mol^{-1}$)
W: weight percentage of a component

c: concentration of a component ($g\ cm^{-3}$)
f: weight fraction of a component
g: grams of a component
h: hydration ($mol\ mol^{-1}$)
i: any solution component
j: any nonmacromolecular solution component
k: apparatus constant of digital density meter
m: molality
n: number of moles
$\bar{v}$:   partial specific volume of a component ($cm^3\ g^{-1}$); cf. Eq. (8)
$\bar{v}_a$:   assumed (average) value for the partial specific volume ($cm^3\ g^{-1}$)
$\bar{v}_c$:   calculated partial specific volume ($cm^3\ g^{-1}$)
$(\bar{v}_c)_{mod.}$: calculated partial specific volume ($cm^3\ g^{-1}$), obtained from modified procedures
v':   isopotential specific volume ($cm^3\ g^{-1}$); cf. Eq. (11)
w: weight molality ($g\ g^{-1}$)

δ: amount of ligand (e.g., detergent) bound to a macromolecule ($g\ g^{-1}$)
μ: chemical potential
ξ: preferential interaction parameter ($g\ g^{-1}$)
ϱ: density ($g\ cm^{-3}$), density of the solution ($g\ cm^{-3}$)
τ: period of oscillation
ϕ: apparent specific volume ($cm^3\ g^{-1}$); cf. Eq. (9)
ϕ':   apparent isopotential specific volume ($cm^3\ g^{-1}$); cf. Eq. (12)
ϕ'_c: calculated isopotential specific volume ($cm^3\ g^{-1}$)

Subscripts

1:      component 1, principal solvent (generally water)
2:      component 2, macromolecule (e.g., a protein)
3:      component 3, a membrane-diffusible solute in the solvent medium (e.g., a salt)
d: detergent
l: ligand
np: nonprotein component
p: protein component

r: amino acid residue
s: complex solvent mixture (e.g., water + salt(s), buffer)

Superscripts

^{0}:   vanishing concentration of the ith component
$'$:   designates isopotential conditions

Abbreviations

BSA: bovine serum albumin
DTE: dithioerythritol
DTT: dithiothreitol
GuHCl: guanidine hydrochloride
2-ME: 2-mercaptoethanol
SDS: sodium dodecyl sulfate

cmc: critical micellar concentration

1 Introduction

The partial specific volume, $\bar{v}_2$, represents an essential property of macromolecules which is of particular significance for applying a variety of physicochemical and biochemical techniques. The most prominent example is the determination of molecular weights of macromolecules in simple two-component solutions from ultracentrifugal data, for which the buoyancy term, $(1-\bar{v}_2\varrho_1)$, is required [1]. In practice often an apparent specific volume, ϕ_2, is used instead of $\bar{v}_2$. Especially at high values of solvent density, ϱ_1, the calculation of molecular weights is extremely sensitive to the adopted value of the specific volume.

   Experiments are often performed under widely different conditions (e.g., concentration of macromolecule, temperature, pH, presence and concentration of denaturants or other additives such as buffer components, salts). The systems investigated generally contain at least three components. In the work of biochemists, however, the influence of diverse parameters on the specific volume has often been denied or neglected in the past. From the theoretical point of view (thermodynamics of multicomponent solutions, as developed by Casassa and Eisenberg [2, 3, 11, 38] it is clear, however, that this assumption is not correct. This was also corroborated by many experiments on a variety of systems (particularly in the laboratories of Eisenberg, of Timasheff, and of the author; cf., e.g., [105–107, 109, 113, 120, 121, 127, 128, 137, 153]).

   Experiments in multicomponent biopolymer solutions may be performed at constant molality or constant chemical potential of added solvent components. Specific volumes ($\bar{v}_2$, ϕ_2; v_2', ϕ_2') and density increments $(\partial\varrho/\partial c_2)_m$ and $(\partial\varrho/\partial c_2)_\mu$ and quantities derived from increments (definitions cf. Sect. 2) are of manifold interest (cf. [2]), e.g., for determination of true molecular weights in multicomponent systems, calculation of interaction parameters, statements on volume

---

[1] Theoretical treatment of ultracentrifugation involving irreversible thermodynamics demonstrates, that solution density ϱ instead of solvent density ϱ_1 has to be used. However, in general extrapolation to zero solute concentration c_2 is performed, in order to get rid of nonideality terms. Therefore, if this extrapolation is made, or if measurements are performed at low c_2, solvent density can be applied.

changes upon denaturation of proteins, melting of DNA double helices, changes in hydration or electrostriction of solvent by charged solutes etc.

In the case of ultracentrifugal experiments in multicomponent solutions the term $(1-\bar{v}_2\varrho_1)$ should be replaced by the density increment $(\partial\varrho/\partial c_2)_\mu$ (cf. [2]). There is some advantage in expressing $(\partial\varrho/\partial c_2)_\mu$ in a form analogous to the two-component buoyancy term by $(1-v'_2\varrho_s)$, where v'_2, the isopotential specific volume of the biopolymer, and ϱ_s, the density of the complex solvent mixture, are formally substituted for $\bar{v}_2$ and ϱ_1. In practice, v'_2 is often replaced by the apparent quantity ϕ'_2, the apparent isopotential specific volume; v'_2 and ϕ'_2 are only operational quantities which include solvent interactions with the macromolecular component.

Use of a constant value for the specific volume, determined in or calculated for a two-component system, has led to numerous misinterpretations in the past, when experiments were performed in multicomponent solutions. This holds especially for polyelectrolytes, such as nucleic acids or proteins, in high concentrations of salts or denaturants where large volume changes may occur (e.g., [104, 107, 113, 178]). Values of $(\partial\varrho/\partial c_2)_\mu$ may be quite different from the values of $(\partial\varrho/\partial c_2)_m$ at the same c_2.

Additional confusion has arisen by use of different nomenclatures in different laboratories and sometimes in different papers of the same authors (different symbols, subscripts, and superscripts of symbols) and incorrect designations of volume quantities, which in general have been measured correctly. Especially isopotential volume quantities (v'_2, ϕ'_2) have been mixed up with isomolal quantities $(\bar{v}_2, \phi_2)$. Confusing apparent volume quantities (ϕ'_2, ϕ_2) with true quantities $(v'_2, \bar{v}_2)$ seems to be of minor importance, at least at low c_2. What is sometimes called a partial specific volume $\bar{v}_2$ in a multicomponent solution, is rather the apparent quantity ϕ'_2.

The present paper gives a survey of specific volumes of many compounds of different nature, obtained by various methods and techniques under a variety of conditions. This compilation of data (about 1500 values) may enable the user to make reliable approximations in the numerous fields of the biological sciences where knowledge of the specific volume is a necessary prerequisite for evaluation of data. There is no doubt, however, that preference should be given to experimental specific volumes, provided exact determinations of this quantity are feasible.

2 Definitions

Detailed definitions and considerations are given in reviews by Eisenberg [2, 3, 11] or by Kupke [9]. Therefore, only a brief outline of the basic relations pertinent to this article will be given.

2.1 Volume and Density

The total volume, V, of a solution at constant temperature T and pressure P may be defined in terms of the partial specific volumes, $\bar{v}$, and the number of grams, g, of each of the N components:

$$V = \sum_{i=1}^{N} \bar{v}_i g_i .$$ (1)

The partial specific volume of the ith component, $\bar{v}_i$, is defined as the change in total volume per unit mass upon adding an infinitesimal amount of component i at constant T, P, and masses in grams, g_j, of all other components j:

$$\bar{v}_i = \left(\frac{\partial V}{\partial g_i} \right)_{T,P,g_j} \qquad (j \neq i) .$$ (2)

The partial molar volume, $\bar{V}$, is defined in an analogous way by substituting the number of grams, g, by the number of moles, n (cf. [565]):

$$\bar{V}_i = \left(\frac{\partial V}{\partial n_i} \right)_{T,P,n_j} \qquad (j \neq i) ,$$ (3)

thus

$$\bar{v}_i = \frac{\bar{V}_i}{M_i} .$$ (4)

The density ϱ of a solution, at particular T and P, is related to the total volume, V, and may also be expressed by the concentration c:

$$\varrho = \frac{\sum_{i=1}^{N} g_i}{V} = \frac{\sum_{i=1}^{N} g_i}{\sum_{i=1}^{N} \bar{v}_i g_i} = \sum_{i=1}^{N} c_i .$$ (5)

When $\bar{v}$ is given in $cm^3 \ g^{-1}$, and c in $g \ cm^{-3}$, the volume expression [cf. Eq. (1)] becomes:

$$\sum_{i=1}^{N} \bar{v}_i c_i = 1 .$$ (6)

2.2 Specific Volumes

To obtain values of the partial specific volume [cf. Eq. (2)], the volume change of a solution upon adding an infinitesimal amount of a definite component i has to be measured. This can be achieved by measuring the densities of a series of solutions, in which only the mass of the ith component is varied.

Differentiation of the solution density [Eq. (5)] with respect to c_i gives at vanishing concentration of the ith component a simple expression (cf. [9]):

$$\left(\frac{\partial \varrho}{\partial c_i} \right)_{g_j}^{o} = (1 - \bar{v}_i^o \varrho_s) \qquad (j \neq i) ,$$ (7)

where ϱ_s refers to the density of the solvent, and superscript o to vanishing c_i.

The limiting value at $c_2 \rightarrow 0$ is usually taken as the partial specific volume, $\bar{v}_2$, of the macromolecular component 2. Rearrangement of Eq. (7) and evaluation of the limiting slope $(\partial\varrho/\partial c_2)^0_m$ leads to:

$$\bar{v}^0_2 = \frac{1}{\varrho_s}\left[1-\left(\frac{\partial\varrho}{\partial c_2}\right)^0_m\right], \tag{8}$$

where subscript m refers to constant molality of the nonmacromolecular components in a multicomponent system.

Both the densities ϱ of a series of solutions and the density ϱ_s of the solvent itself are then plotted vs. c_2. In general a straight line results at low c_2. In practice, an apparent quantity is often measured at finite c_2. The apparent specific volume, ϕ_2, is then derived from the slope of the straight line between density ϱ at finite c_2 and density ϱ_s of the solvent:

$$\phi_2 = \frac{1}{\varrho_s}\left[1-\left(\frac{\varrho-\varrho_s}{c_2}\right)_m\right]. \tag{9}$$

If $(\partial\varrho/\partial c_2)_m$ is constant, then ϕ_2 equals $\bar{v}^0_2$ and $\bar{v}_2$ at any c_2. If it is not constant, ϕ_2 approximates $\bar{v}^0_2$ only at vanishing c_2; this may also be taken from:

$$\bar{v}_i = \phi_i + g_i\left(\frac{\partial\phi_i}{\partial g_i}\right)_{g_j} \qquad (j\neq i). \tag{10}$$

The apparent specific volume ϕ_2, measured at constant molality of the added solvent components, is a useful quantity, if it can be determined. In practice, however, another quantity, ϕ'_2, obtained at dialysis equilibrium, is usually determined. The main reason is that thermodynamics of multicomponent systems [11] tell us that most experiments (especially in ultracentrifugation or small-angle X-ray scattering) should be done on solutions which have been dialyzed to equilibrium against the solvent.

At equilibrium the molalities of the diffusible components on the two sides of a membrane are not necessarily the same (e.g., due to Donnan effects, preferential binding, etc.). But in spite of the fact that at equilibrium the mass ratios of the diffusible components may differ, the chemical potentials μ of each component are equal in the two phases. In this case a plot of the densities of a series of solutions vs. c_2 yields $(\partial\varrho/\partial c_2)_\mu$ instead of $(\partial\varrho/\partial c_2)_m$.

The isopotential specific volume v' (the prime designating isopotential conditions)[2] at $c_2\rightarrow 0$, v'^0_2, is defined in analogy to Eq.(8):

$$v'^0_2 = \frac{1}{\varrho_s}\left[1-\left(\frac{\partial\varrho}{\partial c_2}\right)^0_\mu\right]. \tag{11}$$

In practice, $(\partial\varrho/\partial c_2)_\mu$ tends to be constant at low c_2. Therefore an apparent quantity, the apparent isopotential specific volume, ϕ'_2, measured at constant chemical potential of the added solvent components, may be obtained from the density difference between solution, containing component 2 at finite concentration c_2, and solvent after equilibrium dialysis:

$$\phi'_2 = \frac{1}{\varrho_s}\left[1-\left(\frac{\varrho-\varrho_s}{c_2}\right)_\mu\right]. \tag{12}$$

---

[2] To avoid confusion, the reader is reminded that some authors call this quantity ϕ', apparent specific volume.

There is no simple relation between v_2' and ϕ_2', as given for $\bar{v}_2$ and ϕ_2 via Eq. (10). Since under isopotential conditions the molalities of the diffusible components are not held constant, v_2' and ϕ_2' are not partial specific volumes in the thermodynamic sense and have no precise physical meaning; however, their definition is operationally very useful. In a two-component system $\bar{v}_2$ and v_2' should be identical as long as Donnan effects are absent. But in the presence of high concentrations of a third component, c_3, large differences between these two volume quantities may occur.

In this connection the question arises how the isopotential specific volume and the true partial specific volume are connected. As shown by Casassa and Eisenberg [11], at vanishing c_2 the difference between isopotential and isomolal slopes is given by:

$$\left(\frac{\partial \varrho}{\partial c_2}\right)_\mu^0 - \left(\frac{\partial \varrho}{\partial c_2}\right)_m^0 = (1 - v_2'^0 \varrho_s) - (1 - \bar{v}_2^0 \varrho_s)$$

$$= (1 - \bar{v}_2^0 \varrho_s) + \xi_1^0 (1 - \bar{v}_1 \varrho_s) - (1 - \bar{v}_2^0 \varrho_s)$$

$$= \xi_1^0 (1 - \bar{v}_1 \varrho_s) = \xi_3^0 (1 - \bar{v}_3 \varrho_s), \tag{13}$$

where ξ_1 and ξ_3 represent preferential interaction parameters; $\xi_1 = (\partial w_1 / \partial w_2)_\mu$ and $\xi_3 = (\partial w_3 / \partial w_2)_\mu$, where ξ_1 (or ξ_3) is the number of grams of component 1 (or 3) which must be added per gram of component 2 to maintain components 1 and 3 at constant chemical potential. $\bar{v}_1$ and $\bar{v}_3$ are partial specific volumes of components 1 and 3 respectively. The parameters ξ_1 and ξ_3 are connected; by using $c_1 + c_3 = \varrho_s$ and $c_1 \bar{v}_1 + c_3 \bar{v}_3 = 1$ we obtain at vanishing c_2: $\xi_1 = -\xi_3 / w_3$, where w_3, equal to c_1 / c_3, is the weight molality, in g of component 3 per g of component 1 (cf. [3]).

In terms of apparent volume quantities the following relation between ϕ_2' and ϕ_2 holds:

$$\xi_j = \left(\frac{\phi_2 - \phi_2'}{1/\varrho_s - \bar{v}_j}\right) \qquad (j = 1, 3). \tag{14}$$

Thus ϕ_2' can be calculated if ϕ_2 and preferential binding have been determined. More detailed treatments concerning interaction parameters reflecting the distribution of diffusible solutes have been reviewed by Eisenberg [2, 3, 11], Kupke [9], and Lee et al. [108].

Partial and apparent specific volumes of third components, $\bar{v}_3$ and ϕ_3, can be calculated at all c_3 from the measured or known dependence of the density of solutions of component 3, ϱ_s, on c_3, in the absence of component 2 (cf. [8, 113, 121, 142, 154, 178]). The same procedure as for macromolecules at constant chemical molality [cf. Eqs. (8) and (9)] may be applied. For aqueous solutions holds:

$$\bar{v}_3^0 = \frac{1}{\varrho_1}\left[1 - \left(\frac{\partial \varrho_s}{\partial c_3}\right)^0\right] \tag{15}$$

$$\phi_3 = \frac{1}{\varrho_1}\left[1 - \left(\frac{\varrho_s - \varrho_1}{c_3}\right)\right]. \tag{16}$$

Mixtures of species being in dynamic equilibrium in solution (e.g., buffer components like K_2HPO_4 / KH_2PO_4) may be considered as a single thermodynamic component (cf. [178]).

3 Determination of Specific Volumes

3.1 Calculation

3.1.1 Traube's Additivity Principle

According to Traube [12, 77] the partial specific volume of any (low-molecular compound in dilute aqueous solution may be calculated assuming additivity of molar volumes of atoms and atomic groups (given for 15 °C). There is, however, no strict additivity: corrections for covolume, electrostriction, ring formation etc. must be taken into account (cf. also [1, 78, 110, 118, 160, 325, 518, 565, 566]). Some examples are given in the following tables, e.g., the value for sialic acid, $\bar{v}_c =$ 0.58–0.59 cm^3 g^{-1}[168, 571], which is in accord with the experimental value of 0.584 cm^3 g^{-1} [133].

Traube's principle of additivity may also be applied to macromolecules or groups of molecules, using the partial specific volumes of the low-molecular constituents without further corrections. The most prominent example of this kind is the well-known method of Cohn and Edsall [1] for calculating the partial specific volumes of proteins. It should be noted that the additivity principle is not restricted to proteins, but may be applied to other molecules including conjugated or associated systems. These calculations generally neglect any kind of accompanying interaction. The partial specific volumes of constituents are generally assumed to be the same as they would be in the isolated state, though this is not an a priori assumption.

3.1.2 Native Nonconjugated Proteins

Provided the amino acid composition is known, the partial specific volume of a native protein (or a large peptide) can be calculated according to Cohn and Edsall [1] from the partial specific volumes of the individual amino acid residues (given for 25 °C).

$$\bar{v}_c = \frac{\sum\limits_{i=1}^{N} W_i \bar{v}_i}{\sum\limits_{i=1}^{N} W_i} = \frac{\sum\limits_{i=1}^{N} N_i M_i \bar{v}_i}{\sum\limits_{i=1}^{N} N_i M_i}, \tag{17}$$

where $\bar{v}_c$ is the calculated partial specific volume, W_i is the weight percentage of the ith amino acid residue, N_i is the number of residues, M_i is the molecular weight of residue (molecular weight of the amino acid minus 18), $\bar{v}_i$ is the partial specific volume of residue.

Values $\bar{v}_r$ for the amino acid residues (Table 1) were obtained by Cohn and Edsall [1] using calculated and experimental apparent molar volumes of amino acid residues. Calculated volumes were obtained according to Traube's rule on the basis of the apparent atomic and apparent molar volumes of the atomic groups of amino acid residues; so-called experimental values of amino acid residues were obtained by subtracting a constant correction factor (taking account of the water eliminated) from the experimentally observed apparent molar volumes of amino acids. As is shown in the table, $\bar{v}_r$ varies from 0.60 (asp) to 0.90

(ile, leu). It should be noted that the value given originally by Cohn and Edsall [1] for cystine (0.61 cm^3 g^{-1}) was corrected later on to 0.63 cm^3 g^{-1} [73, 151]. Proteins with a high content of asp, asn, cys, ser, gly, glu, gln, his ($\bar{v}_r < 0.7$ cm^3 g^{-1}) (cf. Table 1) should produce a low value of $\bar{v}_c$, those with a high content of ile, leu, val, lys ($\bar{v}_r > 0.8$ cm^3 g^{-1}) should produce a high $\bar{v}_c$-value.

Results obtained for various proteins range from 0.69–0.76 cm^3 g^{-1}; an example of calculation is given, for example, in [405]. Although the Cohn-Edsall method is only an approximation, it has been found for many proteins that there is good agreement of calculated values with experimentally determined values, especially at or near neutral pH (e.g., [105, 149, 151, 157, 405]). This simple procedure, of course, gives only reliable results in the case of nonconjugated proteins, and may only be applied for use in two-component solutions. Appreciable deviations were found at high and low pH, after unfolding in denaturing solvents, or in the presence of high concentrations of some other additives.

A modified Cohn-Edsall procedure has been suggested by Zamyatnin [74–76, 118], taking only experimental data for the apparent molar volumes of amino acids for computation of $\bar{v}_r$ (cf. Table 1). The method assumes that the interaction of amino acid residues with the solvent is identical, independent of whether they occur in an amino acid or in a protein; interaction of the amino acid residues with each other in a protein are not taken into account. So far the calculated value, $(\bar{v}_c)_{mod.}$, rather represents the unfolded polypeptide chain (cf. [91]). Indeed, the values given by Zamyatnin for some proteins [118] are slightly smaller than the values obtainable from the conventional Cohn-Edsall method. Recently Zamyatnin [566] has published refined $\bar{v}_r$-values (also included in Table 1), which are based on new experimental data. With the exception of arg the new $\bar{v}_r$-values are smaller than the values reported earlier. At present no statement on the validity of the new $\bar{v}_r$-values can be made.

3.1.3 Native Conjugated Proteins

In the case of native conjugated proteins, the relative amounts and partial specific volumes of nonprotein components must be taken into account (cf. e.g., [9, 37, 110, 125, 173]). The partial specific volume of a conjugated protein can be calculated assuming the validity of an additivity procedure:

$$\bar{v}_c = \sum_{i=1}^{N} f_i \bar{v}_i = f_p \bar{v}_p + \sum f_{np} \bar{v}_{np}, \tag{18}$$

where f_p and f_{np} are the weight fractions of protein or nonprotein components, respectively, obtained from a compositional analysis, and $\bar{v}_p$ and $\bar{v}_{np}$ their assumed, calculated, or experimentally determined partial specific volumes; $\bar{v}_p$ may be calculated from the amino acid increments according to Cohn and Edsall.

It is obvious that nonprotein components have a significant influence on the value of $\bar{v}_c$ only if considerable amounts ($> 5\%$) of them are present in the conjugated protein. Deviations, of course, only play a role when $\bar{v}_p \neq \bar{v}_{np}$, as in conjugated systems consisting of proteins ($\bar{v}_a \approx 0.735$ cm^3 g^{-1}), carbohydrates ($\bar{v}_a \approx 0.61$ cm^3 g^{-1}), lipids ($\bar{v}_a \approx 1$ cm^3 g^{-1}), or nucleic acids ($\bar{v}_a \approx 0.54$ cm^3 g^{-1}), where $\bar{v}_a$ is the assumed (average) value for the partial specific volume. The influence of a single metal atom or, e.g., a cofactor, or even a small number of these ligands can generally be neglected in the calculation of an approximate $\bar{v}_c$-value; however typical metallo-, phospho-, glyco-, or nucleoproteins yield $\bar{v}_c$-values $< \bar{v}_p$, typical lipoproteins $\bar{v}_c$-values $> \bar{v}_p$.

As an example, $\bar{v}_c$ of the bacteriophage T7 (*E. coli*), which has approximately 51% DNA content [476], can be cited. With the above $\bar{v}_a$-values for nucleic acid and protein, a value of $\bar{v}_c = 0.636$ cm^3 g^{-1} can be calculated. This value agrees well with the experimentally determined value of 0.639 cm^3 g^{-1} [475, 476] (cf. Table 10). Similarly, for 30 S ribosomes (*E. coli*), having 70% RNA content [131], the calculated $\bar{v}_c = 0.599$ cm^3 g^{-1} is in good agreement with the experimental value of 0.601 cm^3 g^{-1} [132].

3.1.4 Protein Complexes

Partial specific volumes of complexes, for example, of lipid associated proteins (as found in membranes and serum lipoproteins), of protein-detergent complexes [110, 160, 161, 376, 390], can be calculated either by Eq. (18) or alternatively by

$$\bar{v}_c(\text{complex}) = \frac{\bar{v}_p + \delta_d \bar{v}_d + \sum \delta_l \bar{v}_l}{1 + \delta_d + \sum \delta_l},\qquad(19)$$

where $\bar{v}_p$, $\bar{v}_d$, $\bar{v}_l$ are the partial specific volumes of protein, detergent, and any other bound ligand (e.g., of lipid), respectively; δ_d, δ_l are the corresponding amounts in grams per gram of protein. The molecular weight of the complex, M(complex), is equivalent to $M_p \cdot (1 + \delta_d + \sum \delta_l)$, and may be obtained, for example, by sedimentation equilibrium. The partial specific volumes of bound ligands, e.g., of detergents or lipids, are generally assumed to be the same as in pure detergent micelles or lipid bilayers [160, 161]. A comparison between theoretical and several experimental values has been carried out for cytochrome P-450 as a typical membrane protein by Wendel et al. ([91], cf. Table 6); discrepancies between calculated and experimental values were attributed to preferential solvation effects. Proteins in the presence of SDS have been treated in ([373, 399]; cf. the $\bar{v}_c$-value for the apoferritin-SDS complex in Table 7).

3.1.5 Denatured Proteins

Special methods have been developed to calculate the isopotential specific volumes, ϕ'_c, of proteins denatured in 6 M GuHCl or 8 M urea on the basis of the amino acid composition [129, 130, 162, 165]. As an extension of the Cohn-Edsall procedure, interactions of the proteins with solvent components, such as preferential interaction with water or the denaturant, are taken into account. The total hydration is calculated from the hydration of the constituent amino acids ([81], cf. Table 1). To obtain the extent of denaturant binding, it is assumed that each aromatic amino acid side chain and each pair of peptide bonds interact with one molecule of either GuHCl or urea. Due to these assumptions, calculations fail if proteins containing considerable amounts of nonprotein components are considered. In the references [129, 130, 162, 165] calculated data have been compared with experimental results.

3.2 Experimental Determination

3.2.1 Methods

The classical procedure for measuring densities of liquids is pycnometry. The container ("pycnometer") is a glass vessel of fixed volume (usually 2–25 cm^3). At a given temperature the volume of the container can be determined by filling the pycnometer with a liquid of known density (water) and weighing. Then the liquid of unknown density is weighed. The method is time-consuming and requires considerable amounts of material.

More sophisticated methods comprise a variety of techniques such as density-gradient columns, isopycnic temperature method, magnetic densimetry, the Cahn electrobalance, analytical ultracentrifugation in the presence of D_2O and H_2O, buoyancy measurements of dialysis bags, digital densimetry etc. A survey of the methods mentioned is given in [9]; detailed descriptions of the various techniques are given in [4–8, 87, 106, 108, 110, 112, 117, 405, 406, 593].

The vast majority of values of specific volumes has been obtained from applications of pycnometry, analytical ultracentrifugation, or digital densimetry. Particularly development of high precision digital density meters has introduced ease of performance and increased precision.

3.2.2 Digital Densimetry

The digital density meter, developed in Kratky's laboratory in Graz [5, 10, 84–86], is manufactured by the Paar Co. (Graz, Austria). The measurement is based on the determination of the natural frequency of an electronically excited mechanical oscillator ("mechanical oscillator technique"). The oscillator consists of a hollow U-shaped glass tube with a precisely defined volume into which the samples are injected. The effective mass of the oscillator is thus composed of its own mass and the mass contributed by the sample under investigation. Neither sample weight nor volume need to be known exactly. However, a minimum of approximately 0.7 cm^3 of the sample is required for proper filling.

The determination of a density difference $\Delta\varrho$ between two samples is based on the simplified equation

$$\Delta\varrho = k\Delta\tau^2, \tag{20}$$

where k is an apparatus constant, τ is the period of oscillation. The constant k can be determined by measuring two samples of known density (e.g., water and air). For measurements below room temperature, dry nitrogen can be recommended instead of atmospheric air [105], to avoid condensation of water in the oscillator tube. From the density difference between any two samples (e.g., solution and solvent, or solvent mixture and water) the unknown densities can be calculated using k [Eq. (20)]. For high precision measurements of specific volumes of macromolecules use of relatively concentrated sample solutions is advisable (≈ 5–20 mg cm^{-3}). More technical details can be found in various references (e.g. [4, 5, 108, 109]).

In general, the accuracy of densities and/or specific volumes is not limited by the accuracy of present density meters, but is usually more seriously affected by other parameters. Possible errors may arise from temperature fluctuations, water condensation in the oscillator tube, presence of air bubbles in the solutions, insufficient dialysis time, inaccuracy of sample concentration, high sample viscosity etc. Experience shows that in biochemical work the most critical quantity is the sample concentration. However, also high viscosity of the solutions (>2 mPas) may cause a systematic density meter error, depending on the type of instrument used; in this case small corrections become necessary [138]. When working with multicomponent systems, changes of absorption coefficients may occur and have to be taken into account (cf. e.g., [24, 104–106, 113, 127]).

Recent constructions of density meters employ two physically identical oscillators. Assuming a thermostat accuracy of ± 0.01 °C, these density meters produce a maximum error of $\pm 1.5 \times 10^{-6}$ g cm^{-3}; the error in sample density due to its thermal expansion has to be added to the error of the instrument. The usual precision for water is $\pm 3.5 \times 10^{-6}$ g cm^{-3}.

4 Tabulation and Interpretation of Data

4.1 Specific Volumes

Tables 2–10 summarize specific volumes of a variety of nonmacromolecular and macromolecular compounds of biological interest. To emphasize similarities, compounds are grouped into classes. Since *small* amounts of, e.g., hemo-, flavo-, glyco-, lipo-, metallo-, phospho-moieties in conjugated biopolymers do not significantly influence the value of the specific volume of the macromolecule, these compounds are usually not cited separately.

The data in this compilation had to be selected from a large number of publications, the prime goal of which generally was *not* the determination of specific volumes. Therefore the necessary care has not always been taken by the authors to measure under well-defined conditions. Furthermore, experimental details have often not been cited properly, particularly with regard to isomolal or isopotential conditions. In such cases no assignments of the specific volume quantities have been made.

Therefore, the present compilation necessarily suffers from some quality differences in the data cited; however, obviously wrong values have been omitted. Usually, values reported by different authors were considered instead of taking average values. This approach was particularly important if values for samples from different sources were obtained or if measurements were performed under different conditions. In those few cases where authors cited accuracies of their data, the accuracy was within the range of ± 0.001 to 0.003 cm^3 g^{-1} (e.g., [105, 108, 109]); small amounts of material or the presence of other compounds may cause greater inaccuracies, however, generally the error is $< \pm 0.01$ cm^3 g^{-1} (e.g. [104]).

The occurrence of different values for the specific volume of a given compound in the literature does not necessarily imply that incorrect determinations of specific volumes are the cause. Frequently discrepancies arise from differences in the concentration determination (e.g., use of different absorption coefficients, different degree of purity and different carbohydrate content of the samples etc.). Kupke [9, 173, 540] pointed out that different sample preparations may lead to differences in the specific volume in a given solvent (e.g., different samples of chromatographically purified ribonuclease led to values ranging from 0.692 to 0.713 $cm^3 g^{-1}$). The most appropriate value of the specific volume for a given macromolecule is obtained when density measurements are made on samples under solvent and sample conditions identical to those used in the experiments for which knowledge of the specific volume is required. This point is also stressed by Eisenberg (e.g. [153]).

The tables comprise both calculated values ($\bar{v}_c$, ϕ'_c), indicated by the subscript c, and experimental values of specific volumes under a variety of conditions. Isomolal quantities are denoted by $\bar{v}$ or ϕ; the isopotential quantities, v' or ϕ', are designated by a prime. Unless it is misleading, the original designations given by the authors are cited. Solvent conditions, like water or dilute buffer, are generally not specified in the tables; for details the reader is referred to the original literature. In rare cases, assumed values for the partial specific volume, $\bar{v}_a$, which are frequently used, have been included.

The specific volume quantities listed in Tables 3–10 are valid for two-component or multicomponent systems, respectively. Biochemical work in general requires the presence of buffers or dilute salt solutions to prevent denaturation of the biopolymers. Dilute aqueous solutions containing low concentrations of third components ($c_3 \leq 0.1$ M, cf. [106, 175]) may be considered as two-component systems. However, care should be taken that data selected refer to the same temperature; if this is not the case a temperature correction must be applied (cf. Sects. 4.2.1.2 and 4.2.2).

For practical purposes in two-component solutions the differences between $\bar{v}_a$, $\bar{v}_c$, $\bar{v}^{(0)}$, $v'^{(0)}$, $\phi^{(0)}$, $\phi'^{(0)}$, or the quantities without designation, become insignificant provided these quantities have been determined in the absence of special solvent conditions (e.g., $c_3 \leq 0.1$ M, no extremes of pH, no specific ligands). This applies in most cases to protein solutions when working close to the isoelectric point (cf. [2, 120, 126, 128, 134, 140]). Nucleic acid solutions furnish an exception. For them the proper quantity generally is $v'^{(0)}$ or $\phi'^{(0)}$ and not $\bar{v}^{(0)}$ or $\phi^{(0)}$ (cf. Sect. 4.2.2 and Fig. 3).

When dealing with multicomponent systems, preferably the isopotential quantities $v'^{(0)}$ or $\phi'^{(0)}$ should be adopted, provided they have been determined under exactly comparable conditions; for proteins in 6 M GuHCl or 8 M urea ϕ'_c-values or estimates may also be used. Special problems (calculations of true partial specific volumes or of interaction parameters) may necessitate the use of the isomolal quantities $\bar{v}^{(0)}$ or $\phi^{(0)}$. Indiscriminate use of so-called "constant" values for the specific volume independent of the experimental conditions may lead to serious errors.

Other Reference Sources. Specific volume values of biopolymers, not cited in Tables 3–10, may be found in reviews on the corresponding subjects (physicochemical data) or in papers where these quantities are needed for the evaluation of data (especially papers on ultracentrifugation or small-angle scattering). A series of diverse data, especially of proteins, is compiled in the *CRC Handbook of Biochemistry* [526]; a few values can also be found in various text books of physical biochemistry or related fields; some values on plasma proteins are compiled in [528, 561].

Specific volumes of small molecules (buffer components, salts, sugars, denaturants etc.), not cited in Table 2, may also be obtained from the corresponding tabulated partial molar volumes, preferably at infinite dilution in water (cf. [534, 535, 565, 594]), by dividing these quantities by the corresponding molecular weights [Eq. (4)]. Partial and apparent specific volumes of small molecules may also be calculated from Eqs. (15) and (16), using the tabulated density data of water and aqueous solutions of various substances (e.g., *CRC Handbook of Chemistry and Physics* [527], *International Critical Tables* [533]; densities of aqueous solutions of GuHCl and urea are given by Kawahara and Tanford [198]. Specific volumes of a variety of detergents, bile salts, and lipids are compiled in [110]. A recent compilation of partial specific volumes of amino acids and small peptides is given in [566].

4.1.1 Small Molecules

The partial and apparent specific volumes of small molecules in aqueous solution, generally termed $\bar{v}_3$ or ϕ_3, vary in a wide range, depending on the chemical nature of the compound; the values of Table 2 generally range from $\approx 0.1–1.1$ cm^3 g^{-1}. It should be noted that, in contrast to biopolymers, there is generally a pronounced dependence of the specific volume on the concentration, c_3, of these substances (cf., e.g., NaCl and CsCl [107]).

The specific volumes of amino acids are, of course, different from those of the amino acid residues (cf., e.g., the values for ala in Tables 1 and 2). Specific volumes of carbohydrates are in the range $\approx 0.58–0.68$ cm^3 g^{-1} ($\bar{v}_a$ of simple carbohydrates is $\approx 0.60–0.62$). Inorganic salts range from $\approx 0.1–0.5$ cm^3 g^{-1} (a frequently used value is $\bar{v}_a = 0.3$ cm^3 g^{-1}). Denaturants and detergents range from $\approx 0.6–1.1$ cm^3 g^{-1}, lipids (Table 9) from $\approx 0.85–1.1$ cm^3 g^{-1}. It is interesting to note that the specific volumes of the denaturants GuHCl and urea are similar to those of nonconjugated proteins ($\bar{v}_3$ both of 6 M GuHCl and 8 M urea is 0.763 cm^3 g^{-1}, cf. Table 2). The specific volumes of detergents may be different above and below their cmc (for SDS the difference is 0.055 cm^3 g^{-1}; (cf. [110, 141, 160]; Table 2); some of these substances are able to form micelles of very high molecular weight (e.g., Triton X-100, cf. [210]).

Knowledge of approximate specific volumes of small molecules (e.g., salts) allows rough estimates of how mere binding or unspecific interaction would influence the overall specific volume. This consideration, of course, does not imply the influence of complicated solvent interactions, structural changes etc.

4.1.2 Polyamino Acids and Peptides

The partial specific volumes of polyamino acids (Table 3) lie in the range of 0.6 to 0.9 cm^3 g^{-1}, depending on the nature of the amino acids and the additional groups (if any). The values are similar but not necessarily identical to those of amino acid residues (Table 1). The values for peptides (Table 3) fall in a narrower range and are similar to those of proteins when they are composed of a variety of different amino acids.

4.1.3 Native Nonconjugated Proteins

In two-component solutions both calculated and experimental partial specific volumes of nonconjugated proteins fall in the range $0.69–0.76\ cm^3\ g^{-1}$ (cf. Tables 4–6), depending on their amino acid composition. The majority exhibits values in the range $0.72–0.75\ cm^3\ g^{-1}$.

As mean values for unknown proteins, values of $\bar{v}_a \approx 0.72$ to $0.75\ cm^3\ g^{-1}$ have been proposed in the literature (e.g. [1, 22, 26, 45, 59, 89, 90, 99, 106, 112, 157, 176, 361, 365, 373, 406, 423, 516, 523]). A comparison of this range (mean value $0.735\ cm^3\ g^{-1}$) with the mean values of the above cited ranges (0.725 or $0.735\ cm^3\ g^{-1}$) shows that $\bar{v}_a$-values of $0.74–0.75\ cm^3\ g^{-1}$ (e.g. [1, 176, 523]) are obviously too high to represent an average protein, especially at low temperatures. Values of 0.72–0.735 would represent better approximations. This consideration is also in accord with a recent paper by Attri and Minton [59], who give 0.720 or $0.724\ cm^3\ g^{-1}$ as mean values of 141 randomly selected proteins at 10 ° or 20 °C, respectively.

Although proteins from different sources or isozymes have different amino acid compositions, their specific volumes do not differ significantly [26], due to the similarity of the mean specific volume values of the constituent amino acids (cf., e.g., values for hemoglobin, lactate dehydrogenase in Table 5). Cleavage of proteins into subunits or large fragments is not accompanied by significant specific volume changes (e.g., values for 1/1–1/20 size of hemocyanin, Table 5). Specific ligands generally do not induce pronounced changes (cf. Sect. 4.2.1.9).

Proteins with unusual amino acid composition (e.g., basic or acidic proteins, cf. Table 6) may show partial specific volumes outside the ranges given above and may have specific solvent requirements, for example, halophilic proteins which need high salt concentration c_3 for maintaining their native structure (cf. halophilic malate dehydrogenase, Table 6).

It is emphasized again that the above values for partial specific volumes are only valid for native nonconjugated proteins in two-component systems. Changes in the environment may cause significant volume changes (cf. Sect. 4.2).

4.1.4 Native Conjugated Proteins

Conjugated proteins containing only small amounts of nonprotein components (e.g., hemoproteins like hemoglobin, myoglobin, cytochrome c, cf. Tables 5 and 6) fall in the same range of partial specific volumes as nonconjugated proteins (cf. Sect. 4.1.3).

Conjugated proteins containing appreciable amounts of nonprotein components (cf. Tables 7–10) show a behavior different from that mentioned above. Partial specific volumes are influenced considerably by the content and nature of the nonprotein moiety. In analogy to the calculated values, $\bar{v}_c$, typical metallo-, phospho-, glyco-, and nucleoproteins show specific volumes $< \bar{v}_p$, lipoproteins have values $> \bar{v}_p$. As representative samples the values of ferritin (Table 7), submaxillary glycoproteins (Table 8), high and low density serum lipoproteins (Table 9), bacteriophages T2–T7 and ribosomes (Table 10) may be considered. In order to see the influence of the nonprotein components, the approximate percentages of protein and nonprotein components are given in some cases in the tables (generally the percentages have been adopted from the same authors who measured the specific volumes).

4.1.5 Denatured Proteins

More or less completely unfolded proteins have specific volumes in the presence of the unfolding agent which have been assumed to differ only slightly from those of native proteins. A few authors expressed the opinion that there is no change (e.g., [485]) or even a slight increase of the specific volume (e.g., [454]) as a consequence of denaturation. Most authors, however, reported a slight decrease of the specific volume (e.g., [105, 113, 153, 185, 193, 232, 455, 456, 458]). As representative examples the experiments of Reisler and Eisenberg [113] on BSA and aldolase (Tables 4 and 5) are mentioned: $\bar{v}^0$ of BSA at 25 °C was found to decrease from 0.734 cm^3 g^{-1} in 0.2 M NaCl to 0.728 cm^3 g^{-1} in 6 M GuHCl/0.1 M 2-mercaptoethanol, the corresponding ϕ'-values decreased from 0.732 to 0.721 cm^3 g^{-1}; $\bar{v}^0$ of aldolase at 25 °C decreased from 0.739 cm^3 g^{-1} in H$_2$O to 0.734 cm^3 g^{-1} in 6 M GuHCl/0.01 M 2-mercaptoethanol, the ϕ'-value in the latter solvent amounted to 0.725 cm^3 g^{-1}.

It was generally assumed in the past that ϕ'-values of proteins in 5–6 M GuHCl are lower by ≈ 0–0.03 cm^3 g^{-1}, i.e., by ≈ 0–4% (e.g., [24, 25, 109, 117, 162, 165, 169, 193, 455–457, 516]) than in the native state. This assumption, probably true for most proteins in the presence of 6 M GuHCl (see Sect. 4.2.1.8), is however only a "rule of thumb". It is definitely incorrect to state that there is no change of the specific volume of proteins in the whole range between 0 and 6 M GuHCl (cf. [105, 106, 113, 153]; Fig. 2; Sect. 4.2.1.8).

In this context the question arises, whether similar generalizations can be made for other denaturing conditions (8 M urea, extremes of pH, heat etc.) or for the situation where the denaturing agent can be removed without renaturation of the polymer. Though also a decrease of specific volumes is frequently observed (cf., e.g., the values for chymotrypsin or chymotrypsinogen in urea, Table 5, cf. [130], Sects. 4.2.1.3, 4.2.1.4, 4.2.1.8), the lack of a significant number of data renders these generalizations still more problematic.

In accord with the experimental values, the isopotential specific volumes, ϕ'_c, calculated for proteins in 6 M GuHCl or 8 M urea solutions, were generally also slightly lower than the values of native proteins (cf. [105, 129, 130, 162, 165]).

4.1.6 Polysaccharides

Similarly to low-molecular carbohydrates (Table 2), the values of partial specific volumes of polysaccharides are expected to range in the vicinity of ≈ 0.60–0.62 cm^3 g^{-1}; this is valid for neutral unmodified polysaccharides like inulin or starch (Table 8). Branching or charged side groups could strongly influence these values (cf., e.g., the value for heparin, a sulfated polysaccharide, of ≈ 0.45 cm^3 g^{-1} [344]).

4.1.7 Lipids, Membranes, and Micelles

The partial specific volumes of lipids amount to ≈ 0.85–1.1 cm^3 g^{-1}, most of them are > 0.92 cm^3 g^{-1} (Table 9). As a mean value ≈ 1 cm^3 g^{-1} may be assumed. Similar results may be derived from the few examples of membranes and micelles.

4.1.8 Polynucleotides and Nucleic Acids

A few examples of polynucleotides with a partial specific volume of ≈ 0.54 cm^3 g^{-1} are cited in Table 10. As pointed out by Cohen and Eisenberg [107], the values for nucleic acids are strongly dependent on the presence of electrolytes in solution; both the nature and the amount of the ions strongly influence the values for the specific volumes: $\bar{v}^0$ of NaDNA at 25 °C was found to be 0.500 cm^3 g^{-1} in water, and $\bar{v}^0$ of CsDNA 0.440 cm^3 g^{-1} (Table 10). For DNA in dilute salt solution (≈ 0.2 M NaCl) values of $\phi' = 0.53$–0.55 cm^3 g^{-1} are reported (cf. [90, 107, 304, 477, 480–482]); a value of $\bar{v}_a = 0.54$ cm^3 g^{-1} [111] is commonly accepted. For RNA generally similar values are reported (cf. Table 10); $\bar{v}_a = 0.54$ cm^3 g^{-1} can also be used for calculations of systems conjugated with RNA.

The values for viruses, partial specific volumes ≈ 0.6–0.8 cm^3 g^{-1} (Table 10), are dependent on the content of the constituents (DNA, RNA, protein, lipid). Values for 30 S – 70 S ribosomes are reported to be ≈ 0.59–0.60 cm^3 g^{-1} [132]; for nucleosomes values of ≈ 0.65–0.66 cm^3 g^{-1} [47, 125, 267, 319] are found. In general, the values of constituents are assumed to be the same as for the isolated substances.

4.2 Specific Volume Changes

As outlined above, specific volumes of macromolecules depend critically on environmental conditions (e.g., solvent composition, temperature, pH). The apparent specific volumes may exhibit drastic changes in the presence of high concentrations of additives (e.g., electrolytes). This behavior is documented in Tables 3–10 for different molecules and a variety of conditions. While these experiments are generally restricted to some special conditions, only a few papers (e.g., [105]) try to make systematic investigations in order to allow generalizations. Since the extent of volume changes depends on the molecular characteristics of the biopolymers, generalizations are, of course, problematic, but obviously useful for rough estimates concerning possible interactions and trends of effects.

Most biochemical experiments are performed in multicomponent solutions (macromolecules in buffer). For evaluation of interaction parameters by density measurements both isomolal and isopotential quantities (cf. Sect. 2.2 and [108]) are required. If, however, only specific volumes are to be determined, isopotential conditions are sufficient and convenient, since in most cases equilibrium dialysis is used in the experiments. Therefore volume changes of the isopotential volume quantities, v'_2 or ϕ'_2, are of great practical interest.

Most experiments have been performed with some special, generally nonconjugated proteins (especially BSA), a few with other substances (e.g., DNA).

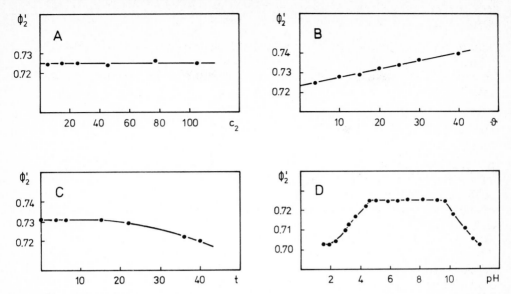

Fig. 1 A–D. Apparent isopotential specific volumes ϕ'_2 (in ml g^{-1}) of BSA in water or dilute aqueous buffer solutions as a function of **A** protein concentration c_2 (mg ml^{-1}); **B** temperature ϑ (°C); **C** time t (h) of storage in water at room temperature; **D** pH. ϕ'_2 was measured at varying c_2 (**A**) or at $c_2 = 10$–20 mg ml^{-1} (**B–D**), and at constant chemical potential with respect to added solvent components at 4 °C (**A, D**), or 4 °–40 °C (**B**), or 20 °C (**C**). [105]

Therefore in the following primarily volume changes of proteins will be considered as a function of various parameters, both in two- and multicomponent solutions. Generalizations have been restricted till now to the isopotential quantity ϕ'_2 [105]. Some typical results, namely on BSA at 4 °C [105], are presented in Figs. 1 and 2.

In this context it should be noted that the direct measurement of volume changes, ΔV, on mixing biopolymers with other substances, as well as the time course of volume changes (kinetic experiments) can also be followed by dilatometry (cf. [9, 501, 502]).

4.2.1 Proteins

4.2.1.1 Protein Concentration. Under isopotential conditions no significant changes of ϕ'_2 were observed with different proteins by means of direct densimetry in the range $\approx 1 < c_2 < 350$ mg cm^{-3} (e.g., [9, 26, 105, 109, 120, 121, 126, 128, 130, 138, 150]; Fig. 1 A); applying additionally ultracentrifugal analysis, c_2 may be extended down to 10^{-4} mg cm^{-3}, thus spanning a range of protein concentration covering 7 orders of magnitude [28, 106]. Under isomolal conditions sometimes slight increases of ϕ'_2 with increasing c_2 have been registered (e.g., [109, 121, 128]). Small peptides or amino acids show a slight linear concentration dependence [26].

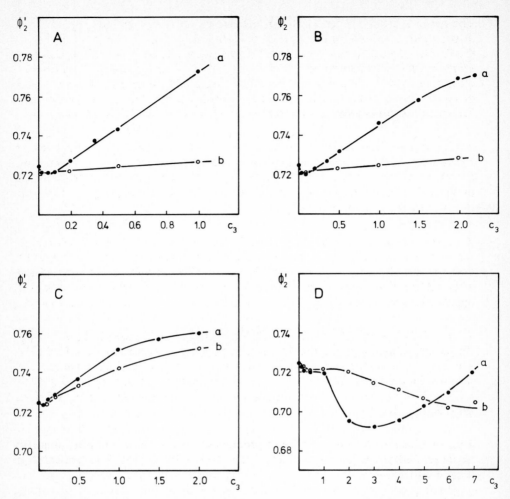

Fig. 2 A–D. Apparent isopotential specific volumes ϕ'_2 (in ml g^{-1}) of BSA in water or various aqueous buffer solutions in the presence of electrolytes, sugars, or denaturants. **A** c_3 = buffer, pH 7.0 (mol l^{-1}): (a) K$_2$HPO$_4$/KH$_2$PO$_4$, (b) Tris/HCl; **B** c_3 = aqueous solution of salt (mol l^{-1}): (a) (NH$_4$)$_2$SO$_4$, (b) NH$_4$Cl; **C** c_3 = aqueous solution of sugar (mol l^{-1}): (a) sucrose, (b) glucose; **D** c_3 = aqueous solution of denaturant (mol l^{-1}): (a) GuHCl, (b) urea. ϕ'_2 was measured at c_2 = 10–20 mg ml^{-1} and constant chemical potential with respect to added solvent components, at 4 °C. [105]

4.2.1.2 Temperature. In the temperature range $\approx 4\,°$–45 °C a linear dependence of specific volumes of different proteins in two-component systems was found. For $\Delta\bar{v}_2/\Delta T$, values of 2.5–10×10^{-4} cm^{3}g^{-1}K^{-1} were reported (e.g., [7, 23, 26, 29, 76, 91, 96, 104, 105, 118, 138, 146, 150, 156, 171, 176, 180, 185, 189, 199, 414, 458, 508, 516, 524]; Fig. 1 B), the majority lying between 3.5–5×10^{-4} cm^3 g^{-1} K^{-1}.

The mean value of the limits of this latter range of 4.25×10^{-4} cm^3 g^{-1}K^{-1} is also in accord with a recent study on some proteins, giving 4.5×10^{-4} cm^3 g^{-1} K^{-1} [105]. The temperature effect is still more pronounced for small peptides or amino acids [26] or denatured proteins ([23, 199, 524]; for heat-denatured ribonuclease: 1.8×10^{-3} cm^{3}g^{-1}K^{-1} [199]). The course of volume changes of proteins at temperatures >45 °C, where thermal transitions and heat denaturation may take place, is more complex and nonlinear [23, 199].

4.2.1.3 Aging, Thermal Denaturation. Aging after long storage of proteins at room temperature or heat denaturation at elevated temperatures is reflected by a decrease of the specific volume ([105]; Fig. 1 C). Effects are more pronounced in water than in buffered solutions. In the case of isoelectric heat aggregation volume increases were observed [97].

4.2.1.4 Acid and Alkaline Denaturation. Denaturation at extremes of pH, resulting in partial unfolding, was accompanied by nonlinear decreases of the specific volume ([98, 105, 120, 181, 182, 505]; Fig. 1 D). Effects may be more pronounced after addition of substances promoting the unfolding process (e.g., reducing agents like 2-mercaptoethanol).

4.2.1.5 Different Buffers. Different buffers cause generally linear increases of ϕ'_2, the effects depending on the buffer components rather than on pH ([105]; Fig. 2 A). Pronounced increases of ϕ'_2 with increasing c_3 were observed in the case of phosphate buffers [27, 105, 106, 178]. The effects could be correlated quantitatively with the results of sedimentation velocity and sedimentation equilibrium experiments [106, 178].

4.2.1.6 Addition of Salts. A series of proteins has been examined in the presence of various salts (e.g., [25, 27, 104, 105, 121, 142, 150, 155, 175, 413, 414, 453, 590]); some experiments were quantitatively correlated with ultracentrifugal studies (e.g., [104]).

   As in the case of buffer ions, aqueous solutions of salts induce linear increases of ϕ'_2 with increasing c_3. In general after a shallow minimum at low c_3, a strong linear increase of ϕ'_2 up to at least $c_3 = 1$ M takes place, which is followed by a plateau or a slight decrease of ϕ'_2 ([105]; Fig. 2 B). A systematic investigation on various alkali and alkali earth salts showed that the effects depend on the nature of both cation and anion. The slopes of the linear increases may be correlated with the Hofmeister series; large effects were found, e.g., for CsCl, $(NH_4)_2SO_4$, Na_2SO_4, $MgSO_4$, $CsSO_4$, Na-tartrate, Na-citrate [83, 105]. Maximum effects amounted to $\Delta\phi'_2 \approx 0.06$ cm^3 g^{-1}.

   The special case of halophilic proteins has been treated in [21, 123, 124, 142, 486].

   Addition of salts under isomolal conditions may cause a quite different behavior (usually $\phi^0_2 \leqq \phi'^0_2$; e.g., [105, 121, 590]). The comparison of quantities under isopotential and isomolal conditions allows the calculation of interaction parameters (e.g., [108, 121, 153, 590]).

4.2.1.7 Addition of Sugars or Polyols. A variety of proteins has been investigated after addition of sugars (glucose, fructose, lactose, sucrose etc.; cf., e.g., [105, 120, 127, 148, 158, 252]) or polyols (glycerol, ethylene glycol, polyethylene glycols, propylene glycol etc.; cf., e.g., [105, 108, 122, 126, 128, 134, 137, 140, 180, 252, 575]). Similar ϕ_2'-effects as found for salts were reported ([105]; Fig. 2 C); the effects, however, were not as prominent. The slopes of the linear increases were postulated to be correlated with the number of hydroxyl groups [105].

Isomolal conditions again induced a behavior different from that under isopotential conditions (usually $\phi_2^0 \leqq \phi_2'^0$; cf. e.g., [108, 120, 122]).

4.2.1.8 Addition of Denaturants. A variety of experiments with proteins in aqueous solutions of GuHCl and urea were performed, primarily at high concentrations of these denaturants (e.g., [24, 25, 98, 105, 106, 108, 109, 113, 117, 142, 153, 154, 163, 170, 185, 191, 193, 457, 458, 488–491, 506–508, 520, 521, 537, 591]); a few calculations were performed for the special case of 6 M GuHCl or 8 M urea (e.g., [105, 129, 130, 162, 165]).

A detailed investigation of the effects in GuHCl or urea ([105, 106]; Fig. 2 D) showed nonlinear changes of ϕ_2' or ϕ_2 with increasing c_3. In general, after a shallow minimum a strong decrease of ϕ_2' is observed which is probably due to denaturation of the protein. At high c_3 specific effects of the additives become predominant (in the case of GuHCl a strong increase of ϕ_2', similar to the effects of salts). Changes of ϕ_2' were found to parallel changes of other molecular parameters [106].

The effects differ for different denaturants. They vary with the presence of additional components (like salts, reducing agents) different pH, or pretreatment (e.g., oxidation with performic acid), as well as for isopotential and isomolal conditions (usually $\phi_2^0 \geqq \phi_2'^0$; cf. [105, 108, 109, 113, 153, 591]). When comparing the values of ϕ_2' in aqueous solution of GuHCl at $c_3 \approx 0$ and $c_3 \approx 6$ M ([105]; Fig. 2 D), it becomes clear that there is only a slight net decrease of ϕ_2' (about 2%). This is obviously due to an accidental compensation of volume changes due to unfolding and increasing ionic strength (cf. also Sect. 4.1.5).

A few experiments were performed using SDS as a denaturing agent [83, 105], showing also nonlinear changes of ϕ_2' or ϕ_2, qualitatively similar to those in the above mentioned denaturants. The effects were different for isopotential and isomolal conditions and could also be influenced by further additives (reducing agents). Use of SDS as a detergent solution in which proteins are not grossly denatured is dealt with in Sect. 3.1.4.

4.2.1.9 Binding of Specific Ligands. A few experiments were performed using specific ligands (e.g., [105, 175, 492, 517]). Generally small changes of the specific volume were reported (cf., e.g., the values for unliganded hemoglobin or methemoglobin with oxy- or CO-hemoglobin, Table 5; or of villin, a Ca^{2+}-modulated regulator protein, in the absence/presence of Ca^{2+}, Table 6). In the case of the brain S-100 protein Ca^{2+}-binding seems to be accompanied by large changes of the specific volume (Table 6; [517]). Binding of NAD^+ or NADH to glyceraldehyde-3-phosphate dehydrogenase or lactate dehydrogenase was accompanied by a slight decrease of ϕ_2; the volume changes increased considerably with raising temperature [105].

4.2.2 Other Substances

Since the pioneering experiments of Cohen and Eisenberg [107] on NaDNA and CsDNA, it is well-known that the specific volumes, $\bar{v}_2^0$ and ϕ_2', of nucleic acids are not constants and their variation with solvent composition is not easily predictable (cf. [111]). Because both proteins and nucleic acids are polyelectrolytes, their behavior under different experimental conditions may be qualitatively similar; the same may be true for systems conjugated with them. For polysaccharides or lipids almost no values concerning volume changes are available.

Experiments concerning volume changes were performed with polynucleotides (e.g. [493, 494]), DNA (e.g. [107, 409, 477, 496–498, 500], ribosomes (e.g. [131]), nucleosomes (e.g. [125, 136]), heparin [344, 585], lipids [581–583], SDS [141]. Reviews treating problems encountered with nucleic acids can be found in papers by Eisenberg [2, 3, 111].

The specific volume of DNA, obtained by dilatometry by Chapman and Sturtevant, increased with temperature in the range 10 °–85 °C [496, 497]. At pH 7.0 no deviations from the monotonically raising curve could be observed at melting temperature (possibly due to compensatory volume effects); the net volume changes at neutral pH accompanying thermal denaturation are obviously very small and do not differ qualitatively from the temperature-dependent volume expansion in the pre-transition region [496]. At pH 11.0 the specific volume increased sigmoidally during the helix-coil transition, the changes at alkaline pH were ascribed to a counterion-dependent loss of protons during the transition [497]. If we approximate the values at pH 7.0 at 10 ° and 40 °C, 40 ° and 85 °C, and 10 ° and 85 °C, which may be derived from the data in [496] (cf. Table 10), by straight lines, temperature coefficients of $\approx 3 \times 10^{-4}$, $\approx 6 \times 10^{-4}$, $\approx 5 \times 10^{-4}$ $cm^3 g^{-1} K^{-1}$ would result; similar results ($3–4 \times 10^{-4}$ $cm^3 g^{-1} K^{-1}$) can be obtained from the data in [500]. At pH 11.0 the slopes in the pre-transition and post-transition ranges (10 °–35 °C, 55 °–80 °C) give also similar results, they amount to $\approx 3–4 \times 10^{-4}$ and $\approx 5–6 \times 10^{-4}$ $cm^3 g^{-1} K^{-1}$, depending slightly on the ionic strength [497]. It is interesting that these temperature coefficients are similar to those of proteins (cf. Sect. 4.2.1.2), but also similar to values found for aqueous solutions of sodium polyacrylates and polyethylenesulfonates [499], and also for polystyrenes and polymethylmethacrylates in organic solvents ([504] and refs. therein). Effects of temperature (5 °–60 °C) on buoyant densities of DNA in CsCl solutions were reported in [498]. Specific volumes of lipids as a function of temperature are treated in [581–583].

In contrast to the results observed for the denaturation of DNA, volume changes were reported to accompany the complex formation of poly A and poly U, as found by dilatometry [494]. The effects of pressure on the helix-coil transitions of the poly A–poly U system, however, were found to be rather small [493].

DNA in different salt solutions was treated in [107, 111, 142, 153, 409, 477, 497, 498]. Properties of NaDNA in NaCl and of CsDNA in CsCl solutions have been described in detail in [3, 107, 111], both $\bar{v}_2^0$ and ϕ_2' increase significantly with increasing c_3. An examination of Fig. 3 also shows that $\bar{v}_2^0$ is not the limit of ϕ_2' in very dilute NaCl solutions (cf. [2, 107, 111]); this is also clear from Eq. (13).

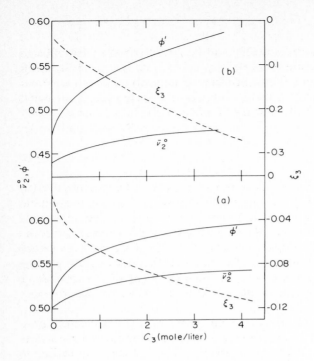

Fig. 3. Specific volumes ($\bar{v}_2$ and ϕ') and preferential interaction parameters ξ_3 in DNA solutions: (a) NaDNA in NaCl; (b) CsDNA in CsCl. From [107]; see [111]

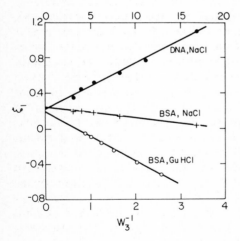

Fig. 4. Preferential interaction (hydration) parameter ξ_1, as a function of reciprocal salt molality, w_3 (g of salt/g of water) for (●) DNA in NaCl and (○) BSA in GuHCl and (+) BSA in NaCl, 23 °C. *Upper scale* NaCl; *lower scale* GuHCl. [142]. $\xi_1 = B_1 - (B_3 - E_3)/w_3$, where B_1 and B_3 denote binding of components 1 (water) and 3 (salt), respectively, and E_3 electrostatic (Donnan) exclusion of component 3

Similarly, ϕ'_2-values of nucleosomes increase with increasing c_3 of NaCl [125]; in the case of SDS [141] and heparin [344, 585] slight increases of the specific volume in the presence of NaCl were also observed.

An interpretation of small-angle X-ray scattering data of DNA solutions has been given in [3, 495, 596]. Small-angle neutron data of tRNA[phe] in aqueous salt solutions are dealt with in [43].

Preferential interaction data of proteins and of nucleic acids have been compared by Eisenberg and co-workers ([142, 153]; cf. also [2, 3, 108, 111]). The authors demonstrated a balance between hydration and solvation for BSA in GuHCl solution, and less pronounced for BSA in NaCl solution, and exclusion of salt for DNA in NaCl solution (Fig. 4). In all three cases the amounts of bound water turned out to be similarly small (0.2–0.24 g g^{-1}); the amount of NaCl bound to BSA is, however, only 1/25 that of GuHCl [142].

5 Conclusions

The values for partial specific volumes of biopolymers vary from about 0.45 to 1.1 cm^3 g^{-1}. In general there is good agreement between values calculated on the basis of compositional increments on one hand and experimental determinations on the other. If there is a choice, preference should be given to values experimentally determined under the conditions applied. Particularly isopotential values obtained after equilibrium dialysis are recommended for experiments in multicomponent solutions; for the calculation of true partial specific volumes or of interaction parameters the determination of the isomolal volume quantities is necessary.

It can be generalized that the absolute value of the partial specific volume is primarily determined by the nature of the constituents (amino acids, monosaccharides, etc.). This volume has also been called a compositional or constitutive volume (cf. [118, 505, 511, 519, 566]). Therefore biopolymers display in the first approximation specific volumes characteristic of their monomers. For rough calculations an assumed or calculated value, or a value from the literature, may be used provided that (1) the conditions of the experiments do not differ significantly from those of a two-component system (absence of special solvent conditions, e.g., $c_3 \leq 0.1$ M, no extremes of pH) and (2) no drastic chemical or structural modifications of the biopolymer occur under the conditions used (cf. [105, 106]).

It should be kept in mind that volume changes may occur under different experimental conditions. These changes have also been interpreted in terms of conformation and solvation increments (cf. [118, 505, 511, 519, 566]).

The generally most useful volume quantity in multicomponent systems is the apparent isopotential specific volume ϕ_2' (cf. [2, 3, 9, 11, 108, 111]). For proteins ϕ_2' has generally the following properties [99, 105]: (1) it exhibits no significant change when protein concentration is changed, (2) there is a predominantly linear increase of ϕ_2' with increasing temperature or increasing concentration of solvent components (buffer components, salts, sugars, polyols), (3) ϕ_2' shows a non-linear decrease upon thermal, acid, or basic denaturation, or in the presence of other denaturing agents (GuHCl, urea, SDS). The extent of changes depends on the experimental conditions, and the specific properties and pretreatments of the biopolymers. Concerning volume changes caused by temperature or addition of salts, results similar to those found for proteins were observed for nucleic acids.

The effects for $\Delta\phi_2'$ may be interpreted as superposition of different volume effects due to charge effects, changes of hydration, preferential ligand binding, folding-unfolding processes, helix-coil transitions, etc. Drastic alterations of ϕ_2' are to be expected with polyelectrolytes (nucleic acids, proteins, or systems conjugated with them) in the presence of large amounts of third components, especially of electrolytes, where strong interactions may occur; effects are more pronounced with nucleic acids than with proteins. Therefore great care must be taken in all cases, where values for specific volumes are used, that have not been determined under the proper conditions.

Knowledge of absolute values of specific volumes is necessary for the application of a variety of techniques, like ultracentrifugation, small-angle X-ray and neutron scattering, both in two- and multicomponent solutions. Moreover, knowledge of constancy or change of specific volumes is a necessary prerequisite for the application of a variety of further methods, including light scattering, viscometric, electrophoretic, chromatographic, calorimetric, crystallographic studies (cf. [2, 3, 106]). A study of volume changes accompanying diverse reactions may also be a helpful tool in understanding biopolymers.

In this context a series of applications can be quoted: determination of molecular weights under native and denaturing conditions, correction of sedimentation velocity coefficients, use of contrast variation methods, penetration experiments, evaluation of SDS gel electrophoresis or gel filtration experiments, calculation of heat capacities, stoichiometry of quaternary structure, interpretation of association or dissociation phenomena, interpretation of biopolymer-ligand interactions (binding and exclusion of specific and non-specific ligands, e.g., of coenzymes, substrates, water, salt), elucidation of crystallization processes, behavior of biopolymers under extremes of physical conditions (e.g., of halophilic, acidophilic, thermophilic enzymes), volume occupancy of biopolymers in connection with transport, regulation, aging phenomena, etc.

Only a beginning has been made in classifying specific volumes and in determining volume changes of biopolymers after alteration of experimental conditions. Therefore, it should be kept in mind that at present explanations and generalizations must be tentative.

Acknowledgments. Experimental work of the author was supported by grants of the Deutsche Forschungsgemeinschaft. The author is much indebted to Prof. H. Eisenberg, Prof. H.-J. Hinz, and Prof. R. Jaenicke for helpful suggestions, and to Mrs. S. Richter for typing assistance.

6 Tables

Table 1. Specific volumes and hydration of amino acid residues[a]

Amino acid	M_r	$\bar{v}_r$ (cm^3 g^{-1})[b]			h_r (mol mol^{-1})[c]	
		Cohn-Edsall [1, 73, 151]	Zamyatnin [118]	[566]	Kuntz [81] pH 6–8[d]	pH 4[e]
Ala	71.1	0.74	0.748	0.732	1.5	1.5
Arg	156.2	0.70	0.666	0.756	3	3
Asn	114.1	0.62	0.619	0.610	2	2
Asp	115.1	0.60	0.579	0.573	6	2
Cys	103.2		0.631	0.630	1	1
2Cys	204.3	0.63	0.676			
Gln	128.1	0.67	0.674	0.667	2	2
Glu	129.1	0.66	0.643	0.605	7.5	2
Gly	57.1	0.64	0.632	0.610	1	1
His	137.2	0.67	0.670	0.659	4	4
Ile	113.2	0.90	0.884	0.876	1	1
Leu	113.2	0.90	0.884	0.876	1	1
Lys	128.2	0.82	0.789	0.775	4.5	4.5
Met	131.2	0.75	0.745	0.739	1	1
Phe	147.2	0.77	0.774	0.766	0	0
Pro	97.1	0.76	0.758	0.748	3	3
Ser	87.1	0.63	0.613	0.596	2	2
Thr	101.1	0.70	0.689	0.676	2	2
Trp	186.2	0.74	0.734	0.728	2	2
Tyr	163.2	0.71	0.712	0.703	3	3
Val	99.1	0.86	0.847	0.831	1	1

[a] M_r, $\bar{v}_r$, and h_r refer to molecular weight, partial specific volume, and hydration of the amino acid residue, respectively.
[b] Values are given for 25 °C.
[c] Values are given for -35 °C ("unfreezable water").
[d] Necessary for calculation of ϕ'_c in 6 M GuHCl (cf. [162, 165]).
[e] Necessary for calculation of ϕ'_c in 8 M urea (cf. [129, 130]).

Table 2. Specific volumes of small molecules

Molecule	Solvent conditions[a]	Temperature (°C)	Specific volume (cm^3 g^{-1})	Reference
Amino acids				
α-Ala				
		25	0.684	[503]
(0.7 M)		20	0.688 ($\bar{v}_3$)	[40]
β-Ala				
(0.7 M)		20	0.666 ($\bar{v}_3$)	[40]
Betaine				
(0.7 M)		20	0.836 ($\bar{v}_3$)	[40]
(2.0 M)		20	0.835 ($\bar{v}_3$)	[40]

Table 2 (continued)

Molecule	Solvent conditions[a]	Temperature (°C)	Specific volume (cm³ g⁻¹)		Reference
Gly					
		25	0.583		[503]
(0.7 M)		20	0.588	$(\bar{v}_3)$	[40]
(2.0 M)		20	0.614	$(\bar{v}_3)$	[40]
Leu		25	0.817		[503]
Val		25	0.770		[503]
(0.5 M)		20	0.762	$(\bar{v}_3)$	[592]
Carbohydrates					
Fructose		20	0.614		[83]
Fucose			0.671	$(\bar{v}_c)$	[571]
Galactose			0.622	$(\bar{v}_c)$	[571]
Glucose					
			0.622	$(\bar{v}_c)$	[571]
		25	0.621		[167]
(0.5 M)		20	0.623	$(\bar{v}_3)$	[120]
(3 M)		20	0.638	$(\bar{v}_3)$	[120]
Hexose			0.613	$(\bar{v}_c)$	[133, 169]
Hexosamine			0.666		[133]
Lactose					
(0.1 M)		20	0.606	$(\bar{v}_3)$	[120]
(0.4 M)		20	0.610	$(\bar{v}_3)$	[120]
Mannose			0.607	$(\bar{v}_c)$	[571]
Methylpentose			0.678	$(\bar{v}_c)$	[169]
N-Acetylgalactose			0.684	$(\bar{v}_c)$	[571]
N-Acetylglucose			0.684	$(\bar{v}_c)$	[571]
N-Acetylhexosamine			0.666	$(\bar{v}_c)$	[169]
N-Acetylneuraminic acid			0.584	$(\bar{v}_c)$	[169]
Raffinose		25	0.608		[167]
Sialic acid					
			0.581	$(\bar{v}_c)$	[571]
			0.590	$(\bar{v}_c)$	[168]
			0.584		[133]
Sucrose					
		1	0.607		[542]
		25	0.618		[167, 542]
(0.05 M)		20	0.613	$(\bar{v}_3)$	[127]
(0.1–0.2 M)		20	0.616	$(\bar{v}_3)$	[127]
(1 M)		20	0.620	$(\bar{v}_3)$	[127]
(1 M)		20	0.620	(ϕ_3)	[83]
Denaturants					
Guanidine acetate					
(1.0 M)		20	0.778	$(\bar{v}_3)$	[591]
Guanidine hydrochloride (GuHCl)					
$(c_3 \rightarrow 0)$		20	0.692	$(\bar{v}_3^0)$	[8, 154]
$(c_3 \rightarrow 0)$		20	0.700		[193]
(low c_3)		25	0.73		[198]
(1.0 M)		20	0.732	$(\bar{v}_3)$	[591]
(1.06 M)		20	0.731		[193]
(2.06 M)		20	0.743		[193]

Table 2 (continued)

Molecule	Solvent conditions[a]	Temperature (°C)	Specific volume (cm³ g⁻¹)	Reference
(2.88 M)		20	0.749	[193]
(3.0 M)		20	0.747 $(\bar{v}_3)$	[591]
(3.00–3.07 M)		25	0.748 $(\bar{v}_3)$	[113]
(4.18 M)		20	0.754	[193]
(4.49–4.52 M)		25	0.757 $(\bar{v}_3)$	[113]
(5.21 M)		20	0.758	[193]
(5.97–6.03 M)		25	0.763 $(\bar{v}_3)$	[113]
(6 M)		20	0.763 $(\bar{v}_3)$	[109, 162, 165]
(6.30 M)		20	0.760	[193]
Guanidine sulfate				
(0.5 M)		20	0.588 $(\bar{v}_3)$	[591]
(1.0 M)		20	0.609 $(\bar{v}_3)$	[591]
(2.0 M)		20	0.631 $(\bar{v}_3)$	[591]
Urea				
$(c_3 \rightarrow 0)$		20	0.735	[83]
(1 M)		20	0.745 (ϕ_3)	[83]
(8 M)		20	0.763	[129, 130]
Detergents				
Brij-56				
(above cmc)		25	0.955	[160]
Dodecyldimethylamine oxide (DDAO)		21	1.122	[586]
Lubrol PX				
(above cmc)			0.958	[161]
Lubrol WX				
(above cmc)		25	0.929	[161, 572]
Sodium deoxycholate (DOC)				
(below and above cmc)		25	0.778	[110, 160, 161]
(above cmc)		20	0.779 $(\bar{v}_3)$	[64]
Sodium dodecyl sulfate (SDS)				
(below cmc)		25	0.813	[383]
(below cmc)		25	0.815	[160]
(above cmc)			0.875	[101]
(above cmc)		25	≈0.86	[141]
(above cmc)		25	0.870	[160, 161]
(above cmc)	Water	25	0.854	[110, 383]
(above cmc)	0.1 M NaCl	25	0.863	[110]
Triton N-101				
(above cmc)		25	0.922	[160, 161]
Triton X-100				
(above cmc)		20	0.9125	[210]
(above cmc)		20	0.93	[587]
(above cmc)		25	0.908	[160, 161]
Tween-80				
(above cmc)		25	0.896	[160, 161]
Inorganic acids				
HCl				
(1 M)		20	0.516 (ϕ_3)	[83]
HNO₃				
		20	0.465	[346]

Table 2 (continued)

Molecule	Solvent conditions[a]	Temperature (°C)	Specific volume (cm³ g⁻¹)	Reference
(1 M)		20	0.464 (ϕ_3)	[83]
H₂SO₄				
(1 M)		20	0.363 (ϕ_3)	[83]
H₃PO₄				
(1 M)		20	0.479 (ϕ_3)	[83]
Inorganic bases				
NaOH				
(1 M)		20	−0.070 (ϕ_3)	[83]
NH₄OH				
(1 M)		20	1.44 (ϕ_3)	[83]
KOH				
(1 M)		20	0.163 (ϕ_3)	[83]
Inorganic salts				
Ca-Acetate				
(1 M)	pH 5.6	20	0.440 ($\bar{v}_3$)	[590]
CaCl₂				
(1 M)	pH 5.6	20	0.216 ($\bar{v}_3$)	[121, 590]
CsCl				
			0.233	[433]
(0.99 M)	H₂O/D₂O Mixture	25	0.2453	[3]
(1 M)		20	0.245 (ϕ_3)	[83]
KCl				
(0–0.23 M)			0.365	[93]
K₂HPO₄/KH₂PO₄				
(1.97 M)	pH 6.97	25	0.203 (ϕ_3)	[178]
KSCN				
(1 M)	pH 5.6	20	0.530 ($\bar{v}_3$)	[121]
Mg-Acetate				
(1 M)	pH 5.6	20	0.313 ($\bar{v}_3$)	[590]
MgCl₂				
(0.5 M)	pH 4.5	20	0.150 ($\bar{v}_3$)	[121, 590]
(1.0 M)	pH 4.5 or 5.6	20	0.176 ($\bar{v}_3$)	[121, 590]
(1.5 M)	pH 4.5	20	0.195 ($\bar{v}_3$)	[590]
(2.0 M)	pH 4.5	20	0.200 ($\bar{v}_3$)	[590]
MgSO₄				
(1 M)	pH 4.5	20	0.136 ($\bar{v}_3$)	[121, 590]
Na-Acetate				
(0.5 M)	pH 4.7	20	0.502 ($\bar{v}_3$)	[121]
(1 M)	pH 4.7 or 5.6	20	0.516 ($\bar{v}_3$)	[121, 590]
NaCl				
		20	0.284 ($\bar{v}_3$)	[43]
(c₃→0)		25	0.2865	[155]
(0.1 M)		25	0.299	[400]
(0.5 M)		25	0.318	[400]
(1 M)	pH 4.5 or 5.6	20	0.331 ($\bar{v}_3$)	[121, 590]
(1 M)			0.335 ($\bar{v}_3$)	[124]
(2 M)			0.357 ($\bar{v}_3$)	[124]
(4 M)			0.388 ($\bar{v}_3$)	[124]

Table 2 (continued)

olecule	Solvent conditions[a]	Temper- ature (°C)	Specific volume (cm^3 g^{-1})	Reference
a$_2$SO$_4$				
(0.5 M)	pH 4.5 or 7.0	20	0.165 ($\bar{v}_3$)	[121]
(1 M)	pH 4.5 or 5.6	20	0.208 ($\bar{v}_3$)	[121, 590]
iCl$_2$				
(1 M)	pH 4.5	20	−0.2 ($\bar{v}_3$)	[590]
iSO$_4$				
(1 M)	pH 4.5	20	−0.021 ($\bar{v}_3$)	[590]
ns				
2 Ca^{2+}			−0.22	[169]
s$^+$			0.16	[169]
$^+$			0.022	[169]
$^+$			−0.13	[169]
a$^+$			−0.061	[169]
			−0.288	[571]
$^-$			0.507	[14]
)$_4^{2-}$			0.310	[571]
pids and constituents				
. Table 9				
rganic acids				
cetic acid				
(0.1 M)		20	0.856 ($\bar{v}_3$)	[339, 340]
(1 M)		20	0.860 ($\bar{v}_3$)	[339, 340]
(1 M)		20	0.862 (ϕ_3)	[83]
(1 M)		20	0.864	[121]
(10 M)		20	0.918 ($\bar{v}_3$)	[339, 340]
ityric acid				
(0.1 M)		20	0.944 ($\bar{v}_3$)	[339]
(1 M)		20	0.967 ($\bar{v}_3$)	[339]
(10 M)		20	1.036 ($\bar{v}_3$)	[339]
ropionic acid				
(0.1 M)		20	0.885 ($\bar{v}_3$)	[339]
(1 M)		20	0.904 ($\bar{v}_3$)	[339]
(10 M)		20	0.991 ($\bar{v}_3$)	[339]
itty acids cf. Table 9				
ılts of some organic acids cf. this table, section Miscellaneous				
iscellaneous				
cetyl-CoA			0.638 ($\bar{v}_c$)	[103]
ctinomycin		5	0.786	[305]
denosine			0.638	[358]
	0.2 M NaCl	20	0.62	[112, 406]
ΓP			0.44 ($\bar{v}_c$)	[337]
ırbamyl phosphate			0.48 ($\bar{v}_c$)	[337, 338]
ΓP			0.44 ($\bar{v}_c$)	[337]
hanol		20	1.18 (ϕ_3)	[83]
MN		25	0.574	[532]

Table 2 (continued)

Molecule	Solvent conditions[a]	Temperature (°C)	Specific volume (cm^3 g^{-1})	Reference
Glycerol				
(10%)		20	0.767 ($\bar{v}_3$)	[128]
(20%)		20	0.768 ($\bar{v}_3$)	[128]
(30%)		20	0.770 ($\bar{v}_3$)	[128]
(40%)		20	0.772 ($\bar{v}_3$)	[128]
2-Methyl-2,4-pentanediol (MPD)				
(50%)		25	1.0565 ($\bar{v}_3$)	[514]
NAD$^+$				
			0.62 ($\bar{v}_c$)	[338, 434]
		20	0.62	[83]
	Aqueous buffer		0.57	[342]
	D$_2$O buffer		0.56	[342]
NADH, sodium salt		20	0.57	[83]
N-(phosphonacetyl)-L-aspartate (PALA)			0.52 ($\bar{v}_c$)	[337]
O-acetyl			0.75 ($\bar{v}_c$)	[169]
Phosphate (not further specified)			0.35 ($\bar{v}_c$)	[337, 338]
Propylene glycol				
(10%)		25	0.923	[575]
(50%)		25	0.940	]575]
Sodium glyoxylate			0.364 ($\bar{v}_c$)	[103]
Sodium pyruvate			0.463 ($\bar{v}_c$)	[103]
Succinate (not further specified)			0.61 ($\bar{v}_c$)	[337, 338]
Trimethylamine N-oxide				
(1 M)		20	0.96 ($\bar{v}_3$)	[592]

[a] Water or dilute buffer.

Table 3. Specific volumes of polyaminoacids and peptides

Molecule	Solvent conditions	Temperature (°C)	Specific volume (cm^3 g^{-1})	Reference
Polyaminoacids				
Poly-β-benzyl-DL-aspartate				
[M ≈ 29 500]	Dimethylformamide	20	0.763	[272]
Poly-β-benzyl-L-aspartate				
[M ≈ 23 000 or 190 000]	Metacresol	20	0.765	[272]
Poly-L-glutamate				
[M = 70 000]	10% PEG 1000, pH 11.0	20	0.700 (ϕ'), 0.490 (ϕ)	[137]
Poly-L-glutamic acid				
[M ≈ 50 000 or 82 500]	Dimethylformamide	20	0.675	[272]
Poly-γ-benzyl-L-glutamate				
[M > 200 000]	Dimethylformamide	20	0.784	[272]
[M = 35 000 or 330 000]	Dimethylformamide	20	0.786	[274, 530]
[M = 1000–20 000]	Dimethylformamide	25	0.791	[275]
[M = 35 000]	Metacresol	25	0.788	[274, 530]
[M = 35 000 or 250 000]	Pyridine	25	0.787	[274, 530]

Table 3 (continued)

olecule	Solvent conditions	Temperature (°C)	Specific volume (cm³ g⁻¹)	Reference
oly-γ-methyl-L-glutamate				
[M = 53 000]	Metacresol	25	0.762	[271]
[M = 53 000]	Pyridine	25	0.763	[271]
oly-L-lysine				
[M = 90 000 or 240 000 or 500 000]	10% PEG 1000, pH 4.7	20	0.890 (ϕ'), 0.790 (ϕ)	[137]
oly-L-lysine, polycation				
[M = 192 000]			0.781 ($\bar{v}_c$)	[14]
oly-L-lysine hydrochloride				
[M = 245 000]	1 M NaCl		0.722	[14]
oly-ε-carbobenzoxy-L-lysine				
[M ≈ 300 000]	Dimethylformamide	20	0.806	[272]
[M ≈ 80 000]	Dimethylformamide		0.847	[529]
oly-DL-phenylalanine				
[M ≈ 60 000]	Toluene	20	0.805	[272]
oly-L-tyrosine				
[M ≈ 370 000]	Dimethylformamide	20	0.720	[272]
ptides				
elittin				
(bee venom)		20	0.716	[152]
sulin				
(bovine)		20	0.735	[151]
		20	0.749	[1, 176, 531]
			0.724 ($\bar{v}_c$)	[151]
rocidine B				
(*Bacillus brevis*)	80% Ethanol		0.793	[334]
			0.746 ($\bar{v}_c$)	[348]

Table 4. Specific volumes of serum albumin, detailed data

olecule	Solvent conditions	Temperature (°C)	Specific volume (cm³ g⁻¹)	Reference
rum albumin				
(bovine)	Water or low c_3	4	0.725 (ϕ')	[105]
	Water or low c_3	20	0.732–0.735 ($\phi'^{(0)}$, $\bar{v}^{(0)}$)	[7, 108, 109, 151, 171, 189, 201, 414]
	Water or low c_3	25	0.734–0.736 ($\phi'^{(0)}$, $\phi^{(0)}$, $\bar{v}^{(0)}$)	[105, 118, 134, 138, 149, 150, 153, 159, 171, 184, 413, 575]
	Water or low c_3	30	0.735–0.7395	[139, 171, 414]
		25	0.734–0.736 ($\bar{v}_c$)	[105, 118, 149, 151, 189]
		25	0.724 ($\bar{v}_c$)$_{mod.}$	[118]

Table 4 (continued)

Molecule	Solvent conditions	Temperature (°C)	Specific volume (cm^3 g^{-1})	Reference
	1 M CaCl$_2$, pH 5.6	20	0.731 (ϕ'^0), 0.734–0.735 (ϕ^0)	[121, 590]
	0.133 M KCl	20	0.734 (ϕ')	[117]
	0.23 M KCl	25	0.735	[150]
	1 M KSCN, pH 5.6	20	0.735 (ϕ'^0), 0.738 (ϕ^0)	[121]
	1 M MgCl$_2$, pH 3.0	20	0.746 (ϕ'^0), 0.734 (ϕ^0)	[590]
	1 M MgCl$_2$, pH 4.5	20	0.739 (ϕ'^0), 0.736 (ϕ^0)	[590]
	1 M MgCl$_2$, pH 5.6	20	0.740 (ϕ'^0), 0.737 (ϕ^0)	[121]
	1 M MgCl$_2$, pH 8.8	20	0.733 (ϕ'^0), 0.735 (ϕ^0)	[590]
	2 M MgCl$_2$, pH 3.0	20	0.741 (ϕ'^0), 0.739 (ϕ^0)	[590]
	2 M MgCl$_2$, pH 4.5	20	0.737 (ϕ'^0), 0.735 (ϕ^0)	[590]
	1 M MgSO$_4$, pH 4.5	20	0.769 (ϕ'^0), 0.734 (ϕ^0)	[121, 590]
	0.02 M Na-Acetate, pH 4.5	20	0.736 (ϕ'^0), 0.735 (ϕ^0)	[121]
	1 M Na-Acetate, pH 5.6	20	0.747 (ϕ'^0), 0.735 (ϕ^0)	[590]
	0.1 M NaCl	20	0.732	[406]
	0.2 M NaCl	20	0.730 (ϕ')	[197]
	0.2 M NaCl	25	0.732 (ϕ'), 0.734 ($\bar{v}^0$)	[113]
	0.2 M NaCl, 0.01 M DTE	25	0.732 (ϕ')	[153]
	1 M NaCl, pH 4.5	20	0.744 (ϕ'^0), 0.734 (ϕ^0)	[121]
	1 M NaCl, pH 5.6	20	0.744 (ϕ'^0), 0.735 (ϕ^0)	[121, 590]
	1–5 M NaCl	23	0.735 ($\bar{v}$)	[142]
	1 M Na$_2$SO$_4$, pH 4.5	20	0.781 (ϕ'^0), 0.735 (ϕ^0)	[121]
	1 M Na$_2$SO$_4$, pH 5.6	20	0.788 (ϕ'^0), 0.735 (ϕ^0)	[121, 590]
	pH 3.0	20	0.722 (ϕ'^0), 0.721 (ϕ^0)	[120]
	0.4 M Lactose, pH 3.0	20	0.753 (ϕ'^0), 0.719 (ϕ^0)	[120]
	0.5 M Glucose, pH 3.0	20	0.734 (ϕ'^0), 0.721 (ϕ^0)	[120]
	2 M Glucose, pH 3.0	20	0.759 (ϕ'^0), 0.728 (ϕ^0)	[120]
	pH 6.0	20	0.728 (ϕ'^0), 0.729 (ϕ^0)	[120]
	0.4 M Lactose, pH 6.0	20	0.742 (ϕ'^0), 0.725 (ϕ^0)	[120]
	0.5 M Glucose, pH 6.0	20	0.736 (ϕ'^0), 0.726 (ϕ^0)	[120]
	2 M Glucose, pH 6.0	20	0.750 (ϕ'^0), 0.727 (ϕ^0)	[120]
	60% Ethylene glycol	25	0.744 (ϕ'^0), 0.732 (ϕ^0)	[134]
	30% Glycerol	15	0.744 (ϕ'^0), 0.726 (ϕ^0)	[134]
	30% Glycerol	25	0.744 (ϕ'^0), 0.729 (ϕ^0)	[134]
	30% Glycerol	35	0.744 (ϕ'^0), 0.731 (ϕ^0)	[134]
	30% Glycerol, pH 2.0	20	0.729 (ϕ'^0), 0.726 (ϕ^0)	[128]
	30% Glycerol, pH 4.0	20	0.749 (ϕ'^0), 0.731 (ϕ^0)	[128]
	30% Glycerol, pH 5.8	20	0.749 (ϕ'^0), 0.732 (ϕ^0)	[128]
	10% PEG 1000	20	0.743 (ϕ'), 0.738 (ϕ)	[137]
	pH 2.0	25	0.728 (ϕ'^0, ϕ^0)	[575]
	10% Propylene glycol, pH 2.0	25	0.724 (ϕ'^0), 0.734 (ϕ^0)	[575]
	20% Propylene glycol, pH 2.0	25	0.726 (ϕ'^0), 0.740 (ϕ^0)	[575]
	50% Propylene glycol, pH 2.0	25	0.735 (ϕ'^0), 0.746 (ϕ^0)	[575]
	30% Sorbitol	25	0.761 (ϕ'^0), 0.738 (ϕ^0)	[134]
	0.7 M α-Ala	20	0.740 (ϕ'^0), 0.735 (ϕ^0)	[40]
	1.4 M α-Ala	20	0.749 (ϕ'^0), 0.737 (ϕ^0)	[40]
	0.7 M Gly	20	0.747 (ϕ'^0), 0.737 (ϕ^0)	[40]
	2.0 M Gly	20	0.758 (ϕ'^0), 0.735 (ϕ^0)	[40]
	1 M Guanidine acetate	20	0.741 (ϕ'^0), 0.739 (ϕ^0)	[591]
	1 M GuHCl	20	0.729 (ϕ'^0), 0.735 (ϕ^0)	[591]

Table 4 (continued)

olecule	Solvent conditions	Temperature (°C)	Specific volume (cm³ g⁻¹)	Reference
	3 M GuHCl	20	0.718 (ϕ'^0), 0.728 (ϕ^0)	[591]
	6 M GuHCl	4	0.710 (ϕ')	[105]
	6 M GuHCl	20	0.717 (ϕ'^0), 0.724 (ϕ^0, $\bar{v}^0$)	[108, 109]
	3 M GuHCl, 0.01 M DTE	25	0.693 (ϕ') , 0.728 ($\bar{v}^0$)	[153]
	6 M GuHCl, 0.01 M DTE	25	0.714 (ϕ') , 0.727 ($\bar{v}^0$)	[153]
	7 M GuHCl, 0.01 M DTE	25	0.725 (ϕ')	[153]
	6 M GuHCl, 0.1 M 2-ME	25	0.720 (ϕ')	[153]
	6 M GuHCl, 0.1 M 2-ME	25	0.721 (ϕ') , 0.728 ($\bar{v}^0$)	[113]
	6 M GuHCl, 0.1 M 2-ME	25	0.725 (ϕ')	[191]
	6 M GuHCl, 0.5 M 2-ME	20	0.726 (ϕ')	[185]
	0.5 M Guanidine sulfate	20	0.729 (ϕ'^0), 0.732 (ϕ^0)	[591]
	1 M Guanidine sulfate	20	0.750 (ϕ'^0), 0.734 (ϕ^0)	[591]
	2 M Guanidine sulfate	20	0.775 (ϕ'^0), 0.734 (ϕ^0)	[591]
(horse)		25	0.735	[149]
(human)		20	0.733	[171]
		25	0.733	[194]
		25	0.736	[149]
(porcine)	0.2 M NaCl	20	0.731 (ϕ')	[197]

Table 5. Specific volumes of selected proteins, detailed data

Molecule	Solvent conditions	Temperature (°C)	Specific volume (cm³ g⁻¹)	Reference
Aldolase (rabbit muscle)		20	0.740–0.742	[151, 189, 521]
	Water		0.742–0.743 $(\bar{v}_c)$	[151, 189]
	Water	20	0.737 $(\bar{v}^0)$	[113]
		25	0.739 $(\bar{v}^0)$	[113]
	pH 2.0	25	0.749	[181]
	pH 12.6	0	0.764	[181]
		20	0.719	[117]
	0.113 M KCl	20	0.742 (ϕ')	[55]
Succinylated	0.5 M KCl, 0.1 M Tris, 0.01 M 2-ME, pH 8.0	20	0.745	[55]
Succinylated	0.5 M KCl, 0.1 M Tris, 0.01 M 2-ME, pH 8.0	20	0.704	[55]
		20	0.734 $(\bar{v}_c)$	
	3 M GuHCl, 0.01 M 2-ME	20	0.70 (ϕ')	[521]
	5 M GuHCl, 0.01 M 2-ME	20	0.70 (ϕ')	[521]
	7 M GuHCl, 0.01 M 2-ME	20	0.73 (ϕ')	[521]
	3 M GuHCl, 0.01 M 2-ME	25	0.703 (ϕ')	[113]
	5 M GuHCl, 0.01 M 2-ME	25	0.714 (ϕ')	[113]
	6 M GuHCl, 0.01 M 2-ME	25	0.725 (ϕ'), 0.734 (v^0)	[113]
	7 M GuHCl, 0.01 M 2-ME	25	0.735 (ϕ')	[113]
	6 M GuHCl, 0.1 M 2-ME	25	0.74 (ϕ')	[191]
	6 M GuHCl, 0.5 M 2-ME	20	0.743 (ϕ')	[185]
α-Chymotrypsin (bovine pancreas)		20	0.738 $(\bar{v}^0)$	[108, 109, 130]
	pH 6.20	27	0.736	[237]
	pH 3.86	27	0.731	[237]
		20	0.732 $(\bar{v}_c)$	[165]
	0.1 M NaCl, 10^{-3} M HCl	20	0.738 (ϕ'^0)	[127]
	1 M Sucrose, 0.1 M NaCl, 10^{-3} M HCl	20	0.766 (ϕ'^0), 0.738 (ϕ^0)	[127]
	0.01 M NaCl, 10^{-3} M HCl, pH 3.0	20	0.736 (ϕ'^0), 0.736 (ϕ^0)	[128]
	30% Glycerol, 0.01 M NaCl, 10^{-3} M HCl, pH 3.0	20	0.748 (ϕ'^0), 0.733 (ϕ^0)	[128]
	6 M GuHCl	20	0.713 (ϕ'^0), 0.732 (ϕ^0, v^0)	[108, 109]
	6 M GuHCl		0.712 (ϕ^0)	[165]
	8 M Urea		0.711 (ϕ'^0), 0.726 (ϕ^0)	[ ]

	Solvent	Temp.	Specific volume	References
Chymotrypsinogen A (bovine pancreas)		20	0.733 ($\bar{v}^0$)	[108, 109, 129, 130, 162]
	Water	25	0.734 ($\bar{v}$)	[163]
			0.731 ($\bar{v}_c$)	[129, 162]
	0.1 M NaCl, 10^{-3} M HCl	20	0.736 (ϕ'^0), 0.734 (ϕ^0)	[127]
	1 M Sucrose, 0.1 M NaCl, 10^{-3} M HCl	20	0.773 (ϕ'^0), 0.736 (ϕ^0)	[127]
	0.01 M HCl, pH 2.0	20	0.732 (ϕ'^0), 0.733 (ϕ^0)	[108, 128]
	30% Glycerol, 0.01 M HCl, pH 2.0	20	0.745 (ϕ'^0), 0.727 (ϕ^0)	[108, 128]
	10% PEG 1000, pH 3.0	20	0.743 (ϕ'), 0.732 (ϕ)	[137]
	6 M GuHCl	20	0.712 (ϕ'^0), 0.729 (ϕ^0, $\bar{v}^0$)	[108, 109, 162]
	6 M GuHCl, 0.1 M 2-ME	25	0.71 (ϕ')	[191]
	6 M GuHCl	20	0.711 (ϕ_c')	[162]
	4 M Urea	25	0.739 ($\bar{v}$)	[163]
	8 M Urea	25	0.734 ($\bar{v}$)	[163]
	8 M Urea	20	0.720 (ϕ'^0), 0.730 (ϕ^0)	[129, 130]
	8 M Urea	20	0.717 (ϕ_c')	[129]
Glyceraldehyde-3-phosphate dehydrogenase (chicken heart)	0.05 M Phosphate	25	0.736 ($\bar{v}_c$)	[178]
	1.07 M Phosphate	25	0.736 (ϕ')	[178]
	1.97 M Phosphate	25	0.768 (ϕ')	[178]
		25	0.812 (ϕ')	[178]
(pig muscle)		5	0.733	[431, 432]
		20	0.737	[438]
	5 M GuHCl	25	0.729 (ϕ')	[192]
(rabbit liver)			0.740 ($\bar{v}_c$)	[435]
(rabbit muscle)		4	0.730	[183]
		5	0.729	[189]
		20	0.737	[183, 430, 438]
		20	0.739	[189]
		40	0.747	[183]
			0.742 ($\bar{v}_c$)	[189]
Succinylated	pH 8.0		0.730	[183]
	pH 12.0		0.730	[183]
Performic acid oxidized	0.133 M KCl	20	0.739 (ϕ')	[117]
	5 M GuHCl	25	0.729 (ϕ')	[192]

Table 5 (continued)

Molecule	Solvent conditions	Temperature (°C)	Specific volume (cm³ g⁻¹)	Reference
(yeast)		4	0.730	[105, 183]
		20	0.737	[183]
		25	0.739	[105]
		40	0.745	[102]
		40	0.747	[183]
		25	0.741 ($\bar{v}_c$)	[105]
Hemocyanin				
(*Astacus leptodactylus*)		4	0.741	[250]
(*Cancer magister*)				
25S species	pH 7.0	25	0.728	[538]
5S species	pH 10.0	25	0.731	[538]
	6 M GuHCl, 0.1 M 2-ME	25	0.700 (ϕ')	[538]
(*Helix pomatia*)				
1/2 size		4–5	0.732	[114, 251, 253]
		4–5	0.727–0.730	[146, 252, 253]
1/10 size		5	0.732	[146]
1/10 size		20	0.722	[255]
1/20 size		20	0.725	[255]
1/2 size	30% Sucrose	4	0.770	[252]
1/2 size	35% Glycerol	4	0.743	[252]
(lobster)		4	0.740	[415]
(*Loligo pealei*)				
1/5 size, 19S species	pH 6.6	25	0.740	[422]
1/10 size, 11S species	pH 10.6	25	0.710	[422]
(*Ommatostrephes sloani pacificus*)				
19.5S		20	0.724	[425]
(*Sepia officinalis*)		5	0.736	[146, 253]
Hemoglobin				
(bovine)				
Oxyhemoglobin		25	0.753	[155]
Methemoglobin		20	0.748	[157]
Methemoglobin		25	0.748	[190]

		Temp (°C)		Ref
Methemoglobin			0.743 ($\bar{v}_c$)	[157]
Methemoglobin	6 M GuHCl, 0.5 M 2-ME	20	0.746 (ϕ')	[185]
(*Chironomus thummi thummi*)				
Methemoglobin		20	0.734	[157]
Methemoglobin			0.736 ($\bar{v}_c$)	[157]
(horse)				
Methemoglobin		20	0.749	[176]
Methemoglobin		20	0.748	[157]
Methemoglobin			0.744 ($\bar{v}_c$)	[157]
(human)				
Unliganded		20	0.749	[176, 415]
Unliganded	2 M NaCl	23	0.779 (ϕ')	[175]
Unliganded	1 M NaI	23	0.757 (ϕ')	[175]
CO-Hemoglobin			0.749	[349]
Methemoglobin		20	0.749	[157]
Methemoglobin		20	0.751	[82]
Methemoglobin			0.745 ($\bar{v}_c$)	[157]
Oxyhemoglobin	0.09 M NaCl	20	0.746 (ϕ')	[175]
Oxyhemoglobin	1 M NaCl	20	0.775 (ϕ')	[175]
Oxyhemoglobin	2 M NaCl	20	0.779 (ϕ')	[175]
Oxyhemoglobin	0.1 M NaI	20	0.751 (ϕ')	[175]
Oxyhemoglobin	1 M NaI	20	0.763 (ϕ')	[175]
Oxyhemoglobin	2 M NaI	20	0.749 (ϕ')	[175]
(*Lumbricidae*)				
60S species	0.1 M Phosphate buffer, pH 6.8	20	0.738 (ϕ')	[421]
3.5 and 10S species	0.1 M Carbonate buffer, pH 9.5	20	0.712 (ϕ')	[421]
(*Lumbricus terrestris*)				
			0.740	[419]
			0.733	[246]
		20	0.731 ($\bar{v}_c$)	[246]
			0.745	[80, 349]
(*Planorbis corneus*)				
(rabbit)				
Methemoglobin		20	0.748	[157]
Methemoglobin			0.740 ($\bar{v}_c$)	[157]
Lactate dehydrogenase (bovine heart) H$_4$				
		20	0.740	[188, 449]
		20	0.741 ($\bar{v}^0$)	[108, 109]
		20	0.747	[179]
		20	0.750	[451]

82 H. Durchschlag

Table 5 (continued)

Molecule	Solvent conditions	Temperature (°C)	Specific volume (cm³ g⁻¹)	Reference
	6 M GuHCl (protein reduced and S-carboxy-methylated)	20	0.736 (ϕ'^0), 0.739 (ϕ^0, $\bar{v}^0$)	[108, 109]
(bovine muscle) M₄	6 M GuHCl, 0.5 M 2-ME	20	0.732 (ϕ')	[185]
(chicken) H₄, M₄	6 M GuHCl	20	0.736 (ϕ'_c)	[536]
(dogfish) M₄		20	0.740	[179]
(human) H₄, H₂M₂, M₄		20	0.740	[179, 440]
			0.74	[440]
(Lactobacillus acidophilus)		20	0.740	[450]
(Lactobacillus casei)			0.74 ($\bar{v}_c$)	[446]
Monomeric form			0.74 ($\bar{v}_c$)	[446]
Tetrameric form	pH 9.2		0.740	[446]
(Lactobacillus curvatus)	pH 5.0		0.748	[446]
			0.74 ($\bar{v}_c$)	[446]
Monomeric form	pH 9.2		0.745	[446]
Tetrameric form	pH 5.0		0.745	[446]
(Lactobacillus plantarum)			0.74 ($\bar{v}_c$)	[446]
(lobster tail)	0.1 M Tris buffer, pH 6.8	4	0.730 (ϕ')	[104]
	1.2 M (NH₄)₂SO₄, 0.1 M Tris buffer, pH 6.8	4	0.767 (ϕ')	[104]
(pig heart) H₄		4	0.734 (ϕ')	[105]
		20	0.740	[174, 440, 443, 444, 447, 449]
		25	0.742 (ϕ')	[105]
		25	0.749 ($\bar{v}_c$)	[105]
	pH 2.3	21	0.736	[448]
	0.1 M Phosphate buffer, pH 7.0	4	0.734 (ϕ')	[104]
	1.2 M (NH₄)₂SO₄, 0.1 M phosphate buffer, pH 7.0	4	0.764 (ϕ')	[104]
	$c_3 \rightarrow 0$, Phosphate buffer, pH 7.0	4	0.729 (ϕ')	[106]
	0.1 M Phosphate buffer, pH 7.0	4	0.734 (ϕ')	[106]

(pig muscle) M_4			
0.2 M Phosphate buffer, pH 7.0	4	0.737 (ϕ')	[106]
0.5 M Phosphate buffer, pH 7.0	4	0.753 (ϕ')	[106]
1.0 M Phosphate buffer, pH 7.0	4	0.772 (ϕ')	[106]
0.8 M Citrate	20	0.747	[174]
	20	0.740	[174, 440, 447]
0.8 M Citrate	20	0.747	[174]
β-Lactoglobulin (bovine milk)			
	20	0.750 ($\bar{v}^0$)	[108, 109]
	20	0.751 ($\bar{v}^0$)	[130, 151, 284]
pH 5.0 or 9.5		0.751	[349]
		0.746 ($\bar{v}_c$)	[151]
1 M MgCl$_2$, pH 3.0	20	0.761 (ϕ'^0), 0.745 (ϕ^0)	[590]
1 M MgCl$_2$, pH 5.1	20	0.750 (ϕ'^0), 0.748 (ϕ^0)	[590]
1 M MgCl$_2$, pH 7.0	20	0.749 (ϕ'^0), 0.748 (ϕ^0)	[590]
2 M MgCl$_2$, pH 7.9	20	0.745 (ϕ'^0), 0.747 (ϕ^0)	[590]
1 M MgSO$_4$, pH 3.0	20	0.795 (ϕ'^0), 0.744 (ϕ^0)	[590]
1 M MgSO$_4$, pH 5.1	20	0.785 (ϕ'^0), 0.745 (ϕ^0)	[590]
0.01 M HCl, pH 2.0	20	0.750 (ϕ'^0), 0.750 (ϕ^0)	[128]
30% Glycerol, 0.01 M HCl, pH 2.0	20	0.760 (ϕ'^0), 0.749 (ϕ^0)	[128]
10% PEG 1000, pH 3.0	20	0.752 (ϕ'), 0.750 (ϕ)	[137]
6 M GuHCl	20	0.719 (ϕ'^0), 0.728 (ϕ^0, $\bar{v}^0$)	[108, 109]
6 M GuHCl, 0.1 M 2-ME	25	0.74 (ϕ')	[191]
8 M Urea	20	0.719 (ϕ'^0), 0.738 (ϕ^0)	[130]
Lysozyme (hen's egg white)			
	20	0.702 ($\bar{v}^0$)	[108, 109, 130]
	20	0.722	[151, 236, 325, 415]
	20	0.726	[147]
		0.730	[148]
	20	0.733 ($\bar{v}$)	[144, 146]
	25	0.714, 0.723	[149]
	25	0.726 ($\bar{v}$)	[145]
	25	0.710–0.720 ($\bar{v}_c$)	[118, 144, 151, 235, 485]

Table 5 (continued)

Molecule	Solvent conditions	Temperature (°C)	Specific volume (cm³ g⁻¹)	Reference
	0.15 M KCl	20	$0.703\ (\bar{v})$	[41]
	1 M NaCl, pH 4.5	20	$0.723\ (\phi'^0),\ 0.707\ (\phi^0)$	[121, 590]
	1 M Mg-Acetate, pH 5.6	20	$0.733\ (\phi'^0),\ 0.708\ (\phi^0)$	[590]
	1 M MgCl$_2$, pH 4.5	20	$0.721\ (\phi'^0),\ 0.712\ (\phi^0)$	[590]
	1 M MgSO$_4$, pH 4.5	20	0.747–$0.748\ (\phi'^0),$ 0.709–$0.710\ (\phi^0)$	[590]
	30% Glycerol, pH 5.8 pH 3.0	20	$0.720\ (\phi'^0),\ 0.709\ (\phi^0)$	[122, 128]
	2.8% PEG 400, pH 3.0	20	$0.704\ (\phi'),\ 0.702\ (\phi)$	[137]
	44.8% PEG 400, pH 3.0	20	$0.708\ (\phi),\ 0.704\ (\phi)$	[137]
	10% PEG 1000, pH 3.0	20	$0.752\ (\phi),\ 0.705\ (\phi)$	[137]
	10% PEG 1000, pH 7.0	20	$0.729\ (\phi),\ 0.705\ (\phi)$	[137]
	10% PEG 1000, pH 10.0	20	$0.723\ (\phi),\ 0.700\ (\phi)$	[137]
	3.75% PEG 4000, pH 3.0	20	$0.699\ (\phi),\ 0.699\ (\phi)$	[137]
	0.7 M Gly	20	$0.744\ (\phi),\ 0.702\ (\phi)$	[40]
	2.0 M Gly	20	$0.710\ (\phi'^0),\ 0.697\ (\phi^0)$	[40]
	6 M GuHCl	20	$0.726\ (\phi'^0),\ 0.700\ (\phi^0)$	[108, 109]
	8 M Urea	20	$0.694\ (\phi'^0),\ 0.704\ (\phi^0,\ \bar{v}^0)$	[130]
		20	$0.700\ (\phi'^0),\ 0.707\ (\phi^0)$	[539]
(human milk)		20	0.721	
Ribonuclease A (bovine pancreas)				
Different preparations		20	$0.696\ (\bar{v}^0)$	[108, 109, 130]
		20	0.692–0.713	[173, 540]
		20–25	0.693–0.709	[117, 151, 154, 193, 200, 245, 329, 355, 396, 414, 508, 509]
		34	0.701–0.712	[199, 414]
		66	0.712	[199]
			0.703–$0.711\ (\bar{v}_c)$	[118, 151, 405]

Condition	Temp (°C)	Specific volume	Ref.
1–412.5 bar, pH 2.03, 2.16, 9.1, 9.7	20	0.707	[541]
0.1 M NaCl	20	0.703	[406]
0.15 M KCl		0.705 ($\bar{v}$)	[170]
0.04 M Gly, pH 2.9	20	0.693 (ϕ'^0), 0.693 (ϕ^0)	[127]
1 M Sucrose, 0.04 M gly, pH 2.9	20	0.743 (ϕ'^0), 0.692 (ϕ^0)	[127]
0.04 M Gly, pH 2.9	20	0.699 (ϕ'^0), 0.699 (ϕ^0)	[128]
30% Glycerol, 0.04 M gly, pH 2.9	20	0.708 (ϕ'^0), 0.696 (ϕ^0)	[128]
10% PEG 1000, pH 3.0	20	0.723 (ϕ'), 0.692 (ϕ)	[137]
5 M GuHCl	20	0.685 (ϕ)	[193]
6 M GuHCl	4	0.700	[508]
6 M GuHCl	20	0.694 (ϕ'^0), 0.694 (ϕ^0, $\bar{v}^0$)	[108, 109]
6 M GuHCl	25	0.709 (ϕ')	[117, 508]
6 M GuHCl		0.715 ($\bar{v}$)	[170]
6 M GuHCl, 0.1 M 2-ME	25	0.69 (ϕ')	[191]
6 M GuHCl, 0.1 M 2-ME		0.712 ($\bar{v}$)	[170]
8 M Urea	20	0.695 (ϕ'^0), 0.696 (ϕ^0)	[130]
Tubulin (calf brain)	20	0.735 (ϕ'^0)	[126]
	20	0.736 (ϕ^0, $\bar{v}^0$)	[108, 109, 140, 158, 537]
0.1 M Sucrose	20	0.727 ($\bar{v}_c$)	[537]
1 M Sucrose	20	0.742 (ϕ^0)	[158]
30% Glycerol	20	0.765 (ϕ^0)	[158]
10% PEG 1000, pH 7.0	20	0.754 (ϕ^0)	[126]
10% PEG 4000, pH 7.0	20	0.773 (ϕ'), 0.736 (ϕ)	[137, 140]
6 M GuHCl	20	0.783 (ϕ), 0.736 (ϕ)	[140]
6 M GuHCl	20	0.725 (ϕ^0), 0.736 ($\bar{v}^0$)	[537]
6 M GuHCl (protein reduced and S-carboxy-methylated)	20	0.725 (ϕ'^0), 0.736 (ϕ^0, $\bar{v}^0$)	[108, 109]

Table 6. Specific volumes of diverse proteins

Molecule	Solvent conditions	Temperature (°C)	Specific volume (cm³ g⁻¹)	Reference
ADP/ATP carrier protein (bovine heart mitochondria)			0.73 ($\bar{v}_c$)	[265]
Alcohol dehydrogenase (horse liver)		20	0.750	[187, 454]
	5 M GuHCl	20	0.754	[454]
	6 M GuHCl, 0.5 M 2-ME	20	0.749 (ϕ')	[185]
α-Amylase (pig pancreas)		20	0.70	[543]
Arginase (rat liver)			0.75	[544]
			0.74 ($\bar{v}_c$)	[544]
F₁-ATPase (*E. coli*)			0.737 ($\bar{v}_c$)	[212]
			0.742	[323]
			0.738	[510]
α₂-Subunit (spinach chloroplasts)		20	0.745	[214]
(*Vicia faba* chloroplasts)			0.737	[215]
Band 3 protein (human erythrocyte membranes) [7% carbohydrate content]	0.1 M NaOH, pH 8.7	20	0.740	[335]
			0.740 ($\bar{v}_c$)	[335]
Protein-detergent complex	0.4 g Dimethyl laurylamine oxide g⁻¹ protein	5	0.79	[57]
Protein-detergent complex	0.64 g Nonaethylene glycol lauryl ether g⁻¹ protein	5	0.82	[57]
Basic proteins Clupein sulfate		20	0.584	[17]
Histone H1 (calf thymus)		25	0.730	[13]

f2al-f3 Complex of histones (calf thymus)		20–25	0.733 ($\bar{v}_c$)	[341]
Iridine chloride			0.66	[13, 17]
Protamine				
(Gibbula divaricata)		25	0.678	[13]
(Loligo pealei)		25	0.667	[13]
ϕ_1 Protamine				
(Mytilus edulis)		25	0.701	[13]
Z Protein				
(Cryptochiton stelleri)		25	0.720	[13]
ϕ_0 Protein				
(Holothuria tubulosa)		25	0.711	[13]
ϕ_3 Protein				
(Mytilus edulis)		25	0.738	[13]
Thymine		25	0.682	[13]
Calmodulin (bovine brain)	+ EGTA	25	0.712 (v')	[381]
	+ Ca^{2+}	25	0.707 (v')	[381]
(bovine heart)			0.72 ($\bar{v}_c$)	[547, 548]
			0.71 ($\bar{v}_c$)	[549]
Carboxypeptidase A (bovine pancreas)		20	0.748 ($\bar{v}^0$)	[108, 109]
	6 M GuHCl (protein reduced and S-carboxymethylated)	20	0.735 (ϕ'^0), 0.741 (ϕ^0, $\bar{v}^0$)	[108, 109]
Pro-(carboxypeptidase A) (pig pancreas)				
Monomeric form		25	0.714 ($\bar{v}^0$)	[54]
Monomeric form			0.730 ($\bar{v}_c$)	[54]
Binary complex with pro-(proteinase E)		25	0.707 ($\bar{v}^0$)	[54]
Binary complex with pro-(proteinase E)			0.727 ($\bar{v}_c$)	[54]
Casein (bovine milk)				
α-Casein		20	0.726	[546]
		25	0.728	[73]
β-Casein		20	0.739	[546]
		25	0.741	[73]

Table 6 (continued)

Molecule	Solvent conditions	Temperature (°C)	Specific volume (cm³ g⁻¹)	Reference
Catalase (bovine liver)		20	0.729 ($\bar{v}^0$)	[238]
		20	0.730 ($\bar{v}^0$)	[108, 109]
	6 M GuHCl (protein reduced and S-carboxy-methylated)		0.725 (ϕ'^0), 0.726 (ϕ^0, $\bar{v}^0$)	[108, 109]
Citrate lyase (*Aerobacter aerogenes*)			0.735 ($\bar{v}_c$)	[417]
(*Klebsiella aerogenes*)			0.728	[374]
Citrate synthase (pig heart)			0.733	[71]
			0.74 ($\bar{v}_c$)	[70]
Collagen (cod skin)		20	0.710	[551]
			0.708 ($\bar{v}_c$)	[551]
(calf skin) Tropocollagen			0.706	[552]
(rat skin) Tropocollagen		20	0.703	[404]
X-irradiated (10 kR)		20	0.688–0.693	[404]
Concanavalin A (Jack bean)		20	0.73	[176]
			0.73 ($\bar{v}_c$)	[336]
α-Crystallin (calf lens)		20	0.740	[42]
			0.745	[263]
γ-Crystallin (bovine lens)			0.729 ($\bar{v}_c$)	[353]
Cytochrome c (bovine heart)		22	0.728	[553]
Ferro und ferri form (porcine heart)		20	0.707	[554]
			0.713	[46]

	Temp. (°C)	$\bar{v}$	Ref.
Cytochrome c oxidase			
(bovine heart)	20	0.72	[555]
(*Pseudomonas aeruginosa*)	20	0.73	[556]
Cytochrome P-450			
(rabbit liver microsomes)			
Complex of protein, detergent and phospholipid	20	0.78–0.811	[91, 92]
Complex of protein, detergent and phospholipid		0.754 ($\bar{v}_c$), 0.743 ($\bar{v}_c$)$_{mod.}$	[91]
Protein moiety		0.742 ($\bar{v}_c$), 0.730 ($\bar{v}_c$)$_{mod.}$	[91]
Deoxynucleotidyl transferase			
(calf thymus gland)		0.65	[557]
		0.73 ($\bar{v}_c$)	[557]
DNA-dependent RNA polymerase			
(*E. coli*)			
Core enzyme	5	0.732	[227]
Core enzyme	21	0.739	[227]
Holoenzyme monomer	5	0.733	[227]
Holoenzyme monomer	21	0.743	[227]
Holoenzyme dimer	5	0.735	[227]
Holoenzyme dimer	21	0.742	[227]
DNA polymerase			
(*E. coli*)	20	0.74	[521]
	25	0.745 ($\bar{v}_c$)	[117, 522]
3 M GuHCl, 0.01 M 2-ME	20	0.72 (ϕ')	[521]
5 M GuHCl, 0.01 M 2-ME	20	0.74 (ϕ')	[521]
7 M GuHCl, 0.01 M 2-ME	20	0.76 (ϕ')	[521]
Edestin	20	0.744	[151, 176]
	25	0.724	[149]
6 M GuHCl, 0.5 M 2-ME	20	0.719 ($\bar{v}_c$)	[149, 151]
Enolase			
(rabbit muscle)	1	0.728	[186]
	20	0.723 (ϕ')	[185]
Fatty acid synthetase			
(pig liver)		0.756 (v')	[217]
(pigeon liver)	20	0.744	[558]
(yeast)		0.748	[233]

Table 6 (continued)

Molecule	Solvent conditions	Temperature (°C)	Specific volume (cm³ g⁻¹)	Reference
Fibrinogen (bovine)	0.2 M NaCl	20	0.706 (ϕ')	[197]
		20	0.710	[195, 351]
(human) [≈2.5% carbohydrate content]		20	0.725	[151, 562]
			0.723 ($\bar{v}_c$)	[151]
β-Galactosidase (*E. coli*)			0.76	[559]
			0.727 ($\bar{v}_c$)	[485]
Gelatin		20	0.682	[151]
		30	0.695	[149]
			0.707 ($\bar{v}_c$)	[149, 151]
7S-Globulin (*Canavalia ensiformis*)			0.729	[44]
11S-Globulin (sunflower seed)			0.730	[283]
Glutamate dehydrogenase (bovine liver)	0.2 M Phosphate buffer, pH 7	20	0.749 (ϕ')	[241]
	0.2 M Phosphate buffer, pH 7	20	0.75	[230, 387]
	50% Glycerol	25	0.751 (ϕ')	[386]
	2–7 M GuHCl	25	0.761 (ϕ')	[385]
		25	0.745 (ϕ')	[508]
			0.735	[426]
	5.7 M GuHCl	25	0.726 (ϕ')	[386]
	5.7 M GuHCl	25	0.733 (ϕ')	[385]
	6 M GuHCl	4	0.739 (ϕ')	[508]
	6 M GuHCl	25	0.752 (ϕ')	[508]

Substance	Solvent	Temperature	v̄	Reference
(*Halobacterium marismortui*) Halophilic glutamate dehydrogenase	4 M NaCl	23	0.716 ($\bar{v}_c$)mod.	[123]
			0.707 (ϕ')	[202]
Glutaminase-Asparaginase (*Pseudomonas 7A*)	5.9 M GuHCl		0.735 ($\bar{v}_c$)	[33]
			0.735 (ϕ'_c)	[33]
Glycerol-3-phosphate dehydrogenase (rabbit muscle)		20	0.73	[332]
Glycinin (soybean)		20	0.730	[270]
Hexokinase (yeast)			0.738	[393]
Immunoglobulins IgG [≈2–3% carbohydrate content] (bovine)		20–25	0.720–0.736	[115, 184, 197]
		25	0.734–0.736	[115]
			0.722–0.724 ($\bar{v}_c$)	[115]
(human)		20	0.718–0.741	[39, 116, 195, 326, 526, 528]
	0.15 M NaCl	20	0.739	[528]
		25	0.739	[194]
			0.726 ($\bar{v}_c$)	[115]
(rabbit)	5 M GuHCl	20	0.711 (ϕ')	[506]
H Chain	5 M GuHCl	20	0.720 (ϕ')	[506]
L Chain	5 M GuHCl	20	0.703 (ϕ')	[506]
	6 M GuHCl		0.72 (ϕ')	[456]
IgM [≈10–12% carbohydrate content] (human)		4	0.724	[277]
		20	0.722	[326]
		20	0.723	[528]
IgA [≈8–10% carbohydrate content] (human)		20	0.725	[528]

Table 6 (continued)

Molecule	Solvent conditions	Temperature (°C)	Specific volume (cm³ g⁻¹)	Reference
(rabbit colostrum)	5 M GuHCl		0.703	[563]
			0.685 (ϕ')	[563]
IgD [≈13% carbohydrate content] (human)		20	0.717	[528]
IgE [≈10–12% carbohydrate content] (human)		20	0.713	[528]
Bence Jones protein (human urine)		20	0.736	[281]
			0.738 ($\bar{v}_c$)	[276]
Anti-poly(D-alanyl) antibodies	+ Tetra-D-alanine	5	0.740	[280]
		5	0.746	[280]
Initiation Factor M2A (rabbit reticulocytes)			0.728 ($\bar{v}_c$)	[394]
α-Ketoglutarate dehydrogenase complex (E. coli)				
Dihydrolipoyl transsuccinylase			0.731	[173]
α-Ketoglutarate dehydrogenase			0.737 ($\bar{v}_c$)	[380]
α-Ketoglutarate dehydrogenase			0.728 ($\bar{v}_c$)	[380]
2-Keto-4-hydroxyglutarate aldolase (E. coli)			0.751 ($\bar{v}_c$)	[366]
α-Lactalbumin (bovine milk) [≈12% carbohydrate content]		20	0.735	[234]
			0.704 ($\bar{v}^0$)	[108, 109, 130]
			0.729 ($\bar{v}_c$)	[273]
	6 M GuHCl	20	0.698 (ϕ'^0), 0.701 (ϕ^0, $\bar{v}^0$)	[108, 109]
	8 M Urea	20	0.699 (ϕ'^0), 0.706 (ϕ^0)	[130]

		Temp.	Value	References
α_2-Macroglobulin (human) [$\approx$8% carbohydrate content]		20	0.735	[195, 528]
			0.739	[282]
Malate dehydrogenase (*Halobacterium marismortui*) Halophilic malate dehydrogenase	1 M NaCl	23	0.750 ($\bar{v}_c$)mod.	[123]
	3 M NaCl	23	0.575 (ϕ')	[123]
	5 M NaCl	23	0.658 (ϕ')	[123]
			0.716 (ϕ')	[123]
(pig or horse heart)			0.74 ($\bar{v}_c$)	[388]
(rat liver)			0.734	[588]
(*T. flavus*) Thermophilic malate dehydrogenase			0.745 ($\bar{v}_c$)	[574]
Malate synthase (yeast)		5	0.738 (ϕ')	[103]
			0.735 ($\bar{v}_c$)	[242]
β_2-Microglobulin (human)			0.727 ($\bar{v}_c$)	[528, 560]
Myoglobin (horse) Metmyoglobin		20	0.741	[157, 176, 177]
Metmyoglobin			0.749 ($\bar{v}_c$)	[157]
(sperm whale)	0.1 M Phosphate pH 7.0		0.74	[249]
			0.747–0.748	[406]
Metmyoglobin		20	0.743	[157]
Metmyoglobin		20	0.752 ($\bar{v}_c$)	[157]
Metmyoglobin	1 M NaCl	20	0.762 (ϕ')	[175]
Metmyoglobin	2 M NaCl	20	0.770 (ϕ')	[175]
(not specified)		20	0.743	[553]
CO-Myoglobin	6 M GuHCl, 0.1 M 2-ME	25	0.74 (ϕ')	[191]
		20	0.743	[553]
Myosin (various sources)		20–25	0.71–0.74	[117, 118, 458, 526, 576, 577]

Table 6 (continued)

Molecule	Solvent conditions	Temperature (°C)	Specific volume (cm³ g⁻¹)	Reference
Dehydrated myosin				
(lobster)			0.676	[576]
(rabbit)	5 M GuHCl	25	0.702 (ϕ'), 0.729 ($\bar{v}$)	[457]
		20–26	0.728	[193, 578, 579, 580]
			0.74 ($\bar{v}_c$)	[580]
		25	0.735 ($\bar{v}_c$)	[458, 578]
		25	0.736 ($\bar{v}_c$), 0.724 ($\bar{v}_c$)$_{mod.}$	[118]
	0.5 M KCl	5	0.720 (ϕ')	[458]
	0.5 M KCl	15	0.725 (ϕ')	[458]
	0.5 M KCl	25	0.732 (ϕ')	[458]
	5 M GuHCl	20	0.720 (ϕ')	[193]
	5 M GuHCl	20	0.710 (ϕ')	[35]
Myogen A		20	0.735	[176, 439]
Tropomyosin		25	0.732	[458]
Tropomyosin	0.5 M KCl	25	0.739 ($\bar{v}_c$)	[458]
Tropomyosin	1 M KCl	25	0.736 (ϕ')	[458]
Tropomyosin	pH 2	25	0.739 (ϕ')	[458]
Tropomyosin	8 M Urea	25	0.752	[458]
Tropomyosin		25	0.728 (ϕ')	[458]
Ovalbumin (hen's egg) [≈3% carbohydrate content]				
	>0.2 g H₂O g⁻¹ protein	20–25	0.745–0.750	[149, 151, 156, 184, 349, 474]
	<0.2 g H₂O g⁻¹ protein		0.737–0.738 ($\bar{v}_c$)	[118, 151]
"Dry protein"	6 M GuHCl, 0.5 M 2-ME	25	0.75	[474]
	8 M Urea	25	0.781	[474]
		20	0.746 (ϕ')	[185]
		25	0.738	[149]

Protein	Conditions	Temp (°C)	Value	Ref
Papain (*Papaya latex*)		20	0.719	[148]
	8 M Urea	20	0.729 ($\bar{v}^0$)	[130]
			0.699 (ϕ'^0), 0.710 (ϕ^0)	[130]
Pepsin		5	0.756	[146]
		25	0.725	[245]
			0.725 ($\bar{v}_c$)	[245]
Pepsinogen (pig)		20	0.730 ($\bar{v}^0$)	[130]
	6 M GuHCl, 0.1 M 2-ME	25	0.74 (ϕ')	[191]
	8 M Urea	20	0.729 (ϕ'^0), 0.731 (ϕ^0)	[130]
Phosphatase, alkaline (*E. coli*)			0.731 ($\bar{v}_c$)	[485]
Phosphofructokinase (*Dunaliella salina*)		20	0.740	[216]
	50 mM Fructose 1,6-biphosphate	20	0.735	[216]
(rabbit)			0.732	[219]
(yeast)		20	0.742	[567, 595]
Phosphorylase b (rabbit muscle)		20	0.739	[232]
		30	0.729	[231]
			0.737 ($\bar{v}_c$)	[485]
	7.2 M GuHCl, 2-ME after dialysis	20	0.736 (ϕ')	[232]
Pyruvate decarboxylase (yeast)				
Apoenzyme		5	0.748	[224]
Holoenzyme		5	0.751	[224]
			0.742 ($\bar{v}_c$)	[225]
Pyruvate dehydrogenase complex (*E. coli*)			0.735	[173]
			0.735	[377]
		20	0.734 ($\bar{v}_c$)	[377]

Table 6 (continued)

Molecule	Solvent conditions	Temperature (°C)	Specific volume (cm³ g⁻¹)	Reference
Dihydrolipoyl transacetylase			0.744 ($\bar{v}_c$)	[378]
Dihydrolipoyl transacetylase fragments			0.743–0.745 ($\bar{v}_c$)	[378]
Transacetylase-pyruvate dehydrogenase subcomplex			0.735 ($\bar{v}_c$)	[379]
Transacetylase-flavoprotein subcomplex			0.746 ($\bar{v}_c$)	[379]
Pyruvate kinase				
(pig liver)		10	0.740 ($\bar{v}_c$)	[427]
(rabbit)			0.74	[51, 117]
(yeast)			0.754	[228]
			0.734 ($\bar{v}_c$)	[428, 429]
Riboflavin-binding protein				
(hen's egg white)				
[≈14% carbohydrate content]		20	0.720	[244]
D-Ribulose-1,5-diphosphate carboxylase				
(*Dasycladus clavaeformis*)			0.730	[218]
Rhodopsin				
(bovine retina)			0.71　($\bar{v}_c$)	[587]
Protein-detergent complex	0.916 g DDAO g⁻¹ rhodopsin	21	0.715–0.744 ($\bar{v}_c$)	[586]
	0.916 g DDAO g⁻¹ rhodopsin		0.905–0.910	[586]
DDAO			0.908–0.924 ($\bar{v}_c$)	[586]
Protein-detergent complex	1.46 g Triton X-100 g⁻¹ rhodopsin	21	1.122	[586]
			0.84 ($\bar{v}_c$)	[587]
Triton X-100		20	0.93	[587]
S-100 protein				
(bovine brain)				
[a highly acidic protein]		20	0.707	[517]
	+1 mM Ca^{2+}	20	0.74	[517]

Spectrin (human erythrocytes)	0.1 M NaCl	11	0.73 (ϕ')	[266]
Superoxide dismutase (bovine erythrocytes)		25	0.736	[53]
		25	0.743	[52]
Thrombin (bovine)		25	0.69	[36]
Prothrombin			0.70	[269]
Autoprothrombin II		25	0.708	[36]
(human) Prothrombin [≈10% carbohydrate content]		21	0.713	[268]
			0.710 ($\bar{v}_c$)	[48]
tRNA ligase (*E. coli*) Lysine: tRNA ligase			0.733 ($\bar{v}_c$)	[222]
(yeast) Lysine: tRNA ligase			0.732	[221]
Lysine: tRNA ligase			0.734 ($\bar{v}_c$)	[222]
Valine: tRNA ligase			0.741 ($\bar{v}_c$)	[222]
tRNA synthetase (*E. coli*) Methionyl-tRNA synthetase			0.729 ($\bar{v}_c$)	[220]
Methionyl-tRNA synthetase, trypsin modified			0.728 ($\bar{v}_c$)	[220]
Tyrosyl-tRNA synthetase			0.74 ($\bar{v}_c$)	[208]
Tyrosyl-tRNA synthetase			0.730 ($\bar{v}_c$)	[424]
(yeast) Phenylalanyl-tRNA synthetase			0.730 ($\bar{v}_c$)	[391]
Phenylalanyl-tRNA synthetase			0.735 ($\bar{v}_c$)	[424]
Valyl-tRNA synthetase			0.74	[213]
Valyl-tRNA synthetase			0.737 ($\bar{v}_c$)	[56]
Trypsinogen (bovine pancreas)		25	0.73	[58]

Table 6 (continued)

Molecule	Solvent conditions	Temperature (°C)	Specific volume (cm³ g⁻¹)	Reference
Trypsin inhibitor (lima bean)		20	$0.699\ (\bar{v}^0)$	[108, 109, 130]
	6 M GuHCl	20	$0.698\ (\phi'^0),\ 0.699\ (\phi^0,\ \bar{v}^0)$	[108, 109]
	8 M Urea	20	$0.691\ (\phi'^0),\ 0.700\ (\phi^0)$	[130]
(soybean)		25	$0.735–0.745$	[67]
	1 M KCl	20	0.698	[68]
Tryptophan synthase (*E. coli*)				
$\alpha_2\beta_2$ Complex			$0.755\ (\bar{v}^0)$	[49]
$\alpha_2\beta_2$ Complex			$0.741\ (\bar{v}_c)$	[49]
$\alpha\beta_2$ Complex			$0.740\ (\bar{v}_c)$	[49]
α-Subunit			$0.745\ (\bar{v}_c)$	[49, 209]
β_2-Subunit		20	0.738	[50]
β_2-Subunit			$0.738\ (\bar{v}_c)$	[49, 209]
Type B toxin (*Clostridium botulinum*)		25	$0.74\ \ (\bar{v}_c)$	[550]
			$0.73\ \ (\bar{v}_c)$	[550]
Urease (Jack bean)	pH 2 or 7	20	0.73	[176]
	6 M GuHCl		0.733	[173]
			$0.722\ (\phi')$	[66]
Villin		20	$0.737\ (\phi^0,\ \bar{v}^0)$	[492]
	+Ca²⁺	20	$0.733\ (\phi^0,\ \bar{v}^0)$	[492]
			$0.73\ \ (\bar{v}_c)$	[492]

Typical conjugated proteins cf. Tables 7–10

Table 7. Specific volumes of hemoproteins, metalloproteins, and phosphoproteins

Molecule	Solvent conditions	Temperature (°C)	Specific volume (cm³ g⁻¹)	Reference
Hemoproteins				
Hemoglobin cf. Table 5				
Catalase, cytochromes, myoglobin cf. Table 6				
Peroxidase cf. Table 8				
Metalloproteins				
Ceruloplasmin				
(human)				
[≈8% carbohydrate content,				
≈0.3% copper content]		20	0.713	[206, 561]
		20	0.714	[226]
		20	0.715	[229]
Apoceruloplasmin		20	0.728	[206]
Ferredoxin				
[≈0.8–4.5% iron content, depending				
on the source]				
(*Clostridium pasteurianum*)			0.63	[407]
	0.1 M NaCl	20	0.61	[406]
(*Clostridium tartarivorum*)		20	0.62	[79]
			0.61 ($\bar{v}_c$)	[79]
Apoferredoxin		20	0.70	[79]
Apoferredoxin			0.71 ($\bar{v}_c$)	[79]
(*Halobacterium marismortui*)				
Halophilic ferredoxin	4.3 M NaCl	25	0.717 (ϕ')	[203]
Ferritin				
[≈20% iron content,				
≙38% iron hydroxide content]				
(horse spleen)			0.589	[359]
Apoferritin		1	0.738	[359]
Apoferritin		20	0.747	[359]
Apoferritin			0.731 ($\bar{v}_c$)	[399]
Apoferritin-SDS complex	1.4 g SDS g⁻¹ protein		0.8275 ($\bar{v}_c$)	[399]
Iron hydroxide micelles			0.260 ($\bar{v}_c$)	[359]
Iron hydroxide micelles			0.286 ($\bar{v}_c$)	[401]
Nitrogenase				
(*Azotobacter vinelandii*)				
Fe Protein			0.72 ($\bar{v}_c$)	[94]
Mo–Fe Protein			0.73	[94]
(*Clostridium pasteurianum*)				
Mo–Fe Protein			0.72	[88]
Mo–Fe Protein			0.735 ($\bar{v}_c$)	[211]
Nonheme iron protein				
(*Azotobacter vinelandii*)			0.684	[95]

Table 7 (continued)

Molecule	Solvent conditions	Temperature (°C)	Specific volume (cm³ g⁻¹)	Reference
Transferrin = siderophilin (human) [≈6% carbohydrate content]				
Apotransferrin			0.716	[195]
Apotransferrin			0.72	[119]
Apotransferrin		25	0.721	[150]
Apotransferrin		37	0.725	[194]
Apotransferrin			0.72 ($\bar{v}_c$)	[196]
Diverse metalloproteins cf. also Tables 5, 6, 8				
Phosphoproteins				
Phosvitin [≈10% phosphate content] (hen's egg yolk)			0.545	[72]
Calcium complex			0.475	[72]
Magnesium complex			0.403	[72]
Casein, ovalbumin, pepsin cf. Table 6				

Table 8. Specific volumes of glycoproteins and polysaccharides

Molecule	Solvent conditions	Temperature (°C)	Specific volume (cm³ g⁻¹)	Reference
Glycoproteins				
α_1-*Acid glycoprotein* = orosomucoid = α_1-seromucoid (human) [≈41% carbohydrate content]			0.675	[172]
			0.688	[195]
		25	0.704	[166]
			0.672 ($\bar{v}_c$)	[166]
			0.693 ($\bar{v}_c$)	[168]
			0.700 ($\bar{v}_c$)	[262, 589]
Biantennary form			0.717 ($\bar{v}_c$)	[589]
Tri/tetraantennary form			0.702 ($\bar{v}_c$)	[589]
Protein			0.762 ($\bar{v}_c$)	[262]
Carbohydrate			0.621 ($\bar{v}_c$)	[262]
Desialized form		25	0.704	[166]
Desialized form			0.690 ($\bar{v}_c$)	[166]
α_1-*Antitrypsin* = α_1-trypsin inhibitor = α_1-glycoprotein (human) [>30% carbohydrate content]		20	0.646	[69]
[≈11.5% carbohydrate content]			0.728 ($\bar{v}_c$)	[303]

Table 8 (continued)

olecule	Solvent conditions	Temper-ature (°C)	Specific volume (cm^3 g^{-1})	Reference
ood group substance A				
(human)				
[≈63% carbohydrate content]		20	0.69	[169, 254]
			0.677 ($\bar{v}_c$)	[169]
ood group substance Le^a				
(human)				
[≈75–90% carbohydrate content]			0.62–0.64	[30]
		20	0.63	[169, 264]
			0.638 ($\bar{v}_c$)	[169]
utyrylcholinesterase				
(horse serum)				
[≈20% carbohydrate content]			0.688	[327]
			0.71	[327]
		25	0.723 (ϕ)	[34]
			0.712 ($\bar{v}_c$)	[327]
			0.710 ($\bar{v}_c$)	[34]
	6 M GuHCl		0.707 (ϕ'_c)	[34]
rvical glycoprotein				
(bovine cervical mucin)				
[≈74% carbohydrate content]		20	0.63–0.64	[169, 278]
			0.630 ($\bar{v}_c$)	[169]
(human cervical mucin)				
[≈70–85% carbohydrate content]	6 M GuHCl	20	0.67 (ϕ')	[243]
Subunits	6 M GuHCl	20	0.66 (ϕ')	[243]
T-Domains	6 M GuHCl	20	0.65 (ϕ')	[243]
rtical-bone glycoprotein				
(bovine)		20	0.64–0.68	[296]
[≈40% carbohydrate content]				
tuin				
(fetal calf serum or cow's fetus)				
[≈23% carbohydrate content]			0.692–0.714	[247, 526]
		20	0.696	[169, 369]
		20	0.70	[370]
		20	0.712	[322]
3.4S component		20	0.702	[312]
20S component		20	0.733	[312]
			0.718 ($\bar{v}_c$)	[169]
4-β-D-Glucan cellobiohydrolase				
[≈13% carbohydrate content]			0.699 ($\bar{v}_c$)	[368]
ucose oxidase				
(*Aspergillus niger*)				
[≈16.5% carbohydrate content]		20	0.711	[372]
(*Penicillium notatum*)		20	0.727	[372]
yco-α-lactalbumin				
(bovine milk)				
[≈9% carbohydrate content]			0.70	[367]
ycophorin A				
(human erythrocytes)			0.68 ($\bar{v}_c$)	[573]
[≈50–60 carbohydrate content]				

Table 8 (continued)

Molecule	Solvent conditions	Temperature (°C)	Specific volume (cm³ g⁻¹)	Reference
Ovomucoid				
(egg white)				
[≈23% carbohydrate content]		25	0.685	[483]
			0.70	[368]
			0.697 ($\bar{v}_c$)	[371]
Peroxidase				
(horseradish)				
[≈16.5% carbohydrate content]		20	0.699	[279]
			0.716 ($\bar{v}_c$)	[367]
Submaxillary glycoprotein				
(bovine)				
[≈44% carbohydrate content]			0.65	[169]
			0.658 ($\bar{v}_c$)	[169]
(ovine)				
[≈41% carbohydrate content]			0.685	[169]
			0.661 ($\bar{v}_c$)	[169]
(porcine)				
[≈54% carbohydrate content]			≈0.66	[515]
Tamm-Horsfall glycoprotein				
(human urine)				
[≈28% carbohydrate content]		20	0.685	[328, 331]
		20	0.705	[331]
Transcortin = corticosteroid binding globulin				
(human)				
[≈26% carbohydrate content]		20	0.708 ($\bar{v}_c$)	[300, 528]
Band 3 protein, fibrinogen, immunoglobulins, α-lactalbumin, α₂-macroglobulin, ovalbumin, prothrombin, riboflavin binding protein etc. cf. Table 6, ceruloplasmin, transferrin cf. Table 7				
Polysaccharides				
Amylose				
		25	0.61[a]	[570]
	2 M NaOH	25	0.64[a]	[570]
Cellulose				
	Cuprammonium		0.64	[176]
Cellulose acetate	Acetone		0.68	[176]
Cellulose nitrate				
[≈11–12% N]	Acetone	21	0.55–0.57	[248]
Hemicellulose A				
(*Chlorella pyrenoidosa*)		20	0.65	[347]
Methyl cellulose				
[≈29% methoxy groups]	Water		0.72	[176]
Chondroitin sulfate				
(pig laryngeal cartilage)			0.662 ($\bar{v}_c$)	[571]
Na₂-Salt			0.534 ($\bar{v}_c$)	[571]
Na₂-Salt		6	0.53 ($\bar{v}^0$)	[571]

Table 8 (continued)

olecule	Solvent conditions	Temperature (°C)	Specific volume (cm³ g⁻¹)	Reference
ondroitin sulfate C				
	CaCl₂ solution		0.55	[297]
	NaCl solution		0.59	[297]
ondromucoprotein				
(bovine nasal septum)				
[≈19% protein,				
≈81% chondroitin sulfate]	CaCl₂ solution		0.58 ($\bar{v}_c$)	[297]
	NaCl solution		0.61 ($\bar{v}_c$)	[297]
lactosamino-glycan				
(bovine cornea)		25	0.52–0.55	[526, 598]
ucosamino-glycan				
(bovine cornea)		25	0.47–0.55	[526, 598]
eparin				
[a sulfated mucopolysaccharide]			0.41–0.50	[cf. 344]
	0.1 M NaCl	25	0.42	[333]
	Water		0.45	[344]
	0.5 M NaCl		0.47	[344]
	1 M HCl		0.47	[344]
yaluronic acid				
			0.669 ($\bar{v}_c$)	[571]
Na-Salt			0.588 ($\bar{v}_c$)	[571]
Na-Salt		6	0.58 ($\bar{v}^0$)	[571]
(bovine)			0.68 ($\bar{v}_a$)	[346]
			0.66	[343, 345]
(rooster)				
Na-Salt	Water	20	0.557 ($\bar{v}^0$)	[569]
Na-Salt	Water	25	0.556 ($\bar{v}$)	[585]
Na-Salt	0.005 M NaCl	25	0.562 ($\bar{v}$)	[585]
Na-Salt	0.05 M NaCl	25	0.569 ($\bar{v}$)	[585]
Na-Salt	0.15 M NaCl	25	0.590 ($\bar{v}$)	[585]
ulin		20	0.601	[350]
oteoglycan				
(bovine femoral-head cartilage)	4 M GuHCl	20	0.53	[62]
(bovine nasal cartilage)				
[≈87% chondroitin sulfate,	0.5–4 M GuHCl		0.55 ($\bar{v}_c$)	[63]
≈ 6% keratan sulfate,				
≈ 7% protein]				
(pig laryngeal cartilage)				
Hyaluronate binding region				
[31% carbohydrate,				
69% protein]			0.663 ($\bar{v}_c$)	[571]
		6	0.66 ($\bar{v}^0$)	[571]
Protein			0.733 ($\bar{v}_c$),	[571]
Protein			0.721 ($\bar{v}_c$)$_{mod.}$	[571]
Carbohydrate			0.537 ($\bar{v}_c$)	[571]
tarch	Water		0.60	[176]

[a] Calculated from $1/\varrho$.

Table 9. Specific volumes of lipoproteins, lipids, membranes, and micelles

Molecule	Solvent conditions	Temperature (°C)	Specific volume (cm^3 g^{-1})	Reference
Lipoproteins				
Very high density lipoproteins, VHDL [$\varrho > 1.21$ g cm^{-3}]			<0.83[a]	cf. [362]
VHDL subfractions (human serum)				
VHDL$_1$ [$\varrho = 1.21–1.25$ g cm^{-3}] [average content: 62.4% protein, 28.0% phospholipids, 0.3% cholesterol, 3.2% cholesteryl esters, 4.6% triglycerides, 0.6% unesterified fatty acids]			0.80–0.83[a]	[362]
VHDL$_1$			0.866[a]	cf. [362]
VHDL$_2$ [$\varrho > 1.25$ g cm^{-3}] [average content: 98.45% protein, 0.83% phospholipids, 0.02% cholesterol, 0.05% cholesteryl esters, 0.05% triglycerides, 0.60% unesterified fatty acids]			<0.80[a]	cf. [362]
High density lipoproteins, HDL [$\varrho = 1.063–1.21$ g cm^{-3}] [average content: 45–55% protein, 30% phospholipids, 5% cholesterol, 20% cholesteryl esters, 3% triglycerides, <1% carbohydrate]			0.83–0.94[a]	cf. [100]
Protein in HDL			0.740 ($\bar{v}_a$)	[133]
HDL subfractions: (human serum)				
HDL$_2$ [41.2% protein content][b]		4	0.914[a]	[100, 204]
HDL$_2$		20	0.889	[133]
HDL$_2$		20	0.905	[356]
HDL$_3$ [55.0% protein content][b]		4	0.848[a]	[100, 204]
HDL$_3$		20	0.859	[133]
HDL$_3$		20	0.867	[356]
LpA		4	0.859	[261]
LpA		25	0.871	[261]
LpA$_2$ [42.0% protein content][b]		4	0.903	[100, 204]
LpA$_3$ [58.0% protein content][b]		4	0.859	[100, 204]
LpC [35.0% protein content][b]		4	0.905	[100, 204]

Table 9 (continued)

olecule	Solvent conditions	Temper-ature (°C)	Specific volume (cm³ g⁻¹)	Reference
(porcine serum)				
HDL₃ [44% protein content][b]		4	0.88[a]	[100, 204, 357]
		25	0.880	[597]
w density lipoproteins, LDL				
[ϱ = 1.006–1.063 g cm⁻³]			0.94–0.99[a]	cf. [100]
[average content:				
25% protein,				
22% phospholipids,				
8% cholesterol,				
37% cholesteryl esters,				
10% triglycerides,				
≈1% carbohydrate]				
(human serum)				
LDL		20	0.952–0.956	[133]
Protein in LDL			0.725 ($\bar{v}_a$)	[133]
LDL		37	0.98 (ϕ')	[257]
LDL	D₂O-containing buffer	37	0.99 (ϕ')	[257]
Protein			0.74 ($\bar{v}_a$)	[257]
Phosphatidylcholine			1.00 ($\bar{v}_a$)	[257]
Phosphatidylcholine-N(CD₃)₃			1.00 ($\bar{v}_a$)	[257]
Cholesterol			0.99 ($\bar{v}_a$)	[257]
Cholesteryl ester			1.08 ($\bar{v}_a$)	[257]
Triglyceride			1.10 ($\bar{v}_a$)	[257]
LDL		4	0.934	[258]
LDL, partially trypsin digested		4	0.944	[258]
ApoLDL	7.6 M GuHCl	25	0.703 (ϕ')	[65]
(human serum, normal subjects)				
LDL		25	0.965–0.969	[360]
(human serum, hyperlipemic subjects)				
LDL		25	0.962–0.971	[360]
DL subfractions				
(human serum)				
LpB [≈21% protein content][b]		4	0.952	[205]
		20	0.956	[205]
		37	0.977	[205]
	11% Sucrose	4	0.967 (ϕ')	[205], cf. [3]
	25% Sucrose	4	0.981 (ϕ')	[205], cf. [3]
	55% Sucrose	4	0.994 (ϕ')	[205], cf. [3]
Protein			0.72 ($\bar{v}_a$)	[205]
Phospholipids			0.98 ($\bar{v}_a$)	[205]
Cholesterol			0.97 ($\bar{v}_a$)	[205]
Cholesteryl ester			1.04 ($\bar{v}_a$)	[205]
Triglyceride			1.09 ($\bar{v}_a$)	[205]
ApoB		20	0.725 ($\bar{v}$)	[64]
ApoB	6 M GuHCl	20	0.722 (ϕ'_c)	[64]

Table 9 (continued)

Molecule	Solvent conditions	Temper-ature (°C)	Specific volume (cm³ g⁻¹)	Reference
ApoB	6 M GuHCl, 0.02 M DTT	25	0.72 (ϕ')	[32]
Complex with DOC	0.42 g DOC g⁻¹ ApoB	20	0.741 $(\bar{v}_c)$	[64]
Complex with DOC	0.64 g DOC g⁻¹ ApoB	20	0.746 $(\bar{v}_c)$	[64]
Sodium deoxycholate (DOC)		20	0.779 $(\bar{v}_3)$	[64]
(porcine serum)				
LDL₁[b]				
$[\varrho=1.02{-}1.05\ \mathrm{g\ cm^{-3}}]$		4	0.939 (ϕ')	[256]
		37	0.965 (ϕ')	[256]
LDL₂[b]				
$[\varrho=1.05{-}1.09\ \mathrm{g\ cm^{-3}}]$		4	0.920 (ϕ')	[256]
		37	0.943 (ϕ')	[256]
(rhesus monkey serum)				
LDL₂				
$[\varrho=1.020{-}1.050\ \mathrm{g\ cm^{-3}}]$		21	0.963	[143]
Very low density lipoproteins, VLDL				
$[\varrho=0.95{-}1.006\ \mathrm{g\ cm^{-3}}]$			0.99–1.05[a]	cf. [100]
[average content:				
<10% protein,				
15–20% phospholipids,				
<10% cholesterol,				
5% cholesteryl esters,				
50–70% triglycerides,				
< 1% carbohydrate]				
ApoC-III			0.723 $(\bar{v}_c)$	[390]
Pathological lipoprotein X, LpX				
(human serum)				
[ϱ in the range of LDL]		20	0.967	[259]
[average content:				
5% protein,				
61% phospholipids,				
26% cholesterol,				
4% cholesteryl esters,				
4% glycerides]				

Lipids and constituents

Carnitines				
C₁₂ Carnitine		25	0.970	[110]
C₁₆ Carnitine		25	1.002	[110]
Fatty acids				
Caprylic acid		20	1.098[a]	[403]
		25	1.100[a]	[403]
Capric acid		25	0.983[a]	[403]
Lauric acid		35	0.990[a]	[403]
		40	0.995[a]	[403]
		45	1.141[a]	[403]
		50	1.148[a]	[403]
Glycerides				
Monoglycerides				
C₁₀ Glycerol		25	0.93	[110, 375]
C₁₂ Glycerol		45	0.96	[110, 375]

Table 9 (continued)

olecule	Solvent conditions	Temperature (°C)	Specific volume (cm^3 g^{-1})	Reference
Triglycerides				
			1.093	[133]
			1.09–1.10 ($\bar{v}_a$)	[205, 257]
ospholipids				
		20	0.981	[93]
			0.970	[133]
			0.98 ($\bar{v}_a$)	[205]
Monoacylphospholipids				
C$_{16}$ Lysophosphatidylcholine				
(egg yolk)			0.921	[110, 363]
		20	0.93 ($\bar{v}_c$)	[110, 161]
Diacylphospholipids				
Phosphatidylcholine				
(egg yolk)		20	0.981	[93, 110, 161]
		20	0.984	[382, 384]
di-C$_6$ Phosphatidylcholine		25	0.865	[110, 161]
di-C$_7$ Phosphatidylcholine		25	0.888	[110, 161]
di-C$_{10}$ Phosphatidylcholine		25	0.927 ($\bar{v}_c$)	[110]
di-C$_{12}$ Phosphatidylcholine		25	0.945 ($\bar{v}_c$)	[110, 161]
di-C$_{14}$ Phosphatidylcholine		23	0.963	[376]
		28	0.972	[376]
di-C$_{16}$ Phosphatidylcholine		25	0.976 ($\bar{v}_c$)	[110, 161]
Phosphatidylethanolamine		20	0.965 ($\bar{v}_c$)	[110, 160]
Phosphatidylglycerol		20	1.015 ($\bar{v}_c$)	[110, 160]
Phosphatidylserine		20	0.93 ($\bar{v}_c$)	[110, 160]
Sphingomyelin		20	1.005 ($\bar{v}_c$)	[110, 160]
(beef brain)	50 mM KCl	20	0.966	[382]
eroids				
Cholesterol			0.938	[205]
			0.968	[133]
			0.97–0.99 ($\bar{v}_a$)	[205, 257]
[below cmc]	Water		0.988 ($\bar{v}_c$)	[513]
[below cmc]	Benzene		1.021	[513]
[above cmc]	Water	20	0.949	[110, 513]
Cholesteryl ester			1.042	[260]
			1.044	[133]
			1.04–1.08 ($\bar{v}_a$)	[205, 257]
Digitonin				
		20	0.738	[110, 364]
Rhodopsin-digitonin complex		20	0.766	[364]
embranes, micelles, vesicles				
le salt/lecithin mixed micelles				
1:1 Micelles	0.15 M NaCl	37	0.912–0.919	[324]
3:1 Micelles	0.15 M NaCl	37	0.875	[324]
hexadecyl phosphate micelles				
[simple membrane model]		20	0.856	[402]

Table 9 (continued)

Molecule	Solvent conditions	Temperature (°C)	Specific volume (cm^3 g^{-1})	Reference
Ganglioside micelles (ox brain) [a glycolipid]		20	0.78	[397]
Phosphatidylcholine vesicles (egg yolk)		20	0.9814	[93]
	50 mM KCl	20	0.9840	[382]
	0.1 M KCl	20	0.9848	[392]
	KCl	20	0.9883 (ϕ')	[93]
	NaCl	20	0.9886 (ϕ')	[93]
di-C$_{14}$ Phosphatidylcholin vesicles		10	0.933 (ϕ')	[100]
		20	0.952 (ϕ')	[100]
		23	0.963	[376]
		28	0.969	[389]
		28	0.972	[376]
		30	0.978 (ϕ')	[100]
di-C$_{14}$ Phosphatidylcholin vesicles/apoC-III complex	0.08 g Protein g^{-1} lipid		0.954 ($\bar{v}_c$)	[376]
	0.23 g Protein g^{-1} lipid		0.962	[390]
	0.23 g Protein g^{-1} lipid		0.936 ($\bar{v}_c$)	[390]
	0.25 g Protein g^{-1} lipid		0.905–0.923	[376]
	0.26 g Protein g^{-1} lipid	28	0.905	[389]
	0.26 g Protein g^{-1} lipid		0.921 ($\bar{v}_c$)	[376]
ApoC-III			0.723 ($\bar{v}_c$)	]390]
di-C$_{16}$ Phosphatidylcholine liposomes	10^{-4} mol Melittin mol^{-1}	30	0.942 (ϕ)	[152]
	10^{-3} mol Melittin mol^{-1}	30	0.960 (ϕ)	[152]
	10^{-2} mol Melittin mol^{-1}	30	0.950 (ϕ)	[152]
	10^{-1} mol Melittin mol^{-1}	30	0.936 (ϕ)	[152]
Diverse detergents cf. Table 2				

[a] Calculated from $1/\varrho$.
[b] Content is given in the reference cited.

Table 10. Specific volumes of polynucleotides, nucleic acids, and nucleoproteins

Molecule	Solvent conditions	Temperature (°C)	Specific volume (cm^3 g^{-1})	Reference
Polynucleotides				
Poly(A)	0.2 M NaCl	5	0.531 (ϕ')	[584]
Poly(U)	0.2 M NaCl	5	0.527 (ϕ')	[584]
Poly(U) · poly(A)	0.2 M NaCl	5	0.529 (ϕ')	[584]
Poly rC	0.1 M NaCl	25	0.505–0.509 (v')	[466]
Poly[d(A-s^{4}U) · d(A-s^{4}U)]			0.538 ($\bar{v}_c$)	[318]
Poly[d(A-s^{4}T) · d(A-s^{4}T)]	NaCl (c$_3$→0)	20	0.54 (v')	[317]
Poly[d(A-T) · d(A-T)]			0.548 ($\bar{v}_c$)	[318]

Table 10 (continued)

olecule	Solvent conditions	Temperature (°C)	Specific volume (cm^3 g^{-1})	Reference
ucleic acids				
NA				
(calf thymus)				
	Low c_3, pH 7.0	10	0.551	[496]
	Low c_3, pH 7.0, after heating to 85 °C and cooling	10	0.554	[496]
	Low c_3, pH 7.0	40	0.561[a]	[496]
	Low c_3, pH 7.0	60	0.572[a]	[496]
	Low c_3, pH 7.0	90	0.590[a]	[496]
	Low c_3, pH 11.0	10	0.563[a]	[497]
;DNA				
(bacteriophage T4, *E. coli*)	CsCl ($c_3 \rightarrow 0$)	25	0.446	[411]
	CsCl (high c_3)	25	0.471	[409, 410, 411]
			0.479	[477]
(bacteriophage λ, *E. coli*)			0.462	[478]
(calf thymus)	CsCl ($c_3 \rightarrow 0$)	25	0.440 ($\bar{v}^0$)	[107]
	0.20 M CsCl	25	0.446 ($\bar{v}^0$)	[107]
	0.97 M CsCl	25	0.460 ($\bar{v}^0$)	[107]
	1.86 M CsCl	25	0.471 ($\bar{v}^0$)	[107]
	2.325 M CsCl	25	0.467 ($\bar{v}^0$)	[107]
	3.44 M CsCl	25	0.471 ($\bar{v}^0$)	[107]
	CsCl ($c_3 \rightarrow 0$)	25	$\approx$0.470 (ϕ')[a]	[111]
	0.20 M CsCl	25	0.503 (ϕ')[b]	[107]
	0.99 M CsCl	25	0.530 (ϕ')[b]	[107]
	2.29 M CsCl	25	0.563 (ϕ')[b]	[107]
	3.67 M CsCl	25	0.587 (ϕ')[b]	[107]
	5.00 M CsCl	25	0.584 (ϕ')[b]	[107]
	7.46 M CsCl	25	0.566 (ϕ')[b]	[107]
DNA				
(bacteriophage T4, *E. coli*)			0.562	[477]
DNA				
(bacteriophage T4, *E. coli*)	Li-Silicotungstate (high c_3)	25	0.561	[410, 477]
(calf thymus)	0.2 M LiCl	25	0.555 (ϕ')[b]	[107]
			0.56	[320]
Solid			0.58	[480]
(Col E$_1$ plasmid, *E. coli*)				
	0.2 M LiCl	20	0.562 (ϕ')	[568]
	2 M LiCl	20	0.596 (ϕ')	[568]
	5 M LiCl	20	0.615 (ϕ')	[568]
	9 M LiCl	20	0.621 (ϕ')	[568]
H$_4$DNA				
(bacteriophage T4, *E. coli*)			0.613	[477]
aDNA	NaCl (low c_3)		0.55 ($\bar{v}_a$)	[107, 477, 480, 481, 482]
(bacteriophage T4, *E. coli*)			0.556	[477]
(bacteriophage λ, *E. coli*)			0.54	[478]
(calf thymus)	NaCl ($c_3 \rightarrow 0$)	25	0.499 ($\bar{v}^0$)	[107]
	0.20 M NaCl	25	0.503 ($\bar{v}^0$)	[107]

Table 10 (continued)

Molecule	Solvent conditions	Temperature (°C)	Specific volume (cm^3 g^{-1})	Reference
	0.976 M NaCl	25	0.528 ($\bar{v}^0$)	[107]
	3.16 M NaCl	25	0.539 ($\bar{v}^0$)	[107]
	4.19 M NaCl	25	0.543 ($\bar{v}^0$)	[107]
	NaCl ($c_3 \to 0$)	25	$\approx$0.514 (ϕ')[a]	[111]
	0.20 M NaCl	25	0.540 (ϕ')[b]	[107]
	1.00 M NaCl	25	0.565 (ϕ')[b]	[107]
	2.00 M NaCl	25	0.579 (ϕ')[b]	[107]
	3.00 M NaCl	25	0.591 (ϕ')[b]	[107]
	4.00 M NaCl	25	0.598 (ϕ')[b]	[107]
	4.99 M NaCl	25	0.595 (ϕ')[b]	[107]
	0.1 M M NaClO$_4$ or Na$_2$SO$_4$	4	0.537 (ϕ')	[302]
	0.18 M NaCl	20	0.530 (ϕ')	[304]
	0.18 M NaCl, +0.175 mol actinomycin C$_3$ mol^{-1} nucleotide pair	20	0.598 (ϕ')	[304]
	NaCl (low c_3)	20	0.56 (ϕ')	[306]
RbDNA				
(bacteriophage T4, *E. coli*)			0.516	[477]
(calf thymus)	0.2 M RbCl	25	0.514 (ϕ')[b]	[107]
RNA				
CsRNA			0.43 ($\bar{v}_a$)	[31]
NaRNA			0.53 ($\bar{v}_a$)	[31]
tRNA			0.53 ($\bar{v}_a$)	[525]
tRNA, unfractionated				
(*E. coli*)		25	0.51	[441]
(yeast)		20	0.531	[420, 436]
tRNA^{met}				
(*E. coli*)			0.50	[207]
tRNA^{phe}				
(*E. coli*)			0.53	[408]
(yeast)		17	0.54	[223]
		20	0.560	[307]
tRNA^{ser}				
(yeast)		25	0.536 (v')	[308]
tRNA^{tyr}				
(*E. coli*)			0.53	[208]
tRNA^{val}				
(yeast)			0.54	[213]
Ribosomal RNA			0.52–0.58 ($\bar{v}_a$)	[287, 293, 45 460, 463, 469, 525]
		20	0.55 ($\bar{v}_a$)	[459, 460]
(*E. coli*)			0.525	[295]
			0.544	[301]
	0.1 M NaCl, pH 4.6	25	0.57	[437]
16S-RNA	0.2 M NaCl	5	0.523 (ϕ')	[584]
16S-RNA	0.1 M KCl	4	0.577 (ϕ')	[464]
16S-RNA	0.2 M KCl	5	0.580 (ϕ')	[584]

Table 10 (continued)

Molecule	Solvent conditions	Temperature (°C)	Specific volume (cm³ g⁻¹)	Reference
16S-RNA, 13S fragment			0.537	[294]
(Jensen sarcoma)		25	0.53	[442]
(rat liver)			0.53 ($\bar{v}_a$)	[19, 20]
Viral RNA				
			0.53–0.58 ($\bar{v}_a$)	[37, 313, 468, 525]
(alfalfa mosaic virus)	0.15 M NaCl	20	0.46–0.47 (ϕ')	[412]
(bacteriophage MS2, *E. coli*)	0.1 M NaCl	5	0.457 (ϕ')	[315]
	0.1 M NaCl	20	0.495	[164]
Formaldehyde-treated	0.1 M NaCl	20	0.440	[164]
(tobacco mosaic virus)			0.578	[445]
		26	0.578	[452]
(turnip yellow mosaic virus)			0.509	[173, 416]

Viruses

Molecule	Solvent conditions	Temperature (°C)	Specific volume (cm³ g⁻¹)	Reference
Alfalfa mosaic virus [content: ≈16% RNA, ≈84% protein]				
	0.1 M NaCl	20	0.703 (ϕ')	[412]
RNA	0.15 M NaCl	20	0.46–0.47 (ϕ')	[412]
Bacteriophage fd [content: ≈12% DNA, ≈88% protein]		20	0.71	[465]
Bacteriophage fr (*E. coli*) [content: ≈30% RNA, ≈70% protein]			0.673	[316]
		20	0.69	[465]
Bacteriophage G (*B. megatherium*) [content: ≈50% DNA, ≈50% protein]			0.67	[467]
Bacteriophage MS2 (*E. coli*) [content: ≈31% RNA, ≈69% protein]			0.685ᶜ	[479]
			0.703ᶜ	[470]
RNA	0.1 M NaCl	5	0.457 (ϕ')	[315]
RNA	0.1 M NaCl	20	0.495	[164]
RNA, formaldehyde-treated	0.1 M NaCl	20	0.440	[164]
Bacteriophage PM2 (*Pseudomonas*) [content: ≈13.5% supercoiled circular DNA, ≈13 % lipid, ≈73.5% protein]				
	1 M NaCl	25	0.771 (ϕ')	[475]

Table 10 (continued)

Molecule	Solvent conditions	Temper-ature (°C)	Specific volume (cm^3 g^{-1})	Reference
Bacteriophage Qβ				
(*E. coli*)				
[content: ≈30% RNA,				
≈70% protein]	0.15 M NaCl	25	0.668 (ϕ')	[475]
			0.695^c	[470]
Bacteriophage R17				
(*E. coli*)				
[content: ≈30% RNA,				
≈70% protein]			0.673	[316]
	0.15 M NaCl	25	0.689 (ϕ')	[475]
			0.67 ($\bar{v}_c$)	[37]
RNA			0.53 ($\bar{v}_a$)	[37]
Bacteriophage T2				
(*E. coli*)				
[content: ≈55% DNA,				
≈45% protein]			0.66	[354]
Bacteriophage T4				
(*E. coli*)				
[content: ≈55% DNA,				
≈45% protein]	0.1 M NaCl	26	0.618 (ϕ')	[476]
DNA			0.540 ($\bar{v}_a$)	[111]
Bacteriophage T5				
(*E. coli*)				
[content: ≈62% DNA,				
≈38% protein]	0.1 M NaCl	26	0.658 (ϕ')	[476]
DNA			0.540 ($\bar{v}_a$)	[111]
Bacteriophage T7				
(*E. coli*)				
[content: ≈51% DNA,				
≈49% protein]	0.1 M NaCl	26	0.639 (ϕ')	[475, 476]
			0.669 ($\bar{v}_a$)	[60, 61]
			0.636 ($\bar{v}_c$)	[83]
DNA			0.54 ($\bar{v}_a$)	[83]
DNA			0.540 ($\bar{v}_a$)	[111]
Protein			0.735 ($\bar{v}_a$)	[83]
Bacteriophage λ				
(*E. coli*)				
[content: ≈63% DNA,				
≈37% protein]			0.61 ($\bar{v}_c$)	[478]
DNA			0.54	[478]
Protein			0.725	[354, 478]
Bromegrass mosaic virus (BMV)				
[content: ≈21% RNA,				
≈79% protein]			0.708 ($\bar{v}_c$)	[313]

Table 10 (continued)

lecule	Solvent conditions	Temperature (°C)	Specific volume (cm³ g⁻¹)	Reference
RNA			0.55 ($\bar{v}_a$)	[313]
Protein			0.751 ($\bar{v}_c$)	[313]
Protein		20	0.725	[309]
g virus 3 (FV3) [content: ≈20% DNA, ≈10% lipid, ≈70% protein]				
			0.713	[310]
DNA			0.50 ($\bar{v}_a$)	[310]
Lipid			1 ($\bar{v}_a$)	[310]
Protein			0.73 ($\bar{v}_a$)	[310]
ato latent mosaic virus (PLMV) [content: ≈ 6% RNA, ≈94% protein]				
			0.73	[484]
thern bean mosaic virus (SBMV) [content: ≈21% RNA, ≈79% protein]				
			0.691	[173]
		28	0.696	[239]
acco mosaic virus (TMV) [content: ≈ 5% RNA, ≈95% protein]				
			0.726	[173]
		20	0.727	[398]
		20	0.73	[352, 395, 487, 512]
RNA			0.578	[445]
RNA		26	0.578	[452]
Protein, unpolymerized		0	0.728	[180]
Protein, unpolymerized	25% Glycerol	0	0.733	[180]
Protein, polymerized		20	0.737	[180]
Protein, polymerized	25% Glycerol	20	0.742	[180]
mato bushy stunt virus (TBSV) [content: ≈16.5% RNA, ≈83.5% protein]				
			0.706	[173, 475]
			0.71	[311]
rnip yellow mosaic virus (TYMV) [content: ≈33.5% RNA, ≈66.5% protein]				
			0.661	[173, 416]
		25	0.665	[170]
			0.666	[240, 314]
	1–412.5 bar, pH 7	20	0.663	[541]
RNA			0.509	[173, 416]
RNA			0.55 ($\bar{v}_a$)	[468]
Protein			0.734	[173, 416]
Protein			0.740	[240, 314]
Protein	1–412.5 bar, pH 7	20	0.727	[541]

Table 10 (continued)

Molecule	Solvent conditions	Temperature (°C)	Specific volume (cm^3 g^{-1})	Reference
Ribosomes				
Ribosomes and ribosomal subunits				
(*E. coli*)				
[content: ≈60–70% RNA,				
≈30–40% protein,				
depending on				
pretreatment]		4	0.59	[132]
		25	0.60	[132]
			0.64	[545]
			0.63–0.66 ($\bar{v}_a$)	[442, 461, 473, 545]
30S		4	0.591	[131, 132]
30S		25	0.601	[132]
30S			0.599 ($\bar{v}_c$)	[83]
Protein			0.735 ($\bar{v}_a$)	[83]
RNA			0.54 ($\bar{v}_a$)	[83]
30S, unfolded		4	0.586–0.618 (ϕ')	[131]
50S			0.59	[295, 299, ?]
50S		4	0.585	[132]
50S		10	0.597–0.600 ($\bar{v}$)	[135]
50S		25	0.592	[132]
70S		4	0.596	[132]
70S		25	0.606	[132]
70S			0.64	[472]
(diverse eukaryotes)			0.61–0.674	[471]
80S			0.66–0.67	[472]
(*Paramecium aurelia*)				
[content: ≈57% RNA,				
≈43% protein]				
80S			0.601–0.607	[16]
(rat liver)				
[content: ≈45% RNA,				
≈55% protein]				
80S		25	0.664	[18]
Large subunit			0.623 ($\bar{v}_c$)	[19, 20]
Small subunit			0.633–0.645 ($\bar{v}_c$)	[19, 20]
Ribosomal proteins				
(*E. coli*)			0.74 ($\bar{v}_a$)	[292, 293, ? 459, 461, 469]
Proteins in 30S			0.742 ($\bar{v}_c$)	[461]
Proteins in 50S			0.745 ($\bar{v}_c$)	[461]
Proteins in 70S			0.743 ($\bar{v}_c$)	[461]
Proteins in 70S	8 M Urea	25	0.737 (ϕ')	[462]
S1–S21			0.716–0.746 ($\bar{v}_c$)	[564]
S1			0.737–0.738 ($\bar{v}_c$)	[298, 564]
S4		20	0.725 (ϕ')	[290]

Table 10 (continued)

olecule	Solvent conditions	Temperature (°C)	Specific volume (cm³ g⁻¹)	Reference
S4			0.73–0.74 ($\bar{v}_c$)	[285, 289, 418, 564]
S5			0.740 ($\bar{v}_c$)	[330]
S8			0.746 ($\bar{v}_c$)	[330]
L7			0.735–0.740	[291]
L12			0.735–0.740	[291]
L18			0.739 ($\bar{v}_c$)	[286]
L25			0.730 ($\bar{v}_c$)	[286]
L7/L12–L10 complex		21	0.726 (ϕ')	[288]
rat liver)			0.74 ($\bar{v}_c$)	[19, 20]
bosomal RNA				
			0.52–0.58 ($\bar{v}_a$)	[287, 293, 459, 460, 463, 469, 525]
		20	0.55 ($\bar{v}_a$)	[459, 460]
E. coli)			0.525	[295]
			0.544	[301]
		25	0.57	[437]
16S-RNA	0.2 M NaCl	5	0.523 (ϕ')	[584]
16S-RNA	0.1 M KCl	4	0.577 (ϕ')	[464]
16S-RNA	0.2 M KCl	5	0.580 (ϕ')	[584]
16S-RNA, 13S fragment			0.537	[294]
Jensen sarcoma)		25	0.53	[442]
rat liver)			0.53 ($\bar{v}_a$)	[19, 20]

romatin

ontent: ≈40–50% DNA, ≈50–60% protein				
			0.63 ($\bar{v}_c$)	[15]
Core particle			0.65 ($\bar{v}_c$)	[15]
DNA			0.54 ($\bar{v}_a$)	[15]
Protein			0.72 ($\bar{v}_a$)	[15]
cleosome				
chicken erythrocytes)				
Core particle	0.1 M NaCl		0.646 (ϕ'_c)	[47, 125]
Core particle	0.7 M NaCl		0.660 (ϕ'_c)	[47, 125]
Histone			0.75	[47, 125]
		25	0.661 (ϕ')	[267, 319]
	4 M Urea	25	0.672 (ϕ')	[319]
	8 M Urea	25	0.677 (ϕ')	[319]
pea buds)			0.650 ($\bar{v}_c$)	[321]
DNA			0.555 ($\bar{v}_a$)	[321]
Histone			0.745 ($\bar{v}_c$)	[321]
rat thymus)				
Core particle			0.660 ($\bar{v}_c$)	[136]

[a] Taken from the corresponding figure.
[b] Calculated from given $(\partial\varrho/\partial c_2)^0_\mu$ and ϱ values.
[c] Calculated from $1/\varrho$.

References

1. Cohn EJ, Edsall JT (1943) In: Cohn EJ, Edsall JT (eds) Proteins, amino acids and peptides as ions and dipolar ions. Reinhold, New York, pp 370–381, 428–431 (1965) reprint by Hafner, New York
2. Eisenberg H (1976) Biological macromolecules and polyelectrolytes in solution. Clarendon, Oxford
3. Eisenberg H (1981) Q Rev Biophys 14:141–172
4. Elder JP (1979) Methods Enzymol 61:12–25
5. Kratky O, Leopold H, Stabinger H (1973) Methods Enzymol 27:98–110
6. Kupke DW, Beams JW (1972) Methods Enzymol 26:74–107
7. Hunter MJ (1978) Methods Enzymol 48:23–29
8. Kupke DW, Crouch TH (1978) Methods Enzymol 48:29–68
9. Kupke DW (1973) In: Leach SJ (ed) Physical principles and techniques of protein chemistry, part C. Academic Press, New York, pp 1–75
10. Kratky O, Leopold H, Stabinger H (1969) Z angew Physik 27:273–277
11. Casassa EF, Eisenberg H (1964) Adv Protein Chem 19:287–395
12. Traube J (1899) Samml Chem Vortr 4:255–332
13. Ausió J, Subirana JA (1982) Biochemistry 21:5910–5918
14. Applequist J, Doty P (1962) In: Stahmann MA (ed) Polyamino acids, polypeptides and proteins. The University of Wisconsin Press, Madison, pp 161–177
15. Spencer M, Staynov DZ (1980) Biophys J 30:307–316
16. Reisner AH, Rowe J, Macindoe HM (1968) J Mol Biol 32:587–610
17. Gehatia M, Hashimoto C (1963) Biochim Biophys Acta 69:212–221
18. Hamilton MG, Cavalieri LF, Petermann ML (1962) J Biol Chem 237:1155–1159
19. Hamilton MG, Ruth ME (1969) Biochemistry 8:851–856
20. Hamilton MG, Pavlovec A, Petermann ML (1971) Biochemistry 10:3424–3427
21. Pundak S, Aloni H, Eisenberg H (1981) Eur J Biochem 118:471–477
22. O'Brien RD, Timpone CA, Gibson RE (1978) Anal Biochem 86:602–615
23. Bull HB, Breese K (1973) Biopolymers 12:2351–2358
24. Noelken ME, Timasheff SN (1967) J Biol Chem 242:5080–5085
25. Hade EPK, Tanford C (1967) J Am Chem Soc 89:5034–5040
26. Durchschlag H, Högel J, Schuster R, Jaenicke R (1980) Hoppe-Seyler's Z Physiol Chem 361:237–238
27. Durchschlag H, Schuster R, Farkas P, Jaenicke R (1980) Hoppe-Seyler's Z Physiol Chem 361:238–239
28. Durchschlag H (1983) Hoppe-Seyler's Z Physiol Chem 364:1257–1258
29. Jaenicke R (1964) In: Rauen HM (ed) Biochemisches Taschenbuch, 2nd edn, vol 2. Springer, Berlin Göttingen Heidelberg New York, pp 746–767
30. Bhaskar KR, Creeth JM (1974) Biochem J 143:669–679
31. Bruner R, Vinograd J (1965) Biochim Biophys Acta 108:18–29
32. Patterson BW, Schumaker VN, Fisher WR (1984) Anal Biochem 136:347–351
33. Holcenberg JS, Teller DC (1976) J Biol Chem 251:5375–5380
34. Teng T-L, Harpst JA, Lee JC, Zinn A, Carlson DM (1976) Arch Biochem Biophys 176:71–81
35. Gazith J, Himmelfarb S, Harrington WF (1970) J Biol Chem 245:15–22
36. Harmison CR, Seegers WH (1962) J Biol Chem 237:3074–3076
37. Enger MD, Stubbs EA, Mitra S, Kaesberg P (1963) Proc Natl Acad Sci USA 49:857–860
38. Casassa EF, Eisenberg H (1961) J Phys Chem 65:427–433
39. Heimburger N, Heide K, Haupt H, Schultze HE (1964) Clin Chim Acta 10:293–307
40. Arakawa T, Timasheff SN (1983) Arch Biochem Biophys 224:169–177
41. Sophianopoulos AJ, Rhodes CK, Holcomb DN, van Holde KE (1962) J Biol Chem 237:1107–1112
42. Andries C, Guedens W, Clauwaert J, Geerts H (1983) Biophys J 43:345–354
43. Li ZQ, Giegé R, Jacrot B, Oberthür R, Thierry J-C, Zaccai G (1983) Biochemistry 22:4380–4388
44. Plietz P, Damaschun G, Müller JJ, Schlesier B (1983) FEBS Lett 162:43–46

45. Harding SE, Rowe AJ (1983) Biopolymers 22:1813–1829
46. Timchenko AA, Denesyuk AI, Fedorov BA (1981) Biofizika 26:32–36
47. McGhee JD, Felsenfeld G, Eisenberg H (1980) Biophys J 32:261–270
48. Lanchantin GF, Hart DW, Friedmann JA, Saavedra NV, Mehl JW (1968) J Biol Chem 243:5479–5485
49. Lane AN, Kirschner K (1983) Eur J Biochem 129:675–684
50. Hathaway GM (1972) J Biol Chem 247:1440–1444
51. Warner RC (1958) Arch Biochem Biophys 78:494–496
52. Martel P, Powell BM, Zepp Johnston RA, Petkau A (1983) Biochim Biophys Acta 747:78–85
53. Wood E, Dalgleish D, Bannister W (1971) Eur J Biochem 18:187–193
54. Martínez MC, Nieuwenhuysen P, Clauwaert J, Cuchillo CM (1983) Biochem J 215:23–27
55. Hass LF (1964) Biochemistry 3:535–541
56. Dietrich A, de Marcillac GD, Pouyet J, Giegé R (1978) Biochim Biophys Acta 521:597–605
57. Pappert G, Schubert D (1983) Biochim Biophys Acta 730:32–40
58. Tietze F (1953) J Biol Chem 204:1–11
59. Attri AK, Minton AP (1983) Anal Biochem 133:142–152
60. Rontó G, Agamalyan MM, Drabkin GM, Feigin LA, Lvov YM (1983) Biophys J 43:309–314
61. Boyarintseva AK, Dembo AT, Tikhonenko TI, Feigin LA (1973) Dokl Akad Nauk SSSR 212:487–489
62. Lyon M, Greenwood J, Sheehan JK, Nieduszynski IA (1983) Biochem J 213:355–362
63. Hascall VC, Sajdera SW (1970) J Biol Chem 245:4920–4930
64. Oeswein JQ, Chun PW (1983) Biophys Chem 18:35–51
65. Smith R, Dawson JR, Tanford C (1972) J Biol Chem 247:3376–3381
66. Reithel FJ, Robbins JE, Gorin G (1964) Arch Biochem Biophys 108:409–413
67. Rackis JJ, Sasame HA, Mann RK, Anderson RL, Smith AK (1962) Arch Biochem Biophys 98:471–478
68. Wu YV, Scheraga HA (1962) Biochemistry 1:698–705
69. Bundy HF, Mehl JW (1959) J Biol Chem 234:1124–1128
70. Singh M, Brooks GC, Srere PA (1970) J Biol Chem 245:4636–4640
71. Wu J-Y, Yang JT (1970) J Biol Chem 245:212–218
72. Joubert FJ, Cook WH (1958) Can J Biochem 36:399–408
73. McMeekin TL, Groves ML, Hipp NJ (1949) J Am Chem Soc 71:3298–3300
74. Zamyatnin AA (1969) Stud Biophys 17:165–172
75. Zamyatnin AA (1970) Stud Biophys 24/25:53–60
76. Zamyatnin AA (1971) Biofizika 16:163–171
77. Traube J (1896) Liebigs Ann Chem 290:43–122
78. Cohn EJ, McMeekin TL, Edsall JT, Blanchard MH (1934) J Am Chem Soc 56:784–794
79. Devanathan T, Akagi JM, Hersh RT, Himes RH (1969) J Biol Chem 244:2846–2853
80. Svedberg T, Eriksson-Quensel IB (1934) J Am Chem Soc 56:1700–1706
81. Kuntz ID (1971) J Am Chem Soc 93:514–516
82. Behlke J (1971) Stud Biophys 28:79–84
83. Durchschlag H et al., unpublished work
84. Stabinger H, Leopold H, Kratky O (1967) Monatsh Chem 98:436–438
85. Leopold H (1970) Elektronik 19:297–302
86. Leopold H (1970) Elektronik 19:411–416
87. Bauer N, Lewin SZ (1959) In: Weissberger A (ed) Physical methods of organic chemistry, 3rd edn, vol 1, part 1, chapter IV. Wiley Interscience, New York, pp 131–190
88. Dalton H, Morris JA, Ward MA, Mortenson LE (1971) Biochemistry 10:2066–2072
89. Smith MH (1970) In: Sober HA (ed) CRC Handbook of biochemistry, 2nd edn. Chemical Rubber, Cleveland, pp C3–C35
90. Jacrot B, Zaccai G (1981) Biopolymers 20:2413–2426
91. Wendel I, Behlke J, Jänig G-R (1983) Biomed Biochim Acta 42:623–631
92. Wendel I, Behlke J, Jänig G-R (1983) Biomed Biochim Acta 42:633–640
93. Huang C-H, Charlton JP (1971) J Biol Chem 246:2555–2560

94. Swisher RH, Landt M, Reithel FJ (1975) Biochem Biophys Res Commun 66:1476–1482
95. Shethna YI (1970) Biochim Biophys Acta 205:58–62
96. Mueller H (1965) J Biol Chem 240:3816–3828
97. Jaenicke R (1971) Eur J Biochem 21:110–115
98. Kauzmann W (1958) Biochim Biophys Acta 28:87–91
99. Durchschlag H (1980) Abstract 5th Int Conf Small-Angle Scattering, Berlin 1980, pp 135–136
100. Laggner P (1982) In: Glatter O, Kratky O (eds) Small angle X-ray scattering. Academic Press, London, pp 329–359
101. Kratky O, Müller K (1982) In: Glatter O, Kratky O (eds) Small angle X-ray scattering. Academic Press, London, pp 499–510
102. Durchschlag H, Puchwein G, Kratky O, Schuster I, Kirschner K (1971) Eur J Biochem 19:9–22
103. Zipper P, Durchschlag H (1978) Eur J Biochem 87:85–99
104. Tuengler P, Long GL, Durchschlag H (1979) Anal Biochem 98:481–484
105. Durchschlag H, Jaenicke R (1982) Biochem Biophys Res Commun 108:1074–1079
106. Durchschlag H, Jaenicke R (1983) Int J Biol Macromol 5:143–148
107. Cohen G, Eisenberg H (1968) Biopolymers 6:1077–1100
108. Lee JC, Gekko K, Timasheff SN (1979) Methods Enzymol 61:26–49
109. Lee JC, Timasheff SN (1974) Biochemistry 13:257–265
110. Steele JCH Jr, Tanford C, Reynolds JA (1978) Methods Enzymol 48:11–23
111. Eisenberg H (1974) In: Ts'o POP (ed) Basic principles in nucleic acid chemistry, vol 2. Academic Press, New York, pp 171–264
112. Edelstein SJ, Schachman HK (1973) Methods Enzymol 27:82–98
113. Reisler E, Eisenberg H (1969) Biochemistry 8:4572–4578
114. Pilz I, Kratky O, Moring-Claesson I (1970) Z Naturforsch Teil B Chem Biochem Biophys Biol Chem 25:600–606
115. Kickhöfen B, Hammer DK, Scheel D (1968) Hoppe-Seyler's Z Physiol Chem 349:1755–1773
116. Pilz I, Puchwein G, Kratky O, Herbst M, Haager O, Gall WE, Edelman GM (1970) Biochemistry 9:211–219
117. Sakura JD, Reithel FJ (1972) Methods Enzymol 26:107–119
118. Zamyatnin AA (1972) Prog Biophys Mol Biol 24:107–123
119. Bezkorovainy A, Rafelson ME Jr (1964) Arch Biochem Biophys 107:302–304
120. Arakawa T, Timasheff SN (1982) Biochemistry 21:6536–6544
121. Arakawa T, Timasheff SN (1982) Biochemistry 21:6545–6552
122. Gekko K (1982) J Biochem (Tokyo) 91:1197–1204
123. Pundak S, Eisenberg H (1981) Eur J Biochem 118:463–470
124. Reich MH, Kam Z, Eisenberg H (1982) Biochemistry 21:5189–5195
125. Eisenberg H, Felsenfeld G (1981) J Mol Biol 150:537–555
126. Na GC, Timasheff SN (1981) J Mol Biol 151:165–178
127. Lee JC, Timasheff SN (1981) J Biol Chem 256:7193–7201
128. Gekko K, Timasheff SN (1981) Biochemistry 20:4667–4676
129. Prakash V, Timasheff SN (1981) Anal Biochem 117:330–335
130. Prakash V, Loucheux C, Scheufele S, Gorbunoff MJ, Timasheff SN (1981) Arch Biochem Biophys 210:455–464
131. Blair DP, Heilmann L, Hill WE (1981) Biophys Chem 14:81–89
132. Hill WE, Rossetti GP, van Holde KE (1969) J Mol Biol 44:263–277
133. Oeswein JQ, Chun PW (1981) Biophys Chem 14:233–245
134. Gekko K, Morikawa T (1981) J Biochem (Tokyo) 90:39–50
135. Tardieu A, Vachette P, Gulik A, le Maire M (1981) Biochemistry 20:4399–4406
136. Karawajew K, Böttger M, Eichhorn I, Lindigkeit R, Osipova TN, Pospelov VA, Vorob'ev VI (1981) Acta Biol Med Ger 40:105–114
137. Lee JC, Lee LLY (1981) J Biol Chem 256:625–631
138. Bernhardt J, Pauly H (1980) J Phys Chem 84:158–162
139. Bull HB, Breese K (1979) Arch Biochem Biophys 197:199–204
140. Lee JC, Lee LLY (1979) Biochemistry 24:5518–5526

141. Doughty DA (1979) J Phys Chem 83:2621–2628
142. Eisenberg H, Haik Y, Ifft JB, Leicht W, Mevarech M, Pundak S (1978) In: Caplan SR, Ginzburg M (eds) Energetics and structure of halophilic microorganisms. Elsevier/North-Holland Biomedical, Amsterdam, pp 13–32
143. Aggerbeck L, Yates M, Tardieu A, Luzzati V (1978) J Appl Cryst 11:466–472
144. Floßdorf J, Süßenbach U (1978) Makromol Chem 179:1061–1068
145. Deonier RC, Williams JW (1970) Biochemistry 9:4260–4267
146. Pilz I, Czerwenka G (1973) Makromol Chem 170:185–190
147. Millero FJ, Ward GK, Chetirkin P (1976) J Biol Chem 251:4001–4004
148. Schausberger A, Pilz I (1977) Makromol Chem 178:211–225
149. Charlwood PA (1957) J Am Chem Soc 79:776–781
150. Hunter MJ (1967) J Phys Chem 71:3717–3721
151. McMeekin TL, Marshall K (1952) Science 116:142–143
152. Posch M, Rakusch U, Mollay C, Laggner P (1983) J Biol Chem 258:1761–1766
153. Reisler E, Haik Y, Eisenberg H (1977) Biochemistry 16:197–203
154. Crouch TH, Kupke DW (1977) Biochemistry 16:2586–2593
155. Bernhardt J, Pauly H (1977) J Phys Chem 81:1290–1295
156. Bull HB, Breese K (1968) J Phys Chem 72:1817–1819
157. Behlke J, Wandt I (1973) Acta Biol Med Ger 31:383–388
158. Lee JC, Frigon RP, Timasheff SN (1975) Ann NY Acad Sci 253:284–291
159. Bernhardt J, Pauly H (1975) J Phys Chem 79:584–590
160. Tanford C, Nozaki Y, Reynolds JA, Makino S (1974) Biochemistry 13:2369–2376
161. Tanford C, Reynolds JA (1976) Biochim Biophys Acta 457:133–170
162. Lee JC, Timasheff SN (1974) Arch Biochem Biophys 165:268–273
163. Škerjanc J, Doleček V, Lapanje S (1970) Eur J Biochem 17:160–164
164. Slegers H, Clauwaert J, Fiers W (1973) Biopolymers 12:2033–2044
165. Lee JC, Timasheff SN (1979) Methods Enzymol 61:49–57
166. Kawahara K, Ikenaka T, Nimberg RB, Schmid K (1973) Biochim Biophys Acta 295:505–513
167. Webber RV (1956) J Am Chem Soc 78:536–541
168. Popenoe EA, Drew RM (1957) J Biol Chem 228:673–683
169. Gibbons RA (1972) In: Gottschalk A (ed) Glycoproteins, part A. Elsevier, Amsterdam, pp 31–140
170. Kupke DW, Hodgins MG, Beams JW (1972) Proc Natl Acad Sci USA 69:2258–2262
171. Hunter MJ (1966) J Phys Chem 70:3285–3292
172. Smith EL, Brown DM, Weimer HE, Winzler RJ (1950) J Biol Chem 185:569–575
173. Kupke DW (1966) Fed Proc 25:990–992
174. Jaenicke R, Koberstein R, Teuscher B (1971) Eur J Biochem 23:150–159
175. Kellett GL (1971) J Mol Biol 59:401–424
176. Svedberg T, Pedersen KO (1940) The ultracentrifuge. Oxford University Press, London
177. Theorell H (1934) Biochem Z 268:46–54
178. Aune KC, Timasheff SN (1970) Biochemistry 9:1481–1484
179. Pesce A, McKay RH, Stolzenbach F, Cahn RD, Kaplan NO (1964) J Biol Chem 239:1753–1761
180. Jaenicke R, Lauffer MA (1969) Biochemistry 8:3083–3092
181. Sia CL, Horecker BL (1968) Arch Biochem Biophys 123:186–194
182. Sine HE, Hass LF (1967) J Am Chem Soc 89:1749–1751
183. Jaenicke R, Schmid D, Knof S (1968) Biochemistry 7:919–926
184. Dayhoff MO, Perlmann GE, MacInnes DA (1952) J Am Chem Soc 74:2515–2517
185. Castellino FJ, Barker R (1968) Biochemistry 7:2207–2217
186. Holt A, Wold F (1961) J Biol Chem 236:3227–3231
187. Ehrenberg A, Dalziel K (1958) Acta Chem Scand 12:465–469
188. Appella E, Markert CL (1961) Biochem Biophys Res Commun 6:171–176
189. Taylor JF, Lowry C, Keller PJ (1956) Biochim Biophys Acta 20:109–117
190. Kirshner AG, Tanford C (1964) Biochemistry 3:291–296
191. Tanford C, Kawahara K, Lapanje S (1967) J Am Chem Soc 89:729–736
192. Harrington WF, Karr GM (1965) J Mol Biol 13:885–893

193. Kielley WW, Harrington WF (1960) Biochim Biophys Acta 41:401–421
194. Oncley JL, Scatchard G, Brown A (1947) J Phys Chem 51:184–198
195. Schultze HE, Schmidtberger R, Haupt H (1958) Biochem Z 329:490–507
196. Bezkorovainy A, Rafelson ME Jr, Likhite V (1963) Arch Biochem Biophys 103:371–378
197. Koenig VL (1950) Arch Biochem 25:241–245
198. Kawahara K, Tanford C (1966) J Biol Chem 241:3228–3232
199. Holcomb DN, van Holde KE (1962) J Phys Chem 66:1999–2006
200. Van Holde KE, Baldwin RL (1958) J Phys Chem 62:734–743
201. Stauff J, Jaenicke R (1961) Koll Z 175:1–14
202. Leicht W, Werber MM, Eisenberg H (1978) Biochemistry 17:4004–4010
203. Werber MM, Mevarech M (1978) Arch Biochem Biophys 187:447–456
204. Laggner P, Müller KW (1978) Q Rev Biophys 11:371–425
205. Müller K, Laggner P, Glatter O, Kostner G (1978) Eur J Biochem 82:73–90
206. Kasper CB, Deutsch HF (1963) J Biol Chem 238:2325–2337
207. Dessen P, Blanquet S, Zaccai G, Jacrot B (1978) J Mol Biol 126:293–313
208. Dessen P, Zaccai G, Blanquet S (1982) J Mol Biol 159:651–664
209. Wilhelm P, Pilz I, Lane AN, Kirschner K (1982) Eur J Biochem 129:51–56
210. Paradies HH (1980) J Phys Chem 84:599–607
211. Meyer J, Zaccai G (1981) Biochem Biophys Res Commun 98:43–50
212. Satre M, Zaccai G (1979) FEBS Lett 102:244–248
213. Zaccai G, Morin P, Jacrot B, Moras D, Thierry J-C, Giegé R (1979) J Mol Biol 129:483–500
214. Paradies HH, Zimmermann J, Schmidt UD (1978) J Biol Chem 253:8972–8979
215. Süss K-H, Damaschun H, Damaschun G, Zirwer D (1978) FEBS Lett 87:265–268
216. Paradies HH, Vettermann W (1978) Arch Biochem Biophys 191:169–181
217. Folkhard W, Felser B, Pilz I, Kratky O, Dutler H, Vogel H (1977) Eur J Biochem 81:173–178
218. Paradies HH, Zimmer B, Werz G (1977) Biochem Biophys Res Commun 74:397–404
219. Paradies HH, Vettermann W (1976) Biochem Biophys Res Commun 71:520–526
220. Gulik A, Monteilhet C, Dessen P, Fayat G (1976) Eur J Biochem 64:295–300
221. Österberg R, Sjöberg B, Rymo L, Lagerkvist U (1973) J Mol Biol 77:153–158
222. Rymo L, Lundvik L, Lagerkvist U (1972) J Biol Chem 247:3888–3899
223. Pilz I, Kratky O, Cramer F, von der Haar F, Schlimme E (1970) Eur J Biochem 15:401–409
224. Pilz I, Ullrich J (1973) Eur J Biochem 34:256–261
225. Ullrich J, Kempfle M (1969) FEBS Lett 4:273–274
226. Morell AG, van den Hamer CJA, Scheinberg IH (1969) J Biol Chem 244:3494–3496
227. Pilz I, Kratky O, Rabussay D (1972) Eur J Biochem 28:205–220
228. Müller K, Kratky O, Röschlau P, Hess B (1972) Hoppe-Seyler's Z Physiol Chem 353:803–809
229. Magdoff-Fairchild B, Lovell FM, Low BW (1969) J Biol Chem 244:3497–3499
230. Sund H, Pilz I, Herbst M (1969) Eur J Biochem 7:517–525
231. Puchwein G, Kratky O, Gölker CF, Helmreich E (1970) Biochemistry 9:4691–4698
232. Seery VL, Fischer EH, Teller DC (1967) Biochemistry 6:3315–3327
233. Pilz I, Herbst M, Kratky O, Oesterhelt D, Lynen F (1970) Eur J Biochem 13:55–64
234. Gordon WG, Semmett WF (1953) J Am Chem Soc 75:328–330
235. Krigbaum WR, Kügler FR (1970) Biochemistry 9:1216–1223
236. Wetter LR, Deutsch HF (1951) J Biol Chem 192:237–242
237. Schwert GW, Kaufman S (1951) J Biol Chem 190:807–816
238. Malmon AG (1957) Biochim Biophys Acta 26:233–240
239. Miller GL, Price WC (1946) Arch Biochem 10:467–477
240. Markham R (1951) Discuss Faraday Soc 11:221–227
241. Reisler E, Pouyet J, Eisenberg H (1970) Biochemistry 9:3095–3102
242. Schmid G, Durchschlag H, Biedermann G, Eggerer H, Jaenicke R (1974) Biochem Biophys Res Commun 58:419–426
243. Carlstedt I, Lindgren H, Sheehan JK (1983) Biochem J 213:427–435
244. Kumosinski TF, Pessen H, Farrell HM Jr (1982) Arch Biochem Biophys 214:714–725

245. McMeekin TL, Wilensky M, Groves ML (1962) Biochem Biophys Res Commun 7:151–156
246. Shlom JM, Vinogradov SN (1973) J Biol Chem 248:7904–7912
247. Pedersen KO (1947) J Phys Chem 51:164–171
248. Zipper P, Krigbaum WR, Kratky O (1969) Kolloid-Z Z Polymere 235:1281–1287
249. Ibel K, Stuhrmann HB (1975) J Mol Biol 93:255–265
250. Pilz I, Goral K, Hoylaerts M, Witters R, Lontie R (1980) Eur J Biochem 105:539–543
251. Berger J, Pilz I, Witters R, Lontie R (1977) Eur J Biochem 80:79–82
252. Berger J, Pilz I, Witters R, Lontie R (1976) Z Naturforsch Sect C Biosci 31:238–244
253. Pilz I, Engelborghs Y, Witters R, Lontie R (1974) Eur J Biochem 42:195–202
254. Morgan WTJ, Pusztai A (1961) Biochem J 81:648–658
255. Pilz I, Walder K, Siezen R (1974) Z Naturforsch Sect C Biosci 29:116–121
256. Jürgens G, Knipping GMJ, Zipper P, Kayushina R, Degovics G, Laggner P (1981) Biochemistry 20:3231–3237
257. Laggner P, Kostner GM, Rakusch U, Worcester D (1981) J Biol Chem 256:11832–11839
258. Laggner P, Chapman MJ, Goldstein S (1978) Biochem Biophys Res Commun 82:1332–1339
259. Laggner P, Glatter O, Müller K, Kratky O, Kostner G, Holasek A (1977) Eur J Biochem 77:165–171
260. Janiak MJ, Loomis CR, Shipley GG, Small DM (1974) J Mol Biol 86:325–339
261. Laggner P, Kratky O, Kostner G, Sattler J, Holasek A (1972) FEBS Lett 27:53–57
262. Li Z-Q, Perkins SJ, Loucheux-Lefebvre MH (1983) Eur J Biochem 130:275–279
263. Siezen RJ, Berger H (1978) Eur J Biochem 91:397–405
264. Pusztai A, Morgan WTJ (1961) Biochem J 78:135–146
265. Block MR, Zaccai G, Lauquin GJM, Vignais PV (1982) Biochem Biophys Res Commun 109:471–477
266. Reich MH, Kam Z, Eisenberg H, Worcester D, Ungewickell E, Gratzer WB (1982) Biophys Chem 16:307–316
267. Olins AL, Carlson RD, Wright EB, Olins DE (1976) Nucleic Acids Res 3:3271–3291
268. Österberg R, Sjöberg B, Österberg P, Stenflo J (1980) Biochemistry 19:2283–2286
269. Lamy F, Waugh DF (1953) J Biol Chem 203:489–499
270. Badley RA, Atkinson D, Hauser H, Oldani D, Green JP, Stubbs JM (1975) Biochim Biophys Acta 412:214–228
271. Ishimuro Y, Yamaguchi S, Hamada F, Nakajima A (1981) Biopolymers 20:2499–2508
272. De Loźe C, Saludjian P, Kovacs AJ (1964) Biopolymers 2:43–49
273. Gordon WG, Ziegler J (1955) Arch Biochem Biophys 57:80–86
274. Luzzati V, Cesari M, Spach G, Masson F, Vincent JM (1961) J Mol Biol 3:566–584
275. Mitchell JC, Woodward AE, Doty P (1957) J Am Chem Soc 79:3955–3960
276. Schiffer M, Stevens FJ, Westholm FA, Kim SS, Carlson RD (1982) Biochemistry 21:2874–2878
277. Wilhelm P, Pilz I, Goral K, Palm W (1980) Int J Biol Macromol 2:13–16
278. Gibbons RA, Glover FA (1959) Biochem J 73:217–225
279. Cecil R, Ogston AG (1951) Biochem J 49:105–106
280. Pilz I, Kratky O, Licht A, Sela M (1973) Biochemistry 12:4998–5005
281. Holasek A, Pascher I, Hauser H (1961) Monatsh Chem 92:463–467
282. Branegård B, Österberg R, Sjöberg B (1980) Int J Biol Macromol 2:321–323
283. Behlke J, Schwenke KD (1976) Stud Biophys 59:55–60
284. Pedersen KO (1936) Biochem J 30:961–970
285. Serdyuk IN, Sarkisyan MA, Gogia ZV (1981) FEBS Lett 129:55–58
286. Österberg R (1979) Eur J Biochem 97:463–469
287. Österberg R, Sjöberg B, Garrett RA (1976) Eur J Biochem 68:481–487
288. Österberg R, Sjöberg B, Pettersson I, Liljas A, Kurland CG (1977) FEBS Lett 73:22–24
289. Österberg R, Sjöberg B, Garrett RA, Littlechild J (1977) FEBS Lett 73:25–28
290. Paradies HH, Franz A (1976) Eur J Biochem 67:23–29
291. Wong K-P, Paradies HH (1974) Biochem Biophys Res Commun 61:178–184
292. Ramakrishnan VR, Yabuki S, Sillers I-Y, Schindler DG, Engelman DM, Moore PB (1981) J Mol Biol 153:739–760

293. Serdyuk IN, Grenader AK, Zaccai G (1979) J Mol Biol 135:691–707
294. Serdyuk IN, Shpungin JL, Zaccai G (1980) J Mol Biol 137:109–121
295. Crichton RR, Engelman DM, Haas J, Koch MHJ, Moore PB, Parfait R, Stuhrmann HB (1977) Proc Natl Acad Sci USA 74:5547–5550
296. Williams PA, Peacocke AR (1965) Biochim Biophys Acta 101:327–335
297. Eyring EJ, Yang JT (1968) J Biol Chem 243:1306–1311
298. Laughrea M, Moore PB (1977) J Mol Biol 112:399–421
299. Stuhrmann HB, Haas J, Ibel K, de Wolf B, Koch MHJ, Parfait R, Crichton RR (1976) Proc Natl Acad Sci USA 73:2379–2383
300. Muldoon TG, Westphal U (1967) J Biol Chem 242:5636–5643
301. Moore PB, Engelman DM, Schoenborn BP (1975) J Mol Biol 91:101–120
302. Zipper P, Ribitsch G, Schurz J, Bünemann H (1982) Z Naturforsch Sect C Biosci 37:824–832
303. Crawford IP (1973) Arch Biochem Biophys 156:215–222
304. Zipper P, Bünemann H (1975) Eur J Biochem 51:3–17
305. Müller W, Emme I (1965) Z Naturforsch Teil B Chem Biochem Biophys Biol 20:835–841
306. Wawra H, Müller W, Kratky O (1970) Makromol Chem 139:83–102
307. Nilsson L, Rigler R, Laggner P (1982) Proc Natl Acad Sci USA 79:5891–5895
308. Pilz I, Malnig F, Kratky O, von der Haar F (1977) Eur J Biochem 75:35–41
309. Cuillel M, Zulauf M, Jacrot B (1983) J Mol Biol 164:589–603
310. Cuillel M, Tripier F, Braunwald J, Jacrot B (1979) Virology 99:277–285
311. Harrison SC (1969) J Mol Biol 42:457–483
312. Turner KJ (1963) Biochim Biophys Acta 69:518–523
313. Bockstahler LE, Kaesberg P (1962) Biophys J 2:1–9
314. Schmidt P, Kaesberg P, Beeman WW (1954) Biochim Biophys Acta 14:1–11
315. Zipper P, Folkhard W, Clauwaert J (1975) FEBS Lett 56:283–287
316. Zipper P, Kratky O, Herrmann R, Hohn T (1971) Eur J Biochem 18:1–9
317. Gottschalk EM, Kopp E, Lezius AG (1971) Eur J Biochem 24:168–182
318. Zipper P (1973) Eur J Biochem 39:493–498
319. Olins DE, Bryan PN, Harrington RE, Hill WE, Olins AL (1977) Nucleic Acids Res 4:1911–1931
320. Langridge R, Marvin DA, Seeds WE, Wilson HR, Hooper CW, Wilkins MHF, Hamilton LD (1960) J Mol Biol 2:38–64
321. Brutlag D, Schlehuber C, Bonner J (1969) Biochemistry 8:3214–3218
322. Deutsch HF (1954) J Biol Chem 208:669–678
323. Paradies HH, Schmidt UD (1979) J Biol Chem 254:5257–5263
324. Müller K (1981) Biochemistry 20:404–414
325. Edsall JT (1953) In: Neurath H, Bailey K (eds) The proteins 1st edn, vol 1, part B. Academic Press, New York, pp 549–726
326. Miller F, Metzger H (1965) J Biol Chem 240:3325–3333
327. Main AR, Soucie WG, Buxton IL, Arinc E (1974) Biochem J 143:733–744
328. Tamm I, Bugher JC, Horsfall FL Jr (1955) J Biol Chem 212:125–133
329. Kondo T, Kawakami M (1981) Anal Biochem 117:374–381
330. Tindall SH, Aune KC (1981) Biochemistry 20:4861–4866
331. Pesce AJ (1976) In: Fasman GD (ed) CRC Handbook of biochemistry and molecular biology, 3rd edn. Proteins, vol 2. CRC, Cleveland, p 304
332. Batke J, Asbóth G, Lakatos S, Schmitt B, Cohen R (1980) Eur J Biochem 107:389–394
333. Laurent TC (1961) Arch Biochem Biophys 92:224–231
334. Paradies HH (1979) Biochem Biophys Res Commun 88:810–817
335. Dorst H-J, Schubert D (1979) Hoppe-Seyler's Z Physiol Chem 360:1605–1618
336. Agrawal BBL, Goldstein IJ (1968) Arch Biochem Biophys 124:218–229
337. Howlett GJ, Schachman HK (1977) Biochemistry 16:5077–5091
338. Kirschner MW, Schachman HK (1971) Biochemistry 10:1900–1919
339. Szuchet S, Yphantis DA (1976) Arch Biochem Biophys 173:495–516
340. Szuchet S, Yphantis DA (1973) Biochemistry 12:5115–5127
341. Roark DE, Geoghegan TE, Keller GH (1974) Biochem Biophys Res Commun 59:542–547

342. Smith GD, Schachman HK (1973) Biochemistry 12:3789–3801
343. Varga L, Pietruszkiewicz A, Ryan M (1959) Biochim Biophys Acta 32:155–165
344. Lasker SE, Stivala SS (1966) Arch Biochem Biophys 115:360–372
345. Varga L (1955) J Biol Chem 217:651–658
346. Cleland RL (1977) Arch Biochem Biophys 180:57–68
347. Olaitan SA, Northcote DH (1962) Biochem J 82:509–519
348. Williams RC Jr, Yphantis DA, Craig LC (1972) Biochemistry 11:70–77
349. Lamm O, Polson A (1936) Biochem J 30:528–541
350. Phelps CF (1965) Biochem J 95:41–47
351. Shulman S (1953) J Am Chem Soc 75:5846–5852
352. Boedtker H, Simmons NS (1958) J Am Chem Soc 80:2550–2556
353. Chiou S-H (1984) J Biochem (Tokyo) 95:75–82
354. Bloomfield V, van Holde KE, Dalton WO (1967) Biopolymers 5:149–159
355. Harrington WF, Schellman JA (1956) Comp Rend Trav Lab Carlsberg Sér Chim 30:21–43
356. Hazelwood RN (1958) J Am Chem Soc 80:2152–2156
357. Atkinson D, Davis MAF, Leslie RB (1974) Proc R Soc Lond B Biol Sci 186:165–180
358. Hanlon S, Lamers K, Lauterbach G, Johnson R, Schachman HK (1962) Arch Biochem Biophys 99:157–174
359. Rothen A (1944) J Biol Chem 152:679–693
360. Fisher WR, Hammond MG, Warmke GL (1972) Biochemistry 11:519–525
361. Martin RG, Ames BN (1961) J Biol Chem 236:1372–1379
362. Alaupovic P, Sanbar SS, Furman RH, Sullivan ML, Walraven SL (1966) Biochemistry 5:4044–4053
363. Perrin JH, Saunders L (1964) Biochim Biophys Acta 84:216–217
364. Hubbard R (1954) J Gen Physiol 37:381–399
365. Llewellyn DJ, Smith GD (1981) Arch Biochem Biophys 207:63–68
366. Wang JK, Dekker EE, Lewinski ND, Winter HC (1981) J Biol Chem 256:1793–1800
367. Leach BS, Collawn JF Jr, Fish WW (1980) Biochemistry 19:5734–5741
368. Leach BS, Collawn JF Jr, Fish WW (1980) Biochemistry 19:5741–5747
369. Spiro RG (1960) J Biol Chem 235:2860–2869
370. Green WA, Kay CM (1963) Arch Biochem Biophys 102:359–366
371. Davis JG, Mapes CJ, Donovan JW (1971) Biochemistry 10:39–42
372. Bodmann O, Walter M (1965) Biochim Biophys Acta 110:496–506
373. Reynolds JA, Tanford C (1970) J Biol Chem 245:5161–5165
374. Bowen TJ, Mortimer MG (1971) Eur J Biochem 23:262–266
375. Reiss-Husson F (1967) J Mol Biol 25:363–382
376. Aune KC, Gallagher JG, Gotto AM Jr, Morrisett JD (1977) Biochemistry 16:2151–2156
377. Danson MJ, Hale G, Johnson P, Perham RN, Smith J, Spragg P (1979) J Mol Biol 129:603–617
378. Bleile DM, Munk P, Oliver RM, Reed LJ (1979) Proc Natl Acad Sci USA 76:4385–4389
379. Reed LJ, Pettit FH, Eley MH, Hamilton L, Collins JH, Oliver RM (1975) Proc Natl Acad Sci USA 72:3068–3072
380. Pettit FH, Hamilton L, Munk P, Namihira G, Eley MH, Willms CR, Reed LJ (1973) J Biol Chem 248:5282–5290
381. Crouch TH, Klee CB (1980) Biochemistry 19:3692–3698
382. Barenholz Y, Gibbes D, Litman BJ, Goll J, Thompson TE, Carlson FD (1977) Biochemistry 16:2806–2810
383. Corkill JM, Goodman JF, Walker T (1967) Trans Faraday Soc 63:768–772
384. Hauser H, Irons L (1972) Hoppe-Seyler's Z Physiol Chem 353:1579–1590
385. Eisenberg H, Josephs R, Reisler E, Schellman JA (1977) Biopolymers 16:2773–2783
386. Eisenberg H, Tomkins GM (1968) J Mol Biol 31:37–49
387. Olson JA, Anfinsen CB (1952) J Biol Chem 197:67–79
388. Thorne CJR, Kaplan NO (1963) J Biol Chem 238:1861–1868
389. Laggner P, Gotto AM Jr, Morrisett JD (1979) Biochemistry 18:164–171
390. Morrisett JD, Gallagher JG, Aune KC, Gotto AM Jr (1974) Biochemistry 13:4765–4771
391. Fasiolo F, Remy P, Pouyet J, Ebel J-P (1974) Eur J Biochem 50:227–236

392. Huang C-H, Lee L-P (1973) J Am Chem Soc 95:234–239
393. Easterby JS, Rosemeyer MA (1972) Eur J Biochem 28:241–252
394. Merrick WC, Kemper WM, Anderson WF (1975) J Biol Chem 250:5556–5562
395. Schachman HK, Lauffer MA (1949) J Am Chem Soc 71:536–541
396. Richards FM, Wyckoff HW (1971) In: Boyer PD (ed) The enzymes, vol 4, 3rd edn. Academic Press, New York, pp 647–806
397. Gammack DB (1963) Biochem J 88:373–383
398. Jaenicke R, Lauffer MA (1969) Biochemistry 8:3077–3082
399. Bryce CFA, Crichton RR (1971) J Biol Chem 246:4198–4205
400. Anacker EW, Rush RM, Johnson JS (1964) J Phys Chem 68:81–93
401. Fischbach FA, Anderegg JW (1965) J Mol Biol 14:458–473
402. Paradies HH (1981) Biochem Biophys Res Commun 101:1096–1101
403. Markley KS (1975) In: Fasman GD (ed) CRC Handbook of biochemistry and molecular biology, 3rd edn. Lipids, carbohydrates, steroids. CRC, Cleveland, p 492
404. Majewska MR, Dancewicz AM (1977) Stud Biophys 63:65–74
405. Schachman HK (1957) Methods Enzymol 4:32–103
406. Edelstein SJ, Schachman HK (1967) J Biol Chem 242:306–311
407. Lovenberg W, Buchanan BB, Rabinowitz JC (1963) J Biol Chem 238:3899–3913
408. Dessen P, Ducruix A, Hountondji C, May RP, Blanquet S (1983) Biochemistry 22:281–284
409. Schmid CW, Hearst JE (1971) Biopolymers 10:1901–1924
410. Hearst JE, Vinograd J (1961) Proc Natl Acad Sci USA 47:1005–1014
411. Hearst JE, Vinograd J (1961) Proc Natl Acad Sci USA 47:825–830
412. Heijtink RA, Houwing CJ, Jaspars EMJ (1977) Biochemistry 16:4684–4693
413. Ifft JB, Vinograd J (1962) J Phys Chem 66:1990–1998
414. Cox DJ, Schumaker VN (1961) J Am Chem Soc 83:2433–2438
415. Cox DJ, Schumaker VN (1961) J Am Chem Soc 83:2439–2445
416. Kaper JM, Litjens EC (1966) Biochemistry 5:1612–1617
417. Bowen TJ, Rogers LJ (1965) Nature 205:1316–1317
418. Serdyuk IN, Gogia ZV, Venyaminov SY, Khechinashvili NN, Bushuev VN, Spirin AS (1980) J Mol Biol 137:93–107
419. Svedberg T, Eriksson I-B (1933) J Am Chem Soc 55:2834–2841
420. Lindahl T, Henley DD, Fresco JR (1965) J Am Chem Soc 87:4961–4963
421. David MM, Daniel E (1974) J Mol Biol 87:89–101
422. Van Holde KE, Cohen LB (1964) Biochemistry 3:1803–1808
423. Svedberg T (1930) Koll Z 51:10–24
424. Krauss G, Pingoud A, Boehme D, Riesner D, Peters F, Maass G (1975) Eur J Biochem 55:517–529
425. Omura T, Fujita T, Yamada F, Yamamoto S (1961) J Biochem (Tokyo) 50:400–404
426. Cassman M, Schachman HK (1971) Biochemistry 10:1015–1024
427. Kutzbach C, Bischofberger H, Hess B, Zimmermann-Telschow H (1973) Hoppe-Seyler's Z Physiol Chem 354:1473–1489
428. Bischofberger H, Hess B, Röschlau P, Wieker H-J, Zimmermann-Telschow H (1970) Hoppe-Seyler's Z Physiol Chem 351:401–408
429. Hunsley JR, Suelter CH (1969) J Biol Chem 244:4815–4818
430. Dietz G, Woenckhaus C, Jaenicke R, Schuster I (1977) Z Naturforsch Sect C Biosci 32:85–92
431. Lakatos S, Závodszky P (1976) FEBS Lett 63:145–148
432. Lakatos S, Závodszky P, Elödi P (1972) FEBS Lett 20:324–326
433. Sloan DL, Velick SF (1973) J Biol Chem 248:5419–5423
434. Fuller Noel JK, Schumaker VN (1972) J Mol Biol 68:523–532
435. Smith CM, Velick SF (1972) J Biol Chem 247:273–284
436. Henley DD, Lindahl T, Fresco JR (1966) Proc Natl Acad Sci USA 55:191–198
437. Kurland CG (1960) J Mol Biol 2:83–91
438. Elödi P (1958) Acta Physiol Acad Sci Hung 13:199–206
439. Gralén N (1939) Biochem J 33:1342–1345
440. Holbrook JJ, Liljas A, Steindel SJ, Rossmann MG (1975) In: Boyer PD (ed) The enzymes, vol 11, 3rd edn. Academic Press, New York, pp 191–292

441. Tissières A (1959) J Mol Biol 1:365–374
442. Petermann ML, Pavlovec A (1966) Biochim Biophys Acta 114:264–276
443. Jaenicke R, Pfleiderer G (1962) Biochim Biophys Acta 60:615–629
444. Jaenicke R, Knof S (1968) Eur J Biochem 4:157–163
445. Boedtker H (1960) J Mol Biol 2:171–188
446. Mayr U, Hensel R, Kandler O (1980) Eur J Biochem 110:527–538
447. Jaenicke R, Gregori E, Laepple M (1979) Biophys Struct Mechanism 6:57–65
448. Jaenicke R (1974) Eur J Biochem 46:149–155
449. Jaenicke R (1964) Biochim Biophys Acta 85:186–201
450. Jaenicke R (1963) Biochem Z 338:614–627
451. Markert CL, Appella E (1961) Ann NY Acad Sci 94:678–690
452. Cohen SS, Stanley WM (1942) J Biol Chem 144:589–598
453. Mire M (1969) Biochim Biophys Acta 181:35–44
454. Green RW, McKay RH (1969) J Biol Chem 244:5034–5043
455. Marler E, Tanford C (1964) J Biol Chem 239:4217–4218
456. Marler E, Nelson CA, Tanford C (1964) Biochemistry 3:279–284
457. Woods EF, Himmelfarb S, Harrington WF (1963) J Biol Chem 238:2374–2385
458. Kay CM (1960) Biochim Biophys Acta 38:420–427
459. Serdyuk IN (1979) Methods Enzymol 59:750–775
460. Serdyuk IN (1979) Mol Biol (Moscow) 13:965–982
461. Spahr PF (1962) J Mol Biol 4:395–406
462. Möller W, Chrambach A (1967) J Mol Biol 23:377–390
463. Stanley WM Jr, Bock RM (1965) Biochemistry 4:1302–1311
464. Ortega JP, Hill WE (1973) Biochemistry 12:3241–3243
465. Marvin DA, Hoffmann-Berling H (1963) Nature 197:517–518
466. Gulik A, Inoue H, Luzzati V (1970) J Mol Biol 53:221–238
467. Murphy JS, Philipson L (1962) J Gen Physiol 45:155–168
468. Haselkorn R (1962) J Mol Biol 4:357–367
469. Koch MHJ, Stuhrmann HB (1979) Methods Enzymol 59:670–706
470. Oberby LR, Barlow GH, Doi RH, Jacob M, Spiegelman S (1966) J Bacteriol 91:442–448
471. Van Holde KE, Hill WE (1974) In: Nomura M, Tissières A, Lengyel P (eds) Ribosomes. Cold Spring Harbor Laboratory, New York, pp 53–91
472. Spirin AS, Gavrilova LP (1969) The ribosome. Springer, Berlin Heidelberg New York
473. Kayushina RL, Feigin LA (1969) Biofizika 14:957–962
474. Bull HB, Breese K (1968) Arch Biochem Biophys 128:497–502
475. Camerini-Otero RD, Pusey PN, Koppel DE, Schaefer DW, Franklin RM (1974) Biochemistry 13:960–970
476. Bancroft FC, Freifelder D (1970) J Mol Biol 54:537–546
477. Hearst JE (1962) J Mol Biol 4:415–417
478. Dyson RD, van Holde KE (1967) Virology 33:559–566
479. Strauss JH Jr, Sinsheimer RL (1963) J Mol Biol 7:43–54
480. Langridge R, Wilson HR, Hooper CW, Wilkins MHF, Hamilton LD (1960) J Mol Biol 2:19–37
481. Kahler H (1948) J Phys Chem 52:676–689
482. Tennent HG, Vilbrandt CF (1943) J Am Chem Soc 65:424–428
483. Fredericq E, Deutsch HF (1949) J Biol Chem 181:499–510
484. Lauffer MA, Cartwright TE (1952) Arch Biochem Biophys 38:371–375
485. Ullmann A, Goldberg ME, Perrin D, Monod J (1968) Biochemistry 7:261–265
486. Werber MM, Mevarech M, Leicht W, Eisenberg H (1978) In: Caplan SR, Ginzburg M (eds) Energetics and structure of halophilic microorganisms. Elsevier/North-Holland Biomedical, Amsterdam, pp 427–445
487. Lauffer MA (1944) J Am Chem Soc 66:1188–1194
488. Škerjanc J, Lapanje S (1972) Eur J Biochem 25:49–53
489. Lapanje S, Škerjanc J (1971) Biochem Biophys Res Commun 43:682–687
490. Katz S (1968) Biochim Biophys Acta 154:468–477
491. Katz S, Ferris TG (1966) Biochemistry 5:3246–3253
492. Hesterberg LK, Weber K (1983) J Biol Chem 258:359–364
493. Hughes F, Steiner RF (1966) Biopolymers 4:1081–1090

494. Noguchi H, Arya SK, Yang JT (1971) Biopolymers 10:2491–2498
495. Eisenberg H, Cohen G (1968) J Mol Biol 37:355–362; erratum (1969) J Mol Biol 42:607
496. Chapman RE Jr, Sturtevant JM (1969) Biopolymers 7:527–537
497. Chapman RE Jr, Sturtevant JM (1970) Biopolymers 9:445–457
498. Vinograd J, Greenwald R, Hearst JE (1965) Biopolymers 3:109–114
499. Ise N, Okubo T (1968) J Am Chem Soc 90:4527–4533
500. Gray HB Jr, Hearst JE (1968) J Mol Biol 35:111–129
501. Katz S (1972) Methods Enzymol 26:395–406
502. Johnson FH, Eyring H, Pollisar MJ (1954) In: The kinetic basis of molecular biology. Wiley, New York, pp 335–344
503. Daniel J, Cohn EJ (1936) J Am Chem Soc 58:415–423
504. Schulz GV, Hoffmann M (1957) Makromol Chem 23:220–232
505. Rasper J, Kauzmann W (1962) J Am Chem Soc 84:1771–1777
506. Small PA Jr, Lamm ME (1966) Biochemistry 5:259–267
507. Lamm ME, Small PA Jr (1966) Biochemistry 5:267–276
508. Reithel FJ, Sakura JD (1963) J Phys Chem 67:2497–2498
509. Clarke AM, Kupke DW, Beams JW (1963) J Phys Chem 67:929–930
510. Paradies HH (1982) FEBS Lett 137:265–270
511. Kauzmann W (1959) Adv Protein Chem 14:1–63
512. Bawden FC, Pirie NW (1937) Proc R Soc Lond B Biol Sci 123:274–320
513. Haberland ME, Reynolds JA (1973) Proc Natl Acad Sci USA 70:2313–2316
514. Pittz EP, Timasheff SN (1978) Biochemistry 17:615–623
515. Shogren R, Jamieson AM, Blackwell J, Cheng PW, Dearborn DG, Boat TF (1983) Biopolymers 22:1657–1675
516. Bowen TJ (1970) An introduction to ultracentrifugation. Wiley Interscience, London
517. Mani RS, Kay CM (1984) FEBS Lett 166:258–262
518. Kremann P (1928) Mechanische Eigenschaften flüssiger Stoffe. Akadem Verlagsanstalt, Leipzig, pp 154–157, 192–194
519. Lapanje S (1978) Physicochemical aspects of protein denaturation. Wiley Interscience, New York
520. Katz S, Denis J (1970) Biochim Biophys Acta 207:331–339
521. Thomas JO, Edelstein SJ (1971) Biochemistry 10:477–482
522. Jovin TM, Englund PT, Bertsch LL (1969) J Biol Chem 244:2996–3008
523. Tanford C (1961) Physical chemistry of macromolecules. Wiley, New York
524. Bull HB (1976) In: Fasman GD (ed) CRC Handbook of biochemistry and molecular biology, 3rd edn. Proteins, vol 3. CRC, Cleveland, p 595
525. Boedtker H (1976) In: Fasman GD (ed) CRC Handbook of biochemistry and molecular biology, 3rd edn. Nucleic acids, vol 2. CRC, Cleveland, pp 405–407
526. Sober HA (ed) (1970) CRC Handbook of biochemistry, 2nd edn. Chemical Rubber, Cleveland, pp C3–C39, E10, J278. Fasman GD (ed) (1975) Handbook of biochemistry and molecular biology, 3rd edn. Lipids, carbohydrates, steroids, p 492; (1976) Proteins, vol 2, pp 242–253, 302, 304, 305; (1976) Nucleic acids, vol 2, pp 405–407, 482–484; (1976) Physical and chemical data, vol 1. CRC, Cleveland, p 404
527. Weast RC (ed) (1982) CRC Handbook of chemistry and physics, 63rd edn. CRC, Boca Raton
528. Masson PL (1976) In: Fasman GD (ed) CRC Handbook of biochemistry and molecular biology, 3rd edn. Proteins, vol 2. CRC, Cleveland, pp 242–253
529. Daniel E, Katchalski E (1962) In: Stahmann MA (ed) Polyamino acids, polypeptides, and proteins. The University of Wisconsin Press, Madison, pp 183–193
530. Luzzati V, Cesari M, Spach G, Masson F, Vincent JM (1962) In: Stahmann MA (ed) Polyamino acids, polypeptides, and proteins. The University of Wisconsin Press, Madison, pp 121–130
531. Sjögren B, Svedberg T (1931) J Am Chem Soc 53:2657–2661
532. Gibson QH, Massey V, Atherton NM (1962) Biochem J 85:369–383
533. Washburn EW (ed) (1928) Int Crit Tables Numerical Data, Physics, Chemistry and Technology, vol 3. McGraw-Hill Book, New York
534. Millero FJ (1971) Chem Rev 71:147–176

535. Millero FJ (1972) In: Horne RA (ed) Water and aqueous solutions. Wiley Interscience, New York, pp 519–595
536. Fosmire GJ, Timasheff SN (1972) Biochemistry 11:2455–2460
537. Lee JC, Frigon RP, Timasheff SN (1973) J Biol Chem 248:7253–7262
538. Ellerton HD, Carpenter DE, van Holde KE (1970) Biochemistry 9:2225–2232
539. Parry RM Jr, Chandan RC, Shahani KM (1969) Arch Biochem Biophys 130:59–65
540. Ulrich DV, Kupke DW, Beams JW (1964) Proc Natl Acad Sci USA 52:349–356
541. Fahey PF, Kupke DW, Beams JW (1969) Proc Natl Acad Sci USA 63:548–555
542. Gosting LJ, Morris MS (1949) J Am Chem Soc 71:1998–2006
543. Danielsson C-E (1947) Nature 160:899
544. Hirsch-Kolb H, Greenberg DM (1968) J Biol Chem 243:6123–6129
545. Tissières A, Watson JD, Schlessinger D, Hollingworth BR (1959) J Mol Biol 1:221–233
546. Sullivan RA, Fitzpatrick MM, Stanton EK, Annino R, Kissel G, Palermiti F (1955) Arch Biochem Biophys 55:455–468
547. Lin YM, Lin YP, Cheung WY (1974) J Biol Chem 249:4943–4954
548. Watterson DM, Harrelson WG Jr, Keller PM, Sharief F, Vanaman TC (1976) J Biol Chem 251:4501–4513
549. Teo TS, Wang TH, Wang JH (1973) J Biol Chem 248:588–595
550. Beers WH, Reich E (1969) J Biol Chem 244:4473–4479
551. Young EG, Lorimer JW (1961) Arch Biochem Biophys 92:183–190
552. Rice RV, Casassa EF, Kerwin RE, Maser MD (1964) Arch Biochem Biophys 105:409–423
553. Ehrenberg A (1957) Acta Chem Scand 11:1257–1270
554. Nozaki M (1960) J Biochem (Tokyo) 47:592–599
555. Love B, Chan SHP, Stotz E (1970) J Biol Chem 245:6664–6668
556. Horio T, Higashi T, Yamanaka T, Matsubara H, Okunuki K (1961) J Biol Chem 236:944–951
557. Chang LMS, Bollum FJ (1971) J Biol Chem 246:909–916
558. Yang PC, Bock RM, Hsu RY, Porter JW (1965) Biochim Biophys Acta 110:608–615
559. Sund H, Weber K (1963) Biochem Z 337:24–34
560. Berggård I, Bearn AG (1968) J Biol Chem 243:4095–4103
561. Schultze HE, Heremans JF (1966) Molecular biology of human proteins with special reference to plasma proteins, vol 1. Elsevier Amsterdam, pp 173–235
562. Armstrong SH Jr, Budka MJE, Morrison KC, Hasson M (1947) J Am Chem Soc 69:1747–1753
563. Cebra JJ, Small PA Jr (1967) Biochemistry 6:503–512
564. Donner D (1976) In: Fasman GD (ed) CRC Handbook of biochemistry and molecular biology, 3rd edn. Nucleic acids, vol 2. CRC, Cleveland, pp 482–484
565. Høiland H (1986) This volume
566. Zamyatnin AA (1984) Annu Rev Biophys Bioeng 13:145–165
567. Plietz P, Damaschun G, Damaschun H, Kopperschläger G, Kröber R, Müller JJ (1977) Stud Biophys 65:9–22
568. Borochov N, Eisenberg H (1984) Biopolymers 23:1757–1769
569. Cleland RL (1984) Biopolymers 23:647–666
570. Ebert B, Elmgren H (1984) Biopolymers 23:2543–2557
571. Perkins SJ, Miller A, Hardingham TE, Muir H (1981) J Mol Biol 150:69–95
572. Reynolds JA, Tanford C (1976) Proc Natl Acad Sci USA 73:4467–4470
573. Dohnal JC, Potempa LA Garvin JE (1980) Biochim Biophys Acta 621:255–264
574. Iijima S, Saiki T, Beppu T (1984) J Biochem (Tokyo) 95:1273–1281
575. Gekko K, Koga S (1984) Biochim Biophys Acta 786:151–160
576. Garrigos M, Morel JE, Garcia de la Torre J (1983) Biochemistry 22:4961–4969
577. Garcia de la Torre J, Bloomfield VA (1980) Biochemistry 19:5118–5123
578. Parrish RG, Mommaerts WFHM (1954) J Biol Chem 209:901–913
579. Lowey S, Slayter HS, Weeds AG, Baker H (1969) J Mol Biol 42:1–29
580. Yang JT, Wu C-SC (1977) Biochemistry 16:5785–5789
581. Nagle JF, Wilkinson DA (1978) Biophys J 23:159–175
582. Wilkinson DA, Nagle JF (1981) Biochemistry 20:187–192

583. Seddon JM, Cevc G, Kaye RD, Marsh D (1984) Biochemistry 23:2634–2644
584. Pearce TC, Rowe AJ, Turnock G (1975) J Mol Biol 97:193–205
585. Davies A, Gormally J, Wyn-Jones E, Wedlock DJ, Phillips GO (1982) Int J Biol Macromol 4:436–438
586. Sardet C, Tardieu A, Luzzati V (1976) J Mol Biol 105:383–407
587. Osborne HB, Sardet C, Helenius A (1974) Eur J Biochem 44:383–390
588. Mann KG, Vestling CS (1969) Biochemistry 8:1105–1109
589. Perkins SJ, Kerckaert J-P, Loucheux-Lefebvre MH (1985) Eur J Biochem 147:525–531
590. Arakawa T, Timasheff SN (1984) Biochemistry 23:5912–5923
591. Arakawa T, Timasheff SN (1984) Biochemistry 23:5924–5929
592. Arakawa T, Timasheff SN (1985) Biophys J 47:411–414
593. Stevens CL, Lauffer MA (1965) Biochemistry 4:31–37
594. Cesàro A (1986) This volume
595. Kopperschläger G, Bär J, Nissler K, Hofmann E (1977) Eur J Biochem 81:317–325
596. Timasheff SN (1973) Adv Chem Ser 125:327–342
597. Cox AC, Tanford C (1968) J Biol Chem 243:3083–3087
598. Laurent TC, Anseth A (1961) Exp Eye Res 1:99–105

Chapter 4 Partial Molar Compressibilities of Organic Solutes in Water

HARALD HØILAND

Symbols

v:	specific volume of solution $[cm^3 g^{-1}]$
v*:	specific volume of pure solvent $[cm^3 g^{-1}]$
V_2:	partial molar volume
V_ϕ:	apparent molar volume
K_2:	partial molar compressibility
K_ϕ:	apparent molar compressibility
K_2^0:	partial molar compressibility at infinite dilution
$K_{\phi, s}$:	isentropic apparent molar compressibility
$K_{s, NaA}^0$:	isentropic partial molar compressibility of a sodium carboxylate at infinite dilution
$K_{s, HA}^0$:	isentropic partial molar compressibility of a carboxylic acid at infinite dilution
$K_{2, el}^0$:	electrostriction contribution to partial molar compressibility of solute at infinite dilution
$K_{2, int}$:	intrinsic partial molar compressibility
ΔK^0:	compressibilities of ionisation
κ:	isothermal compressibility of the solution $[bar^{-1}]$
κ^*:	isothermal compressibility of pure solvent
E_2^0:	partial molar expansibility of solute at infinite dilution
α:	coefficient of thermal expansion $[K^{-1}]$
$C_{p, 2}^0$:	partial molar heat capacity of solute at infinite dilution
σ:	heat capacity per unit volume
m:	molality of solute [mol/kg pure solvent]
η_H:	hydration number
x_1:	weight fractions of solvent $[g g^{-1}]$
x_2:	weight fractions of solute $[g g^{-1}]$
ϱ:	density of liquid
ϱ^*:	density of pure solvent $[g cm^{-3}]$
U:	speed of sound
δ_0:	difference between isothermal and isentropic compressibilities

1 Introduction

In recent years numerous data concerning partial molar quantities of aqueous solutes have accumulated. The aim of such studies have been to elucidate solute-water interactions, the structure of water, its effect on the conformation of biological molecules, and hydration in connection with reaction rates. Especially partial molar heat capacities and partial molar volumes have been widely studied. Since there is presently no rigorous statistical thermodynamic theory to account for the observed quantities, much effort has been directed at establishing empirical relations. One such relationship concerns the additivity of partial molar heat capacities and partial molar volumes.

The partial molar volumes of organic solutes at infinite dilution have proved to be additive with reasonably good accuracy (see this chapter). However, partial molar volumes may be divided in an intrinsic part and a part due to solute-solvent interactions. The intrinsic part is by far the largest. It could be that the observed additivity simply reflects additivity of intrinsic volumes, and that this oversha- dows contributions from solute-solvent interactions. One way to study this more closely is offered by partial molar compressibilities. The intrinsic volume can be regarded as incompressible, and the partial molar compressibility will thus reflect only solute-solvent interactions. A stringent test of the additivity relations is at hand, and the partial molar compressibility should prove a powerful tool in the study of aqueous solutions.

2 Definitions

The partial molar compressibility can be defined as:

$$K_2 = -(\partial V_2/\partial P)_T, \tag{1}$$

where V_2 is the partial molar volume. However, it is not easy to measure the par- tial molar volume accurately as a function of pressure. It may be easier to calcu- late the apparent molar compressibility:

$$K_\phi = -(\partial V_\phi/\partial P)_T = \frac{1000}{m\varrho^*}(\kappa - \kappa^*) + \kappa V_\phi, \tag{2}$$

Here V_ϕ is the apparent molar volume, κ and κ^* are the isothermal compressibili- ties of the solution and the pure solvent, respectively, i.e., $\kappa = -\frac{1}{V}(\partial V/\partial P)_T$, ϱ^* is the density of the pure solvent. The factor 1000 appears when densities and compressibilities are given as $g\,cm^{-3}$ (which is often the case) instead of $kg\,m^{-3}$.

In order to obtain accurate apparent molar compressibilities, it must be pos- sible to measure the isothermal compressibilities of solution and solvent with great precision. This is not an easy task, and as far as isothermal partial molar compressibilities are concerned, it may be easier to employ Eq. (1) as it stands. However, one may measure the speed of sound through any liquid with very great precision [1, 2]. The speed of sound is related to the isentropic compressibility through the Laplace equation:

$$\kappa_s = 1/u^2\varrho, \tag{3}$$

where u is the speed of sound and ϱ the density of the liquid. On this basis it is possible to obtain isentropic apparent molar compressibilities, $K_{\phi,s}$. Ideally, one would prefer the isothermal apparent compressibilities, and it is possible to con- vert the isentropic values since:

$$\kappa_T = \kappa_s + \alpha^2 T/\sigma, \tag{4}$$

where α is the coefficient of thermal expansion, $1/V \, (\partial v/\partial T)_p$, and σ the heat capacity per unit of volume, $(C_{p,2}/V_2)$.

The relation between partial and apparent molar compressibilities is as follows:

$$K_2 = K_\phi + m(\partial K_\phi/\partial m). \tag{5}$$

From this equation it can be seen that the partial and apparent molar compressibilities become equal at infinite dilution. It is possible to equate the two quantities provided that the apparent molar compressibility is known as a function of the solute molality. Similar equations can be derived on the molarity scale.

The relationship between apparent molar compressibility and concentration depends on the solute. If it is a nonelectrolyte a linear function has proved to be adequate:

$$K_\phi = K_2^0 + b_k m, \tag{6}$$

where K_2^0 is the partial molar compressibility at infinite dilution, and b_k an adjustable parameter. Equation (6) applies whether one deals with isothermal or isentropic compressibilities.

Electrolytes behave differently, and by differentiating the Redlich-Meyer equation [3] with respect to pressure, the following expression is obtained:

$$K_\phi = K_2^0 + S_k c^{1/2} + b_k c. \tag{7}$$

Here S_k is the theoretical limiting slope. In principle it can be calculated from the Debye-Hükel limiting law, but this requires knowledge of the second derivative of the dielectric constant of the solvent with respect to pressure. If we deal with isentropic apparent molar compressibilities, it will be even more complicated to determine $S_{k,s}$. It therefore appears more advantageous to follow Sakurai's approach [4]. By measuring the isentropic apparent molar compressibilities of simple 1:1 electrolytes an empirical value of $S_{k,s}$ was established. The average value of $S_{k,s}$ from all measurements is $(5.6 \pm 0.7) \times 10^{-4}$ cm^3 mol$^{-3/2}$ dm$^{-1/2}$ bar^{-1}. This value appeared to be fairly constant over a temperature range from 5 ° to 45 °C. It also agrees fairly well with other such investigations. Millero and coworkers [5] found $S_{k,s}$ values in the range between 1.74×10^{-4} and 7.79×10^{-4} cm^3 mol$^{-3/2}$dm$^{-1/2}$bar^{-1}. The units of $S_{k,s}$ may seem unnecessarily complicated as given here, but apparent molar compressibilities are generally given as cm^{3}mol^{-1}bar^{-1}. By tabulating $S_{k,s}$ values in the units given above one may use Eq. (7) directly without having to worry about the units. (1 bar $= 10^5$ Pa).

Equation (4) is useful for converting isentropic apparent molar compressibilities to isothermal apparent molar compressibilities. However, Desnoyers and Philip [6] have derived another equation that can be used at infinite dilution:

$$K_2^0 = K_{2,s}^0 + \delta_0(2 E_2^0/\alpha_0 - C_{p,2}^0/\sigma_0), \tag{8}$$

where E_2^0 and $C_{p,2}^0$ are the partial molar expansibility and heat capacity of the solute at infinite dilution. δ_0 is the difference between the isothermal and isentropic compressibilities, $(\kappa\text{-}\kappa_s)$. α and σ are the same as in Eq. (4). The subscript zero refers to pure solvent. The expansibility term of Eq. (8) is usually the important one. In many instances the heat capacity term is less than the experimental uncertainty [7]. It may be easier to use Eq. (8), but both Eqs. (4) and (8) require additional data. These may not be available, and in many cases one must settle for the isentropic partial molar compressibilities. This may not be entirely satisfactory, but valuable information about solute-solvent interactions may still be obtained.

3 Tabulation of Experimental Results

In the following, partial molar compressibilities of organic solutes in water will be presented. The values given are at infinite dilution and 298.15 K. All available literature has been examined, and the data have been critically evaluated. Where more than one set of data could be found, the recommended value is equal to the mean value calculated from the various sets. The error limits have been based on the difference between different sets of data compared to the error limits given in each paper. There are generally too few data to calculate standard deviations for the partial molar compressibility of a solute. In fact, in several instances only one set of data could be found. These data have been tabulated as reported in the original paper including error limits if given. If possible the concentration dependence of the apparent molar compressibilities has been tabulated according to Eq. (6) or (7).

Inorganic electrolytes and polymers have not been included. However, at the end a brief discussion of compressibilities of proteins has been included.

4 Alcohols, Diols, and Ethers

The partial molar compressibilities at infinite dilution are positive for the lower alcohol homolog decreasing with increasing size and eventually becoming negative. The obvious interpretation is that water around hydrophobic groups is less compressible than in the bulk. As the number of hydrophobic methylene groups increases, the partial molar compressibility will decrease.

The partial molar compressibilities of the ethers in Table 1 are all positive, and size effects seem rather irregular. The oxygen atoms are, however, positioned at

Table 1. Isentropic and isothermal partial molar compressibilities of alcohols, diols, and ethers at infinite dilution in aqueous solution; recommended values at 298.15 K [a]

	$\dfrac{10^4 \cdot K_{2,s}^0}{cm^3\,mol^{-1}\,bar^{-1}}$ $(10^4 \cdot b_k)$ [b]	$\dfrac{10^4 \cdot K_2^0}{cm^3\,mol^{-1}\,bar^{-1}}$	Reference
Methanol	12.5 ± 0.1 ($-$ 0.3)	12.6	[19–23]
Ethanol	9.9 ± 0.1 ($-$ 1.6)	10.0	[19, 20, 22, 23]
1-Propanol	6.2 ± 0.3 (0.7)	7.3	[22, 24]
2-Propanol	6.1 ± 0.3 ($-$ 0.3)	7.1	[20, 22, 24]
1-Butanol	4.6 ± 0.2 (1.6)	5.2	[22–24]
2-Butanol	3.5 ± 0.2 (0.9)	4.2	[23, 24, 29]

Table 1 (continued)

	$\dfrac{10^4 \cdot K_{2,s}^0}{cm^3\,mol^{-1}\,bar^{-1}}$ $(10^4 \cdot b_k)$[b]	$\dfrac{10^4 \cdot K_2^0}{cm^3\,mol^{-1}\,bar^{-1}}$	Reference
2-Methyl-1-propanol	9.4±0.2		[23]
2-Methyl-2-propanol	4.6±0.7 (−14.4)		[20, 29, 44]
1-Pentanol	2.4±0.3 (1.8)	4.3	[22, 24]
2-Pentanol	1.0±0.3 (1.4)	1.8	[7, 24]
3-Pentanol	3.8		[29]
1-Hexanol	0.5±0.3	0.9	[7, 24]
2-Hexanol	− 1.2±0.3	− 0.4	[7, 24]
3-Hexanol	2.8	5.5	[29]
4-Heptanol	−11.2		[29]
Cyclopentanol	3.1	4.1	[29]
Cyclohexanol	2.6	4.3	[29]
Cycloheptanol	2.2	4.8	[29]
1,2-Ethanediol	4.7 (0.7)		[22]
1,2-Propanediol	2.4±0.3 (0.2)		[24]
1,3-Propanediol	9.5±0.3 (0.3)	10.2	[7, 22, 24]
1,3-Butanediol	5.7±0.3 (0.2)		[24]
1,4-Butanediol	9.2±0.3 (0.0)	9.9	[7, 22, 24]
2,3-Butanediol	6.9±0.3 (0.2)		[24]
1,5-Pentanediol	8.0±0.2 (2.8)	8.9	[7, 22, 24]
2,4-Pentanediol	4.0±0.3 (2.4)		[24]
1,6-Hexanediol	6.6 (6.7)		[22]
1,7-Heptanediol	6.0±0.3		[24]
2-Methoxyethanol	6.0±0.5 (2.8)	7.6	[31]
2-Ethoxyethanol	3.4±0.5 (2.8)	5.3	[31]
2-Propoxyethanol	1.5±0.5 (5.7)	4.0	[31]
2-Butoxyethanol	− 0.8±0.5 (1.1)	2.9	[20, 31]
Diethyleneglycol	4.3±0.5 (3)		[31]
Triethylene glycol	2.3±0.5 (3)		[31]
Tetraethylene glycol	1.1±0.5 (7)		[31]
Tetrahydrofuran-2-ol	4.4		[43]
Tetrahydropyran-2-ol	− 2.0		[43]
Trimethyleneoxide	10.1	12.1	[29]
Tetrahydrofuran	8.0	9.8	[29, 43]
Tetrahydropyran	12.0	14.7	[29]
	6.4		[43]
1,3-Dioxolane	7.1	10.4	[29]
1,3-Dioxane	10.0	13.8	[29]
1,4-Dioxane	9.3		[43]
1,3-Dioxepane	8.1	11.6	[29]
1,3,5-Trioxane	2.3	6.0	[29]
Dimethoxymethane	16.8	19.4	[29]
Diethoxymethane	23.8	26.6	[29]
1,2-Dimethoxyethane	13.5	15.8	[29]
	8.6±0.5 (2)		[31]
1,1'-Oxybis(2-methoxyethane)	6.3±0.5 (4)		[31]
2,5,8,11-Tetraoxadodecane	4.4±0.5 (8)		[31]
2,5,8,11,14-Pentaoxapentadecane	2.8±0.5 (13)		[31]

[a] Data at other temperatures can be found in [7, 22, 24, 31].
[b] In units of $cm^3\,mol^{-2}\,kg\,bar^{-1}$.

various places in the ether molecule, and this may disrupt the hydrophobic hydration sheath, masking regular size effects, as observed with alcohols.

It should also be noted that although the compressibility of hydrophobic hydration sheaths appears less than bulk water, it is far more compressible that
water around ions or charged groups. The partial molar compressibilities of
simple $1:1$ inorganic electrolytes is of the order -50×10^{-4} cm^{3}mol^{-1}bar^{-1}
[5].

Several investigations of aqueous alcohol and ether solutions have concentrated on the sound velocity as a function of concentration over the entire solubility region [8–20]. Most of these have not calculated the partial molar compressibilities, but concentrated on the sound velocity as such. The general trend
seems to be equal for ethers and alcohols. The sound velocity increases with concentration until a maximum is reached. Presently there is no molecular theory to
account for the observed concentration dependence, but the results have been explained satisfactorily in terms of clathrate-like or "iceberg" structures. Baum-
gartner and Atkinson[10] and Endo [11] have suggested rather well-defined clathrate structures with 17 water molecules per alcohol or ether molecule.

The molar compressibilities of pure alcohols increase with increasing size of
the alcohol. The difference $K_2^0 - K^*$, the excess molar compressibility, seems to be
negative in all cases [19, 22]. The excess molar compressibility will be a measure
of the hydrophobic hydration, and again most clathrate or "iceberg" models can
account for the sign of the excess molar compressibility and the trends of this
quantity as a function of solute size [22, 25, 26].

Lara and Desnoyers [20] have investigated the concentration dependence of
the totally water-soluble alcohols methanol, ethanol, and 2-propanol. They all
have negative slopes in the dilute region up to a certain concentration. For methanol the isentropic apparent molar compressibility then increases monotonically
towards the isentropic molar compressibility of pure methanol. This can be taken
as evidence for methanol participating in the water structure by forming hydrogen bonds with water [29, 27]. Kaulgud and Rao [23] have arrived at the same
conclusion by using the two-state model of Hall [28] to interpret the results. The
compressibility is divided into a structural term, κ_{str}, and a bulk term, κ. Methanol apparently causes very little change in κ, which indicates that methanol dis-
solves without affecting the free water molecules and thus becomes part of the
water lattice.

The isentropic partial molar compressibilities of ethanol and 2-propanol also
reach minimum values at certain concentrations. However, above the minima the
isentropic partial molar compressibilities increase sharply. Lara and Desnoyers
[20] argue that such sharp increases are evidence for microphase transitions. Partial molar volume and heat capacity data are taken in support of this view. The
microphases could be nonsolvated alcohol molecules or even aggregates similar
to micelles. Beyond the transition region, the isentropic partial molar compressibilities increase moderately with concentration, and the solution properties are essentially like those of ordinary nonaqueous solutions.

The isentropic partial molar compressibilities of alcohols and ethers increase
rapidly with temperature [7, 22, 23, 29]. This rapid increase can be explained by
a two-state model for water in the hydration sheath. One state is water hydrogen-

bonded to the alcohol hydroxyl group, or, for ethers, free high density water around the ether oxygens [29, 30]. The other state is water around hydrophobic groups. As the temperature increase, the water structure around hydrophobic groups is gradually destroyed, increasing the partial molar compressibility.

Cabani et al. [29] have studied the possibilities of using partial molar compressibilities to distinguish hydrophobic hydration effects from other types of hydration. They conclude that although partial molar compressibilities are sensitive to solute-solvent interactions, it is not possible to make such distinctions. However, Høiland [7] has shown that this may be possible if group values are calculated. At least for the alcohols it appears that the group molar compressibilities of hydrophobic groups increase with temperature, while for hydrophilic groups the partial molar compressibilities remain constant or decrease with increasing temperature.

5 Carboxylic Acids and Sodium Carboxylates

The ionization (dissociation) equilibria of carboxylic acids in water are important in relation to their partial molar compressibilities. Like the volumes of ionization, the compressibilities of ionization, ΔK^0, in itself require further comment. The compressibilities of ionization can be obtained by two independent methods of measurement. First, isentropic compressibilities can be measured and ΔK^0 can be calculated:

$$\Delta K_s^0 = K_{s,\,NaA}^0 - K_{s,\,HA}^0 - (K_{s,\,Na}^0 - K_{s,\,H}^0). \tag{9}$$

Here $K_{s,\,NaA}^0$, and $K_{s,\,HA}^0$ are the isentropic partial molar compressibilities of the sodium carboxylates and carboxylic acids at infinite dilution respectively. The difference between the isentropic partial molar compressibilities of the sodium ion and the proton at infinite dilution, $K_{s,\,Na^+}^0 - K_{s,\,H^+}^0$, can be calculated from literature data [5, 34], -42.2×10^{-4} cm^{3}mol^{-1} bar^{-1} at 298.15 K. ΔK^0 can also be calculated by measuring the equilibrium constant as a function of pressure:

$$\Delta K^0 = RT(\partial^2 \ln k / \partial P^2). \tag{10}$$

Values calculated from the two methods are in fair agreement [35], but it is obvious that in order to employ Eq. (10) with confidence the equilibrium constant data must be very accurate.

ΔK^0 is always negative [35], as expected from electrostriction theory [38, 39]. It decreases with increasing size of the carboxylic acid, ranging from -9×10^{-4} cm^{3}mol^{-1}bar^{-1} for formic acid to -30×10^{-4} cm^{3}mol^{-1}bar^{-1} for pentanoic and hexanoic acid. For the series of 2-hydroxycarboxylic acids ΔK^0 is constant, -26×10^{-4} cm^{3}mol^{-1}bar^{-1}, from 2-hydroxy butanoic acid onwards. This confirms the results from volumetric measurements that the negative charge of the carboxylate group has no effect on the hydration sheath of the solute beyond the γ-carbon.

Lown et al. [40] observed a linear relationship between ΔV_{el}^0 and ΔK_{el}^0 by studying a number of acids and bases including the self-ionization of water. The electrostriction theory predicts linearity between ΔV_{el}^0 and ΔK_{el}^0 where the subscript el is for electrostriction [41, 42]:

$$\Delta K_{el}^0 = -(\partial \Delta V_{el}^0/\partial P) = \left(\frac{1\partial\varepsilon}{\varepsilon\partial P} - \frac{\partial^2\varepsilon}{\partial P^2}\frac{\partial P}{\partial\varepsilon}\right)V_{el}^0 = 1.99 \times 10^{-4}\,\Delta V_{el}^0. \qquad (11)$$

The empirical slope found by Lown et al. [40], 2.1×10^{-4} bar^{-1}, is in good agreement with the theoretical slope of Eq. (11). However, by using data for all the carboxylic acids presented here, the agreement is not quite so good, the slope proving to be 2.9×10^{-4} bar^{-1}. Still it shows that the volume and compressibilities of ionization can be correlated with reasonable accuracy via a simple electrostriction model.

Due to the ionization equilibria, the partial molar compressibility of the acids, as directly calculated from densities, will include a contribution from ionized species. It is, however, possible to correct for this by the following equation:

$$K_\phi - \alpha\Delta K^0 = K_2^0 + b_k m, \qquad (12)$$

where K_ϕ is the directly calculated apparent molar compressibility, α is the degree of ionization, and b_k an empirical constant that is comparable to b_k of Eq. (6). K_2^0 is the interesting quantity, i.e., the partial molar compressibility of the unionized acid at infinite dilution. Since ΔK^0 depends on K_2^0, and both are unknown, it is necessary to carry out an iteration procedure in order to determine both quantities. The data in Table 2 have all been corrected for ionization.

Table 2. Isentropic partial molar compressibilities of carboxylic acids and sodium carboxylates at infinite dilution in aqueous solution; recommended values at 298.15 K [a] ($K_{2,s}^0$ in cm^3 mol^{-1} bar^{-1})

	$10^4 \cdot K_{2,s}^0 (10^4 \cdot b_k)$ [b]	$10^4 \cdot K_{2,s}^0 (10^4 \cdot b_k)$ [c]	$10^4 \cdot K_{2,s}^0 (10^4 \cdot b_k)$ [c]	Reference
Formic acid	3.8 ± 0.5 (0.0)	-47.7 ± 0.5 (5.9)		[32]
Acetic	6.9 ± 0.5 (1.5)	-60.0 ± 0.5 (12.2)		[33–35]
Propanoic	6.4 ± 0.5 (5.5)	-67.5 ± 0.5 (16.9)		[35]
Butanoic	6.2 ± 0.5 (7.5)	-71.2 ± 0.5 (18.6)		[35]
Pentanoic	5.5 ± 0.5 (13.5)	-73.4 ± 0.5 (22.7)		[35]
Hexanoic	4.2 ± 0.5 (24)	-75.2 ± 0.5 (27.5)		[35]
2-Hydroxyacetic	5.4 ± 0.5 (-7.5)	-63.3 ± 0.5 (14.4)		[35]
2-Hydroxypropanoic	3.6 ± 0.5 (-5.0)	-65.9 ± 0.5		[35]
2-Hydroxybutanoic	3.0 ± 0.5 (-1.5)	-67.0 ± 0.5 (15.0)		[35]
2-Hydroxy-2-methylpropanoic	4.2 ± 0.5 (0.0)	-60.0 ± 0.5 (-18.1)		[35]
2-Hydroxypentanoic	1.4 ± 0.5 (-3.0)	-69.5 ± 0.5 (15.0)		[35]
2-Hydroxy-2-methylbutanoic	5.2 ± 0.5 (1.0)	-68.1 ± 0.5 (25.6)		[35]
2-Hydroxyhexanoic	0.7 ± 0.5 (4.8)	-70.6 ± 0.5 (23.8)		[35]
Galacturonic	-25.9 ± 0.5 (1.7)	-90.5 ± 0.5 (18.8)		[36]
Glucuronic	-17.8 ± 0.5 (0.5)	-88.6 ± 0.5 (75)		[36]
Butanedioic	5.8 ± 0.4 (0.3)	-62.0 ± 0.5 (14.1)	-132.8 ± 0.5 (35.1)	[32, 37]
Pentanedioic	7.4 ± 0.5 (4.4)	-64.6 ± 0.5 (20.8)	-137.5 ± 0.5 (36.8)	[32]

[a] Data at other temperatures can be found in [35].
[b] In units of cm^3 mol^{-2} kg bar^{-1}.
[c] In units of cm^3 mol$^{-3/2}$ dm$^{1/2}$ bar^{-1}.

6 Amines, Amides, and Ureas

Amines dissolved in water hydrolyze. This affects the partial molar compressibilities of the amine, and it is necessary to correct for the effects of hydrolysis. Kaulgud et al. [45] have proposed the following equation:

$$\frac{K^*_{\phi,s}}{(1-\alpha)} = K^0_{\phi,s(BH_2O)} + (1-\alpha)cS_k, \tag{13}$$

where

$$K^*_{\phi,s} = K^{obs}_{\phi,s} - \alpha K^0_{\phi,s(BH^+OH^-)}. \tag{14}$$

In these equations $K_{\phi,s(BH_2O)}$ and $K_{\phi,s(BH^+OH^-)}$ are the apparent molar compressibilities of a hypothetical neutral aquoamine complex and the corresponding ammonium hydroxide, respectively. α is the degree of hydrolysis. Equation (13) can be linearized by multiplying both sides by $(1-\alpha)c$:

$$K^*_{\phi,s}c = (1-\alpha)\,c\,[K^0_{\phi,s(BH_2O)} + (1-\alpha)cS_k]. \tag{15}$$

If $(1-\alpha)c\,S_{k,s}$ is small, plots of $K^*_{\phi,s}c$ versus $(1-\alpha)c$ should be linear. Such plots have been made by Kaulgud et al. [45]. However, marked deviations from linearity have resulted especially at low temperatures, 5 ° and 15 °C. Only for ammonia and methaneamine were linear plots obtained at all temperatures. Kaulgud et al. [45] argues that the observed deviation from linearity is due to solute-solute interactions. At low concentrations a linear region is observed. It seems possible to carry out the extrapolation to obtain $K^0_{\phi,s(BH_2O)}$ and consequently determine the isentropic partial molar compressibility of the amines at infinite dilution. Another possibility is to avoid hydrolysis by adding a small amount of NaOH or KOH. Conway and Verrall [25] thus measured isentropic apparent molar compressibilities of amines in 0.025 N aqueous KOH solutions. This is easier in the sense that no correction functions are needed. On the other hand, the KOH added may lead to errors due to possible interactions between these ions and the amines. In Table 3 the data for the amines have been taken from Kaulgud et al., corrected by using Eq. (15).

It is surprising to see that the isentropic partial molar compressibilities of neutral amines increase with increasing alkyl chain length at 25 °C. This is contrary to what is observed for alcohols. However, at higher temperatures, 45 °C, $K_{2,s}$ of alcohols exhibit the same behavior, though the increase per CH_2-group seems less. It is further observed that $(dK^0_{2,s}/dT)$ is positive for all amines at lower temperatures. At higher temperatures only the higher homologs retain positive values of $(dK^0_{2,s}/dT)$. Kaulgud et al. [45] conclude that this shows that amines behave as structure stabilizers when the hydrophobic chain is large, while the lower homologs disturb the structure, presumably due to the NH_2 group. By comparison with alcohols, it seems that the NH_2 group disturbs the water structure around the neighboring hydrophobic groups much more than the OH-group.

The isentropic partial molar compressibilities of the tetra-alkylammonium salts at infinite dilution decrease with increasing hydrophobicity. Conway and Verrall [25] have interpreted this in terms of increased water structure around the hydrophobic groups. Mathieson and Conway [34] have also measured the isentropic partial molar compressibilities of the tetra-alkylammonium salts in D_2O. The partial molar compressibilities of these ions become progressively more negative in D_2O compared to H_2O the larger the size of the organic ion. This behavior is consistent with the trend in H_2O itself as interpreted by Conway and Verrall [20]. The larger tetra-alkylammonium ions promote the water structure to a larger extent than the smaller ones. This effect becomes more pronounced in D_2O, so that $K_{\phi,s}$ is more negative in D_2O than in H_2O.

As can be seen from Table 3, compressibility data on amines are rather scarce. In the few cases where comparisons can be made the agreement between various series is not particularly good. For instance, Kaulgud et al. [45] present 21.06×10^{-4} $cm^3 mol^{-1} bar^{-1}$ for N-methylmethaneamine, while Conway and Verrall [25] found the value to be 9×10^{-4} $cm^3 mol^{-1} bar^{-1}$. Large differences also appear for N,N-dimethylmethaneamine. True, Conway and Verrall have carried out their measurements in 0.025 N aqueous NaOH, but this cannot be the cause for such large discrepancies. So when using the data of Table 3, these discrepancies should be kept in mind.

Table 3. Isentropic partial molar compressibilities of amines, ammonium salts, amides, and ureas at infinite dilution in water; recommended values at 298.15 K [a]

	$\dfrac{10^4 \cdot K^0_{2,s}(B)}{cm^3 mol^{-1} bar^{-1}}$	$\dfrac{10^4 \cdot K^0_{2,s}(BHCl)}{cm^3 mol^{-1} bar^{-1}}$	Reference
Ammonia	4.22	-20.2 ± 0.2	[34, 45]
Methaneamine	7.72	-22.0	[25, 45]
Ethaneamine	13.32	-18.2 ± 0.2	[34, 45]
1-Propaneamine	13.4		[45]
1-Butaneamine	15.42		[45]
2-Methyl-2-propaneamine	12.92		[45]
N-Methylmethaneamine	20.8	-20.2	[25, 45]
N,N-Dimethylmethaneamine	15.52	-18.1	[25, 45]
N-Ethylethaneamine		-14.4	[25]
N,N-Diethylethaneamine		-12.8	[25]
Tetramethylammoniumbromide		-4.1	[25, 34]
Tetraethylammoniumbromide		-4.6	[25, 34]
Tetrapropylammoniumbromide		-9.6	[25, 34]
Tetrabutylammoniumbromide		-17.3	[25, 34]
Pyrrolidine	-3.5		[46]
Piperidine	-12.0	-22.0 ± 0.25	[46, 57]
1-Methylpiperidine [b]	0.3 ± 0.25	-20.9 ± 0.25	[57]
Morpholine	4.0		[46]
Pyridine	2.2		[47]
Acetamide	5.94 ± 0.18		[37]
Thioacetamide	8.35 ± 0.24		[37]
Lactamide	4.32		[58]
Glycolamide	2.74		[58]
Urea	-3.20 ± 0.08		[37]
1,3-Dimethylurea	-0.24 ± 0.13		[37]
1,1,3,3-Tetramethylurea	8.58 ± 0.15		[37]
Thiourea	-2.34 ± 0.03		[37]

[a] Data at other temperatures in references [46, 48].
[b] In 0.1 N KOH.

7 Carbohydrates

In water the aldoses are almost entirely represented by their cyclic forms [49, 50]. Galactose, glucose, and mannose have a pyranose C1 conformation with the majority of OH groups in an equatorial configuration [51]. The distance between the groups in this configuration compares well with distances between water molecules [52, 53]. On this basis Tait et al. [54] have proposed a specific hydration model where these hexoses are part of the tetrahedral arrangement of water molecules. The ultrasonic studies of aldose hydration by Juszkiewicz [55] seem to agree, since the calculated hydration numbers at 0 °C correspond to the mean number of equatorial hydroxyl groups of glucose, mannose, xylose, saccharose, and lactose. Further support for the model can be derived from the dielectric relaxation and NMR measurements of Tait et al. [54]. From these measurements it was impossible to distinguish between the hydration properties of galactose, glucose, and mannose.

The specific hydration model thus has found considerable support from experiment. If it was correct in a strict sense, one would expect that the partial molar compressibilities of galactose, glucose, and mannose were equal. However, by inspecting the data, one finds significant differences. This does not necessarily render the specific hydration model invalid, but it strongly suggests differences in the hydration properties of the sugars. How significant these differences are remains an open question, but the partial molar compressibilities should, in principle, reflect the hydration number. Shio [56] has in fact used compressibilities cal-

Table 4. Isentropic partial molar compressibilities of carbohydrates at infinite dilution in water; recommended values at 298.15 K

	$\dfrac{10^4 \cdot K^0_{2,s}}{cm^3\,mol^{-1}\,bar^{-1}}$ $(b_k)^a$	Reference
Arabinose	-19.3 ± 0.5 (4.3)	[36]
Ribose	-12.5 ± 0.2 (3.1)	[36, 37, 43]
Xylose	-12.9 ± 0.5 (3.1)	[36]
D-Galactose	-20.8 ± 0.5 (5.2)	[36]
D-Glucose	-17.6 ± 0.3 (3.7)	[36, 37, 43]
D-Mannose	-16.0 ± 0.5 (4.2)	[36]
Lactose	-30.4 ± 1.0 (13.5)	[36]
Maltose	-23.7 ± 1.0 (13.6)	[36]
Sucrose	-18.5 ± 0.5 (8.9)	[36, 37]
Melezitose	-24.6 ± 1.0 (25.2)	[36]
Raffinose	-30.9 ± 1.0 (37.2)	[36]
Methyl-α-glucopyranoside	-13.0	[43]
Methyl-β-glucopyranoside	-5.9	[43]
Arabitol	-10.0 ± 0.5 (3.6)	[36]
Ribitol	-9.6 ± 0.5 (3.4)	[36]
Xylitol	-12.0 ± 0.5 (3.3)	[36]
Galactitol	-14.6 ± 0.5 (4.8)	[36]
Glucitol	-14.0 ± 0.5 (3.7)	[36]
Mannitol	-14.9 ± 0.5 (4.7)	[36]

^{a} In units of $cm^3\,mol^{-2}\,kg\,bar^{-1}$.

culated from ultrasound measurements to determine hydration numbers of several saccharides. Galactose and mannose were not included in these studies, but by using his method of calculation one arrives at the following hydration numbers: 4.1, 3.5, and 3.1 for galactose, glucose, and mannose respectively.

It is also interesting to compare the isentropic partial molar compressibilities of di- and trisaccharides. The partial molar volumes increase with increasing solute size, and usually the compressibilities exhibit the same kind of behavior. However, in this case there is no significant difference in the isentropic parital molar compressibilities of di- and trisaccharides. $K^0_{2,s}$ of melezitose equals maltose, and $K^0_{2,s}$ values of lactose and raffinose are also practically equal. The results suggest that specific hydration takes place and that the size of the solute is not the predominant factor. However, if we inspect the b_k values, it increases significantly with solute size. That is, solute-solute interactions depend very much on the solute size, as expected (Table 4).

8 Amino Acids and Dipeptides (Table 5)

Few studies have been carried out dealing with partial molar compressibilities of amino acids. Millero et al. [60] and Cabani et al. [61] have provided most of the data. Unfortunately, these two sets of data do not agree very well. The data of Cabani et al. [61] are less negative than these of Millero et al. [60], for DL-alanine the difference is 3.4×10^{-4} cm^{3}mol^{-1}bar^{-1}. Cabani et al. [61] suggest that the discrepancy is due to different experimental conditions. They measured the speed of sound at a frequency of 15 MHz, while Millero used 2 MHz. However, the author has measured $\phi_{k,s}$ for glycine and diglycine with a frequency of 13 MHz. The glycine result $(-26.5 \pm 0.5) \times 10^{-4}$ cm^{3}mol^{-1}bar^{-1} agrees well with Millero's, and for diglycine the result was $(-41.5 \pm 0.5) \times 10^{-4}$ cm^{3}mol^{-1}bar^{-1}, 6×10^{-4} cm^{3}mol^{-1}bar^{-1} more negative than reported by Cabani et al. [61]. Thus, it seems questionable if the frequency dependence can account for the differences. We have therefore recommended the data of Millero et al. [60], as agreeing with other literature data. However, for several solutes the only data are those of Cabani et al. [61], and they have been included.

Millero et al. [60] have calculated hydration numbers of amino acids from the following equation:

$$n_H = \frac{-K^0_{2,el}}{\kappa^* V^*}, \tag{16}$$

$K^0_{2,el}$ is the electrostriction contribution to the partial molar compressibility:

$$K^0_{2,el} = K^0_2 - K_{2,int}, \tag{17}$$

where $K_{2,int}$ is the intrinsic partial molar compressibility. To a first approximation $K_{2,int}$ can be regarded as zero. The calculated values of n_H range from 2.9 to 4.5 with an average of 3.6 ± 0.5.

Cabani et al. [61] have calculated the changes in isentropic partial molar compressibilities for reaction of internal and external proton exchange, i.e., zwitterion formation and reaction of the type glycolamide glycine. It turns out that there is a significant difference between the two, though the calculation is based on a few examples only. Still it seems that interactions between the NH_3^+ and COO^- groups in neighboring positions increase the compressibility compared to a situation where the two groups are separated by methylene groups.

9 Proteins

The apparent specific compressibility is defined as:

$$K_\phi = (v \, \kappa - v^* \kappa^* x_1)/x_2. \tag{18}$$

Here κ and κ^* are the compressibilities of the solution and solvent respectively. v and v^* are the specific volumes ($1/\varrho$) of solution and solvent and x_1 and x_2 the weight fractions of solute and solvent respectively.

Compressibility data on proteins in solution are rather few. Usually only the compressibilities, κ, are presented, or the apparent specific compressibilities have been calculated. Fahey et al. [63] used a direct densitometric method at pressures up to 400 atm. Apparently they observed no change in the partial specific volumes of RNase, turnip yellow mosaic virus, and its capsid protein. Hearst et al. [64, 65] showed that a plot of the compositional buoyant density of a macromolecule versus the pressure at band center of a centrifuge was linear. The slope was termed the standard pressure coefficient, ψ. ψ-values of approximately 22×10^{-6} atm^{-1} were obtained for T-4 bacteriophage DNA. Similar values of ψ have been obtained for nucleic acids [66]. Sharp et al. [67] used the same method of measurement on human γ-immunoglobulin, bovine serum mercaptalbumin, and egg albumin in CsCl solution. The ψ-values for egg albumin and mercaptalbumin were positive, though smaller than for nucleic acids. However, the ψ-value of γ-immunoglobulin was negative, suggesting an unusual internal structure of the protein. Compression of the protein voids or cavities less accessible to solvation could account for this observation. Such a compression would result in extrusion of the interacting water, the result being that preferential hydration decreases with pressure.

Apparent specific compressibilities from ultrasound measurements still seem the most accurate method. Miyahara [68] has carried out measurements on egg albumin in water, Jacobson [69] on serum albumin, oxyhemoglobin, and hemocyanine. Gekko and Noguchi [70] have measured compressibilities of 14 globular proteins in water, and Millero et al. [71] have studied the apparent specific compressibility of aqueous lysozyme solutions. It appears that the isentropic compressibilities of the globular proteins are all positive. It has also been observed that ribonuclease A, one of the most hydrophilic proteins, exhibits a small isentropic compressibility compared to other more hydrophobic proteins. It is also quite clear that one cannot add compressibilities of the relevant amino acids and thus obtain the partial specific compressibility of the protein.

The partial specific compressibility can be divided into two contributions [70]. One is due to the compression of void volumes, i.e., the void volumes generated by the random close packing of the units constituting the protein. The other contribution is due to the compressibility of the hydration sheath:

$$\kappa_s = \kappa_{s, \, void} + \kappa_{s, \, sol}. \tag{19}$$

Table 5. Isentropic and isothermal partial molar compressibilities of amino acids and dipeptides at infinite dilution in water; recommended values at 298.15 K

	$\dfrac{10^4 \cdot K_{2,s}^0}{cm^3\,mol^{-1}\,bar^{-1}}$ $(b_k)^a$	$\dfrac{10^4 \cdot K_2^0}{cm^3\,mol^{-1}\,bar^{-1}}$	Reference
Glycine	−27.0 ±0.2 (4.6)	−23.4±0.4	[58–62, 79]
DL-α-Alanine	−25.0 ±0.5 (4.1)	−21.9±0.5	[58–62]
D-α-Alanine	−25.53±0.28 (5.0)		[60]
L-α-Alanine	−25.56±0.60 (4.8)		[60, 79]
β-Alanine	−26,36±0.11		[79]
L-Valine	−30.62±0.13 (8.4)		[60, 79]
DL-Valine	−24.0	−21.4	[61]
L-Leucine	−31.78±0.56 (−13.9)		[60]
L-Proline	−23.25±0.11 (5.8)		[60]
Phenylalanine	−34.54±1.48 (36.3)		[60]
D-Tryptophan	−30.24±0.23 (−41.1)		[60]
DL-Methionine	−31.18±0.25 (14.3)		[60]
L-Serine	−29.88±0.10		[79]
L-Threonine	−31.23±0.10		[79]
L-Cysteine	−32.82±0.26 (7.3)		[60]
DL-Aspartic acid	−33.12±0.07 (47.8)		[60]
Glutamic acid	−36.17±0.38 (16.1)		[60]
L-Arginine	−26.62±0.20 (12.1)		[60]
Histidine	−31.84±0.29 (−13.9)		[60]
DL-2-Aminobutanoic acid	−21.8 ±0.2	−19.2±0.2	[61]
DL-3-Aminobutanoic acid	−18.7 ±0.6	−16.8±0.8	[61]
2-Amino-2-methylpropanoic acid	−23.46±0.27		[79]
4-Aminobutanoic acid	−27.0 ±1.4	−24.5±0.6	[61]
5-Aminopentanoic acid	−27.3 ±2.0	−25.2±1.5	[61]
6-Aminohexanoic acid	−29.2 ±0.3	−26.1±0.4	[61]
Glycylglycine dkp	−11.1 ±0.2	− 4.5±1.6	[61]
Alanylalanine dkp	3.8 ±0.4	7.5±1.0	[61]
Sacrosylsarcosine dkp	3.2 ±0.6	7.0±0.6	[61]
Diglycine	−35.5 ±0.5	−31.9±0.6	[61, 62]

a In units of $cm^3\,mol^{-2}\,kg\,bar^{-1}$.

The void term is expected to contribute positively and the solvation term negatively. A large positive value of κ_s can thus be expected for a protein with large voids or little hydration. The observed values of κ_s suggest that the void term of proteins is large, so large that it outweighs the hydration effects. Gekko and Noguchi [70] have also shown that there is a linear relationship between the isentropic partial specific compressibility and the polarity parameter of Bigelow [76]. Exceptions to the rule are bovine serum albumine, α-chymotrypsinogen, and trypsin.

The pressure-induced denaturation of proteins has also been studied [72–75]. The conclusion after these studies is that the specific compressibility of denatured protein is larger than that of the native by about 1.5×10^6 bar^{-1}. Sharp et al. [67] argue that this is the opposite of what was to be expected if the denaturation process exposes hydrophobic groups to the solvent. Instead they find it likely that even small increases in the void volume upon denaturation could overcome an increase in the solvation term [Eq. (19)]. This would lead to an increase in the compressibility of the denatured protein.

As is evident from this short summary of protein compressibilities in solution, few investigations have been carried out so far. However, it appears that compressibility studies might prove fruitful. Perhaps one should advocate studies of volumes and compressibilities at various temperatures. It seems that such studies might be of special significance.

10 Nucelobases and Nucleosides

Recently the isentropic partial molar compressibilities of some nucleobases and nucleosides have been published [80]. The data are presented in Table 6.

The nucleobases and the nucleosides self-associate in a vertical form, often termed stacking. The apparent molar compressibilities are thus no longer linear with concentration. A curvature is observed reflecting the equilibrium between nucleobases or nucleosides in a monomeric form and in the stack. The changes in the isentropic partial molar compressibility as these compounds associate can be calculated as:

$$\Delta K^0_{stack} = K^0_{2, stack} - K^0_{2, mono}. \tag{20}$$

Here $K^0_{2, stack}$ is the isentropic partial molar compressibility at infinite dilution of a nucleobase or a nucleoside in the aggregated state (in the stack), and $K^0_{2, mono}$ that of the monomeric species.

In order to calculate ΔK^0_{stack}, the equilibrium constant must be known. Unfortunately this is not available for every step as the stack grows, only average values can be determined. This means that one must rely on models for calculating the equilibrium constant. Two models are generally used; the sequential isodesmic model (SEK) and the attenuated constant model (AK), see Chapter 2.3, this volume. Both sets of data are presented in Table 7.

Normally, negative volume differences for a process lead to negative compressibility changes and vice versa [35]. This is not so for the nucleobases. They exhibit negative ΔV^0_{stack} and positive ΔK^0_{stack} values. Furthermore the ΔK^0_{stack}

Table 6. Isentropic partial molar compressibilities and changes in the partial molar compressibilities of stacking of nucleobases and nucleosides at infinite dilution in water, values at 298.15 K. (AK) calculated from the attenuated constant model, (SEK) from the sequential equal constant model

	$\dfrac{10^4 \cdot K^0_{2, s}}{cm^3 \, mol^{-1} \, bar^{-1}}$	$\dfrac{10^4 \cdot K^0_{stack}(AK)}{cm^3 \, mol^{-1} \, bar^{-1}}$	$\dfrac{10^4 \cdot K^0_{stak}(SEK)}{cm^3 \, mol^{-1} \, bar^{-1}}$	Reference
Purine	1.5	5.8	7.6	[80]
Caffeine	5.2	7.0	8.8	[80]
Cytidine	−18.2	17.0	19.0	[80]
Uridine	−17.0	71	81	[80]
Thymidine	− 4.0		30	[80]

values of the nucleobases are unusually large. This has been explained in terms of an intrinsic compressibility of the stack, so that the distance between the associated base units is pressure-dependent.

11 Summary

From data on partial molar compressibilities of homologous series of solutes it is possible to estimate the increment per methylene group. Cabani et al. [61] have calculated an average value of $(-1.9\ 1.1) \times 10^{-4}\ cm^3 mol^{-1} bar^{-1}$ from 33 solutes of various homologous series. However, the methylene group value varies considerably from series to series and apparently becomes positive for amines. The methylene group value is also far from being constant within each homologous series. Some isentropic partial molar compressibilities of functional groups have been presented in Table 7. It can be seen that the methyl group value also varies, but at least it seems to be positive in all cases.

From volumes and compressibilities of ionization it has been shown that the charge of a carboxylate or amino group influences the hydration of neighboring groups as far as the γ-carbon. Charges bring about significant changes in the hydration, and these are therefore easily measured. It can be seen as changes both in the partial molar volume and the partial molar compressibility. A noncharged polar group, on the other hand, does not seem to bring about dramatic changes in the hydration sheath of neighboring groups. Volumetric data indicate that only the nearest methylene group is affected, and even this only marginally. However, the poor additivity of partial molar compressibilities could be explained if even noncharged polar groups influence the hydrophobic hydration sheath of neighboring methylene groups. If this is so, the amino group certainly disrupts the hydration sheath differently from the hydroxyl group of alcohols.

Edward et al. [77, 78] have suggested the following equation to account for the partial molar volumes of organic solutes:

$$V_2^0 = \tfrac{4}{3}\pi(r_w + \Delta)^3 + n_i\sigma_i, \tag{21}$$

Table 7. Isentropic partial molar compressibilities of various functional groups at infinite dilution in water at 298.15 K [a]

	$\dfrac{K_{2,s}^0 \times 10^4}{cm^3\,mol^{-1}\,bar^{-1}}$	Source
$-CH_3$	0.8	Alcohols, diols [7, 24]
	2.0 ± 0.2	Methylamines [57]
	1.6	Alanine-glycine [60]
$-CH_2OH$	6.5	Alcohols, diols [7, 24]
$-CH_2-$	-1.6 ± 0.6	Alcohols, diols [7, 24]
	-1.2 ± 0.8	Tetraalkylammonium salts [57]
	2.0	Butaneamine-propaneamine [45]
	-1.9 ± 1.1	Average of 33 solutes [61]
$-CHOH-$	2.2	2-Alcohols [24]

[a] Millero et al. [60] have tabulated other group values of special relevance to amino acids.

where r_w is the van der Waals radius, Δ the increment due to the void volume around the solute, and σ_i a shrinkage parameter necessary for polar groups. This equation can be differentiated with respect to pressure:

$$K_2^0 = 4\pi(r_w + \Delta)^2(\partial\Delta/\partial P) + n_i(\partial\sigma_i/\partial P). \tag{22}$$

Here it has been assumed that the van der Waals radius is independent of pressure. The values of $(\partial\Delta/\partial P)$ and $(\partial\sigma_i/\partial P)$ are not known, but it is possible to estimate them from the experimental data. However, if this is attempted, different values are obtained for different solutes. The reason seems obvious. The hydration sheath is not uniformly compressed, as follows from Eq. (21) or (22), and is demonstrated by the values of the group molar compressibilities. It thus appears that the simple Eq. (21) that works reasonably well with partial molar volumes is too simple to stand the more severe test of partial molar compressibilities. The partial molar compressibility measures solute-solvent interactions directly, i.e., no intrinsic contributions appear for monomers. Specific interactions between various functional groups seem to influence the hydration properties. Strictly speaking it thus seems impossible to rationalize the partial molar compressibilities in terms of simple additivity schemes However, the amount of data presently available is not overwhelming, and it seems worthwhile to continue compressibility measurements. Perhaps the temperature dependence of the partial molar compressibilities may be informative. As mentioned before, it has been shown [7] that alcohol data suggest that hydrophobic and hydrophilic group values vary differently with temperature. Moreover, the data of Cabani et al. [29], Nakajima et al. [22], and Kaulgud et al. [23] show that the rate of change is very different for neutral and ionic solutes. Therefore, investigations of the partial molar compressibilities as functions of temperature may be expected to provide important information about solute-solvent interactions.

References

1. Greenspan M, Tschiegg CE (1957) J Res Natl Bur Stand 59:249
2. Garnsey R, Boe RJ, Mahoney R, Litovitz TA (1969) J Chem Phys 50:5222
3. Redlich O, Meyer DM (1964) Chem Rev 64:221
4. Sakurai M, Komatsu T, Nakagawa T (1981) Bull Chem Soc Jpn 54:643
5. Millero FJ, Ward GK, Chetirkin PV (1977) J Acoust Soc Am 61:1492
6. Desnoyers JE, Philip PR (1972) Can J Chem 50:1094
7. Høiland H (1980) J Solution Chem 11:857
8. Yasunaga T, Hirata Y, Kawano Y, Miura M (1964) Bull Chem Soc Jpn 37:867
9. Yasunaga T, Usui I, Iwata K, Miura M (1964) Bull Chem Soc Jpn 37:1658
10. Baumgarten EK, Atkinson G (1971) J Phys Chem 75:2336
11. Endo H (1973) Bull Chem Soc Jpn 46:1586
12. Malenkov GG (1966) Zh Strukt Khim 7:331
13. Pesce B, Giaconimi A (1940) Ric Sci 11:619
14. Nozdrev VF (1963) Application of ultrasonics in molecular physics. Gordon and Breach, New York
15. Bruun SG, Sørensen PG, Hvidt Å (1974) Acta Chem Scand A 28:1047
16. Dale WDT, Flavelle PA, Kruus P (1976) Can J Chem 54:355
17. Emery J, Gasse S (1979) Acustica 43:205
18. Gasse S, Emery J (1980) J Chem Phys 77:263

19. Kiyohara O, Benson GC (1981) J Solution Chem 10:281
20. Lara J, Desnoyers JE (1981) Solution Chem 10:465
21. Garnsey R, Mahoney R, Litovitz TA (1964) J Chem Phys 64:2073
22. Nakajima T, Komatsu T, Nakagawa T (1975) Bull Chem Soc Jpn 48:788
23. Kaulgud MV, Rao SM (1979) J Chem Soc Faraday Trans I 75:2237
24. Høiland H, Vikingstad E (1976) Acta Chem Scand A 30:692
25. Conway BE, Verrall RE (1966) J Phys Chem 70:3952
26. Verrall RE, Conway BE (1966) J Phys Chem 70:3961
27. Franks F, Ives DJG (1966) Q Rev Chem Soc Lond 20:1
28. Hall L (1948) Phys Rev 73:775
29. Cabani S, Conti G, Matteoli E (1979) J Solution Chem 8:11
30. Cabani S, Conti G, Matteoli E (1976) J Solution Chem 5:751
31. Harada S, Nakajima T, Komatsu T, Nakagawa T (1978) J Solution Chem 7:463
32. Høiland H, unpublished results
33. Owen BB, Brinkley SR (1941) Chem Rev 29:461
34. Mathieson JG, Conway BE (1974) J Chem Soc Faraday Trans I 70:752
35. Høiland H, Vikingstad E (1976) J Chem Soc Faraday Trans I 72:1441
36. Høiland H, Holvik H (1978) J Solution Chem 7:587
37. Lo Surdo A, Shin C, Millero FJ (1978) J Chem Eng Data 23:197
38. Drude P, Nernst W (1894) Z Phys Chem 15:79
39. Desnoyers JE, Conway BE, Verrall RE (1965) J Chem Phys 43:243
40. Lown DA, Thirsk HA, Wynne-Jones WFK (1968) Trans Faraday Soc 64:2073
41. Born M (1920) Z Physiol 1:45
42. Hamann SD (1974) In: Conway BE, Bockriss JO'M (eds) Modern aspects of electrochemistry, vol 9. Plenum, New York
43. Franks F, Ravenhill JR, Reid DS (1972) J Solution Chem 1:3
44. Patil KJ, Raut DN (1980) Indian J Pure Appl Phys 18:499
45. Kaulgud MV, Shrivastava A, Awode MR (1980) Indian J Pure Appl Phys 18:864
46. Kaulgud MV, Patil KJ (1976) J Phys Chem 80:138
47. Conway BE, Laliberte LH (1968) In: Covington AK, Jones P (eds) Hydrogen-bonded solvent systems. Taylor and Francis, London
48. Kaulgud MV, Patil KJ (1974) J Phys Chem 78:714
49. Cantor SM, Penniston QP (1940) J Am Chem Soc 62:2113
50. Los JM, Simpson LB, Wiesener K (1956) J Am Chem Soc 78:1564
51. Angyal SJ (1969) Angew Chem Int Ed Engl 81:172
52. Warner DT (1962) Nature 196:1055
53. Danford MD, Levy HA (1962) J Am Chem Soc 84:3965
54. Tait MJ, Suggett A, Franks F, Ablett S, Quickenden PA (1972) J Solution Chem 1:131
55. Juszkiewicz A (1981) Arch Acoust 6:307
56. Shio H (1958) J Am Chem Soc 80:70
57. Laliberté LH, Conway BE (1970) J Phys Chem 74:4116
58. Gucker FT, Allen TW (1942) J Am Chem Soc 64:191
59. Gucker FT, Haag RM (1953) J Acoust Soc Am 25:470
60. Millero FJ, Lo Surdo A, Shin C (1978) J Phys Chem 82:784
61. Cabani S, Conti G, Matteoli E, Tiné MR (1981) J Chem Soc Faraday Trans I 77:2385
62. Yayanos AA (1972) J Phys Chem 76:1783
63. Fahey PF, Kupke DW, Beams JW (1969) Proc Natl Acad Sci USA 63:548
64. Heart JE, Ifft JB, Vinograd J (1961) Proc Natl Acad Sci USA 47:1015
65. Heart JE, Vinograd J (1961) Proc Natl Acad Sci USA 47:999
66. Bauer W, Prindaville F, Vinograd J (1971) Biopolymers 10:2615
67. Sharp DS, Fujita N, Kinzie K, Ifft JB (1978) Biopolymers 17:817
68. Miyahara Y (1956) Bull Chem Soc Jpn 29:741
69. Jacobson B (1950) Ark Kemi 2:177
70. Gekko K, Noguchi H (1979) J Phys Chem 83:2706
71. Millero FJ, Ward GK, Chetirkin P (1976) J Biol Chem 251:4001
72. Anderson GR (1963) Ark Kemi 20:513
73. Zipp A, Kauzmann W (1973) Biochemistry 12:4217

74. Hawley SA (1971) Biochemistry 10:2436
75. Brandts JF, Oliveira RJ, Westort C (1970) Biochemistry 9:1083
76. Bigelow CC (1967) J Theor Biol 16:187
77. Edward JT, Farrell PG (1975) Can J Chem 53:2965
78. Edward JT, Farrell PG, Shahidi F (1977) J Chem Soc Faraday Trans I 5:705
79. Ogawa T, Yasuda M, Mizutani K (1984) Bull Chem Soc Jpn 57:662
80. Høiland H, Skauge A, Stokkeland I (1985) J Phys Chem 88:6350

Chapter 5 Heat Capacities of Biological Macromolecules

GEORGE M. MREVLISHVILI

Symbols

c:	heat capacity
c*(T):	variation of heat capacity with temperature determined by measuring finite temperature changes caused by finite amounts of thermal energy at different temperatures. c*(T) is in the limit of $\Delta T \to 0$ equal to c(T).
c_p:	isobaric specific heat capacity $[\text{J g}^{-1}\text{K}^{-1}]$
C_p:	isobaric molar heat capacity $[\text{J mol}^{-1}\text{K}^{-1}]$
$c_{p,1}$:	isobaric partial specific heat capacity of solvent
$c_{p,2}$:	isobaric partial specific heat capacity of solute
$c_{p,s}$:	heat capacity in solid state
$\delta c_{p,s}$:	average difference between specific heat capacities of various proteins in the solid state
Δc_p:	average difference between the specific heat capacities of the unfolded and native state for a given protein
$\Delta c_p(\text{ion})$:	heat capacity contribution resulting from ionization

$$\frac{-(G^0 - H_0^0)}{T} = \int_0^T \frac{C_p}{T}dt - \frac{1}{T}\int_0^T C_p dt : \text{Gibbs energy function}$$

G^0:	standard Gibbs energy
H_0^0:	standard enthalpy at 0 K

It is not specified here, whether G and H_0^0 represent molar or specific quantities or refer to another amount of material (such as 100 g). These definitions are made in the tables. A conversion factor of 1 cal = 4.186 J has been used throughout this chapter.

1 Introduction

Thermal measurements of biological objects are gaining increasing importance in molecular-biological studies (see, e.g., reviews [1–9]. This is due both to recent developments in the field of thermophysical instruments (including scanning differential microcalorimeters [10–18], and to advances in configurational statistics and statistical mechanics of macromolecules, the statistical-thermodynamic theory of phase transitions in biopolymers, and the theory of solutions of macromolecules.

The modern methods of calorimetry prove extremely useful and informative when applied to studying the properties of biopolymers and their synthetic analogs over a broad temperature range. This includes the liquid-helium range, i.e., temperatures far below the region of biological function of the macromolecules. Such an approach is valuable, since it reveals characteristic features of the material under extreme environmental conditions (low and high temperature, high pressures, etc.).

Heat capacity studies have been performed on a wide variety of biopolymers at different temperatures (see [19–51].

This review attempts to compile and classify available experimental data on the heat capacity of amino acids, polypeptides, proteins, and nucleic acids. Analysis of the literature data shows that the accuracy with which the heat capacity of biopolymers over a wide temperature range has been measured is rather high. Yet modern technology will permit substantial improvements of calorimetric measurements in the future.

2 Calorimetric Measurements

Currently, most measurements of heat capacity are made by either of the following methods (Fig. 1):

a) The "heat-pulse" method, which consists of providing a known amount of thermal energy (ΔQ) to an isolated specimen and measuring the resulting temperature rise ΔT. The heat capacity is defined by

$$C = \lim_{\Delta T \to 0} \frac{\Delta Q}{\Delta T}.$$

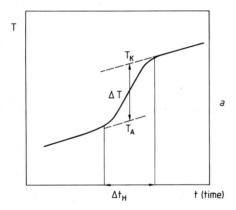

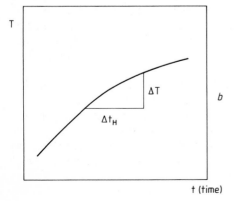

Fig. 1a, b. Principles of measurements of heat capacity: **a** Heat-pulse (step-heating) method:

$$C = \frac{\Delta Q}{\Delta T} = \frac{I_H \cdot U_H \cdot \Delta T}{T_K - T_A} = C \frac{(T_K + T_A)}{2}$$

b Continuous-heating method:

$$C(T) = \frac{I_H \cdot U_H}{(dT/dt)}$$

I_H heating current; U_H voltage across heater resistance; T_K final temperature (after heating); T_A initial temperature (before heating); Δt_H heating time; ΔT temperature increment

In measuring the heat capacity one usually applies the approximation

$$C^*(T) = \frac{\Delta Q}{\Delta T}.$$

Here T is the midpoint of the temperature interval ΔT. Usually one measures C_p, the heat capacity at constant pressure.

b) The "continuous-heating" method, in which one supplies heat to the specimen at a constant rate $dQ/dt = P$ and records the rate of temperature rise of the specimen. In this case the heat capacity is calculated by the formula

$$C = P \bigg/ \left(\frac{dT}{dt}\right).$$

Heat capacity studies in molecular biophysics and biology in the temperature range 0 ° to 100 °C have been mainly performed using continuous heating in so-called scanning calorimeters (see reviews [2–9]).

The continuous-heating method has a number of advantages. It is most conenient for studying heats of intramolecular phase transitions and in general for characterization of the temperature dependence of the heat capacity, since it permits high temperature resolution (small ΔT).

Modern scanning differential microcalorimeters operating in a continuous-heating regime allow precise measurements from microgram amounts, which is very important for biological studies where frequently only small amounts of material are available. The sensitivity of modern scanning microcalorimeters is about 10^{-7} W.

An advantage of the heat-pulse method is that the specimen is practically in a state of thermal equilibrium before the onset of heating and after it has finished [25, 29, 33, 45]. Hence, the quantity $\Delta Q/\Delta T$ characterizes the properties of the specimen in a state of equilibrium. When applying continuous heating, appropriate tests, such as application of different heating rates, have to be made in order to check whether the system is in equilibrium. The requirement of equilibrium conditions for calorimetric measurements becomes especially important for analysis of the shape of the heat capacity vs. temperature curves at low temperatures and for calculation of absolute values of the thermodynamic functions (change in enthalpy, entropy, Gibbs energy).

Specific heat is a fundamental thermodynamic quantity which permits calculation of the entropy S from heat capacity data according to the formula: $C = T(dS/dT)$. It is also connected with the quantum-level energies E_i of the system via the partition function Z:

$$S = k \cdot T \frac{\partial (T \ln Z)}{\partial T}; \quad Z = \sum_i \exp\left(-\frac{E_i}{kT}\right),$$

where k is the Boltzmann constant.

Knowing the temperature dependence of the heat capacity in a wide temperature range, one can determine the absolute values of the thermodynamic functions:

1. Entropy

$$S_T - S_0 = \int_0^{T_1} (C_p/T)dT + \varDelta H_1/T_1 + \int_{T_1}^{T_2} (C_p/T)dT$$

$$+ \varDelta H_2/T_2 + \ldots + \int_{T_n}^{T} (C_p/T)dT$$

2. Enthalpy

$$H_T - H_0^0 = \int_0^{T_1} C_p dT + \varDelta H_1 + \int_{T_1}^{T_2} C_p dT + \varDelta H_2 + \ldots + \int_{T_n}^{T} C_p dT$$

3. Gibbs energy

$$(G^0 - H_0^0)/T = (H_T - H_0^0)/T - (S_T - S_0^0),$$

where $\varDelta H_i$ and T_i are enthalpy and temperature of isothermal phase transitions, respectively, and $\varDelta S_i = \varDelta H_i/T_i$ refers to the entropy change at the i th phase transition.

For illustration purposes the low-temperature calorimeter [41, 45] of the Institute of Physics of the Academy of Sciences of the Georgian SSR is exhibited in Fig. 2. A distinctive feature of the calorimeter is that both methods, the heat-pulse method and the continuous-heating method can be employed. Both modes of application are controlled by one electronic system [16, 17].

The left-hand side of Fig. 2 shows schematically an absolute adiabatic calorimeter, which for low temperature measurements will be immersed into a cryostat

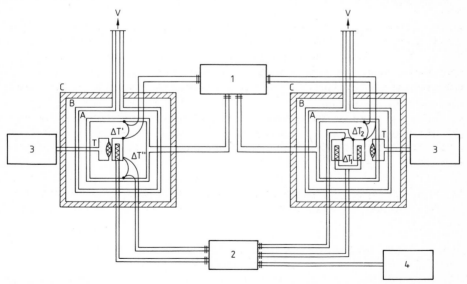

Fig. 2. Diagram of the calorimeter which allows study of thermal properties of biological macromolecules in the step-heating and continuous-heating mode over a broad range of temperatures. *A* adiabatic shield; *B* vacuuam jacket; *C* cryostat; *ΔT* detectors for temperature differences; *V* vacuum line [44, 45]; *1* shield thermoregulating systems; *2* cell thermoregulating system; *3* temperature indicating systems; *4* recorder

Table 1. Characteristics of commercial scanning microcalorimeters

Quality	Characteristic	Unit	Perkin Elmer DSC-2	Dupont 910 DSC	Daini Seikoshi SSC-50	Acad. Sci USSR DASM-4
Operation range	Volume of the cell	ml	0.03	0.03	0.07	0.5
	Temperature range	K	10–1000	100–1000	120–400	278–430
	Heating rates	K/min	0.3–320	0.5–100	0.01–5.0	0.1–2.0
Accurary	Relative error in the heat capacity determination	%	1.0	1.0	0.5	0.005

The table shows that heat capacity measurements using small samples or dilute solutions can be performed with a mean square error of below 1% over a wide temperature range.

containing liquid nitrogen or helium; the right-hand side of Fig. 2 is a schematic diagram of a vacuum-jacketed adiabatic differential scanning microcalorimeter. The new generation of adiabatic calorimeters is equipped with computer-controlled adiabatic shields and on-line data collection and treatment [review 18].

Table 1 gives some characteristics of different commercially available scanning microcalorimeters. Among these are no adiabatic heat pulse microcalorimeters which operate in the temperature range 4 to 400 K.

3 Experimental Results

3.1 Heat Capacity and Thermodynamic Properties of Amino Acid Residues over the Temperature Range 1.5–350 K

Data on the thermodynamic properties for practically all the amino acid residues [19–24] in the solid state over the temperature range 10–350 K (for L-alanine down to 1.5 K) are available.

Table 2 gives molar thermodynamic parameters: the isobaric heat capacity, C_p, the absolute entropy, S^0, and change of enthalpy $(H^0-H^0_0)/T$ and Gibbs energy $-(G^0-H^0_0)/T$ for all the amino acids at the temperature 298.15 K.

3.2 Heat Capacity and Thermodynamic Properties of Polypeptides over the Temperature Range 1.5–300 K

Precision measurements of the heat capacity of poly-L-alanine (PLA) in two conformations: α-helical and β-pleated, have been made by Delhaes and his associates [32] in the Paul Pascal Research Center (Talence, France) in 1970, and by Finegold and Cude [26–29] at the University of Colorado, USA in 1972 (in the range 1.5–20 K). They were repeated by the French group in 1975, but over a far broader temperature range (1.5–300 K) [33]. The specimens contained 85±15% of polypeptide chains in α-form of 8515% in the β-form. They calculated the heat capacity by the classical Debye model for different values of the parameter τ (the

Table 2. Thermodynamic values for amino acids in a solid state at 298.15 K
$(J \, mol^{-1} \, K^{-1})$. [19–24]

Amino acid	C_p	$S^0 = \int_0^{298}(C_p/T)dT$	$\dfrac{H^0 - H_0^0}{T}$	$\dfrac{-(G^0 - H_0^0)}{T}$
Gly	99.25	103.56	54.29	49.27
Ala	122.31	129.26	76.23	62.04
Val	168.91	178.95	92.55	86.4
Leu	201.05	211.9	106.12	105.78
Ileu	188.37	208.09	104.23	103.85
Asp	155.26	170.2	86.61	83.85
Glu	175.14	188.29	96.53	91.76
Ser	135.63	149.23	76.02	73.21
Thr	166.18	–	–	–
Lys	239.02	264.6	132.78	131.82
Arg	261.08	286.45	144.79	141.86
His	249.65	276.23	138.81	137.43
Phe	203.10	213.74	107.2	106.53
Tyr	216.54	214.11	11.98	102.1
Try	238.27	251.16	124.87	126.29
Cys	262.04	280.71	144.08	136.67
Met	290.17	231.57	117.79	113.78
Pro	151.24	164.13	82.01	102.13

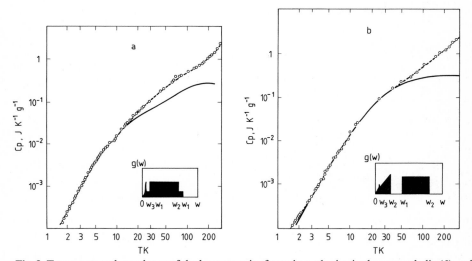

Fig. 3. Temperature dependence of the heat capacity for poly-L-alanine in the pure α-helix (*3*) and pure β-structure (**b**) conformations. *Solid lines* calculated by the Tarasov model. The vibrational spectrum for the two forms is also given as an *insert* [33]. *w* vibration frequency

number of vibrators). Figure 3 shows the temperature dependence of the heat capacity of PLA for the "pure" α- and β-conformations in the temperature range 1–300 K.

Fanconi and Feingold [30] at the Institute for Materials Research of the National Bureau of Standards (Washington) and at Drexel University (Philadel-

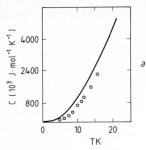 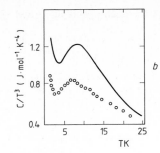

Fig. 4. Experimental (*dots*) and calculated relationships (*solid lines*) $C_p = f(T)$ (**a**) and $C_p/T^3 = f(T)$ (**b**) for polyglycine II [30]

phia) have obtained interesting results in studying the low-temperature heat capacity of homopolymers. They studied the heat capacity of polyglycine II. Figure 4 shows the experimental and calculated variations of the heat capacity $C_p = f(T)$ and $C_p/T^3 = f(T)$ for PG-II. The presented temperature dependence shows fair qualitative agreement of the experimental and calculated values. Although no satisfactory explanation can be offered, the authors consider it improbable that the quantitative differences stem from experimental artifacts. The maximum at a temperature near 8 K was first noted for homopolymers. It corresponds to a frequency of ~ 5 cm^{-1}, and its physical origin is still unclear. In any case it can be stated that the PG-II conformation leads to features in the low-temperature heat capacity that have not been observed for poly-L-alanine in either the α- or the β-conformations. Hence they directly reflect the characteristics of packing of the polypeptide chains and the nature of the bonds of the triple-helix conformation of PG-II.

Thus heat capacity data of polypeptides [26–30, 32, 33] reveal peculiarities which reflect characteristic structural changes as a function of temperature. With the help of reference systems of known structure, it may be possible in the future to derive some structural characteristics of unknown polypeptides on the basis of heat capacity data.

3.3 Heat Capacity and Thermodynamic Properties of Globular Proteins over the Temperature Range 10–350 K

Globular proteins were among the first biopolymers studied by heat capacity measurements. These studies focused on the thermodynamic parameters characterizing the phase transitions proteins can undergo in dilute aqueous solution (see [1–9]). Further interest concerned the state of the water of hydration in solutions of biopolymers. This was studied by low-temperature calorimetry (in the temperature range from $-50°$ to $20 °C$ [58]). The first precision calorimetric measurements of the heat capacity of globular proteins in the solid state over a broad temperature range (10–350 K) were performed in Hutchens' laboratory in 1968–1970 [25].

The procedures of dehydrating and measuring the degree of hydration (see also chapter 11) of a biopolymer, as well as the problem of determining the nativity of the biopolymer after these procedures have been applied as has been discussed in detail [25]. The material was filled into the calorimeter, and the dished cap was soldered in place. The calorimeter was weighed in the presence of air and then flushed three times with dry helium. The weight of the hydrated sample was calculated from the weight of the filled (i.e., sample plus helium) sealed calorimeter minus the weight of the empty calorimeter, cap, and solder, applying corrections for the buoyancy of helium in air. The sample weight was determined to ~0.04% accuracy. After a series of heat capacity measurements, the cryostat was disassembled, the helium-filled calorimeter was weighed, and, after opening of the pinhole, sample drying began. Water removed from the sample was measured in two ways: (a) by weight loss of the calorimeter and (b) by the weight of the water which condensed in a freeze trap immersed in liquid nitrogen. By measuring the freezing point, it was shown that the removed fluid was pure water. The weight of the dry sample is calculated from the weight of the empty calorimeter, plus cap and solder, and a helium buoyancy correction.

Following these measurements, there was a period – 20 days during which the material was drying in vacuum at $<10^{-6}$ mm HG at room temperature (23 °– 25 °C). Approximately 80% of the water was removed in 2 days. Exponentially decreasing amounts were removed thereafter. In this way unhydrous materials were obtained.

Figure 5 shows the temperature dependence of the heat capacity of these proteins over a broad temperature range (10–350 K). It can be seen from the diagram that the temperature dependence of the solid state heat capacity is the same as for all measured globular proteins. It is worth mentioning that between approximately 40 K and 300 K the heat capacity is to a good approximation a linear function of temperature.

All thermodynamic functions of interest were calculated from the heat capacity data. Tables 3–10 summarize data on the thermodynamic properties of some globular proteins in the "solid" state for a wide range of temperatures.

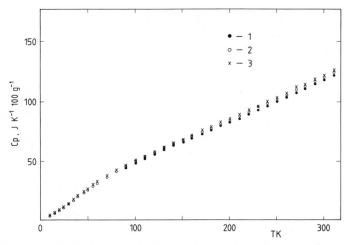

Fig. 5. Temperature dependence of the heat capacity for Zn-insulin (*1*), chymotrypsinogen-A (*2*) and serum albumin (*3*) [25, 39, 61].

Table 3. Thermodynamic properties of anhydrous bovine serum albumin $(J\ K^{-1}\ 100\ g^{-1})$ [61]

T, K	C_p^0	S^0	$\dfrac{H^0 - H_0^0}{T}$	$\dfrac{-(G^0 - H_0^0)}{T}$
10	1.854	0.707	0.188	0.188
15	4.236	1.892	1.348	0.544
20	7.062	3.491	2.415	1.076
25	10.14	5.396	3.688	1.708
30	14.33	7.526	5.069	2.461
35	16.61	9.833	6.518	3.315
40	19.72	12.26	8.008	4.249
45	22.72	14.75	9.502	5.249
50	25.72	17.30	11.00	6.300
55	28.62	19.89	12.49	7.397
60	31.38	22.50	13.98	8.523
70	36.64	27.74	16.88	10.86
80	41.61	32.96	19.69	13.26
90	46.25	38.13	22.42	15.71
100	50.57	43.21	25.04	18.18
110	54.71	48.22	27.56	20.91
120	58.77	53.16	29.99	23.19
130	62.66	58.02	32.35	25.68
140	66.47	62.83	34.66	28.16
150	70.28	67.52	36.90	30.63
160	74.09	72.21	39.11	33.09
170	77.86	76.81	41.28	35.52
180	81.63	81.33	43.41	37.94
190	85.35	85.85	45.50	40.34
200	89.12	90.33	47.59	42.74
210	92.84	94.77	49.69	45.12
220	96.53	99.17	51.74	47.47
230	100.30	103.56	53.75	49.81
240	104.06	107.91	55.76	52.16
250	107.96	112.23	57.77	54.46
260	111.93	116.54	59.78	56.76
270	115.95	120.85	61.78	59.06
280	120.01	124.99	63.79	61.32
290	124.20	129.43	65.80	63.63
300	128.38	133.70	67.81	65.89
310	132.61	137.97	69.82	68.15

It can be seen from Table 10 that the values of the specific heat capacity and entropy data at 298.15 K differ little for the different proteins studied so far. Moreover, since the values of the standard thermodynamic properties for all the amino acid residues and the amino acid compositions of all the listed proteins are known one can calculate the entropy change in the hypothetical reaction: N amino acid residues↔protein + (N+1)·H_2O. This makes it possible to estimate the entropy change in forming a peptide link by comparing the calculated and experimental data [25].

Bull and Breese [36] and Suurkuusk [34, 35] measured the specific heat capacity of proteins in the solid state at different water contents. Partial specific heat

Table 4. Thermodynamic properties of hydrated (2.14% H_2O) bovine serum albumin (J K^{-1} 100 g^{-1}) [61]

T, K	C_p^0	S^0	$\dfrac{H^0 - H_0^0}{T}$	$\dfrac{-(G^0 - H_0^0)}{T}$
10	1.679	0.619	0.456	0.163
15	4.039	1.733	1.243	0.486
20	6.965	3.278	2.294	0.984
25	10.540	5.237	3.600	1.637
30	13.709	7.438	5.019	2.419
35	16.970	9.795	6.492	3.303
40	20.118	12.27	7.999	4.270
45	23.186	14.81	9.519	5.299
50	26.208	17.42	11.04	6.379
55	29.105	20.05	12.55	7.501
60	31.918	22.70	14.04	8.657
70	37.306	28.03	16.99	11.05
80	42.36	33.35	19.84	13.50
90	47.09	38.62	22.61	16.00
100	51.49	43.79	25.28	18.52
110	55.67	48.89	27.85	21.06
120	59.78	53.92	30.34	23.58
130	63.71	58.86	32.76	26.11
140	67.69	63.75	35.11	28.62
150	71.71	68.52	37.42	31.12
160	75.77	73.30	39.68	33.61
170	79.78	78.03	41.94	36.08
180	83.72	82.67	43.95	38.54
190	87.74	87.32	46.36	40.99
200	91.76	91.92	48.53	43.35
210	95.78	96.49	50.65	45.84
220	99.79	101.07	52.78	48.22
230	103.85	105.57	54.92	50.65
240	107.91	110.09	57.05	53.04
250	111.98	114.57	59.15	55.38
260	116.12	119.05	61.28	57.77
270	120.39	123.49	63.38	60.11
280	124.57	127.97	65.51	62.45
290	128.84	132.40	67.60	64.80
300	133.11	136.84	69.74	67.10
310	137.34	141.28	71.83	69.45
273.15	121.69	124.91	64.04	60.86
298.15	132.32	136.00	69.32	66.68
310.15	137.38	141.32	71.87	69.49

capacity was derived from the experimental specific heat capacity, C_p, by means of Eq. (1):

$$(1 + W_2) \cdot C_p = C_{p,1} + W_2 \cdot C_{p,2}, \tag{1}$$

where $C_{p,1}$ and $C_{p,2}$ are the partial specific heats of the solvent and solute, respectively, and W_2 is the weight ratio between the masses of solute and solvent. In these measurements the protein is considered as the solvent. A plot of $(1 + W_2) \cdot C_p$ against W_2 gives $C_{p,1}$ as the ordinate intercept, and $C_{p,2}$ as the slope,

Table 5. Thermodynamic properties of anhydrous denatured bovine serum albumin $(J K^{-1} 100 g^{-1})$ [61]

T, K	C_p^0	S^0	$\dfrac{H^0 - H_0^0}{T}$	$\dfrac{-(G^0 - H_0^0)}{T}$
10	1.867	0.975	0.565	0.23
15	4.274	1.984	1.381	0.603
20	7.179	3.604	2.461	1.143
25	10.36	5.542	3.717	1.825
30	13.62	7.719	5.098	2.62
35	16.87	10.06	6.547	3.516
40	20.00	12.52	8.037	4.487
45	23.06	15.06	9.536	5.521
50	26.09	17.64	11.038	6.605
55	29.08	20.27	12.545	7.727
60	31.93	22.92	14.044	8.883
70	37.11	28.24	16.974	11.269
80	41.90	33.51	19.796	13.722
90	46.46	38.72	22.508	16.208
100	50.82	43.83	25.124	18.716
110	54.96	48.89	27.648	21.231
120	59.02	53.83	30.093	23.743
130	62.96	58.73	32.471	26.246
140	66.81	63.54	34.786	28.737
150	70.66	68.27	37.050	31.211
160	74.47	72.92	39.269	33.672
170	78.24	77.57	41.450	36.121
180	81.96	82.129	43.58	38.55
190	85.73	86.69	45.71	40.37
200	89.54	91.17	47.80	43.37
210	93.35	95.65	49.90	45.75
220	97.20	100.04	51.95	48.10
230	101.09	104.48	54.00	50.48
240	105.07	108.88	56.05	52.83
250	109.13	113.23	58.10	55.13
260	113.27	117.58	60.11	57.47
270	117.50	121.94	62.16	59.78
280	121.77	126.29	64.25	62.09
290	126.08	130.64	66.31	64.34
300	130.44	135.00	68.36	66.64
310	134.83	139.35	70.41	68.90
273.15	118.84	123.44	62.83	60.49
298.15	129.60	134.16	67.98	66.22
310.15	134.87	139.39	70.45	68.94

if the curve is linear in the range studied. From Fig. 6, which shows the results for lysozyme, it can be seen that a small negative deviation from a straight line was observed in the range $W_2 = 0.05$ to $W_2 = 0.2$. Similar patterns were found for chymotrypsinogen and ovalbumin (see also Fig. 7).

The comparison of the specific heat capacities of different proteins with those of their component amino acids indicates that it is possible to calculate specific heat capacities for proteins from their amino acid composition within a few per-

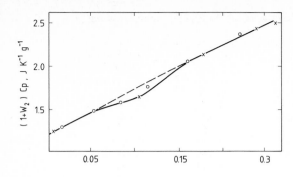

Fig. 6. Specific heat of solid lysozyme as a function of the water content [35] w according to Mrevlishvili, unpublished data

Table 6. Thermodynamic properties of anhydrous chymotrypsinogen $(J K^{-1} 100 g^{-1})$ [25]

T, K	C_p^0	S^0	$\dfrac{H^0 - H_0^0}{T}$	$\dfrac{-(G^0 - H_0^0)}{T}$
10	1.875	0.778	0.557	0.222
15	4.345	1.980	1.390	0.594
20	7.292	3.629	2.491	1.139
25	10.53	5.597	3.772	1.829
30	13.88	7.815	5.178	2.637
35	17.15	10.21	6.656	3.550
40	20.34	12.70	8.167	4.533
45	23.39	15.27	9.691	5.584
50	26.41	17.90	11.21	6.685
55	29,35	20.55	12.73	7.824
60	32.21	23.23	14.23	8.996
70	37.50	28.60	17.18	11.42
80	42.28	33.92	20.03	13.90
90	46.88	39.17	22.76	16.41
100	51.15	44.33	25.38	18.95
110	54.88	49.39	27.92	21.49
120	59.36	54.38	30.37	24.02
130	63.33	59.32	32.75	26,55
140	67.18	64.13	35.07	29.06
150	70.99	68.90	37.34	31.55
160	74.76	73.59	39.56	34.04
170	78.61	78.24	41.75	36.50
180	82.46	82.84	43.91	38.95
190	86.27	87.40	46.05	41.388
200	90.12	91.92	48.14	43.79
210	94.02	96.44	50.23	46.21
220	97.83	100.88	52.32	48.60
230	101.72	105.32	54.38	50.94
240	105.61	109.72	56.43	53.29
250	109.55	114.11	58.48	55.63
260	113.61	118.51	60.53	57.98
270	117.63	122.86	62.45	60.32
280	121.73	127.21	64.59	62.62
290	125.91	131.57	66.64	64.92
300	130.14	135.88	68.69	67.23
310	134.33	140.23	70.74	69.49

Table 7. Thermodynamic properties of hydrated (10.7% H_2O) chymotrypsinogen ($J\,K^{-1}\,100\,g^{-1}$) [25]

T, K	C_p^0	S^0	$\dfrac{H^0-H_0^0}{T}$	$\dfrac{-(G^0-H_0^0)}{T}$
10	1.779	0.745	0.532	0.213
15	4.186	1.892	1.327	0.565
20	7.313	3.512	2.419	1.092
25	10.85	5.521	3.751	1.771
30	14.44	7.815	5.232	2.583
35	18.07	10.31	6.806	3.508
40	21.62	12.96	8.439	4.521
45	25.04	15.71	10.09	5.613
50	28.30	18.51	11.75	6.760
55	31.47	21.36	13.40	7.958
60	34.56	24.23	15.04	9.197
70	40.47	30.01	18.25	11.75
80	45.96	35.77	21.38	14.40
90	51.15	41.49	24.40	17.09
100	56.01	47.13	27.32	19.82
110	60.74	52.70	30.14	22.55
120	65.47	58.18	32.88	25.29
130	70.16	63.63	35.53	28.03
140	74.85	68.98	38.21	30.76
150	79.53	74.30	40.81	33.49
160	84.26	79.58	43.37	36.20
170	88.99	84.85	45.92	38.91
180	93.81	90.04	48.43	41.61
190	98.75	95.27	50.94	44.29
200	103.90	100.46	53.50	46.97
210	109.17	105.65	56.01	49.65
220	114.61	110.84	58.56	52.32
230	120.22	116.08	61.12	54.96
240	125.91	121.31	63.67	57.60
250	131.65	126.58	66.31	60,28
260	137.47	131.86	68.90	62.92
270	143.37	137.13	71.58	65.55
280	149.31	142.45	74.22	68.23
290	155.30	147.81	76.94	70.87
300	161.41	153.16	79.66	73.55
310	167.19	158.57	82.38	76.18
273.15	145.42	138.81	72.42	66.39
298.15	160.49	152.16	79.16	73.05
310.15	167.48	158.65	82.05	76.60

cent [25]. Using Eq. (2) the specific heat capacity, c_p^0, for a protein with n amino acids, molecular weight M, and i peptide chains can be calculated [35].

$$C_p^0 = \frac{\Sigma C_p(\text{amino acid}) - (n-i) \cdot 34.7}{M} \tag{2}$$

ΣC_p (amino acid) is the sum of the heat capacities of the amino acids, and $(n-i) \cdot 34.7$ ($J\,K^{-1}\,mol^{-1}$) is a correction for the formation of (n-i) peptide bonds. From the fact that the specific heat capacity values, c_p^0, for amino acids [19–24]

Table 8. Thermodynamic properties of anhydrous bovine zinc insulin $(J K^{-1} 100 g^{-1})$ [25]

T, K	C_p^0	S^0	$\dfrac{H^0 - H_0^0}{T}$	$\dfrac{-(G^0 - H_0^0)}{T}$
10	1.998	0.820	0.594	0.226
15	4.429	2.068	1.448	0.619
20	7.309	3.370	2.550	1.180
25	10.42	5.693	3.809	1.884
30	13.65	7.878	5.176	2.696
35	16.87	10.22	6.618	3.604
40	19.96	12.68	8.096	4.584
45	22.91	15.20	9.578	5.622
50	25.75	17.77	11.05	6.710
55	28.49	20.35	12.52	7.832
60	31.15	22.94	13.96	8.983
70	36.23	28.13	16.76	11.35
80	41.02	33.28	19.51	13.77
90	45.59	38.38	22.16	16.22
100	49.90	43.41	24.72	18.69
110	53.96	48.35	27.20	21.16
120	57.81	53.20	29.59	23.63
130	61.53	57.98	31.90	26.09
140	65.30	62.71	34.15	28.54
150	69.07	67.31	36.35	30.97
160	72.84	71.91	38.52	33.38
170	76.56	76.44	40.65	35.78
180	80.25	80.91	42.74	38.16
190	83.85	85.35	44.83	40.53
200	87.44	89.75	46.84	42.86
210	91.09	94.10	48.89	45.21
220	94.47	98.41	50.86	47.55
230	98.62	102.72	52.87	49.86
240	102.47	106.99	54.84	52.12
250	106.37	111.26	56.85	54.42
260	110.26	115.49	58.81	56.68
270	114.19	119.72	60.78	58.94
280	118.09	123.95	62.79	61.20
290	122.11	128.17	64.76	63.42
300	126.17	132.36	66.72	65.64
310	130.31	136.59	68.69	67.86
273.15	115.41	121.06	61.41	59.65
298.15	125.41	131.61	66.35	65.22
310.15	130.35	136.63	68.73	67.90

(see Table 2) differ only slightly from each other and from those derived from Eq. (2), it was concluded that c_p^0 values for most proteins can be expected to be close to $1.2 \, J \, K^{-1} \, g^{-1}$.

An attempt was made to analyze the obtained C_p^0 values by use of model compound data and simple additivity rules. The different contributions to the C_p^0 values were separated into several groups [35]: (A) back bone, (B) polar groups, (C) nonpolar groups and (D) ionization. The group (B) consists of two C_p effects: the C_p^0 values for nonsolvated, nonpolar groups and a contribution account-

Table 9. Thermodynamic properties of hydrated (4.0% H_2O) bovine zinc insulin ($J\,K^{-1}\,100\,g^{-1}$) [25]

T, K	C_p^0	S^0	T, K	C_p^0	S^0
10	1.934	0.800	160	75.26	73.80
15	4.341	2.018	170	79.24	78.49
20	7.288	3.663	180	83.18	83.14
25	10.50	5.630	190	87.07	87.74
30	13.89	7.845	200	91.00	92.30
35	17.24	10.24	210	94.98	96.82
40	20.46	12.75	220	99.00	101.34
45	23.57	15.34	230	103.06	105.82
50	26.56	17.98	240	107.12	110.35
55	29.44	20.65	250	111.26	114.78
60	32.21	23.33	260	115.49	119.22
70	37.48	28.70	270	119.80	123.65
80	42.40	34.03	280	124.16	128.09
90	46.97	39.29	290	128.51	132.53
100	51.28	44.45	300	132.91	136.97
110	55.42	49.56	310	137.30	141.36
120	59.44	54.54	273.15	121.18	125.04
130	63.42	59.44	298.15	132.07	136.13
140	67.35	64.30	310.15	137.34	141.44
150	71.33	69.07			

Table 10. Thermodynamic properties of protein in a solid state at 298.15 K ($J\,K^{-1}\,g^{-1}$)

Protein	C_p^0	S^0	$\dfrac{H^0-H_0^0}{T}$	$\dfrac{-(G^0-H_0^0)}{T}$
Zn-Insulin	1.254	1.316	0.663	0.663 [a]
Serum Albumin	1.275	1.329	0.675	0.654 [a]
A-Chymotrypsinogen	1.293	1.351	0.681	0.669 [a]
Chymotrypsinogen	1.223 [b]	–	–	–
Lysozyme	1.192 [b]	–	–	–
Ovalbumin	1.231 [b]	–	–	–
Collagen	1.210	1.240	0.631	0.609 [c]

[a] According to Hutchens et al. [25, 61].   [c] According to Mrevlishvili [45].
[b] According to Suurkuusk [35].

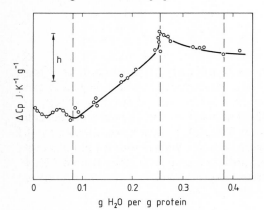

Fig. 7. Apparent specific heat capacity of lysozyme (for 298 K) as a function of water content (unit $h = 0.2\ J\ K^{-1}\ g^{-1}$.) This function is a measure of the excess heat capacity of the system per gram protein [50]

ing for the transfer of the groups to aqueous solution, $\alpha \Delta C_p$, where α is the value for the degree of solvation of the nonpolar groups ($\alpha = 1$, at complete solvation) [35]. The estimated molar contributions from the different groups are summarized in Table 11. The $C_{p,2}$ can be written as the sum of all the different contributions [35]:

$$C_{p,2} = C_p(\text{back bone}) + C^0_{p,2}(\text{polar})$$
$$+ C^0_p(\text{nonpolar}) + \alpha \Delta C_p(\text{solv.}) + \Delta C_p(\text{ion}). \tag{3}$$

Table 11. Contributions from the different groups to the total heat capacity of a protein in aqueous solution [35]

A

Contributions from the back bone: C^0_p (back-bone)
C^0_p (back-bone) $= n \cdot 99.20 - (n-i)34.7 \, (\text{J k}^{-1} \text{mol}^{-1})$

B

Contributions from solvated polar groups: $C^0_{p,2}$ (polar)

Alcohol	$C^0_{p,2}$ (polar) $= 64 \, \text{J K}^{-1} \text{mol}^{-1}$
Alcohol (Phenol)	$- \ 30$
Amino	58
Carboxylic	87
Amido	86
Guanidino	140
Imidazole	110

C

Contributions from nonpolar groups:	$C^0_{p,2} =$
	$C^0_{p,2}(\text{nonpolar}) + C_p(\text{solv})$
Noncyclic aliphatic (CH_2 increment)	$C^0_{p,2}(\text{nonpolar}) = \ 23.5 \, \text{J K}^{-1} \text{mol}^{-1}$
Cyclic aliphatic (CH_2 increment)	17.3
Aromatic ($-CH=$ increment)	13.5
Benzene	80.8
Indole	115.9
Monosulfide	114.3
Disulfide	17.4
Noncyclic aliphatic (CH_2 increment) $C_p(\text{solv}) =$	$65.7 \, \text{J K}^{-1} \text{mol}^{-1}$
Cyclic aliphatic (CH_2 increment)	54.5
Aromatic ($-CH=$ increment)	43.2

D

Contributions from ionization:	$\Delta C_p(\text{ion})$
	$\Delta C_p(\text{ion}) = 167 \, (\text{J K}^{-1} \text{mol}^{-1}$

In order to fit the experimental results with data calculated from Eq. (3) (cf. Table 12) α values of 0.27 and 0.20 were found for lysozyme and chymotrypsinogen respectively. Thus an attempt has been made to calculate the entropy for proteins in the solid state from their amino acid composition and from experimental data on the heat capacity of amino acids and proteins in the solid state. However, evidently this approach is inapplicable in practice for calculating the entropy of proteins in dilute aqueous solutions, in which stabilization of the spatial structure of the protein is attained through a balance of the forces of the intramolecular and intermolecular bonds with the solvent.

Table 12. Total contributions from the different groups to the heat capacity of lysozyme and chymotrypsinogen[a] $(J K^{-1} mol^{-1})$ [35]

	Lysozyme	Chymotrypsinogen
Observed $C_{p,2}^0$	21 385	39 243
Calculated contributions from:		
Back bone, C_p^0 (back bone)	8 354	15 835
Polar groups, C_p^0 (polar)	5 289	8 040
Nonpolar groups, C_p^0 (nonpolar)	6 927	12 867
Solvation of nonpolar groups, ΔC_p (solv)	18 669	35 613
Ionisation, ΔC_p (ion)	− 4 175	− 4 509
Complete solvated $C_{p,2}$ (solv) $(\alpha = 1)$	35 064	67 846
Degree of solvation (α)	0.27	0.20

Subsequent experiments [45, 59] showed that the deviations in the heat capacities for different globular proteins in the solid state $(\delta C^P_{,s})$ are far smaller than the difference between the heat capacities of a given protein existing in the helical and coil states in dilute aqueous solutions (ΔC_p). Thus, $\delta C_{p,s} = \pm 0.04 \, J K^{-1} g^{-1}$, whereas $\Delta c_p = +0.4 \, J K^{-1} g^{-1}$ [59].

Analysis of the experimental data [44, 45] shows that the formation of the tertiary structure of any biopolymer depends on the formation of a few hydration layers on the macromolecules. In other words, each biopolymer has its critical value of the volume fraction of solvent at which it adopts its "equilibrium"-ordered form [44, 45].

Hence, in measurements of the heat capacity of biopolymers in the solid state, one must carefully control the hydration of the macromolecule. The measurements should be performed under conditions in which the biopolymer contains the minimal amount of water required to maintain the given conformation in the native state. The quantity of this water (the so-called bound-unfreezable water) can be established by measuring thermal effects in the region of the ice-water phase transition, by means of the low-temperature scanning microcalorimetry (see Table 13).

This will be illustrated in the following sections by heat capacity data of collagen and DNA.

3.4 Thermodynamic Properties of Fibrous Collagen in the Helical and Coil States at Temperatures 4–400 K

Studies have been performed on the connective-tissue protein collagen in the helical and coil states over the temperature range 4–400 K [39, 42, 43].

The collagen molecule consists of three polypeptide chains that form a triple superhelix stabilized by intramolecular hydrogen bonds. In addition, the collagen macromolecule is stabilized by intrinsic water in the triple helix [45]. Hence, in determining the thermodynamic properties of helical and coiled collagen not only the order-disorder transition of the polypeptide chains themselves, but also the order-disorder transition in the solvent (water) must be taken into account [45].

Table 13. Microcalorimetric data on hydration of biomacromolecules [46, 55, 56]

Object	Hydration	
	g H_2O/g polymer	mol H_2O/mol polymer
Globular protein		
Serum albumin	0.31	1120
Ovalbumin	0.32	820
Hemoglobin	0.34	1280
Lysozyme	0.40	311
Ribonuclease	0.32	230
Fibrous protein	± 0.01	
Collagen	0.48	9600
Nucleic acids		
DNA of calf thymus (42% GC)	0.62	24 (per base pair)
DNA of *M. Lysodeikticus* (71% GC)	0.51	20 (per base pair)
poly-[d(a-T)]	0.75	28 (per base pair)
poly-[d(G-C)]	0.41	16 (per base pair)

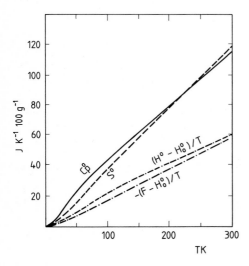

Fig. 8. Temperature dependence of the thermodynamic parameters of native collagen [39, 45]

Figure 8 shows the temperature dependence of the heat capacity of native dehydrated collagen over the temperature range 4–320 K. These results agree well with previous measurements (see Tables 14 and 15).

From the heat capacity data the other thermodynamic functions the entropy S^0, the enthalpy change $(H^0 - H_0^0)/T$, and the Gibbs energy $-(G^0 - H_0^0)/T$ have been calculated. Figure 8 shows the variation with temperature of these quantities. Figure 9 shows the specific heat capacity of collagen as a function of water content. Figure 10 shows specific heat capacity of wool keratin water systems as a function of relative humidity [51]. Figure 11 shows the temperature dependence of the difference between the heat capacity of hydrated (0.35 g H_2O/g protein) denatured (coil) and hydrated native (helix) collagen in the temperature range 4–310 K [39, 45]: $\Delta C_p = C_{coil} - C_{helix}$.

Table 14. Thermodynamic properties of anhydrous collagen $(J\,K^{-1}\,100\,g^{-1})$ [39, 61]

T, K	C_p^0	S^0	$\dfrac{H^0 - H_0^0}{T}$	$\dfrac{-(G^0 - H_0^0)}{T}$
10	1.775	0.632	0.469	0.163
15	4.106	1.783	1.285	0.498
20	6.894	3.336	2.327	1.005
25	10.01	5.207	3.550	1.653
30	13.18	7.313	4.893	2.419
35	16.34	9.582	6.304	3.278
40	19.43	11.97	7.752	4.215
45	22.38	14.42	9.213	5.212
50	25.29	16.94	10.68	6.258
55	28.08	19.48	12.13	7.346
60	30.78	22.04	13.57	8.464
70	35.95	27.18	16.40	10.77
80	40.73	32.29	19.16	13.14
90	45.12	37.35	21.80	15.55
100	49.23	42.32	24.33	17.98
110	53.25	47.18	26.78	20.41
120	57.14	51.99	29.15	22.84
130	60.91	56.72	31.45	25.27
140	64.59	61.37	33.68	27.68
150	68.19	65.93	35.87	30.08
160	71.75	70.45	38.00	32.46
170	75.26	74.93	40.08	34.83
180	78.86	79.32	42.15	37.18
190	82.46	83.68	44.16	39.51
200	86.06	87.99	46.17	41.83
210	89.62	92.26	48.14	44.12
220	93.14	96.53	50.11	46.42
230	96.70	100.76	52.07	48.68
240	100.30	104.94	54.00	50.94
250	103.98	109.09	55.92	53.16
260	107.75	113.27	57.85	55.46
270	111.51	117.37	59.78	57.64
280	115.37	121.52	61.66	59.86
290	119.17	125.62	63.58	62.04
300	122.98	129.72	65.51	64.21
310	126.79	113.83	67.44	66.39
273.15	112.73	118.71	60.36	58.35
298.15	122.27	128.97	65.13	63.84
310.15	126.83	133.91	67.44	66.43

Three regions in the curves are worth mentioning:

1. The difference in heat capacities between the coil and the helix gradually rises from 4 K, reaches a maximum at 15 K, and then begins to decline (Fig. 11 A).
2. The heat capacity of collagen in the coil state becomes larger than that of the native protein at about 70 K. At 120 K, ΔC_p reaches a maximum.
3. At temperature of 230–240 K, the heat capacities of the native and of the denatured protein become equal, then ΔC_p is negative up to 310 K, i.e., up to the

Table 15. Thermodynamic properties of hydrated (13.53% H_2O) collagen $(J K^{-1} 100 g^{-1})$ [39]

T, K	C_p^0	S^0	$\dfrac{H^0 - H_0^0}{T}$	$\dfrac{-(G^0 - H_0^0)}{T}$
10	1.536	0.548	0.406	0.142
15	3.897	1.595	1.159	0.44
20	6.949	3.119	2.210	0.908
25	10.41	5.040	3.504	1.536
30	14.00	7.254	4.952	2.302
35	17.69	9.691	6.509	3.181
40	21.27	12.29	8.133	4.157
45	24.54	14.99	9.774	5.212
50	27.87	17.74	11.42	6.325
55	31.11	20.55	13.06	7.489
60	34.18	23.40	14.7	8.898
70	39.96	29.10	17.90	11.06
80	45.46	34.80	21.00	13.80
90	50.65	40.45	24.01	16.45
100	55.55	45.96	26.84	19.13
110	60.28	51.49	29.66	21.82
120	64.97	56.93	32.41	24.52
130	69.61	62.29	35.09	27.22
140	74.38	67.65	37.72	29.31
150	78.91	72.92	40.31	32.61
160	83.59	78.15	42.86	35.29
170	88.32	83.39	45.42	37.96
180	93.14	88.53	47.93	40.63
190	98.12	93.72	50.44	43.28
200	103.18	98.87	52.95	45.92
210	108.38	104.02	55.46	48.60
220	113.65	109.21	57.98	51.24
230	119.01	114.36	60.53	53.83
240	124.57	119.55	63.08	56.47
250	130.35	124.74	65.64	59.11
260	136.30	129.97	68.23	61.74
270	142.41	135.25	70.74	64.38
280	148.64	140.52	73.55	66.98
290	155.09	145.84	76.23	69.61
300	161.70	151.24	78.99	72.25
310	168.49	156.64	81.75	74.89
273.15	144.37	136.88	71.71	65.18
298.15	160.49	150.23	78.49	71.75
310.15	168.57	156.72	81.79	74.93

temperature of onset of the anomalous growth in the heat capacity involving the melting of the native protein.

A completely different pattern is observed in studying the heat capacity difference for dehydrated specimens in the helical and coiled states. In this case $\Delta C_p = f(T)$ has the form depicted in Fig. 11 B. The difference between the entropy of hydrated collagen in the coil state and that of the protein in the helical state is positive throughout the temperature range. At 298 K it amounts to $\Delta S = 33.6$ J K^{-1} g^{-1} (where $\Delta S = S_{coil} - S_{helix}$).

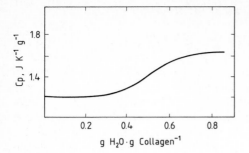

Fig. 9. Specific heat capacity of collagen as a function of water content [44, 45]

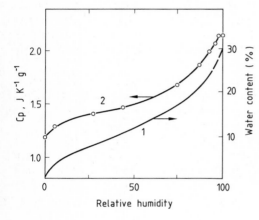

Fig. 10. Plots of *1* equilibrium water content of wool at 293 K against relative humidity (ordinates at right); *2* specific heat for wool keratin against relative humidity (ordinates at *left*) [51]

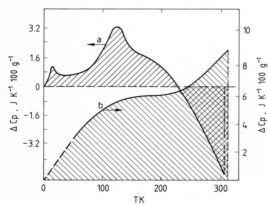

Fig. 11. Temperature dependence of the differences of heat capacities between hydrated collagen specimens (**a**) in the helix and coil states ($\Delta C_p = C_{coil} - C_{helix}$). **b** $\Delta C_p = f(T)$ for dehydrated collagen [39, 45]

It appears to be important that the difference in the thermodynamic properties of a protein in the helical and coiled state stems to a considerable extent from the different character of the interaction of the polypeptide chains with water molecules. This is already manifested at temperatures remote from those of the helix coil transition.

3.5 Heat Capacity of DNA

It is emphasized again that control and knowledge of the degree of solvation is essential for proper analysis of heat capacity measurements at low temperatures. This is also pertinent to studies on DNA. A dependence of the conformation of DNA on water content was demonstrated already in the early X-ray diffraction studies of fibers [52] and is clearly demonstrated in recent studies of single crystals of DNA [53]. Any change in the conformation of DNA that results from changes in the external conditions of the medium arises either from direct action of the perturbing agents on the double helix (e.g., binding of ligands) or via the solvent molecules (change in ionic strength, pH, polarity, pressure, temperature, etc.). Hence the study of the state of the solvent molecules directly adjacent to the DNA macromolecule, i.e., the hydration of the double helix, is especially significant. Heat capacity studies of DNA [44–48] as a function of hydration have yielded data on the mode of interaction of DNA with solvent molecules and on the thermal properties of this macromolecule.

Figure 12 shows the temperature dependence of the heat capacities of native DNA at different water contents. (Here and henceforth we report the heat capacity data averaged over equal temperature intervals.)

Inspection shows that a water content of 18% H_2O does not lead to a deviation from a linear variation of the heat capacity of DNA or to the appearance of heat absorption in the region of the phase transition of ice. This indicates that this amount of water is firmly bound to the DNA molecules. When the water con-

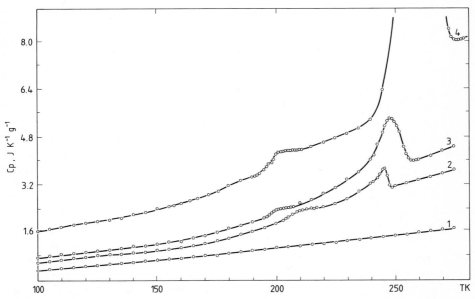

Fig. 12. Temperature dependence of the specific heat capacity of native DNA at different water contents: *1* 18%; *2* 31%; *3* 37%; *4* 61% H_2O by weight [41]

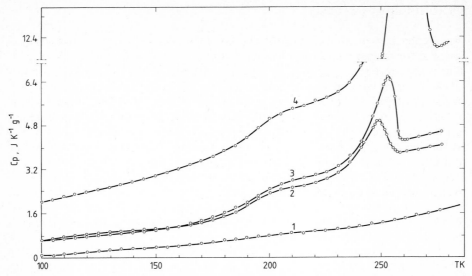

Fig. 13. Temperature dependence of the specific heat capacity of denatured DNA at different water contents: *1* 6%; *2* 35%; *3* 39%; *4* 68% H_2O by weight [41]

tent was increased to 31% (curve 2 in Fig. 12), a slight change in the heat capacity curve was observed in the temperature range 200–220 K, and then a heat absorption due to melting of a certain amount of water that had crystallized upon cooling. The temperature corresponding to the maximum of the heat absorption peak at the given values of water concentration is 245 K. This is 28 K lower than the melting temperature of pure water. Further increase in the water content manifests even more clearly the heat anomaly at 190–210 K. At the same time it shifts the maximum of the heat absorption peak toward 273 K.

Figure 12 shows that the melting of water in a DNA-H_2O system at low water concentration loses the characteristics of a sharp phase transition; it degenerates into a broad transition of low cooperativity. It should be mentioned that the temperature interval of the extended transition and the character of the heat absorption curve depend (at a particular water content) on the conformation of the biopolymer [44, 45]. It has been shown [57] that the reason for the degeneration of the first-order phase transition is not the small amount of solvent (since decomposition into two phases, i.e., a sharp phase transition, is observed even in systems of several hundred particles), but the interaction of the water molecules with the biopolymer and its effect on the structure of the solvent. The temperature dependence of the heat capacity of denatured DNA (curve 1 in Fig. 13) is considerably different from that of native DNA (cuves 2 and 3 in Fig. 13). In particular, the heat anomaly in the temperature range 200–210 K is changed. This anomaly is assumed to reflect transitions involving the vitrification of a certain fraction of the water [44, 45]. Heat capacity of denatured DNA existing in a random coil state varies very smoothly in this temperature interval without a sharp jump. Here the shape of the heat capacity curve (within the limits of experimental error) no

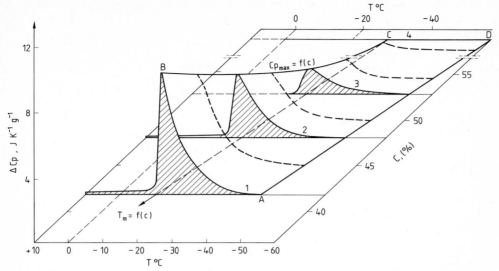

Fig. 14. Apparent excess specific heat capacity (ΔC_p) vs. temperature of poly-[d(A-T)] aqueous solution at ice-water phase transition temperature region [55, 56]. Note: Δc_p here represents a different measurement than in Fig. 11

longer depends on the thermal history of the specimens, and in particular, on the rates of cooling of the solutions.

Recently high sensitivity low temperature scanning microcalorimetry (LTDSC) has been applied to solutions of coiled and helical DNA, to elucidate the interactions of DNA with water and ions (Mrevlishvili et al. [42, 54–56].

Figure 14 shows the microcalorimetric recordings of the heat absorption observed in the region of the ice water phase transition during heating of frozen poly-[d(A-T)]-H_2O solutions of differing H_2O/poly-[d(A-T)] ratio. The microcalorimetric data allowed us to obtain precise results on the phase transition temperature of the water and the enthalpy of the "degenerate" phase transition. The critical values of the polymer concentration at which all the water existing in the system exists in the nonfreezable (bound) hydration state were established.

Figure 15 shows the heat capacity of DNA as a function of water content at 298 K. One sees that complete formation of the hydration shell ($\sim$0.6 g H_2O g^{-1} DNA) changes the thermal behavior of DNA.

Verkin and associates [49] have studied the heat capacity of moist DNA films in the temperature range 4–200 K. The authors compare the experimental heat capacity curve with theoretical predictions of the theory of heat capacity of chain and layer structures at low temperatures [49, 60]. However, for an unambiguous conclusion of the "conservation of the secondary structure of DNA in dry films", further studies at different water contents up to values that correspond to a complete hydration layer on the double helix are required.

Andronikashvili et al. [47, 48] and Mrevlishvili et al. [48] obtained the experimental data on the heat capacity of natural DNA fibers [47, 48] and synthetic polynucleotide solutions [55, 56] at different humidities, within a wide tempera-

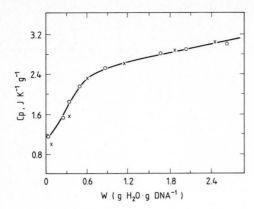

Fig. 15. Specific heat capacity of DNA as a function of water content, w, [36]

Table 16. Low temperature heat capacity of DNA at different water content (n) $(J K^{-1} mol^{-1})$; n in mol H_2O/mol base pair [47]

T, K	C_p n=0–2	C_p n=8–10	C_p n=28	C_p n=42
5	2.10	2.12		
6	2.80	2.51	2.10	3.50
7	3.57	2.80	3.50	4.20
8	4.20	4.22	4.55	6.30
9	5.60	6.31	7.00	7.70
10	6.30	8.40	9.10	9.10
11	7.70	10.50	10.50	11.20
12	10.51	12.60	14.00	14.00
13	12.60	15.40	16.10	16.80
14	14.71	18.20	18.90	20.30
15	16.09	21.00	22.40	24.50
18	25.91	32.20	35.00	39.90
19	30.78	35.70	41.30	45.50
20	34.32	39.20	46.20	51.80
21	38.49	42.00	52.50	58.10
22	40.58	45.50	56.70	63.70
23	44.10	49.00	61.60	68.60
24	49.00	53.90	67.90	74.90
25	53.87	58.10	73.50	80.50
26	58.80	63.00	79.45	86.10
27	63.01	67.20	85.40	93.10
28	68.60	74.20	91.70	99.40
29	72.12	80.50	94.50	104.30
30	75.59	54.70	103.60	111.30
35	94.48	112.00	131.60	142.10
40	114.11	136.50	157.50	172.90
45	136.50	159.60	184.10	200.90
50	161.00	186.20	213.50	228.90
55	183.41	209.30	240.10	256.20
60	198.80	231.70	264.60	277.20
65	212.08	252.00	287.00	294.70
70	226.81	269.50	306.60	315.00
75	242.9	287.00	328.30	337.40

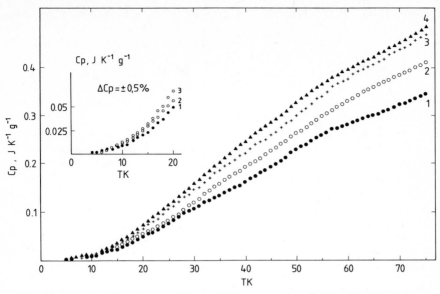

Fig. 16. Temperature dependence of the specific heat capacity of native DNA in the low-temperature region at different water contents [47, 48]: n: mol H_2O/mol base pair. 1 n = 0–2; 2 n = 8–10;
3 = 23; 4 = 42

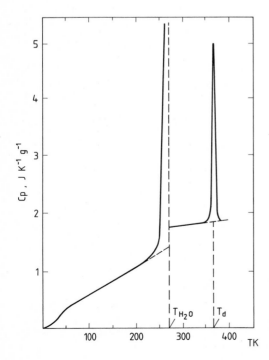

Fig. 17. Temperature dependence of specific heat capacity of DNA in the gel state
in the temperature region 4–400 K [47,
48]. Water content n = 42 mol H_2O/mol
base pair; T_{water}: ice melting temperature
in gels; the parameters of thermal
denaturation of DNA gels (in salt-free
conditions): T_d = 363 K; ΔH_d = 32.7 kJ/
mol base pair; ΔT_d = 4.8 K [47, 48]

ture interval, including the liquid helium temperature region [47, 48]. It has been shown [55, 56] that DNA hydration depends on guanidine cytosine base pairs (GC)-content. The hydration parameters are the following: $n_\Sigma = \{28.0–0.12(\%GC)\}$ and $n_s = \{12.0–0.06(\%GC)\}$, where in $_\Sigma$ is the amount of water molecules in the fully hydrated shell, n_s the number of water molecules involved in the inner ("structural") hydration layers of the molecule; (n is expressed in moles of water per mole of base pair. It has been found that the shape of the low temperature heat capacity curve (i.e., the exponent m in the usual expression $C_p = AT^m$) depends on the degree of DNA hydration (see Fig. 16, Table 16, and details in [47, 48]). Thus, the direct correlation between the DNA hydration parameters (n_Σ, n_s), GC-content and its thermal properties is found. Figure 17 represents the temperature dependence of the heat capacity of a DNA gel in the temperature range 2–400 K [48]. The heat capacity anomaly at 265 K and the heat absorption peak at 363 K are caused by the phase transition in the solvent (water) and by the intramolecular melting of DNA double helix respectively [48].

4 Concluding Remarks

In addition to providing heat capacity data of biopolymers for a wide temperature range, the various unsolved problems of interpretation and physical understanding of c_p-data of biopolymers have been mentioned. There is certainly a lack of understanding of the processes on the basis of quantum mechanics. However, even on a less sophisticated level there are several open questions and particularly a deficiency of good data. Although modern microcalorimetry is able to measure small samples with high precision, the accuracy is often low due to poorly controlled biochemical treatment of the samples.

Proper isolation, purification, and characterization of the specimens (oriented moist films, supermolecular structures of biopolymers, fibers, etc.) is essential for the physical measurements. Often the method of preparation of the specimens is decisive for the quality of the results. Quantitative determination of the ratio water/biopolymer is of utmost importance. The minimal amount of water needed to preserve the native structure of the macromolecules (the so-called bound, or unfreezable water) can be established by measuring the thermal effects in the region of the ice water phase transition.

References

1a. Sturtevant JM (1972) In: Hirs CHW, Timasheff SN (eds) Methods in enzymology, vol 26. Academic Press, New York, p 227
  b. Sturtevant JM (1974) Annu Rev Biophys Bioeng 3:35
2. Andronikashvili EL (1972) Biofizika 17:1068–1178
3. McKnight RP, Karasz FE (1973) Thermochim Acta 5:339
4. Privalov PL (1974) FEBS Lett 40:140–153
5. Rialdi G, Biltonen RL (1972) In: Skinner HA (ed) MTP Int Rev Sci Phys Chem. Butterworths, London, p 103
6. Spink C, Wadsö I (1976) In: Glick D (ed) Methods of biochemical analysis, vol 23. Willey Interscience, New York, p 83

7. Privalov PL (1980) Pure Appl Chem 52:479–497
8. Hinz H-J (1979) In: Jones MN (ed) Biochemical thermodynamics. Elsevier, Amsterdam, p 116–167
9a. Privalov PL (1979) Adv Protein Chem 33:167–241
  b. Privalov PL (1982) Adv Protein Chem 35:1–104
10. Privalov PL, Monaselidze JR, Mrevlishvili GM, Magaldadze VA (1964) Zh Exsp Teor Fiz (USSR) 47:2073–2079
11. Gill SJ, Beck K (1965) Rev Sci Instrum 36:274–276
12. Danford R, Krakauer H, Sturtevant JM (1967) Rev Sci Instrum 38:484–487
13. Watson ES, O'Neil MJ, Justin J, Brenner N (1964) Anal Chem 36:1233–1238
14. Privalov PL, Monaselidze JR (1965) Pribory i Tekhnika Eksperimenta (USSR) 6:174–178
15. Privalov PL, Plotnikov VV, Filimonov VV (1975) J Chem Thermodynam 7:41–47
16. Bakradze NG, Monaselidze JR (1971) Izmeritelnaya Tekhnika 2:58–61
17. Monaselidze JR, Bakradze NG (1973) In: Andronikashvi Li E (ed) Conformational changes of biopolymers in solutions. Nauka, Moscow, p 300
18. Gmelin E (1979) Thermochim Acta 29:1–39
19. Hutchens JO, Cole AG, Robie RA, Stout J (1963) J Biol Chem 238:2407–2412
20. Hutchens JO, Cole AG, Stout J (1963) J Phys Chem 67:1128–1130
21. Cole AG, Hutchens JO, Stout J (1963) J Phys Chem 67:1852–1855
22. Cole AG, Hutchens JO, Stout JW (1963) J Phys Chem 67:2245–2247
23. Hutchens JO, Cole AG, Stout JW (1964) J Biol Chem 239:591–96
24. Hutchens JO, Cole AG, Stout JW (1964) J Biol Chem 239:4194–4196
25. Hutchens JO, Cole AG, Stout JW (1969) J Biol Chem 244:26–32
26. Finegold L, Cude JL (1972) Nature 237:334–336
27. Finegold L, Cude JL (1972) Nature 238:38–40
28. Finegold L, Cude JL (1972) Biopolymers 11:683
29. Finegold L, Cude JL (1972) Biopolymers 11:2483–2491
30. Fanconi B, Finegold L (1975) Science 100:458–459
31. Mizutani U, Massalski TB, McGinness JE, Corry PM (1976) Nature 259:505–507
32. Delhaes P, Daurel M, Dupart E (1972) CR Acad Sci Paris Ser B 274:308–312
33. Daurel M, Delhaes P, Dupart E (1975) Biopolymers 14:801–823
34. Konisek J, Suurkuusk J, Wadsö I (1971) Chem Scr 1:217–273
35. Suurkuusk J (1974) Acta Chem Scand B 28:409–417
36. Bull HB, Breese K (1968) Arch Biochem Biophys 128:497–503
37. Andronikashvili EL, Mrevlishvili GM, Japaridze GS, Sokhadze VM (1973) In: Proc 6th All-Union Conf Calorimetry. Metsniereba, Tbilisi, pp 490–495
38. Andronikashvili EL, Mrevlishvili GM, Japaridze GS, Sokhadze VM (1974) Dokl Akad Nauk SSSR 215:457–461
39. Andronikashvili EL, Mrevlishvili GM, Japaridze GS, Sokhadze VM, Kvavadze KA (1976) Biopolymers 15:1991–2001
40. Andronikashvili EL, Mrevlishvili GM (1976) In: 31st Annu Calorimetry Conf: Abstracts, Argonne Nat Lab USA, p 48
41. Mrevlishvili GM (1977) Biofizika (USSR) 22:180–191
42. Andronikashvili EL, Mrevlishvili GM, Japaridze GS, Sokhadze VM (1979) Int J Quantum Chem 16:367–377
43. Andronikashvili EL, Mrevlishvili GM, Japaridze GS, Sokhadze VM, Tatishvili DA (1981) J Polym Sci Part 0 Makromol Rev 69:11–15
44. Mrevlishvili GM (1977) Biofizika (USSR) 26:233–241
45. Mrevlishvili GM (1979) Sov Phys Usp 22(6):433–455
46. Mrevlishvili GM (1980) In: 6th Int Conf Thermodynam. Main Lectures, Merseburg, GDR, p 39
47. Mrevlishvili GM, Andronikashvili EL, Japaridze GS, Sokhadze VM, Tatishvili DA (1982) Biofizika 27:987–993
48. Mrevlishvili GM, Andronikashvili EL, Japaridze GS, Sokhadze VM, Tatishvili DA (1982) Dokl Akad Nauk SSSR 264:729–732
49. Verkin BI, Sukharevskii BY, Telezhenko YV, Alapina AV, Vorob'eva NN (1977) Sov J Low Temp Phys 3:121–125

50. Carreri G, Gratton E, Yang PH, Rupley JA (1980) Nature 284:572–573
51. Haly AK, Snaith JW (1968) Biopolymers 14:801–812
52. Franklin RF, Gossling RG (1953) Acta Cryst 6:673–680
53. Dickerson RE, Drew HR, Conner BN, Wing RM, Fratini AV, Kopka ML (1982) Science 216:475–485
54. Mrevlishvili GM, Japaridze GS, Sokhadze VM, Bilinska B (1978) Biofizika 4:605–610
55. Mrevlishvili GM, Japaridze GS, Sokhadze VM, Tatishvili DA, Orvelashvili LV (1981) Mol Biol (Mosc) 15:336–343
56. Mrevlishvili GM (1981) Dokl Acad Nauk SSSR 260:761–764
57. Mrevlishvili GM, Syrnikov YP (1974) Stud Biophys 3:155–170
58. Privalov PL, Mrevlishvili GM (1967) Biofizika 12:22–29
59. Sturtevant JM (1977) Proc Natl Acad Sci USA 74:2236–2240
60. Telezhenko YV, Sukharevskii BY (1982) Low Temp Phys (USSR) 8:188–198
61. Hutchens JO (1968) In: The handbook of biochemistry with selected data for biophysicists. Chemical Rubber Company, Cleveland, Ohio, and personal communication TG Mrevlishvili

Chapter 6   **Thermodynamics of Carbohydrate Monomers and Polymers in Aqueous Solution**

Attilio Cesàro

Symbols

J:	any thermodynamic property
ΔJ:	any change in a thermodynamic property
J^E:	any "excess" thermodynamic property
J^o:	any thermodynamic property in a standard state
$\bar{J}$:	any partial molal property
$\bar{J}^o$:	any property at infinite dilution
$\bar{C}_p^o$:	partial molal heat capacity
ΔH_F^o:	heat of formation
$g_{ij}(r)$:	radial distribution function
B^*:	second virial coefficient
j_{ij}:	virial-like coefficients of the property J
$\bar{K}^o$:	partial molar isoentropic compressibility
$\bar{\alpha}^o$:	partial molal isobaric expansibility
m:	molality
ϕ:	osmotic coefficient
R:	concentration ratio
α:	degree of dissociation
β:	degree of binding (complexation, protonation)
ξ:	charge density parameter
k:	Debye screening parameter

1 Introduction

Carbohydrate monomers and polymers are the most ubiquitous natural products, and are the subject of expanding interest because of their physical, biochemical, and industrially useful properties. Among the known polymers, starch and cellulose are frequently cited as the two canonical examples of homopolysaccharides that display different properties due to different linkage geometry. Starch has been subjected to intensive investigation for many years, probably to a greater extent than any other biopolymer, but much of the early work is confusing and unsatisfactory to the point that in 1975 Suggett [1] wrote: "Although many volumes have been written over the years on the chemistry of starches, it is often difficult to separate scientific observations from folklore". In the same year, a note in the J Chem Education [2] revealed that out of 22 popular organic textbooks "only four correctly stated that amylopectin is the water-soluble, and amylose the water-insoluble starch component". The two cited examples perhaps give an idea of the difficulties in selecting good thermodynamic data in the field of carbohydrates. By and large these are due either to the inadequate sensitivity of the techniques, which were developed for other polymers and then extended to carbohydrates, or to the poor chemical characterization of the materials sometimes employed with even advanced physicochemical methods.

The subtle equilibria (i.e., the mixture of several isomers) that exist in solutions of monomeric and oligomeric carbohydrates, and the unavoidable inhomogeneity of polymeric preparations have also contributed to the state of confusion. The first problem is now being solved, through the extensive theoretical and experimental studies that have established the relative stabilities, and hence the equilibrium populations, of alternative conformations and isomers [3, 4]. To overcome the second problem, one should either use standard protocols for purification, which in no way guarantees homogeneous samples [5], or establish a sample bank [1].

From the more specific thermodynamic point of view, it should also be noted that, despite their solubility in water, solutions of even simple sugars have been little studied in the past. This may be a consequence of the widely held belief that sugars form ideal solutions in water and are therefore of little interest. Whatever the reason, up until 1975, the very few thermal studies in solution were limited to the heat of solution of glucose [6, 7], isomerization equilibria [7–9] and some hydrolysis reactions of sugars [10]. The seemingly ideal behavior of aqueous sugar solution has been ascribed to the preference given to isopiestic measurements [11, 12], which provide little or no information about the properties of such solutes in dilute solution [12]. It will be shown that, in fact, many processes involving carbohydrate exhibit a wide range of ΔH values, while ΔG is nearly constant, indicating a large variety of ΔS values.

Isomeric sugar compounds are often cited as typical examples of contrasting behavior. Franks has noted that although galactaric and glucaric acids differ only in the hydroxyl configuration at carbon [3], the former is almost insoluble, while the latter is very soluble in water [13]. Although, in this example, the different stabilities of the crystal structures are more relevant than specific solvation effects (the melting temperatures are 255 ° and 125 °C, respectively), it nonetheless warns against the general tendency to relegate carbohydrate solutes to a single thermodynamic class. It is also becoming apparent that, among a great variety of hydrophilic substances, sugars may exhibit quite peculiar thermal properties. For example, the empirical scheme suggested by Wadso [14] for the estimation of $\bar{C}_p^0$ values for nonionic compounds was shown to work quite well for many substances including simple hydroxyl compounds, but not for sugars. Hence the importance of direct calorimetric data for understanding solute properties and of their relation to solute structure cannot be stressed too strongly.

The scope of this review is not simply to provide tables of the (few) thermodynamic data that are available for carbohydrate systems, but also to organize them into a framework of facts from which consistent models can be derived. It has been too often repeated that thermodynamics cannot provide models, but it is incontestable that it provides data against which different models can be tested. This review, therefore, will also touch upon some of the debates in the carbohy-

---

[1] This need was favorably considered at the European Science Foundation (ESF) Meeting on Polysaccharides held in Uppsala, 1983.

drate field, comparing experimentally available data with those theoretically predicted, or suggesting the need for more data when necessary.

In the following sections, experimental data for carbohydrates in solution are
reported after a theoretically based scheme for the concentration dependence is
outlined, to allow a rationalization of the results. In the third section, attempts
are made to predict the thermodynamic stability of sugar molecules, in order to
stimulate an interest in understanding macroscopic properties at the molecular
level. Mixed interactions are treated in the next two sections: the fourth deals with
inclusion complexes of cyclodextrins and amylose with iodine; and the fifth con-
siders interactions in solution of ionic polysaccharides, treated within the framework of polyelectrolytic theory.

The somewhat arbitrary account of properties is probably justified only by the
variety of compounds (monomers and polymers, ionic and nonionic) which prevents uniformity of presentation. Some physical chemical properties of solutions
containing sugars or sugar polymers have already been reviewed by Franks [15],
Bettelheim [16], Reid [17, 18] and Suggett [1]. A recent guide to current research
is the book on *Solution Properties of Polysaccharides* based on an ACS Sympo-
sium [19]. The reader is also referred to the articles by Franks [20, 21], where fundamental aspects of the thermodynamics of aqueous solutions of hydrophilic solutes are elucidated. Structures and solution properties of polysaccharides have
been described by Rees [22–24], who has been prominent in the popularization
of conformational topologies of carbohydrate chain polymers.

2 Rationale for Concentration Dependence

All theoretical and experimental approaches to the study of thermodynamics of
solutions must recognize the fundamental problem of the concentration dependence of the properties under study. The numerical data taken from any experiment are, indeed, devoid of interest unless the phenomenological properties are
(1) converted into a proper standard framework, and (2) related if possible to a
formal theory of solutions which makes use of molecular parameters for the
understanding of the macroscopic properties.

As regards the first point, the major source of confusion in developing a formalism for the excess properties of solutions comes from the choice of the standard state, which, although arbitrary, must be clearly specified [12]. In principle,
the ideal gas phase is the most sensible convention for the solute reference state
[25, 26], but for the purpose of the present article the more convenient reference
is that of the infinitely dilute solution state, largely adopted for measuring the
concentration dependence. In fact, this choice is almost imperative in view of the
absence of data for the transfer process of the solute from the isolated molecule
in vacuo to the isolated molecule in solution.

The other reference state popular in thermochemistry, i.e., that of pure elements, is not practicable because of the scarcity of heat of combustion data for
sugars. Furthermore, the errors in ΔH_F^0 derived from the combustion data are of
the same order of magnitude as the heat effects in solution.

As regards the second point, we resort to the treatment of solution properties
originally presented by McMillan and Mayer [27] and specifically applied to

aqueous solutions of electrolytes by Mayer [28] and Friedman [29] and to aqueous solutions of nonelectrolytes by Kauzmann and coworkers [30]. In the last decade this virial formalism, extended to many nonelectrolyte aqueous systems, has unified the way of presenting numerical parameters [31–39].

From the theoretical point of view one must realize that a rigorous description of the liquid system is particularly difficult because of the absence of a long-range structural order as well as of complete statistical disorder. That short-range correlations do occur is shown by X-ray and neutron scattering experiments [40]. The nonrandom distribution of molecules may be conveniently described by the radial distribution function $g_{ij}(r)$ [41], which usually shows damped oscillations and decays more or less rapidly. The knowledge of $g_{ij}(r)$ for all pairs of molecules, i.e., solvent-solvent, solvent-solute, solute-solute, would allow a complete characterization of the solution. The words "solvent-structuring" and "solvent-destructuring" have been often used in the past for a qualitative description of the macroscopic change observed in the thermodynamic and nonthermodynamic properties and, more or less appropriately attributed to the unknown changes in the correlation functions [42–46].

Before going on to discuss experimental approaches, we therefore find it necessary to mention briefly the virial-type expansion of solution properties developed by McMillan and Mayer. The essential result of this theory [27] is a rigorous one-to-one correspondence between the equations of imperfect gas theory and dilute solutions of nonelectrolytes: the pressure behavior of the gas corresponds to the osmotic pressure π of the solutions which can be fitted by a virial expansion of the form

$$\frac{\pi}{kT} = \varrho + B^*\varrho^2 + C^*\varrho^3 + \dots, \tag{1}$$

where ϱ is the number of solute particles per unit volume. The virial coefficients are formally identical to those of imperfect gas theory, but instead of the potential U_N, one must use the potential of mean force of N solute molecules in the pure solvent. The theory can also be extended to the development of a distribution function since, although the solutions are represented by an idealized model, one should keep in mind that its statistical-mechanical basis lies ultimately in the McMillan-Mayer theory.

The virial coefficients are therefore described in terms of integrals of the radial distribution function which is related in turn to the exponential term of the molecular pair interaction potential, $W(r)$:

$$B^* = -\frac{1}{2} \int_0^\infty [\langle g(r) \rangle - 1] 4\pi r^2 dr, \tag{2}$$

$$\langle g(r) \rangle = \exp - \left[\frac{\langle W(r) \rangle}{kT}\right], \tag{3}$$

where the $\langle g(r) \rangle$ brackets signify an average over all mutual orientations of the molecules. Although possible in principle, the evaluation of $\langle g(r) \rangle$ from known molecular parameters is, in practice, a formidable task for nonspherical hydrated solutes. Equation (2) shows also that the orientational averaging process may smooth out any discernible character of the $\langle g(r) \rangle$ in the experimentally determined B* values.

Apart from any further use, a heuristic validity of the approach resides in the impulse given to the presentation of the experimental results in terms of the interaction parameters. Thus, any excess thermodynamic properties, J^E, of a solution can be expressed as a virial expansion of molality m, in the Lewis-Randall molality scale [2]:

$$J^E = J - J^0_s - -m \, \bar{J}^0_i = j_{ii}m^2 + j_{iii}m^3 + \dots, \tag{4}$$

where J and J^E are the thermodynamic property and the corresponding excess quantity, respectively, referred to an amount of solution containing m moles of solute i and 1 kg of solvent; J^0_s and $\bar{J}^0_i$ are the standard property of 1 kg of solvent and that of the solute in the infinite dilution state.

By considering the proper units (when using the molality scale), an equation for the excess free energy can be written in a form similar to that of Eq. (1) and experimentally related to the osmotic coefficient ϕ of the solution:

$$G^E = \frac{\pi}{m} - RT = RT(\phi - 1) = g_{ii}m + 2g_{iii}m^2 + \dots, \tag{5}$$

where it should be recalled that $g_{ii}0$ for an ideal solution due to the molality scale [3] $[g_{ii} = -1/2(MRT/1000)$, where M is the molar mass of solvent in kg mol^{-1}].

Similarly, the relative molar apparent enthalpy ϕL is by definition the "excess" enthalpy of the solution and will therefore be expressed in the virial form:

$$H^E = \phi L = h_{ii}m + 2h_{iii}m^2 + \dots \tag{6}$$

It is therefore apparent that the experimentally determined coefficients g_{ii}, g_{iii}, ... h_{ii}, h_{iii} are related to the interaction coefficients of the McMillan-Mayer theory.

Accurate measurements of the osmotic coefficients and of the dilution enthalpies are the only requirement for evaluating the parameters of Eq. (5) and (6). Most of the recent data have now been analyzed in terms of pairwise interactions, and the polynomial coefficients for many carbohydrate solutes have been compared.

Without going into the group-contribution approach by Wood et al. [34–36] and by Barone et al. [37–39], let it here be simply stated that, from the osmotic coefficients and heats of dilution so far reported in the literature the contributions of each functional group to the pair, triplet, etc., interaction coefficients have been estimated and used for predicting the polynomial coefficients of other solutes. The agreement between the predicted and the experimental data can be considered as generally fairly good, but it emphasizes that the true cause of the differences in the behavior of carbohydrates in water is of a stereochemical nature.

---

[2] Even if the McMillan-Mayer theory is constructed at constant volume and therefore the concentration scale to be used is that of molarities, it is most customary to express the experimental properties in terms of molalities. The correlation between the two different scales has been discussed in the literature [47–49].

[3] Stokes [50] provided the equation for the osmotic coefficient of the ideal solution, showing the non zero coefficients in the molality scale, but a mistake in the sign of the coefficient g_{ii} should be noted.

In fact, it must be strongly emphasized that, by their fundamental nature, the group contributions cannot take into account the stereochemical contributions of the individual components. Accurate determinations of these coefficients will therefore allow construction of a reference framework against which the peculiar interactions of any given carbohydrate system may be compared. The fact that, in the polynomial coefficients, all pairwise interactions in the solution are averaged out, including those between water-water and solute-water, implies that mutually compensatory effects reflecting some peculiar aspects of sugar solutions may be obscured, especially if only one thermodynamic property is considered.

3 Experimental Data for Simple Carbohydrates in Solution

The most basic thermodynamic experiments are those in which pure compounds (s, l, or g) are transferred to infinitely dilute solutions. From the calorimetric experiments, ΔH_s^0 and $\Delta \bar{C}_p^0$ values are obtained. The other common property often studied is the density of the solution, which gives the partial molar volume $\bar{V}^0$. These properties, if related to a common standard state (i.e., the isolated molecule in the gas state) should provide useful information of the solute-solvent interactions. In view of the difficulties mentioned in the previous section, the best approach to the study of the solvation of sugars is to use a combination of results from nonthermodynamic techniques [51–59], to complement thermodynamic data, such as partial molar volume, expansibility, etc. As for other properties, additivity rules have been worked out. One use of such procedures is to detect deviations reflecting water-solute interactions which do not exist in the model system used for deriving such additivity parameters. In the case of sugars, the effects of functional groups are overlapping and enhanced by steric compatibility with bulk water structure.

Thermodynamic properties are often studied in the limit of infinite dilution, as the enthalpy of solution, ΔH_s^0, the partial molal heat capacity, $\bar{C}_p^0$, and the partial molal volume, $\bar{V}^0$. Furthermore, partial molal isentropic compressibilities, $\bar{K}^0 = (\partial \bar{V}^0 / \partial P)_s$, and expansibilities, $\alpha^0 = (\partial \bar{V}^0 / \partial T)_p$, have been reported by Franks et al. [60] and by Hoiland and Holvik [61].

It must be recalled that, for processes involving a solid component, the anomeric state must be specified, while solution state should usually refer to the equilibrium mixture of isomers, unless otherwise stated. Before reporting on the preferred values of the above-mentioned properties (selected from the literature values, when several were available), the examination of two model examples is very instructive.

Glucose and sucrose could be taken as examples of well studied compounds in order to compare the values of $\bar{V}^0$, ΔH_s^0, and $\bar{C}_p^0$ obtained by different authors (Table 1). Partial molal volumes have been determined for glucose and sucrose in water at 25 °C by a variety of experimental procedures ranging from the magnetic-float technique to the vibrating tube densitometer. All these methods are quite accurate for low molecular weight solutes and allow the determination of the molar volumes with an error of $< \pm 0.2\%$.

There is no abundance of data for the thermal properties: the ΔH_s^0 for sucrose quoted by Franks and Reid [12] (5.85 kJ mol^{-1}) comes from Table 1 of the Ap-

Table 1. Properties of glucose and sucrose[a]

	V_{vdW} [b]	V^{0} [c]	$\bar{V}^{0}$	ΔH_{s}^{0}	$\bar{C}_{p}^{0}$
α-D-Glucose	88.34	115.3	112.2 ±0.4[d]	10.70±0.04[i]	357 ±12[j]
M = 180.16			117.3 ±0.3[e]	10.82±0.03[k]	331 ± 7[l]
			115[f]	10.71[u]	322.6[m]
			111.9 ±0.3[ghi]	11.00[v]	347 ±3[h]
			111.7[i]		
Selected value			111.7 ±0.3	10.82±0.03	347 ±3
Sucrose	164.3	216.6	211.6 ±0.3[e]	5.85[s]	649.4±2[o]
M = 342.30			210.2 ±0.8[d]	4.91[w]	655[t]
			209.9[g]		
			211.0 ±0.3[i]		
			211.8[o]		
			211.12±0.02[p]		
			211.5 ±0.1[q]		
			212.7[r]		
Selected value			211.3 ±0.2	5.85	649.4±2

[a] Unless otherwise specified data refer to 25 °C. Units: v, V^{0}, $\bar{V}^{0}$ (cm^{3} mol^{-1}); ΔH_{s}^{0} (kJ mol^{-1}); $\bar{C}_{p}^{0}$ (J mol^{-1} K^{-1}). [b] Calculated from atomic increments [62]. [c] Calculated from density of the solid at 17.5° given in *Handbook of Chemistry and Physics* Chemical Rubber Co., Cleveland, 1979. [d] [64]. [e] [61]. [f] [66]. [g] [63]. [h] [60]. [i] From data of [65]. [j] [67] (C_{p}^{0} at 30 °C). [l] [72] (C_{p}^{0} at 30 °C). [m] [75]. [n] [73]. [o] [68] ($\bar{V}^{0}$ at 24°). [p] [69]. [q] [70]. [r] [71]. [s] This value, quoted in [12], is taken as the difference of the heat of formation quoted in [79]. The value of ΔH_{f}^{0} of the solid (531.0 Kcal mol^{-1}) quoted in [79] is different from the selected value of 532.00±0.70 reported in [77]. [t] [74]. [u] [76]. [v] [7]. [w] [78] at 20 °C and final concentration of about 0.12 m.

pendix of Brown's book [79], and does not agree with the other available figure of 4.91 ±0.04 kJ mol^{-1} [78]. On the other hand, the four figures existing for the ΔH_{s}^{0} of α-D-glucose show deviations only slightly larger than the experimental errors and an average value of 10.82±0.03 kJ mol^{-1} can be selected.

Some comment is also needed on the determinations of $\bar{C}_{p}^{0}$. This quantity is often evaluated from calorimetric experiments by measuring the heats of solution at different temperatures ($\Delta \bar{C}_{p}^{0}$) and summing the heat capacities C_{p}^{0} of the pure compounds. Unfortunately, the availability of C_{p}^{0} data for pure compounds is limited, and too often values of C_{p}^{0} are estimated from heat capacity data for similar pure solutes. In principle, the method of direct calorimetry would appear preferable, but some difficulty may arise from the existence of anomeric equilibria in solution. The temperature dependence of the heat of isomerization has also been considered and the small correction to the heat capacity data was evaluated by Wadso [73]. From the data of Table 1 it appears that the $\bar{C}_{p}^{0}$ values of glucose obtained with the two different procedures fall within the expected experimental errors.

Only very recently a number of new thermal data for carbohydrates at finite concentration have appeared in the literature. The major contribution to the data for enthalpies of dilution is due to Barone and coworkers [39, 82, 87, 88] in a series of recent papers where direct calorimetric experiments were carried out for a quite large concentration range. The authors analyzed their results in terms of the poly-

Table 2. Properties of simple sugars

	h_{ii} [a]	g_{ii} [b]	ΔH_s^0 [c]	$\bar{C}_p^0$ [d] (30 °C)	$\bar{\alpha}^0$ [e] (15 °C)	$\bar{K}^0$ [f]	$-\Delta H_F^0$ [g]
D-Arabinose	177±17 [l]		13.24±0.08 [q]	278± 3 [s]			
L-Arabinose	178± 9 [i]			270± 4 [s]			
D-Ribose	202± 8 [h]		13.04±0.14 [q]	271± 2 [s]	0.115	−12.5	
D-Lyxose	243± 7 [h]		10.10±0.01 [q]	(285) [u]			
D-Xylose	339±16 [j]	34 [j,m]	11.98±0.05 [q]	281± 2 [s]		−12.9	
L-Xylose	336± 8 [i]		12.19±0.08 [q]	(291) [u]			
D-Galactose	133± 8 [j]		17.20±0.10 [q]	324±10 [t]	0.150	−20.8	307.4
D-Mannose	207±14 [j]		6.86±0.06 [q]	337± 5 [t]		−16.0	301.9
D-Glucose	343±10 [k]	70 [j,n]	10.82±0.03 [r](α) 4.42±0.07 [q](β)	331± 7 [t]	0.120	−17.0	304.3
D-Fructose	264±18 [j]		8.80±0.10 [q]	352± 8 [t]			
L-Sorbose	395± 9 [i]		6.94±0.03 [q]	(450) [u]			
L-Rhamnose	685±32 [l]	140±34 [l]					
L-Fucose	700±16 [l]	131±36 [l]		(394) [u]			
2-Deoxyribose	468±12 [h]		10.73±0.13 [q]	(333) [u]			
2-Deoxygalactose	442±22 [h]		10.91±0.10 [q]	(386) [u]			
2-Deoxyglucose	592±17 [h]		9.43±0.06 [q]	(381) [u]			
Me-α-Xylose	468±12 [o]	95±20 [p]	1.82±0.05 [q]	(369) [u]			
Me-α-Galactose	900±25 [o]	5±36 [p]	−1.07±0.04 [q]	(436) [u]			
Me-β-Galactose			2.64±0.05 [q]				
Me-α-Glucose	1097±39 [o]	139±18 [p]	3.67±0.07 [q]	(435) [u]	0.115		294.8
Me-β-Glucose			9.18±0.09 [q]		0.045		295.8
Me-α-Mannose	1206±14 [o]	122±36 [p]	9.09±0.07 [q]	(445) [u]			

[a] See Eq. (6); units: $J\,mol^{-1}\,(mol\,kg^{-1})^{-1}$. [b] See Eq. (5); unit: $J\,mol^{-1}\,(mol\,kg^{-1})^{-1}$. [c] Heat of solution; units; $kJ\,mol^{-1}$. [d] Partial molar heat capacity; units: $J\,mol^{-1}\,K^{-1}$. [e] Partial molar expansibility; [60]; units: $cm\,mol^{-1}\,K^{-1}$. [f] Partial molar compressibility; [60]; units: $cm\,mol^{-1}\,bar^{-1}$. [g] Molar enthalpy of formation; [77]; units: $kJ\,mol.$ [h] [39]. [i] [88]. [j] [82]. [k] [34]. [l] [87]. [m] [83]. [n] [84]. [o] [89]. [p] [90]. [q] [67]. [r] From Table 2. [s] [91]. [t] [72]. [u] All data in parenthesis are taken from [67] and are calculated from the temperature coefficient of the heat of solution, ΔC_p^0, and an estimated value for the heat capacity of the solid.

Table 3. Properties of some di- and tri-saccharides

	h_{ii} [a]	g_{ii} [b]	ΔH_s^0 [c]	$\bar{C}_p^0(30\,°C)$ [d]	$\bar{K}^0$ [e]	ΔH_F^0 [f]
Lactose	556.6 [g] 506 ±32 [h]			619 ±16 [l]	−30.4	2238.33±0.88
Sucrose	577.1± 5.9 [i]	183 [h,j]	4.91±0.04 [m] 5.85 [m]	649.4±2 [m]	−17.8	2227.5 ±3.2
Maltose				614.20 [l]	−23.7	2461.5 ±0.92
Melezitose					−24.6	
Raffinose	811 ±50 [e]	331 [h,k]		931 ± 7 [l]	−30.9	

[a] See Eq. (6); units: $J\,mol^{-1}\,(mol\,kg^{-1})$. [b] See Eq. (5); units: $J\,mol^{-1}\,(mol\,kg^{-1})$. [c] Heat of solution; units: $kJ\,mol^{-1}$. [d] Partial molar heat capacity; units: $J\,mol^{-1}\,K^{-1}$. [e] Partial molar compressibility; [60]; units: $cm\,mol^{-1}\,bar^{-1}$. [f] Molar heat of formation; [77]; units: $kJ\,mol^{-1}$. [g] [86]. [h] [82]. [i] [34]. [j] [81]. [k] [80]. [l] [72]. [m] See Table 1.

Table 4. Thermodynamic data for polyols

	a	h_{xx} [b]	$\bar{C}_p^{0}$ [c]	$\Delta\bar{H}_s^{0}$ [d]	$\bar{V}^{0}$ [e]
Ethanediol	(2)	$415^f, 363^g$	193 ± 2^i		54.6^l
Glycerol	(3)	$490^f, 251^g$	240 ± 4^i	5.36^m	70.95^l
Erythritol	(4)	$398^f, 358^h$	310 ± 2^i	22.38^n	87.10^l
Pentaerythrytol	(4)	$393 \pm 2^{g,h}$	301 ± 17^l	22.02^n	
Ribitol	(5)	295 ± 5^h	376 ± 2^i	18.55^n	103.11^l
Xylitol	(5)	80 ± 11^h	346 ± 2^i	22.40^n	102.14^l
Arabitol	(5)	186 ± 4^h	374 ± 3^i	18.71^h	103.3^l
Mannitol	(6)	66 ± 12^h	452 ± 4^i	21.92^i	119.71^l
Sorbitol	(6)	-11 ± 5^f	412 ± 5^i	18.66^i	119.16^l
Galactitol	(6)	-132 ± 50^h	445 ± 16^l	29.22^n	119.3^l
Myo-inositol	(6)	-800 ± 29^h	340 ± 5^i	15.29^i	
Perseytol	(7)	-299 ± 20^h			

[a] Number of OH groups. [b] See Eq. (6); units: $J\,mol^{-1}\,(mol\,kg^{-1})$. [c] Partial molal heat capacity; units: $J\,mol^{-1}\,K^{-1}$. [d] Heat of solution; units $kJ\,mol^{-1}$. [e] Partial molal volume; units: $cm^3\,mol^{-1}$. [f] [95]. [g] [34]. [h] [37]. [i] [73]. [l] [67]. [m] [52]. [n] [68].

nomial expansions and only in a few cases found significant values of h_{iii}. The values of their coefficients h_{ii} and the 95% confidence limits are reported in Tables 2 and 3.

Barone and coworkers have grouped the monomeric carbohydrates in three different ranges of h_{ii} values: a) pentoses and hexoses (values of h_{ii} ranging from 100 to 400 $J\,mol^{-2}$ kg), b) deoxysugars (h_{ii} from 400 to 700 $J\,mol^{-2}$ kg), c) methylglycosides (h_{ii} from 900 to 1200 $J\,mol^{-2}$ kg). They also found that the di- and tri-saccharides studied showed h_{ii} values given approximately by the sum of the h values of the constituent monomers[4]. Table 2 also reports the few g_{ii} coefficients calculated by the same authors on the basis of either their own osmotic coefficient data (isopiestic measurements) or literature data. The scarcity of free energy data [83–88] does not allow any conclusion about similarity of behavior, nonetheless it is possible to notice that all values of g_{ii} are positive, as are also the values of h_{ii}, but for inositol.

The data listed in Tables 2 and 3 are taken as a selection of representative properties and substances. Other calorimetric data of sugar derivatives and related compounds (methyl derivatives and polyols) have been reported by Jasra and Ahluwalia [67], Barone and coworkers [37, 92, 93], Wood and coworkers [34, 94], Lian et al. [73], and Franks and Pedley [95]. Some of these data concerning sugar alcohols are reported in Table 4.

An analysis of the concentration dependence of the available thermodynamic data rules out the simple static hydration model early proposed by some authors and discloses some more specific effects which depend upon the stereochemical environment of the solute, and are dynamically mediated by water.

---

[4] We do not report the values of the coefficients calculated by the same authors with the group contribution method because the arbitrariness in the choice of constituent groups may give results of little significance in the present article. For a longer discussion of the method, the reader is referred to the series of articles by Wood and coworkers [34–36] and by Barone and coworkers [37–39, 82, 87, 88].

Table 5. Partial molar volumes of carbohydrates[a]

	b	c	d	
D-Arabinose			94.00	
L-Arabinose	91.9±0.8	93.2		
D-Ribose		95.2	95.56	95.3[e]
D-Lyxose			95.69	
D-Xilose	94.8±0.4	(95.4)	95.60	
L-Xilose		95.4	95.59	
D-Galactose	111.9±0.3	110.2	110.64	110.7[e]
D-Mannose	111.7±0.5	111.3	111.96	
D-Fructose	110.4±0.4		110.88	
L-Sorbose	110.6±0.4		111.54	
2-Deoxy-ribose	93.8±0.4			
2-Deoxy-glucose	110.4±0.5			
Methyl-α-glucose	132.6±0.4		133.29	133.2[e]
				132.9[e]
Methyl-β-glucose	133.6±0.5		132.09	133.5[e]
Methyl-α-galactose	132.6±0.4			
Methyl-β-galactose	132.9±0.2		131.37	
Methyl-α-mannose	132.9±0.4		132.87	
Methyl-α-xylose			116.36	
Methyl-β-xylose			117.39	
Lactose	207.6±0.1	209.1		
Trehalose	206.9±0.5			
Melibiose	204.0±0.7			
Melezitose	305.8±1.2			
Raffinose	302.2±1.4	307		306.6[f]
Cellobiose	213.6±0.5			211.9[g]
Cellotriose	309.2±1			
Cellotetraose	409.0±1			404.2[g]
Cellopentaose	499.6±1			493.7[g]
Cellohexaose	601.7±2.6			589.6[g]
Maltose	208.8±0.8	210.0		
Maltotriose	304.8±1.2			
α-Cyclodextrin	602.0±1.1			
β-Cyclodextrin	710.8±0.7			
Starch, (Glc)n	n.(97.5±0.6)			

[a] Sucrose and glucose are reported in Table 1. Unit is $cm^3 mol^{-1}$. [b] [64]. [c] [61]. [d] [67]. [e] [60]. [f] [63]. [g] [65].

The values of the partial molar volume at infinite dilution, reported in Table 5, clearly show an overall relation with the van der Waals volumes. The discord among literature data does not permit any conclusions about the stereochemical differences of the hydration interactions.

4 Molecular Calculations of Thermodynamic Properties

4.1 Conformational Properties

After the original simple proposal that the solution properties of sugars in water could be described in terms of ideal equilibria between a series of hydrated solutes [84, 96, 97], further studies by spectroscopic and dynamic methods have provided some evidence for the "stereochemical" model of hydration [9, 51].

The term "specific hydration" has been used mainly to describe the concerted interaction between water molecules and the polar sites of the solute by hydrogen bonds. Because of the nature of the hydrogen bond and its orientation-dependent potential, it has been inferred that "specific hydration" strongly depends upon the detailed stereochemistry of the interacting groups. The possibility that several conformers exist, because of rotation about carbon-carbon bonds [4, 98, 99] adds further complexity to the story and suggests that the population of the various conformational states will change in different solvents [4]. As a consequence, it is very hard to establish, at macroscopic level, an a priori dependence of the physicochemical properties of the carbohydrates in solution on their constitution.

The complexity of the problem arising from conformational ring interconversions (not to speak of anomeric interconversion) also deserves some comment in view of the specificity claimed for the water-carbohydrate interaction. The study of conformations and of compositions of the equilibrium mixtures formed by sugars in solution (and especially in water) is of great importance in carbohydrate chemistry. Properties of carbohydrates differing in the steric arrangements could be accounted for, in principle, by conformational factors. Fortunately, among the 26 different pyranoid rings (see Fig. 1), only a very few may be important for the equilibrium thermodynamics, but their possible role in macromolecular chain conformation is being recognized [100, 101].

Qualitative [102] and semi-quantitative [103, 104] empirical approaches to predict the Gibbs energy differences between monosaccharide conformations have been replaced by more recent, and even ab initio, calculations [105–113], which with a veneer of sophistication have shifted the approximations to another level. Within this field, one group of scientists believes that "the calculations are now refined to a point at which an accurate prediction can usually be made about of the shapes and configurations that will predominate in sugar solutions, as well as the proportions of each form" [24]. For the other group, "there is, as yet, no unified treatment which is capable of accounting for internal energies of monosaccharides, let alone free energies in solutions" and "attempts to incorporate solvation corrections into the conformational energy calculations cannot be regarded as anything better than arm-waving" [15].

In fact, the two different views reflect the very large jump from thermodynamics to interaction mechanisms. The thermodynamic properties are averaged in the energy space and not related therefore to a single structural state, although some states can be more populated than others. The wide gap, geometry versus statistics, originates from the intrinsic difference between the minimum energy approach and that of Gibbs energy space.

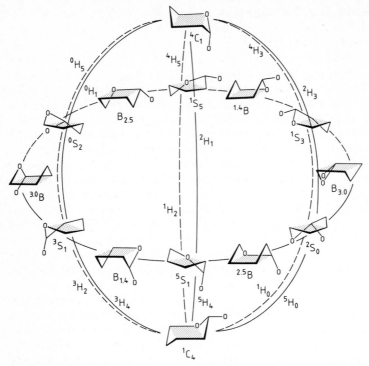

Fig. 1. Conformations of the α-pyranoid ring and their position on the conformational sphere. The positions of the 12 half-chain conformations and the interconversion paths are also indicated

Attempts at energy calculations are too often carried out including intramolecular hydrogen bonding, in addition to van der Waals forces, coulombic interactions, and torsional contributions. Intramolecular hydrogen bonding is unlikely to be important in influencing conformational equilibria in aqueous solutions and usually has to be ignored in the calculations. There is much evidence to indicate that, although intramolecular hydrogen bonding may influence conformational equilibria in aprotic solvents, in aqueous solutions intermolecular hydrogen bonding with water is much more important and probable on a statistical basis.

Once a structure is obtained, either from the coordinates of model systems [114–117] or from more sophisticated calculations, it must be clearly kept in mind that it refers to an in vacuo situation. Therefore the energy differences evaluated for different conformers reflect almost exclusively the intramolecular interactions. Actually, in the case of semi-empirical approaches, a refinement of potential functions and of molecular parameters is usually made to fit the experimental solution properties. This procedure may compensate for the solvation effects, albeit not explicitly.

Some recent studies concerning the stability (structure) of simple sugars have been devoted to calculation of the energy of the individual molecule and from there to the evaluation of the Gibbs energy of solvated conformers. The first type of study provides the internal energy of the molecule in one of the minima of the energy space, with even a rough estimation of the entropy. The second approach calculates the total conformational Gibbs energy of the solute in a particular solvent by adding a term for intermolecular-solvent-solute interactions, G_{solv}, to the Gibbs energy of the solute molecule in the free-space approximation, G_{int}. The G_{solv} term can be obtained through an extension to aqueous solution of the scaled-particle theory [118–121]. Calculations of the G_{solv} term for model sugar molecules and for maltose have been carried out by Tvaroska [122–124] and nicely predict the effect of the solvent on the equilibrium populations of different conformers.

The main validity of these studies resides in the use of molecular parameters for the statistical mechanical evaluation of thermodynamic and other physico-chemical properties. Although the results can be considered still approximate, the promises offered by these studies are to be taken into account for the understanding of the phenomena of our concern.

4.2 Comparison of Experimental Data with Those from Energy Calculations

Calculations of the relative stabilities of different conformers were approached by Angyal by using values of interaction energies between vicinal groups obtained from equilibria of sugars and of borate-cyclitol complexes. The interaction energies considered to be additive and, since obtained from data in aqueous solution, referred to similar conditions. The contribution from the anomeric effect was not taken into account, but obtained in fact from the difference between the calculated and the experimentally found fraction of the α-anomer.

Table 6 reports data taken from the literature concerning the α- and the β-anomers of D-glucose. In the same table the data obtained by Dunfield and

Table 6. Energetic data for the α- and β-anomers for D-glucose

	α	β	α	β
	(Experimental)		(Calculated)	
Conformation of ring [a]	C1	C1	C1	C1
% Anomer H_2O [b]	37%			
$\quad$ D_2O [b]	34%			
$\quad$ DMSO [b]	44%			
% Anomer vacuo [c]			47% (30% C1)	
$\quad$ solvated [c]			33%	
Free energy (C1) [a]	10.0	8.58		
Free energy (1C) [a]	27.4	33.5		
Internal energy [c]			12.8	13.3
Entropy [c]			61.67	64.56
Free energy [c]			− 5.78	− 6.32
ΔH (25°, aq, α→β) [d]	− 1.17			
ΔG (α→β)	− 1.42 [a]		−0.54 [c]	

Units are kJ mol^{-1}, and J mol^{-1} K^{-1} for entropy. [a] [3, 103]. [b] [15]. [c] [106]. [d] [6].

Whittington [106] show that the predicted anomeric composition compares with that measured in DMSO or pyridine, while solvent corrections must be included in order to fit the experimental data in water. It is evident, therefore, that the in vacuo calculations simulate quite well the less polar situation of the nonaqueous solvents. A solvent dependence has been found also by Tvaroska in his studies on the conformational equilibria of maltose and of 2-methoxytetrahydropyran [122–124]. In both cases the Gibbs energy contribution due to the solvent (G_{solv}) was found to affect the population of different rotamers only slightly in many solvents, but in water the effect was larger.

These findings are important in that they are able to explain the contribution of solvation effects to the so-called exo-anomeric effect and to the conformational free energy of dimeric sugars in solution without resorting to undefined "solvation properties" of the solvent with respect to carbohydrates [124]. Since the calculation does not consider explicitly any contribution from direct solvent-solute interactions (that is, "specific interactions"), the good agreement between calculated and experimental values supports the view that the conformational dependence of specific interactions is negligible. Of course, this conclusion, which refers to the case of conformers differing in the torsional glycosidic angle, cannot be extended to different anomers or to ring conformers of glucose, where different forms may offer different interactions with the water molecules. On the other hand, at least in the case of 2-methoxytetrahydropyran, the stability of the anomeric forms was also correctly predicted [123].

5 Inclusion Complexes

5.1 Cyclodextrin Complexes

The essential characteristic of the cyclodextrin molecules (at least α- and β-) is their doughnut shape, which provides a widely unspecific matrix for the inclusion of molecules both in the crystal state and in aqueous solution.

The binding forces between cyclodextrins and guest molecules have been the subject of long controversy and speculation. From time to time, they have been attributed to one or more of the following contributions: nonspecific van der Waals forces, hydrogen bonding, release of "activated" water molecules from the cavity, or gain from the strain energy in the case of α-cyclodextrin.

The inherent ring structure is, of course, decisive, since inclusion complexes are only formed if there is a good spatial fit between the host and the guest molecules.

Many reviews have been written on the cyclodextrins and their inclusion complexes [125–129], where data concerning the energetics of the process can be found[5]. The importance of microcalorimetry for directly measuring the thermodynamic properties of complex formation [127] is recognized.

---

[5] Reference [129] reports a summary of the parameters of the interactions of cyclodextrins in Table 11.XI (p. 453). It should be noted that thermodynamic parameters are incorrectly reported as activation parameters.

Table 7. Thermodynamic parameters for the formation of cyclodextrins inclusion complexes

Substrate	$-\Delta G$ [a]	$-\Delta H$ [a]	ΔS [b]	Reference
1 p-Nitrophenol	12.1 ±1.7	30.6± 6.3	− 62.3	[127]
1′ p-Nitrophenol (β)	17.1 ±9.6	44.0±10.5	− 88	[127]
2 p-Nitrophenolate	19.6	30.1	− 36.4	[130]
3 Perchloric acid	9.2 ±0.4	31.4± 4.2	− 71.2	[127]
4 Sodium perchlorate	7.53±0.4	40.6± 5.0	− 96.3	[127]
5 Anilinium perchlorate	8.4 ±1.2	51.5± 8.0	−146	[127]
6 Benzoic acid	17.1 ±0.4	40.2± 0.4	− 75	[127]
6′ Benzoic acid (β)	12.1 ±2.1	31.8±11.3	− 67	[127]
7 4-Aminobenzoic acid	15.9 ±0.4	48.6± 2.1	−109	[127]
8 2-Aminobenzoic acid	28.4 ±7.5	1.2± 0.4	88	[127]
9 Diisopropyl phosphorofluoridate	4.3	30.5	− 88	[131]
10 m-Chlorophenylacetate (β)	14.1	4.2	33.4	[132]
11 m-Ethylphenylacetate (β)	15.5	19.2	− 12.5	[132]
12 Benzoylacetic acid (β)	13.1	23.9	− 36	[133]
13 Phenol	23.8 ±7.5	7.5± 0.8	54	[127]
13′ Phenol (β)	19.2 ±5.9	10.9± 0.8	29	[127]
14 Indole	44.4 ±0.4	3.3± 0.4	138	[127]
15 Hydrocinnamic acid	17.6 ±0.4	31.4± 0.4	− 46	[127]
16 L-Phenylalanine	23.4 ±8.0	4.6± 1.2	62.8	[127]
17 L-Tyrosine	16.7 ±3.3	4.2± 4.2	42	[127]
18 L-Triptophan	8.4 ±0.4	7.5± 0.4	4	[127]
19 3-Methylbenzoic acid	13.8 ±0.4	48.6± 3.7	−117	[127]
20 2-Nitrophenol	20.4 ±6.3	2.1± 0.4	63	[127]
21 L-Mandelic acid	13.0 ±8.0	20.5±10.4	− 25	[127]
22 Acetic acid	21.8 ±6.7	5.0± 0.4	54	[127]
23 Pyridine	12.6 ±8.0	10.5± 3.7	8.4	[127]
24 3,4,5-Trimethylphenylacetate	13	10.5± 2.9	− 8±12	[134]
25 p-Methylbenzoylacetic acid	15.5	27.6± 1.7	−41± 5	[134]
26 m-Chlorobenzoylacetic acid	14.2	21.8± 4.6	25±14	[134]
27 1-Adamantane carboxylate	17.6 ±8.0	5.0± 1.7	42± 8	[135]
27′ 1-Adamantane carboxylate (β)	18.76±0.4	19.7± 3.3	− 4±12	[135]

[a] Units are kJ mol^{-1}.
[b] Units are J mol^{-1} K^{-1}.

Values of the ΔG, ΔH and ΔS obtained either from direct calorimetric measurement or the temperature dependence of the complex equilibrium constant are reported in Table 7. The ΔH values are always negative, although differing in size, whereas the ΔS values may be positive or negative, indicating that several forces are involved in the complex formation. A plot of ΔH vs. ΔS values reveals that almost all guest compounds fall on a straight line (Fig. 2). This linear relationship (called compensation effect) has been observed frequently for aqueous solutions and has been related to the involvement of water molecules from the solvation co-spheres in complex formation [136]. In the present case it reflects perturbations of both the solvation sphere of the guest molecules and the water molecules grasped by the host cyclodextrin. Although oversimplified, this means that the observed differences in the thermodynamics of binding are largely due to differences in the solvation state of the guest molecules. Inclusion complexes are not limited to organic molecules; anions can also easily fit the cavity. In most cases

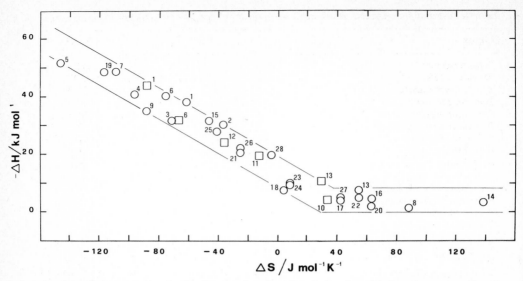

Fig. 2. Enthalpy-entropy plot for the complexation of cyclodextrins with the substrate numbered in Table 6

the interactions between host and guest molecules must be weak because the small molecules have often been found to be statistically disordered within the cavity [137–140].

5.2 Energetics of the Amylose-Iodine Complex

Another well-known example of inclusion complex in aqueous solution is due to the reaction of amylose with iodine in the presence of iodide to give a dark blue complex.

There seems now little doubt that iodine and triiodide reside in the dissolved complex within the annular cavity of a more or less regular helical amylose chain [141]. This picture emerged early from the work of Rundle and coworkers on the crystalline [142] and dissolved [143] complex and has not been seriously challenged by any more recent studies, although the helical character of the dissolved complex does not imply that the complexed polymer adopt a rigid, rod-like conformation. Recent results support a stoichiometry with a fraction of I_3^-, R, in the range 0.2–0.5, and suggest that R may depend somewhat on such variables as mean polymer chain length, iodide and/or salt concentration, and the physical state of the complex (i.e., crystalline or dissolved). At macroscopic level, the complex seems to exist in a variety of different compositions which, however, in the limit of R = 1, do not produce the other known optical properties typical of the blue complex [141, 144].

Concerning the energetics of the complexation reaction, the enthalpy of complexation is found to be constant in the range of reaction conditions which leave

Table 8. Enthalpy of reaction of the amylose-iodine-triiodide complex in solution[a]

Method	Temp.	C_{KI}	β [b]	$-\Delta H$
Potentiometric	1 –21	5×10^{-2}	0.5	64.9
Potentiometric	1.4–20.4	2.4×10^{-3}	0.5	65.7
Spectrophotometric	1 –57	10^{-2}	Variable	87.1
Spectrophotometric	2.7–52	Variable	0.5	40.6
Spectrophotometric	13 –30	10^{-4}	0.5	69.5 ± 2.1
Spectrophotometric	13 –30	10^{-4}	0.5	53.6 ± 2.1
Calorimetric	25	0.15–0.6	$0 \rightarrow 0.2$	70.3 ± 2.0
Calorimetric	25	10^{-2}	$0 \rightarrow 0.2$	71.6 ± 1.5
Calorimetric	25	10^{-2}	$0 \rightarrow 1$	59.0 ± 1.5

[a] From [144] and from references therein quoted.
[b] Degree of saturation.

the corresponding spectroscopic properties unchanged [144], and it varies with amylose chain length in a way that mimics the dependence of absorption properties (λ_{max}) on the degree of polymerization [145]. The rather large enthalpy change (ca. -71 kJ mol of bound I) sustains its largest contribution from the cooperative interactions between iodine units bound in linear chains and a much smaller contribution from interactions of these species with the polymer chain.

Most of the enthalpic data on complex formation (Table 8) were derived from the van't Hoff plot of the apparent equilibrium constant as a function of the temperature, ΔH_{vH} and range from -42 to -87 kJ mol^{-1} (of bound molecular iodine). Unfortunately, despite the number of factors which affect the properties of the complex, too often the experimental conditions are poorly reported. As a further complication, the variable stoichiometry results in a high degree of arbitrariness in defining the mole unit of the bound species. For example, the two values of ΔH_{vH} obtained by Cronan and Schneider [146] (-69.5 ± 2.1 and -53.6 ± 2.1 kJ mol^{-1}, calculated by referring the equilibrium constant to the free iodine and to the free triiodide, respectively) show that the difference between them is almost equal to the enthalpy of triiodide formation, and it is clearly associated with the way the equilibrium constants are expressed. Direct calorimetric measurements have been reported by Takahashi and Ono [147], by Cesàro and Brant [148] and by Cesàro et al. [144]. In the first two cases the integral heat of reaction between $\beta = 0$ to $\beta = 0.2$ was measured, where β is the degree of complexation. The agreement between these data supports a strong contribution from iodine-iodine interaction, since the latter refers to slightly substituted carboxymethylamylose. Given the cooperativity of the complex formation, one would also expect a dependence of the isosteric heat of binding on the degree of complexation as well as on temperature. The direct microcalorimetric determinations [144] have indeed shown that the integral heat of reaction from $\beta = 0$ to 0.2 ($\Delta H_{\beta \rightarrow 0.2}$) is ca. 13 kJ mol^{-1} more negative than the value of $\Delta H_{\beta \rightarrow 1}$ under the same experimental conditions, a fact which is a consequence of the cooperativity of the process. The cases reported show the capability of direct microcalorimetric methods to provide not only more reliable enthalpy changes, compared with the van't Hoff methods, but also valuable hints for the molecular description of the process.

6 Ionic Polysaccharides

6.1 Premise

Ionic polysaccharides are a class of biological polyelectrolytes of very wide interest for their physicochemical properties. Their main physiological role is often related to their ability to entrap large quantities of water through ion-induced structures, which may eventually become macroscopically evident as a gel phase.
Therefore for both scientific and industrial interest, the interactions with water
and with ions have been extensively studied. A detailed review of the thermodynamic data concerning these systems may be unrealistic and useless unless the results are correlated with the particular structure of the polysaccharide and the
data "cleaned up" from the "expected" thermodynamic changes.

The presence of chemical irregularities of polysaccharide chains and the occurrence of conformational changes to form different structures and/or states of
aggregation are formidable complications in the interpretation of experimental
data. The very difference in respect to other biopolymers (like nucleic acids and
proteins or polypeptides) resides in the uniformity of the chain backbone of these
latter polymers, whereas the insertion of other monomers within an homopoly-
meric polysaccharide chain may result in a dramatic change of the chain topol-
ogy. In the following only some basic aspects of the thermochemical characterization of aqueous solution of ionic polysaccharides are considered. Fundamental
macromolecular properties of polysaccharide solutions will not be discussed here,
because, although of thermodynamic origin, they require other kinds of approach. Details of the basic conformational properties and interactions of carbohydrate chains have recently been given in the literature [17–19, 22–24, 149–
151].

6.2 The Polyelectrolytic Contrast Effect

Central to the nature of ionic polymers is the problem of the interactions between
the charged polymer and the counterions. The interactions are far more complicated than those occurring in a simple salt solution, because charge density on the
chain (i.e., local ionic concentration) may reach much higher values, a fact which
explains the use of expressions like "ion-atmosphere" or "ion-binding" for the
counterions accumulated in the polyelectrolyte domain. As for the polymeric
counterpart, it must be clearly kept in mind that, although charges may be localized on specific chemical groups ($-SO_3^-$, $-COO^-$,...), their distance distribution is defined by the Gibbs energy minimum of charge interactions in the conformational energy space. An important consequence is that many ionic polysaccharides in aqueous solution undergo conformational transitions between states
with different degrees of intramolecular order when, for example, the temperature, the ionic strength of the solution, or the charge density of the polymer are
changed. These processes may noticeably change the formal charge density on the
polyelectrolyte chain as monitored by experimental thermodynamic results.

Although molecular polyelectrolyte theories have been proposed since the
early 1950's, and their use has become popular among the practitioners, they have

only recently been fully accepted within the different fields of biopolyelectrolytes. The objective has been to develop a theory which can be regarded as the polyelectrolytic counterpart of the Debye-Hückel theory for simple electrolytes, to allow the popular "polyelectrolyte effect" to be described quantitatively. Theories for the polyelectrolytic behavior of macromolecules were independently developed by many authors [152–161]. In the cylindrical symmetry the distribution of small ions in the field of the polyion has been calculated by suitable application of the Poisson-Boltzmann equation. Although based on different procedures, the results of these approaches converge almost quantitatively. Among all theories, the most attractive one from the formal aspect of the analytical equations is that originally proposed by Oosawa [156] and extensively developed by Manning [157–161]. Under the following assumptions, it provides a simple equation for the electrostatic part of the Helmholtz free energy of a chain in solution:

1. The real chain of finite spatial dimensions is replaced by a cylinder with infinite length-to-radius ratio (linear charge distribution);

2. The actual charge distribution is replaced by a linear charge density with a constant charge spacing, b, along the cylinder axis;

3. The interactions between two or more polyions are neglected (infinite dilution state);

4. The dielectric constant, D, of the solution is taken as equal to that of the bulk solvent;

5. The free ions in solution are treated under the Debye-Hückel approximation;

6. The reference state for any thermodynamic property is that of a solution containing small ions at a concentration equal to the total charge concentration of the polyelectrolyte solution. The contribution of mobile ions to the electrostatic excess free energy of the solution is neglected. Therefore the calculated free energy is an excess free energy relative to the reference state defined above.

Under these conditions the excess electrostatic free energy, F^{el}, can be evaluated in terms of Debye-Hückel potentials, summed over all pairs of charges on the polyion, to give, for the simple case of monovalent counterions,

$$F^{el} = -R\,T\xi^n \ln(1-e^{-kb}), \tag{7}$$

where k is the Debye screening parameter ($k^2 = 4\pi e^2 N(C_+ + C_-)/10^3 DKT$), and ξ is the so-called charge density parameter ($\xi = e^2/DKTb$); ($C_+ + C_-$) is the sum of the concentrations of counterions and co-ions. In Eq. (7), n=1 when $\xi < 1$ and n = −1 when $\xi \geq 1$. The critical condition $\xi = 1$ arises as a convergence requirement in the procedure of potential summation. The consequence is that for $\xi > 1$ a fraction, $r = 1 - \xi$, of counterions per charged group on the chain will "condense" to reduce the effective linear charge density, to $\xi = 1$. For $\xi > 1$ a Gibbs energy of "mixing" term, accounting for the entropy change upon ion condensation, is to be added to the right-hand side of Eq. (7) to give the total excess polyelectrolytic free energy. Full details for the derivation of the complete equations and for the derivation of the enthalpy function are given elsewhere [162–164].

Analytical expressions can be derived from Eq. (7) to evaluate the changes of other state functions for given processes like dilution, mixing with ions, dissociation, etc. Although some of these derivations may suffer from the inadequacy of the assumptions (especially when processes span across the value of $\xi=1$), the theory is nonetheless able to give a consistent set of predictions as far as the energetic behavior of polyelectrolyte solutions is concerned. The possibility of factorizing the purely electrostatic contribution from the nonionic one appears to be a significant achievement in the study of conformational transitions ("contrast effect") and enables one to evaluate the nonionic contribution. As regards the conditions (1) and (2), the charge parameter ξ is usually calculated from the known chain structural parameters. Although the assumption of such a rigid conformation in solution may appear unrealistic in some cases, there is an increasing body of evidence that charged polymers at the local level are extended enough for the application of the theory, despite the common idealization of a random coil structure.

By proper application of the equation for the electrostatic enthalpy to a general process of mixing a polyelectrolyte solution with a solution containing a simple salt, the following general equation is obtained:

$$\Delta H_{mix} = -\frac{1}{2}RT\left(1+\frac{d\ln D}{d\ln T}\right)\xi^n \cdot \left\{2\ln\frac{1-e^{-k_i b}}{1-e^{-k_i b}} + \frac{k_f b}{e^{k_f b}-1} - \frac{k_i b}{e^{k_i b}-1}\right\}, \qquad (8)$$

where C_p and C_s are the molar concentrations of polyelectrolyte and simple salt, respectively, $\lambda = 4\pi e^2 N/10^3 DKT$, $k_{i,f}^2 = \lambda(2C_{s_{i,f}} + \xi^m C_{p_{i,f}})$, $m=0$ for $\xi<1$, $m=-1$ for $\xi \geq 1$ and the subscripts i and f refer to the initial and to the final states, respectively [165]. For the particular case in which $C_{s_i}=0$ and $kb \ll 1$, the equation has been already reported by Boyd et al. [165]. Other equations have been derived for the enthalpy change of dissociation of weak polyacids (which can be of particular interest because many polysaccharides contain carboxylate residues in the chain), and for the electrostatic enthalpy contribution in a conformational tran-

Table 9. Heat of dilution data for ionic polysaccharides[a]

Polymer	ξ[b]	$-(\partial H/\partial \log C)$[c]	$\xi(\partial H/\partial \log C)$[c]	Reference
Alginate	$\begin{pmatrix}1.38^d\\1.64^d\end{pmatrix}$	600– 800	–	[202]
		815–1200	–	[212]
Xanthan	1.12	800	970	[213]
Carboxymethylcellulose	1.15	780	900	[202]
Heparin	1.66	597	990	[203]
	1.75	1030	1800	[203]
	2.10	480	1000	[203]
	2.33	570	1030	[203]
Chondroitin-sulfate	1.49		970	[217–218]
Dextran-sulfate	2.95	400	1180	[202]
Pectate ($\alpha=1$)[e]	1.60	330	530	[211]
Pectate ($\alpha=0.3$)[e]		− 480		[211]
Alginate BA		− 100		[212]
(Dextran $\partial H/\partial C=$ 5 J mol^{-2} L)				[219]
(Glucose $\partial H/\partial C=30$ J mol^{-2} L)				

[a] Taken from the experimental data of the sodium salt of the polymer. [b] Calculated from the structural data. [c] Units: J mol. [d] The two values refer to pure mannuronate and guluronate sequences. [e] Degree of ionization of the polyuronate.

sitions of the polyelectrolyte [164]. Although both contributions are often small when compared with the total heat effect experimentally determined, once again the importance of separating the electrostatic (expected) contribution from the other (sometimes unexpected) ones is to be emphasized.

As far as the experimental literature is concerned, ionic activities have been measured for many polysaccharide-counterion systems [166–196], mainly alginates and pectins, hyaluronates, carrageenans, and ionic derivatives of cellulose and dextrans. Rationalization of these experimental data is not always possible in view of the fact that different techniques intrinsically give figures which are averaged over different ionic distribution distances [197, 198]. Furthermore the changes in the ionic activities are almost dominated by the "expected" ones, with a complex compensation of the entropy and enthalpy contributions, even though these are of different origins. On the other hand, direct determination of the enthalpy changes has proved to be capable of detecting processes other than the ionic interactions. Isothermal microcalorimetric studies of mixing ions with ionic polysaccharides are essentially limited to work carried out by groups in Salford, Rome, and Trieste [199–218].

6.3 Enthalpies of Dilution in Water

The heat of dilution data existing in the literature have been collected in Table 9. This thermodynamic property provides a measure of the changes in the interaction of solvated solute species upon dilution and, as clearly outlined above, it includes the contribution due to changes in the electrostatic interactions between counterions and polyion. Therefore, when cleaned up from the electrostatic interactions, heat of dilution data can provide useful information on both nonelectrostatic interactions and conformational effects. In the same table, for purpose of comparison, are included the only existing data for the enthalpy of dilution of a nonionic polymer (dextran) [219], which has been reported to have a specific heat of dilution independent of the molecular weight, M, (for $M > 10^3$). The negligible nonionic contribution to the enthalpy of dilution of a hydrophilic polymer is thus illustrated.

The equation for dilution with solvents is obtained by setting $C_{s_{i,f}} = 0$ in Eq. (8); in the limit of very dilute solutions [see assumptions (3)–(6)], it reduces to

$$\Delta H_{dil} = -\frac{1}{2} RT \left(1 + \frac{d\ln D}{d\ln T}\right) \xi^n \ln \frac{C_{pf}}{C_{pi}}. \tag{9}$$

The derivative of ΔH_{dil} with respect to log C_p turns out to depend exclusively upon the charge density parameter, ξ, of the polymer.

A normalization of the concentration derivative of the heat of dilution data (Table 9) by the charge density parameter provides in most cases values of $\xi^n(\partial \Delta H_{dil}/\partial \log C_p)$ close to 10^3, as predicted by the theoretical approach. No difference in the behavior can be ascribed to whether the charges on the polymer are carboxylate or sulfate groups, as required by purely electrostatic interactions.

On the contrary, it is not speculative to suggest that additional (conformational) effects occur in the case of diluting partially neutralized pectate ($\alpha = 0.34$)

[211], or a bacterial alginate [212], which contains some heterogeneous sequences of acetylated derivatives.

A closer inspection of the literature reveals that no clear mention is made of the fact that the heat effect recorded in a microcalorimetric experiment must be corrected for the heat of dilution of a simple salt undergoing the same dilution process, as is indicated by condition (6). This correction approaches the Debye-Hückel limiting behavior and becomes negligible at low concentration, because of the different concentration dependence (square root vs. logarithmic). In the work of Pass and Hales [199–204], deviations at high polymer concentration occur in a range where differences among counterions are also noticeable. Behind the problem just raised, it is to be remembered that the validity of the theoretical approach is limited to very dilute solutions; thus it is not clear whether this counterion dependence concerns the dilution of small ions exclusively, the interactions between polyions and counterions, or both. Comparison between experimental data and calculations would have been preferable in a more dilute region.

6.4 Enthalpies of Protonation of Carboxylated Polysaccharides

Most neutral polysaccharides with a regular chain structure are insoluble in water. This behavior cannot be ascribed a priori to the enthalpy of mixing the polymer with the solvent; polysaccharide and water experience a large variety of interactions, so that the predictions of sign and magnitude of ΔH are impossible. If the crystalline rigidity is partially retained in solution, then the entropy change may become sufficiently unfavorable to prevent isotropic solutions. On the other hand, favorable contributions to ΔS are usually significant in the presence of ionic interactions. Therefore, the introduction of fixed charges on the polymer chain is the most efficient way of decreasing the free energy of mixing with water.

The study of the protonation of carboxylate polymers is a potent tool for understanding the energetics of the system and reveals any changes in the interactions occurring upon charge density modulation. The presence of uronate residues in polysaccharide structures has stimulated some studies on the thermodynamics of the protonation (i.e., proton binding process) of the charged carboxylate. Some of these data have been reviewed by Cesàro et al. [213]. They show that proton association to uronates in water is constantly endothermic with values of $\Delta H \simeq 7.5$ kJ (mol $H^+)^{-1}$, except for a few anomalous cases. This result is extremely interesting in view of the fact that the enthalpy of protonation seems to be largely independent of the nature of the monomers. Deviations from the "base line" at 7.5 kJ disclose also a dependence of ΔH upon the degree of protonation β_H of the polymer. At least for two polysaccharides, polygalacturonate and xanthan, the trend to negative values of ΔH is paralleled by sigmoidal changes in other properties and has been ascribed to conformational changes induced by the neutralization of carboxylates.

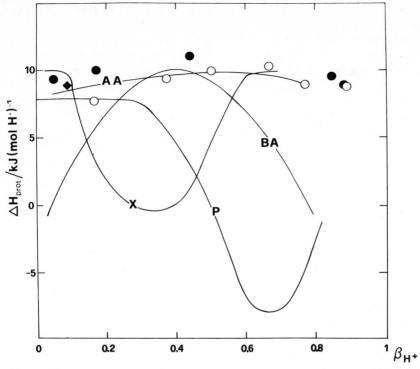

Fig. 3. Enthalpy change on protonation, ΔH, of polyuronates as a function of the degree of protonation, β_H, in water at 25 °: data refer to algal alginate (*AA*), bacterial alginates (*BA*), poly-(guluronate) (○), poly(mannuronate) (●), poly(galacturonate) (*P*), xanthan (*X*) and K-63 (◆); [213]

6.5 Enthalpy of Mixing with Ions

An introductory statement is necessary before analyzing the thermochemical results of ion-polysaccharide interactions. Thermodynamically speaking, "ion-binding" is considered to occur only if the distance distribution function between ions shows a marked minimum between localized ion-pair contact and the free ions. In the absence of any clear indication of this "binding", which occurs in the case of proton association, it is impossible to normalize the observed thermodynamic property change by the amount of ions involved in the process.

Cursory inspection of a thermodynamic property of a polysaccharide solution as a function of the amount of salt added (usually expressed as ratio, R, of the salt-to-polymer concentrations) may reveal interesting features. Again, as in the previously cited examples, it is important to make a preliminary screening of the expected polyelectrolyte behavior (the "contrast effect") from other solution processes [220]. In the case of polyelectrolytes, for example, the "expected" enthalpy change of the system upon mixing a salt solution with the polymer solution is positive and asymptotic in behavior on increasing R [164]. From the thermody-

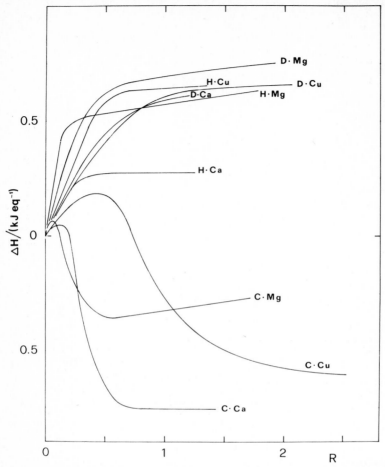

Fig. 4. Enthalpy of interaction of divalent cations (Mg, Ca, Cu) with sulfated polysaccharides as a function of the equivalent concentration ratio of counterion to polymer, R, in water at 25 °: data refer to ι-carrageenan (*C*), heparin (*H*), and dextran sulfate (*D*); [213]

namic point of view, the definition of different types of binding made by Rees [22], although operationally useful for modeling, may cause some confusion, since it neglects the fact that all types of interaction must satisfy the requirements given by the charge density of the polymer. In other words, the description of ion-polymer interaction must proceed first by the identification of the proper value of the integral of the distribution function and then by the subdivision into different modes of interaction [197, 198]. The general result of the polyelectrolyte approach is that nonspecific ion interactions are a direct function of the ionic strength and of the "structural" charge density.

An extension of Manning's approach has been presented to give the "reference" enthalpy change upon mixing a polyelectrolyte with a simple salt [164]. The

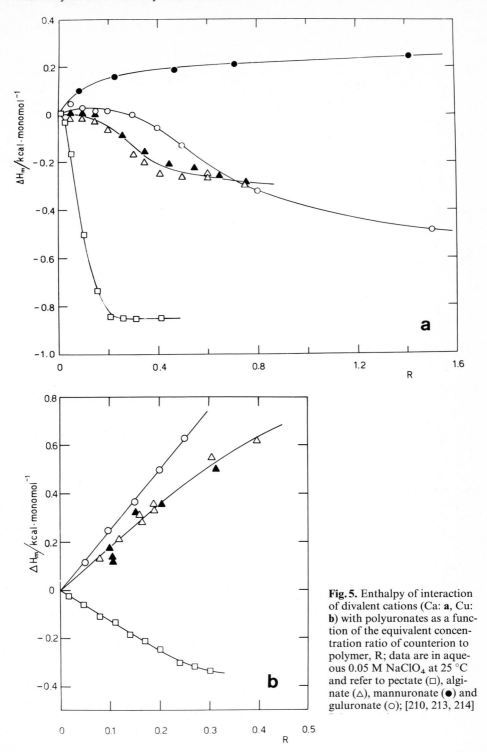

Fig. 5. Enthalpy of interaction of divalent cations (Ca: **a**, Cu: **b**) with polyuronates as a function of the equivalent concentration ratio of counterion to polymer, R; data are in aqueous 0.05 M NaClO₄ at 25 °C and refer to pectate (□), alginate (△), mannuronate (●) and guluronate (○); [210, 213, 214]

endothermic effect predicted is not realised in some cases, which include all the reported cases of ion-induced conformational transitions as monitored, for example, by changes in optical activity [205–214]. A few examples are reported in Figs. 4 and 5 where both endothermic (i.e., "expected") and exothermic (i.e., "conformational transition") enthalpy changes are observed.

7 Conclusions

To conclude this review, it is worth listing some unanswered questions and the contribution one would expect to obtain from a calorimetric approach.

Most of the problems and obstacles in studying polysaccharides concern, as outlined at the beginning, the basic covalent structure and the modifications induced in their secondary and tertiary structures. Bearing in mind the statistical occurrence of these "defects", one would say that, despite structures, thermodynamics of the processes involving the carbohydrate chains may be affected to a smaller extent.

The main result of focusing attention on the energetics of the processes and not only on the structures is, after all, the positive result of the thermodynamic approach. The more or less "specific" interactions occurring between carbohydrate moieties and other molecules (solvent, carbohydrates, or other cell constituents) have still to be fully identified and rationalized before transferring the solid state geometry to the real situation in solution.

Concepts like "hydrophobicity" and "hydrogen bond key-recognition" have been, too often dangerously, exported from the realm of nucleic acids and proteins to the polysaccharide field. Both these mechanisms need to be adapted before being applied to carbohydrate chains, because of the absence of clear hydrophobic clefts and the presence of too many hydrogen bonding sites. Polysaccharides surfaces may be seen more as a fakir nail-bed than a key and lock contact.

The need of such a revision has been stressed also in other fields. Contrasting and scarce evidence has been offered, for example, in the protein-saccharide interaction to let a clear picture emerge [221–223]. Even in the simple case of self-association of caffeine and other purines [224–226] or caffeine-phosphorylase interaction [227], other attractive (van der Waals and dipole-dipole) but canonical hydrophobic forces have been found significant, thus casting doubt upon this habit of generalization.

The current collection of a body of data about the thermodynamics of simple carbohydrate has to be completed and extended to higher structures so that a new "culture" of understanding can develop, devoid of the uncritical transfer of experience from seemingly similar "poly"-mers.

Acknowledgments. The author has enjoyed a protracted interaction with Prof. V. Crescenzi, G. Barone, S. Paoletti and F. Delben. The patient revision of the manuscript by T. J. Painter is gratefully acknowledged. The paper has been prepared with the financial support of Progetto Finalizzato Chimica Fine e Secondaria, C.N.R. Rome and of University of Trieste.

References

1. Suggett A (1973) In: Franks F (ed) Water – a comprehensive treatise, vol 4. Plenum, New York, pp 519–567
2. Green MM, Blankenhorn G, Hart H (1975) J Chem Educ 52:729–730
3. Angyal SJ (1969) Angew Chem Int Ed Engl 8:157–226
4. Stoddard JF (1971) Stereochemistry of carbohydrates. Wiley Interscience, New York
5. Aspinall GO (1982) In: Aspinall GO (ed) The polysaccharides. Academic Press, New York, pp 19–34
6. Sturtevant JM (1941) J Phys Chem 45:127–134
7. Taylor JB, Rowlinson JS (1955) Trans Faraday Soc 51:1183–1192
8. Andersen B, Gronlund F (1965) Acta Chem Scand 19:723–728
9. Kabayama MA, Patterson D, Piche L (1958) Can J Chem 36:557–562
10. Ono S, Takahashi K (1969) In: Brown HD (ed) Biochemical microcalorimetry. Academic Press, New York, pp 99–116
11. Robinson RA, Stokes RH (1959) Electrolyte solutions, 2nd edn. Butterworths, London
12. Franks F, Reid DS (1973) In: Franks F (ed) In: Water – a comprehensive treatise, vol 2. Plenum, New York, pp 323–380
13. Franks F (1967) In: Franks F (ed) Physico-chemical processes in mixed aqueous solvents. Heineman, London, pp 50–70
14. Nichols N, Sköld R, Spink C, Suurkuusk J, Wadso I (1976) J Chem Thermodynam 8:1081–1093
15. Franks F (1979) In: Blanshard JMV, Mitchell JR (eds) Polysaccharides in foods. Butterworths, London, pp 33–50
16. Bettelheim FA (1970) In: Veiss A (ed) Biological polyelectrolytes. Dekker, New York, pp 131–209
17. Reid DS (1979) In: Jones N (ed) Biochemical thermodynamics. Elsevier, Amsterdam, pp 168–184
18. Reid DS (1983) In: Wilson AD, Prosser HJ (eds) Developments in ionic polymers 1. Applied Science, London, pp 269–292
19. Brant DA (ed) (1981) Solution properties of polysaccharides. ACS Symp Ser 150. Am Chem Soc Wash DC
20. Franks F (1979) In: Jones MN (ed) Biochemical thermodynamics. Dekker, New York, pp 15–74
21. Franks F (1977) Philos Trans R Soc Lond B Biol Sci 278:33–56
22. Rees DA (1975) In: Whelan WJ (ed) MTP Int Rev Sci, Biochem Ser 1. Butterworth, London, pp 1–42
23. Rees DA (1977) Polysaccharide shapes, outline studies in biology. Chapman and Hall, London
24. Rees DA, Morris ER, Thom D, Madden JK (1982) In: Aspinall GO (ed) The polysaccharides, vol 1. Academic Press, New York, pp 195–290
25. Ben Naim A (1978) J Phys Chem 82:792–803
26. Friedman HL, Krishnan CV (1973) In: Franks F (ed) Water – a comprehensive treatise, vol 3. Plenum, New York, pp 1–118
27. McMillan WG, Mayer JE (1945) J Chem Phys 13:276–306
28. Mayer JE (1950) J Chem Phys 18:1426–1436
29. Friedman HL (1960) J Chem Phys 32:1351–1362
30. Kozak JJ, Knight WS, Kauzmann W (1968) J Chem Phys 48:675–690
31. Friedman HL, Krishnan CV (1973) J Solution Chem 2:119–138
32. Fortier JL, Leduc PA, Picker P, Desnoyers JE (1973) J Solution Chem 2:467–475
33. Franks F, Pedley M, Reid DS (1976) J Chem Soc Faraday Trans I 72:359–367
34. Savage JJ, Wood RH (1976) J Solution Chem 5:733–750
35. Tasker IR, Wood RH (1982) J Solution Chem 7:469–480
36. Tasker IR, Wood RH (1982) J Phys Chem 86:4040–4045
37. Barone G, Cacace P, Castronuovo G, Elia V (1983) Carbohydr Res 119:1–11
38. Cesàro A, Russo E, Barone G (1982) Int J Pept Protein Res 20:8–15
39. Barone G, Castronuovo G, Doucas D, Elia V, Mattia CA (1983) J Phys Chem 87:1931–1937

40. Enderby JE, Neilson GW (1979) In: Franks F (ed) Water – a comprehensive treatise, vol 6. Plenum, New York, pp 1–46
41. McQuarrie DA (1976) Statistical mechanics. Harper and Row, New York
42. Frank HS, Wen WY (1957) Discuss Faraday Soc 24:133
43. Nèmethy G, Scheraga HA (1962) J Chem Phys 36:3382–3400
44. Franks F (1968) In: Covington AK, Jones P (eds) Hydrogen bonded solvent systems. Taylor and Francis, London, pp 31–47
45. Kavanau JL (1964) Water and solute-water interactions. Holden-Day, San Francisco
46. Holtzer A, Emerson MF (1969) J Phys Chem 73:26–33
47. Friedman HL (1972) J Solution Chem 1:387–412
48. Friedman HL (1972) J Solution Chem 1:413–417
49. Friedman HL (1972) J Solution Chem 1:419–431
50. Stokes RH (1972) Aust J Chem 20:2087–2100
51. Tait MJ, Suggett A, Franks F, Ablett S, Quickenden PA (1972) J Solution Chem 1:131–151
52. Franks F, Reid DS, Suggett A (1973) J Solution Chem 2:99–113
53. Suggett A, Clark AH (1976) J Solution Chem 5:1–15
54. Suggett A, Ablett S, Lillford PJ (1976) J Solution Chem 5:17–31
55. Suggett A (1976) J Solution Chem 5:33–46
56. Harvey JM, Symons MCR, Naftalin RJ (1976) Nature 261:435–436
57. Bociek S, Franks F (1979) J Chem Soc Faraday Trans I 75:262–270
58. Bock K, Lemieux RU (1982) Carbohydr Res 100:63–74
59. Harvey JM, Symons MCR (1978) J Solution Chem 7:571–586
60. Franks F, Ravenhill JR, Reid DS (1972) J Solution Chem 1:3–16
61. Hoiland H, Holvik H (1978) J Solution Chem 7:587–596
62. Edward JT (1970) J Chem Educ 47:261–270
63. Longsworth LG (1955) In: Shedlovsky T (ed) Electrochemistry in biology and medicine. Wiley, New York
64. Shahidi F, Farell PG, Edward JT (1976) J Solution Chem 5:807–816
65. Neal JL, Goring DAI (1970) J Phys Chem 74:658–664
66. Neal JL, Goring DAI (1970) Can J Chem 48:3745–3747
67. Jasra RV, Ahluwalia JC (1982) J Solution Chem 11:325–338
68. Di Paola G, Belleau B (1977) Can J Chem 55:3825–3830
69. Sangster J, Teng T-T, Lenzi F (1976) J Solution Chem 5:575–585
70. Garrod JE, Herrington TM (1970) J Phys Chem 74:363–370
71. Dack MRJ (1976) Aust J Chem 29:779–786
72. Kawaizumi F, Nishio N, Nomura H, Miyahara Y (1981) J Chem Thermodynam 13:89–98
73. Lian Y-N, Chen A-T, Suurkuusk J, Wadso I (1982) Acta Chem Scand A 36:3–7
74. Gucker FT, Pickard HB, Plenck RW (1939) J Am Chem Soc 61:459–470
75. Bonner OD, Cerrutti PL (1976) J Chem Thermodynam 8:105–112
76. Gucker FT, Ayres FD (1937) J Am Chem Soc 59:447–452
77. Cox JD, Pilcher G (1970) Thermochemistry of organic and organometallic compounds. Academic Press, London
78. Barry F (1920) J Am Chem Soc 42:1911–1945
79. Wilhoit RC (1969) In: Brown HD (ed) Biochemical microcalorimetry. Academic Press, New York
80. Ellerton HD, Reinfelds DE, Mulcahy DE, Dunlop PJ (1964) J Phys Chem 68:398–402
81. Robinson RA, Stokes RH (1961) J Phys Chem 65:1954–1958
82. Barone G, Cacace P, Castronuovo G, Elia E (1981) Carbohydr Res 91:101–111
83. Uedaira H, Uedaira H (1970) J Phys Chem 74:1931–1936
84. Stokes RH, Robinson RA (1966) J Phys Chem 70:2126–2131
85. Uedaira H, Uedaira H (1969) Bull Chem Soc Jpn 42:2137–2140
86. Lange E, Markgraf HG (1950) Z Elektrochem 54.73–76
87. Barone G, Cacace P, Castronuovo G, Elia V, Iappelli F (1981) Carbohydr Res 93:11–18
88. Barone G, Cacace P, Gastronuovo G, Elia V (1982) Gazz Chim Ital 112:153–158
89. Ref. 3 in Ref. 88

90. Quoted in Ref. 88
91. Kawaizumi F, Kushida S, Miyahara Y (1981) Bull Chem Soc Jpn 54:2282–2285
92. Barone G, Cacace P, Castronuovo G, Elia V, Lepore U (1983) Carbohydr Res 115:15–22
93. Barone G, Bove B, Castronuovo G, Elia V (1981) J Solution Chem 10:803–809
94. Tasker IR, Wood RH (1983) J Solution Chem 12:315–325
95. Franks F, Pedley MD (1983) J Chem Soc Faraday Trans I 79:2249–2260
96. Scatchard G (1921) J Am Chem Soc 43:2406–2419
97. Scatchard G, Hamer WJ, Wood SE (1938) J Am Chem Soc 60:3061–3070
98. Volkenstein MV (1963) Configurational statistics of polymeric chains (English translation). Wiley Interscience, London
99. Eliel EL, Allinger NL, Angyal SJ, Morrison GA (1965) Conformational analysis. Wiley, New York
100. Goebel KD, Harvie CE, Brant DA (1976) Appl Polymer Symp 28:671–691
101. Brant DA, Hsu B (1983) Carbohydr Res
102. Reeves RE (1951) Adv Carbohydr Chem 6:107–134
103. Angyal SJ (1968) Aust J Chem 21:2737–2746
104. Angyal SJ (1969) Angew Chem Int Ed Engl 8:157–226
105. Rees DA, Smith PJC (1975) J Chem Soc Perkin Trans II 2:830–835
106. Dunfield LG, Whittington SG (1977) J Chem Soc Perkin Trans II:654–658
107. Melberg S, Rasmussen K (1979) J Mol Struct 57:215–239
108. Melberg S, Rasmussen K (1979) Carbohydr Res 69:27–38
109. Melberg S, Rasmussen K (1979) Carbohydr Res 71:25–34
110. Melberg S, Rasmussen K (1980) Carbohydr Res 78:215–224
111. Giacomini M, Pullman B, Maigret B (1970) Theor Chim Acta 19:347–364
112. Saran A, Govil G (1971) Indian J Chem 9:1095–1097
113. Jeffrey GA, Yates JH (1980) Carbohydr Res 79:155–163 (and reference therein)
114. Pensak DA, French AD (1980) Carbohydr Res 87:1–10
115. Arnott S, Scott WE (1972) J Chem Soc Perkin Trans II:324–335
116. French AD, Murphy VG (1973) Carbohydr Res 24:391–406
117. Jeffrey GA, French AD (1978) In: Sutton LE, Truter ME (eds) Structure by diffraction methods, vol 6. Specialist periodical reports. Chem Soc, Lond
118. Pierrotti RA (1965) J Phys Chem 69:281–288
119. Sinagoglu O (1968) In: Pullman B (ed) Molecular association in biology. Academic Press, New York
120. Stillinger FH (1973) J Solution Chem 2:141–158
121. Beveridge DL, Kelly MM, Radna RJ (1974) J Am Chem Soc 96:3769–3778
122. Kozàr T, Tvaroska I (1979) Theor Chim Acta 53:9–19
123. Tvaroska I, Kozàr T (1980) J Am Chem Soc 102:6929–6936
124. Tvaroska I (1982) Biopolymers 21:1887–1897
125. Thoma JA, Stewart L (1965) In: Whistler RL (ed) Starch, chemistry and technology. Academic Press, New York
126. Cramer F, Hettler H (1967) Naturwissenschaften 54:625–632
127. Lewis EA, Hansen LD (1973) J Chem Soc Perkin Trans II:2081–2085
128. Saenger W (1980) Angew Chem Int Ed Engl 19:344–362
129. Fendler JH, Fendler EJ (1975) Catalysis in micellar and macromolecular systems, chap 11. Academic Press, New York
130. Cramer F, Saenger W, Spatz H-C (1967) J Am Chem Soc 89:14–20
131. van Hooidonk C, Breebart-Hansen JCAE (1971) Recl Trav Chim Pays-Bas Belg 90:680–686
132. van Etten RL, Sebastian JF, Clowes GA, Bender ML (1967) J Am Chem Soc 89:3242–3253
133. Straub TS, Bender ML (1972) J Am Chem Soc 94:8881–8888
134. Griffiths DW, Bender ML (1973) Adv Catal 23:209–261
135. Komiyama M, Bender ML (1978) J Am Chem Soc 100:2259–2260
136. Lumry R, Rajender S (1970) Biopolymers 9:1125–1227
137. Hingerty B, Saenger W (1975) Nature 255:396–397
138. Saenger W, Noltenmeyer M, Manor PC, Hingerty B, Klar B (1976) Bioinorg Chem 5:187–195

139. Saenger W (1976) In: Pullman B (ed) Environmental effects on molecular structure and properties. Reidel Publishing Company, Dordrecht Holland/Boston USA
140. Lindner K, Saenger W (1978) Angew Chem Int Ed Engl 17:694–695
141. Banks W, Greenwood CT (1975) Starch and its components. Edinburgh University Press, Edinburgh
142. Rundle RE (1947) J Am Chem Soc 69:1769–1772
143. Rundle RE, Baldwin RR (1943) J Am Chem Soc 65:554–558
144. Cesàro A, Jerian E, Saule S (1980) Biopolymers 19:1491–1506
145. Banks W, Greenwood CT, Kahn KM (1971) Carbohydr Res 17:25–33
146. Cronan CL, Schneider FW (1969) J Phys Chem 73:3990–4004
147. Takahashi K, Ono S (1972) J Biochem (Tokyo) 72:1041–1043
148. Cesàro A, Brant DA (1976) Biopolymers 16:983–1006
149. Brant DA (1976) Q Rev Biophys 9:527–596
150. Brant DA (1982) In: Preiss J (ed) The biochemistry of plants. Academic Press, New York
151. Smidsrød O (1980) In: Varmavuori A (ed) Proc 27th IUPAC. Pergamon, Oxford
152. Lifson S, Katchalsky A (1954) J Polym Sci 13:43–55
153. Alfrey T, Berg PW, Morawetz H (1951) J Polym Sci 7:543–547
154. Fuoss RM, Katchalsky A, Lifson S (1951) Proc Natl Acad Sci USA 37:579–589
155. Hermans JJ, Overbeck JTG (1948) Rec Trav Chim 67:761–776
156. Oosawa F (1971) Polyelectrolytes. Dekker, New York
157. Manning GS (1969) J Chem Phys 51:924–933
158. Manning GS (1977) Biophys Chem 7:95–102
159. Manning GS (1978) Biophys Chem 9:65–70
160. Manning GS (1978) Q Rev Biophys 11:179–246
161. Manning GS (1979) Accounts Chem Res 12:443–449
162. Paoletti S, Delben F, Crescenzi V (1981) J Phys Chem 85:1413–1418
163. Fenyo JC (1981) J Polym Sci Part D Macromol Rev 19:1015–1018
164. Paoletti S, Cesàro A, Delben F, Crescenzi V, Rizzo R (1985) In: Dubin P (ed) Microdomains in polymers. Plenum, New York
165. Boyd GE, Wilson DP, Manning GS (1976) J Phys Chem 80:808–810
166. Podlas TJ, Ander P (1969) Macromol Rev 2:432–436
167. Podlas TJ, Ander P (1970) Macromol Rev 3:154–157
168. Tomasula M, Swanson N, Ander P (1978) In: Schirager RG (ed) ACS Symp Ser 77. Carbohydrate sulfates. Washington, pp 245
169. Kowblansky A, Sasso R, Spagnuola V, Ander P (1977) Macromol Rev 10:78–83
170. Ander P, Gangi G, Kowblansky A (1978) Macromol Rev 11:904–908
171. Ander P in ref. 19, chap 28 and references cited therein
172. Satake L (1972) J Polym Sci Part D Macromol Rev 10:2343–2354
173. Hoguchi H, Gekko K, Makino S (1973) Macromol Rev 6:438–442
174. Pass G, Phillips GO, Wedlock DJ (1977) J Chem Soc Perkin Trans D 2:1229
175. Pass G, Phillips GO, Wedlock DJ, Morley RG (1978) In: Schweiger RG (ed) ACS Symp Ser 77. Carbohydrate sulfates. Washington, p 275
176. Pass G, Phillips GO, Wedlock DJ (1977) Macromol Rev 10:197–201
177. Oman S (1974) Makromol Chem 175:2133–2140
178. Preston BN, Snowden JMcK, Houghton KT (1972) Biopolymers 11:1645–1659
179. Kwak JCT, O'Brien MC, Maclean DA (1975) J Phys Chem 79:2381–2386
180. Joshi YM, Kwak JCT (1979) J Phys Chem 83:1978–1983
181. Magdelenat H, Turq P, Chemla M (1976) Biopolymers 15:175–186
182. Magdelenat H, Turq P, Chemla M (1974) Biopolymers 13:1535–1548
183. Magdelenat H, Turq P, Tivant P, Drifford M in ref. 19, chap 27
184. Magdelenat H, Turq P, Tivant P, Chemla M, Menez R, Drifford M (1979) Biopolymers 18:187–201
185. Rinaudo M, Milas M (1970) Comp Rend Acad Sci Paris Ser C 271:1170–1172
186. Rinaudo M, Milas M (1973) Macromolecules 6:879–881
187. Ravanat G, Rinaudo M (1980) Biopolymers 19:2209–2222
188. Wedlock DJ, Bradshaw F, Phillips GO (1981) Int J Biol Macromol 3:275–277
189. Kohn R, Furda I (1967) Collect Czech Chem Commun 32:1925–1937

190. Kohn R (1975) Pure Appl Chem 42:371–397
191. Kohn R, Malovíkova A (1981) Collect Czech Chem Commun 46:1701–1707
192. Malovíková A, Kohn R (1982) Collect Czech Chem Commun 47:702–708
193. Chakrabarti B, Park JV (1980) CRC Crit Rev Biochem 8:225–313
194. Cleland RL (1979) Biopolymers 18:2673–2681
195. Cleland RL (1982) Macromolecules 15:382–386
196. Cleland RL, Wang JL, Detweiler DH (1982) Macromolecules 15:386–395
197. Manzini G, Cesáro A, Delben F, Paoletti S, Reisenhofer E (1984) Bioelectrochem Bioenerg 12:443–454
198. Reisenhofer E, Cesáro A, Delben F, Manzini G, Paoletti S (1984) Bioelectrochem Bioenerg 12:455–465
199. Hales PW, Pass G (1980) J Chem Soc Faraday Trans I 76:2080–2083
200. Hales PW, Pass G (1981) J Chem Soc Faraday I 77:2009–2013
201. Hales PW, Pass G (1981) Eur Polymer J 17:1289–1292
202. Pass G, Hales PW in ref. 19, chap 24, pp 349–365
203. Hales PW, Pass G (1982) J Chem Res 206–206
204. Hales PW, Pass G (1982) J Chem Soc Faraday Trans I 78:283–287
205. Crescenzi V, Dentini M, Paradossi G, Rizzo R (1979) Polymer Bull 1:771–776
206. Crescenzi V, Dentini M, Paradossi G, Rizzo R (1979) Polymer Bull 1:777–784
207. Crescenzi V, Dentini M, Rizzo R in ref. 19 chap 23
208. Crescenzi V, Airoldi C, Dentini M, Pietrelli L, Rizzo R (1981) Macromol Chem 182:219–223
209. Crescenzi V, Dentini M, Pietrelli L (1981) Period Biol 83:125–128
210. Paoletti S, Cesàro A, Ciana A, Delben F, Manzini G, Crescenzi V in ref. 19, chap 26
211. Cesàro A, Ciana A, Delben F, Manzini G, Paoletti S (1982) Biopolymers 21:431–449
212. Delben F, Cesàro A, Paoletti S, Crescenzi V (1982) Carbohydr Res 100:C46–C50
213. Cesàro A, Paoletti S, Delben F, Crescenzi V, Rizzo R, Dentini M (1982) Gazz Chim Ital 112:115–121
214. Cesàro A, Delben F, Paoletti S (1982) In: Miller B (ed) Thermal analysis. Wiley-Heiden, Chichester, pp 815–821
215. Paoletti S, Cesàro A, Delben F (1983) Carbohydr Res 123:173–178
216. Mita K, Okubo T (1974) J Chem Soc Faraday Trans I 70:1546–1550
217. Tsuge H, Yonese M, Kishimoto H (1979) Bull Chem Soc Jpn 52.2846–2848
218. Yonese M, Tsuge H, Kishimoto H (1981) Bull Chem Soc Jpn 54:20–24
219. Basedow AM, Ebert KH, Feigenbutz W (1980) Makromol Chem 181:1071–1080
220. Angyal SJ (1980) Chem Soc Rev 9:415–428
221. Di Paola G, Belleau B (1978) Can J Chem 56:848–850
222. Banerjee SK, Rupley JA (1975) J Biol Chem 250:8267–8274
223. Goldsmith E, Fletterick R (1983) Pure Appl Chem 55:577–622
224. Gill SJ, Downing M, Sheats GF (1967) Biochemistry 6:272–276
225. Marenchic MG, Sturtevant JM (1973) J Phys Chem 77:544–548
226. Cesàro A, Russo E, Crescenzi V (1976) J Phys Chem 80:335–339
227. Sprang S, Fletterick R, Stern M, Yang D, Madsen N, Sturtevant J (1982) Biochemistry 21:2036–2048

Section III   **Interactions in Solution**

Chapter 7 Thermodynamic Data for Protein-Ligand Interaction

Heiner Wiesinger and Hans Jürgen Hinz

Symbols

ΔG^0:	standard Gibbs energy change
ΔH^0:	standard enthalpy change
ΔS^0:	standard entropy change
ΔC_p^0:	standard molar heat capacity change
$\Delta H^{0\prime}$:	apparent molar enthalpy change
$\Delta G^{0\prime}$:	apparent molar Gibbs energy change
$\Delta S^{0\prime}$:	apparent molar entropy change
$\Delta C_p^{0\prime}$:	apparent molar heat capacity change

1 Introduction

Knowledge of the magnitude of the energy, entropy, and heat capacity changes involved in ligand binding equilibria of biological compounds is essential for a prediction of the thermodynamic properties of the systems under different environmental conditions. Quantitative thermodynamic data will provide the ultimate test for the quality of ab initio theoretical stability calculations of binding phenomena if such treatments become possible in the future. Heat capacity changes are an excellent qualitative indicator of the structural changes (including the hydration changes) that many proteins and ligands undergo on complex formation [1, 2]. Since heat capacity is related to the number of degrees of freedom in the distribution of enthalpy states [3], changes in heat capacity resulting from ligand binding may reflect changes in the static and the dynamic properties of the macromolecule.

The structural changes associated with the energy changes on complex formation can result in permanent shifts of the relative positions of the atoms of the macromolecule large enough to be detectable by X-ray analysis. This is understandable if one realizes that the forces operative between proteins and ligands are of the same order of magnitude as the forces stabilizing the native structure of proteins [4–6]. However, X-ray analysis will only detect the more spectacular structural alterations, since, depending on the characteristic internal macromolecular interactions, energy changes can be distributed among a large number of weak interactions, each of which need not produce sizable changes in the static structure of the respective groups. Therefore changes in heat capacity will be a more sensitive but less uniquely interpretable diagnostic tool for the detection of "structural" changes particularly, since changes in heat capacity may also reflect

changes in the soft internal vibrational and rotational modes of the macromolecule [1], as well as reactions with third components, such as water or buffer ions,
concomitant with ligand binding. Thus, when properly interpreted, the energy
and entropy data can provide more than just thermodynamic insight into the
reaction.

2 Thermodynamic Quantities and Their Measurements

The thermodynamic quantities useful for characterization of a ligand binding
equilibrium are the standard Gibbs energy change, ΔG^0, the standard enthalpy
change, ΔH^0, the standard entropy change, ΔS^0, and the heat capacity change at
constant pressure, ΔC_p^0.

2.1 Gibbs Energy Changes

For biochemical ligand binding equilibria ΔG^0 is usually obtained from the equilibrium constant, K_{eq}, employing the relationship

$$\Delta G^0 = -RT \cdot \ln K_{eq}, \tag{1}$$

where R is the gas constant and T the absolute temperature. ΔG^0 will be given
in J mol^{-1}, if R is expressed in J mol^{-1} K^{-1}. The value used for R $= 8.314$ J mol^{-1}
K^{-1}. In general the numerical value of ΔG^0 will depend on the concentration
units employed in calculating the equilibrium constant, unless $\Sigma n_i = 0$. For clari-
fication, see the excellent chapters in [3] and [7]. While for reactions between inorganic or organic compounds mole fractions are often appropriate concentration units, the most widely used measure of concentration in biochemical studies
is the molarity, i.e., the number of moles of the solute per dm^3 of the solution.

   Although generally formulation of the equilibrium constant would involve the
activities of the reacting species, lack of knowledge of activity coefficients for
most reactions involving biological macromolecules necessitates assuming the activity coefficients to be unity and using the equilibrium concentrations. Therefore
the value of the equilibrium constant and concomitantly that of ΔG^0 may vary
with composition. Since protein ligand equilibria often depend on many factors
such as pH, salts, reducing agents, and buffering compounds, and since their influence on K_{eq} is seldom known, it is appropriate to use apparent equilibrium
constants K_{eq}' and apparent Gibbs energies $\Delta G^{0'}$. It should be mentioned that
this definition is not identical to the convention often used in biochemical research, where a standard Gibbs energy change referring to pH 7 is specified by
a prime.

2.2 Enthalpy Changes

Apparent standard enthalpy changes of ligand binding reactions can be indirectly
obtained from the dependence on temperature of the apparent equilibrium constant K_{eq}' according to the van't Hoff equation

$$d\ln K_{eq}'/d(1/T) = -(\Delta H^{0'}/R), \tag{2}$$

R being the gas constant and T the absolute temperature, or directly by microcalorimetric measurements of the heat effects involved in the binding process. Calorimetric $\Delta H^{0\prime}$-values are obtained as the algebraic sum of all heat effects occurring in the system, whereas van't Hoff enthalpies are the result of more complex summations of all the individual enthalpies through a proper combination of equilibrium constants [8]. The intrinsic dependence on the model used to describe the equilibrium is a possible source of error when applying the van't Hoff method. However, it does not usually play any role for simple ligand binding reactions where the stoichiometry is known [9].

Calorimetric enthalpies are essentially model free quantities. They are determined from the observed overall heat effect Q_{obs} by applying proper corrections for experimental artifacts. Q_{obs} consists in general of three experimentally separable heat contributions

$$Q_{obs} = Q_R + Q_{Dil} + Q_{Mix}. \tag{3}$$

Q_R is the heat of the ligand binding reaction, which may include contributions from linked protonation and deprotonation equilibria, interaction with buffer components etc. Q_{Dil} refers to the heat effects associated with dilution of the reaction components in the mixing process and Q_{Mix} to all physical effects, such as viscous heat, concomitant with flowing or agitating solutions, noncompensatory background heat effects due to nonidentity of sample and reference cell in twin calorimeters, and differences in the wetting of dry surfaces of the cells during rotation. To calculate the heat of the reaction Q_R from the observed heat effect, Q_{obs}, Q_{Dil}, and Q_{Mix} must be determined in separate mixing and dilution experiments without the reaction taking place. The apparent reaction enthalpy per mole of protein, $\Delta H^{0\prime}$, is obtained from Q_R by division by the actual number of moles of protein, n, involved in the reaction:

$$\Delta H^{0\prime} \ (J \ mol^{-1} \ protein) = Q_R/n. \tag{4}$$

Usually not the enthalpy per mole of protein is of thermodynamic interest, but the enthalpy per mole of binding site. In case of identical and independent binding sites, the site-specific enthalpy change is calculated simply by division of $\Delta H^{0\prime}$ by the number of sites, s:

$$\Delta H^{0\prime} (J \ mol^{-1} \ of \ binding \ site) = Q_R/n \cdot s. \tag{5}$$

Determination of site-specific enthalpies for enzymes having enthalpically different classes of identical and independent binding sites or involving cooperation between the sites requires more sophisticated mathematical analyses. Due to the large number of adjustable parameters, these calculations are only meaningful on the basis of extensive thermal titration data and, preferably, in connection with independently determined equilibrium constants [10, 11].

2.3 Changes in Entropy

$\Delta S^{0\prime}$ usually has been calculated from the equation

$$\Delta S^{0\prime} = (\Delta H^{0\prime} - \Delta G^{0\prime})/T, \tag{6}$$

using the experimental $\Delta G^{0\prime}$- and $\Delta H^{0\prime}$-values.

2.4 Changes in Heat Capacity

The majority of enthalpies of protein ligand interactions studied is strongly dependent on temperature. This finding can always be formally described as an apparent change in heat capacity, $\Delta C_p^{0'}$, associated with complex formation. $\Delta C_p^{0'}$ is calculated from the variation with temperature of $\Delta H^{0'}$ according to the equation:

$$\Delta C_p^{0'} = d/dt \, (\Delta H^{0'}).\tag{7}$$

In practically all cases $\Delta C_p^{0'}$ has been found to be temperature-independent within the precision of the measurements.

$\Delta C_p^{0'}$ values can in principle also be estimated from the degree of curvature of a van't Hoff plot according to the equation:

$$\Delta C_p^{0'} = \frac{\partial}{\partial T} \left(R \, \frac{\partial \ln K'_{eq}}{\partial (1/T)} \right).\tag{8}$$

However, the precision required for the values of the equilibrium constant to permit a meaningful second derivative with respect to temperature is rarely attained. Therefore direct microcalorimetric studies of the variation with temperature of the enthalpy of reaction are the most reliable source of heat capacity data.

2.5 Potential Errors in Microcalorimetric Measurements of Protein-Ligand Interactions

The virtually ubiquitous occurrence of heat effects of physical and chemical origin necessitates extreme care being taken when performing microcalorimetric measurements. The importance of accurate dilution and mixing experiments to determine Q_{Dil} and Q_{Mix}, respectively, employing all reaction components except one to prevent the reaction, has been emphasized repeatedly [12]. However, in addition to the correct methodological approach a few other sources of systematic error should be mentioned.

a) The accuracy of microcalorimetric studies on enzyme-ligand interactions can be strongly influenced by the purity of the biological samples. It is of utmost importance to check the homogeneity of the proteins in order to exclude enthalpy contributions from side reactions. Tests of the specific enzymic activity before and after the calorimetric runs should be mandatory to provide a basis for comparison of results obtained with different sample preparations or by different groups.

b) The possible dependence of the thermodynamic parameters of protein-ligand equilibria on the presence and concentration of other components such as buffers, neutral salts, and hydrogen ions, requires careful studies of their influence, since any calorimetric measurement is the algebraic sum of all heat effects.

c) The majority of proteins is multimeric, the subunits being identical or nonidentical. For quantitative thermodynamic studies the exact stoichiometry of the reaction observed in the calorimeter should be known. This requirement is self-evident, in practice, however, often difficult to fulfill. Usually a variety of different techniques ought to be employed to determine the exact number of interaction sites per mole of protein.

d) Biological samples often show low stability. A potential source of error is partial denaturation of the sample during the relatively long duration of calorimetric studies. A good criterion of the native state is a test of the specific activity before and after the measurement, as mentioned above.

2.6 Sensitivity of Microcalorimeters and Calibration Procedures

The average thermopile output of the microcalorimeters used for determination of thermodynamic parameters of biological reactions is in the range of 0.01– 0.1 $\mu V/\mu W$. Electronic amplification of the μV signal is in principle no longer a serious problem, however, the precision of the measurement depends of course critically on the signal-to-noise ratio. Baseline instabilities can often be reduced by careful thermostating of the calorimeter (water thermostat instead of air thermostat). A useful sensitivity can be defined as the minimum detectable heat effect which causes a distinct response from the heat sensing device. This sensitivity lies between 0.4 and 1 $\mu J\ s^{-1}$ for the common flow- and batch-calorimeters. The reproducibility is usually better than 1% (standard deviation, sd) at high rates of heat flow (LKB 10700 flow version: sd $<1\%$ at 450 $\mu J\ s^{-1}$; Beckman 190 flow version: 0.2% at 420 $\mu J\ s^{-1}$) while at low rates of heat flow (1 to 10 $\mu J\ s^{-1}$) it may be only about 10%.

The accuracy of the thermodynamic parameters derived from the calorimetric measurements depends on proper calibration. Electrical calibration via built-in heaters is convenient and generally can be performed with high accuracy. However, it is not a trivial procedure, since positioning of the heater within the reaction cell may influence the result, as has been discussed by [12]. An alternative way of calibration is by chemical reactions. Very well characterized reactions are the neutralization reaction between a strong base and a strong acid, the titration of ultra pure Tris [13], and the dilution of sucrose [14]. Preferentially, microcalorimeters should routinely be calibrated employing electrical and chemical methods.

2.7 Criteria of Selection and Arrangement of Tables

Unlike the situation in general thermodynamics, the majority of thermodynamic parameters for biochemical reactions has often been determined by only one laboratory. Therefore it is sometimes not easy to judge the quality of the data, since in addition to proper execution of thermodynamic and calorimetric measurements correct biochemical treatment of the samples is of fundamental importance. Several of the differences in the thermodynamic quantities, where the same proteins have been studied by various groups, are probably due to differences in the quality of the biological samples employed and not to variances between the thermodynamic measurements themselves. However, unfortunately the samples used sometimes have been unsatisfactorily characterized with regard to homogeneity, specific activity, and the influence of solution components. Thus in the present selection it was attempted to judge, besides the formal correctness of the thermodynamic measurements, also the quality of the samples, as far as it was possible on the basis of the usually brief "materials and methods" sections. Other general criteria were the availability of enthalpy and entropy data in addition to the Gibbs energies. Preference was given to calorimetrically determined enthalpies, when both van't Hoff data and calorimetric values were accessible. Entropy values were calculated according to Eq. (6). Throughout the tables all values refer to one mole of binding site. Where available, $\Delta C_p^{0\prime}$ data have been obtained from the dependence on temperature of calorimetrically determined enthalpies according to Eq. (7).

The arrangement of the results in the tables does not follow a unique scheme. Biochemical aspects and the attempt to facilitate a comparison of thermodynamic parameters of classes of enzymes with the same ligand or different ligands with the same enzyme were considered to be more useful than pure, e.g., alphabetic systematization. No molecular interpretation of the thermodynamic data has been attempted. General ideas concerning mechanistic interpretation of energetic parameters can be found in recent articles [1, 2, 15, 16].

However, comments concerning the individual experiments will be given in connection with the tables. For the compilation original data have been converted from cal to J using 1 cal = 4.184 J. Due to rounding errors, multiplication of the $\Delta S^{0\prime}$-values reported in the original publication by the factor 4.184 have led to results which were slightly different from those obtained by first converting $\Delta G^{0\prime}$ and $\Delta H^{0\prime}$ values to J and then calculating $\Delta S^{0\prime}$ according to Eq. (6). The latter procedure was adopted to be consistent with the Gibbs-Helmholtz equation and the corresponding $\Delta S^{0\prime}$ data have been listed.

Unless otherwise stated, all experiments were performed in commercially available LKB and Beckman microcalorimeters of the batch and flow type, respectively.

Table 1. Thermodynamic parameters for protein-ligand interaction

Entry protein No.	Ligand	$\Delta G^{0\prime}$ (kJ mol^{-1})	$\Delta H^{0\prime}$ (kJ mol^{-1})	$\Delta S^{0\prime}$ (J mol^{-1} K^{-1})	$\Delta C_p^{0\prime}$ (kJ mol^{-1} K^{-1})	Reference
1 Albumin (human plasma)	Warfarin	−30.8	−13.1	59.4		[36]
2 Alcohol dehydrogenase						
Horse liver ADH	NADH	−37.1±0.6	2.5± 3.3	116.0±113.0	−1.70±0.13	[34]
	NADH	−36.8	0.0± 1.7	123.0± 7.0		[35]
	NAD$^+$	−21.6±0.2	− 3.6± 0.9	60.2± 4.0	−0.45	[34]
	NAD$^+$	−19.2±0.4	− 4.2± 0.8	51.0± 4.2		[22]
	AMP	−23.4±0.6	− 42.7± 3.3	− 64.8± 1.3	−1.23±0.13	[34]
	ADP	−20.4±0.3	− 40.2± 2.5	− 66.5± 9.2	−0.56±0.10	[34]
	ADP-ribose	−25.5±0.6	− 36.8± 1.3	− 38.1± 6.3	−0.66±0.05	[34]
	ADP-ribose		− 25.0± 0.8			[35]
Yeast ADH	NADH	−26.4±0.4	− 39.7± 0.8	− 44.8± 4.2		[35]
	NAD$^+$	−18.0±0.4	− 37.2± 1.7	− 64.4± 6.3		[22]
3 Aldolase (rabbit muscle)	Arabinitol 1,5-diphosphate	−41.8±0.8	− 15.9± 1.7	86.3± 4.2		[17]

Table 1 (continued)

try protein	Ligand	$\Delta G^{0\prime}$ (kJ mol^{-1})	$\Delta H^{0\prime}$ (kJ mol^{-1})	$\Delta S^{0\prime}$ (J mol^{-1} K^{-1})	$\Delta C_p^{0\prime}$ (kJ mol^{-1} K^{-1})	Reference
Arabinose binding	L-Arabinose		$-$ 64.1$\pm$ 2.1		$-1.8\ \pm0.2$	[24]
protein (*E. coli*)	D-Galactose		$-$ 62.8$\pm$ 0.6		$-1.6\ \pm0.1$	[24]
Aspartate amino-	PLP(pH6)		$-$ 8.4		-1.55	[18]
transferase	PLP(pH8)		$-$ 33.5		-1.38	[18]
(pig heart)						
Aspartate trans-						
carbamoylase						
(ATCase,						
E. coli)						
c_6r_6	PALA		-210.6			[19]
c_3	PALA		-168.7			[19]
c_6r_6	PALA	-38.1	$-$ 35.6$\pm$ 4.2	8.4		[20]
c_3	PALA	-40.6	$-$ 47.3$\pm$ 2.9	$-$ 22.5		[20]
c_6r_6	CP	-27.6	$-$ 16.7$\pm$ 2.5	36.5		[20]
c_3	CP	-36.0	$-$ 15.1$\pm$ 2.5	70.2		[20]
Chymotrypsin	Indole	-16.7	$-$ 25.1	$-$ 28.5		[25]
	Indole	-18.0	$-$ 63.6$\pm$ 3.4	-153.0		[26]
	N-Acetyl-D-tryptophan	-13.7	$-$ 79.5$\pm$ 2.5	-220.7		[26]
	N-Acetyl-L-tryptophan	-12.4	$-$ 88.7$\pm$ 5.0	-256.2		[27]
	Benzoate	-11.1	$-$ 76.2$\pm$ 4.2	-218.5		[27]
	Hydrocinnamate	-15.5	$-$ 65.7$\pm$ 4.2	-168.4		[27]
	β-Naphtoate	-22.2	$-$ 73.2$\pm$ 4.2	-171.3		[27]
	Phenol	-12.8	$-$ 56.5$\pm$ 2.1	-146.4		[27]
	Indole-3-propionic acid	-18.5 ± 0.2	$-$ 15.3$\pm$ 0.2	10.8$\pm$ 1.2		[28]
Dihydrofolate	NADP	-31.0	$-$ 2.5	95.5		[29]
reductase	NADPH	-37.2	$-$ 14.2	77.3		[29]
(chicken liver)	Folate	-41.0	$-$ 47.3	$-$ 21.1		[29]
	Dihydrofolate	-44.8	$-$ 47.3	$-$ 8.4		[29]
	Methotrexate	-49.0	$-$ 55.7	$-$ 22.5		[29]
Flavin binding	FMN	-31.8	$-$ 84.8	-172.2		[70]
protein	FAD	-27.2	$-$ 43.0	$-$ 62.3		[70]
(egg white)	3-CM-Rf	-46.4	$-$ 87.4$\pm$ 3.3	-142.1		[71]
Flavodoxin	FMN	-46.0	$-120.8\pm$ 6.7	-250.8		[72]
	FAD	-33.8	$-$ 72.7$\pm$ 4.6	-129.6		[72]
	8-COOH-Rf	-38.0	$-$ 58.5$\pm$ 4.2	-204.8		[72]
Glutamate	NADPH	-27.6 ± 0.4	6.7$\pm$ 1.3	115.1$\pm$ 8.4	-0.26 ± 0.25	[30]
dehydrogenase	L-Glu	-11.7 ± 0.8	$-$ 2.5$\pm$ 4.6	30.9$\pm$ 25.1	0.04 ± 0.25	[30]
(beef liver)	Adenosine	-18.0 ± 0.4	$-$ 51.5$\pm$ 2.1	$-112.6\pm$ 5.4	-0.04 ± 0.3	[23]
	AMP	-24.3 ± 0.8	$-$ 32.2$\pm$ 1.7	$-$ 26.8$\pm$ 5.4	$-0.4\ \pm0.2$	[23]
	ADP	-31.0 ± 0.4	$-$ 51.9$\pm$ 0.8	$-$ 70.3$\pm$ 2.9	$-0.5\ \pm0.2$	[23]
	ATP	-23.4 ± 0.8	$-$ 42.7$\pm$ 1.7	$-$ 64.5$\pm$ 5.4	$0.4\ \pm0.2$	[23]
Glutamine						
synthetase						
(E. coli)						
ADP-MgGS	L-Glu	-15.9	$-$ 32.2	$-$ 53.8		[31]
ADP-P$_i$-MgGS	L-Ala	-10.5	7.5	$-$ 59.4		[31]
MnGS	L-Met-S-Sulfoximine	-25.5	$-$ 52.3	$-$ 88.4		[32]
	L-Met-R-Sulfoximine	-18.8	$-$ 48.1	$-$ 96.7		[32]
	L-Glu	-12.6	$-$ 40.6	$-$ 92.5		[32]

Table 1 (continued)

Entry protein No.	Ligand	$\Delta G^{0'}$ (kJ mol^{-1})	$\Delta H^{0'}$ (kJ mol^{-1})	$\Delta S^{0'}$ (J mol^{-1} K^{-1})	$\Delta C_p^{0'}$ (kJ mol^{-1} K^{-1})	Reference
13 Hexokinase (yeast)	D-Glucose	-21.3 ± 0.2	$-\ 3.1\pm\ 3.8$	$61.1\pm\ 13.0$		[33]
14 Isoleucine tRNA synthetase (*E. coli* MRE 600)	L-Ile	-30.1	$-\ 15.5\pm\ 3.8$	49.1	-1.79 ± 0.08	[37]
	L-Leu	-14.6	$-\ 15.5\pm\ 3.8$	$-\ \ 2.8$	-2.00 ± 0.12	[37]
	LVal	-18.4	$-\ 15.5\pm\ 3.8$	9.8	-2.00 ± 0.10	[37]
	L-Isoleucinol	-11.7	$3.8\pm\ 3.8$	52.0	-1.52 ± 0.15	[37]
	tRNA^{Ile}	-39.8 ± 1.7	$-\ 11.1\pm\ 3.8$	$96.0\pm\ 25.0$	-2.26 ± 0.50	[38]
15 Lactate dehydrogenase						
Beef heart LDH	NADH	-36.8	$-\ 40.6\pm\ 0.8$	$12.6\pm\ 4.2$		[21]
	NAD$^+$	-18.0 ± 0.4	$-\ 35.6\pm\ 1.7$	$-\ 59.4\pm\ 5.9$		[22]
Pig heart LDH	NADH	-30.9 ± 0.3	$-\ 44.4\pm\ 1.3$	$-\ 45.2\pm\ 5.0$	-0.70 ± 0.04	[45]
	NAD$^+$	-19.9 ± 0.8	$-\ 25.5\pm\ 0.8$	$-\ 18.8\pm\ 0.3$	-0.35 ± 0.03	[45]
	Adenosine	-12.1 ± 0.2	$-\ 25.5\pm\ 1.3$	$-\ 44.8\pm\ 5.0$	0.01 ± 0.04	[46]
	ADP	-16.9 ± 0.3	$-\ 24.7\pm\ 1.7$	$-\ 26.4\pm\ 6.0$	-0.31 ± 0.06	[45]
	ADP-ribose	-20.8 ± 0.1	$-\ 26.8\pm\ 6.8$	$-\ 20.5\pm\ 1.7$	-0.41 ± 0.03	[45]
	AMP	-16.7 ± 0.4	$-\ 13.4\pm\ 0.8$	$10.9\pm\ 4.0$	-0.43 ± 0.03	[45]
Pig muscle LDH	NADH	28.9 ± 0.3	$-\ 31.6\pm\ 2.1$	$-\ \ 9.0\pm\ 8.4$	-1.26 ± 0.04	[47 48]
	NAD$^+$	-18.8 ± 0.3	$-\ 27.6\pm\ 2.5$	$-\ 29.3\pm\ 13.4$	-0.51 ± 0.05	[46]
	Adenosine	-11.9	$-\ 27.4\pm\ 4.8$	$-\ 51.9$	-1.09 ± 0.09	[49]
	AMP	-14.6 ± 0.4	$-\ 16.9\pm\ 3.0$	$-\ \ 7.5\pm\ 11$	-0.79 ± 0.06	[46]
	ADP	-14.5 ± 0.4	$-\ 21.9\pm\ 2.4$	$-\ 24.4\pm\ 9.0$	-0.88 ± 0.06	[46]
	ADP-ribose	-17.0 ± 0.2	$-\ 32.6\pm\ 2.3$	$-\ 52.2\pm\ 8.0$	-1.44 ± 0.07	[46]
Rabbit muscle LDH	NADH	-31.0	$-\ 28.9\pm\ 0.4$	$-\ \ 7.1\pm\ 2.9$		[21]
	NAD$^+$	-15.9 ± 0.4	$-\ 26.4\pm\ 0.8$	$-\ 34.7\pm\ 4.2$		[22]
16 Lac repressor	IPTG(pH7)	-30.8	$-\ 15.6$	51.0		[39]
	IPTG(pH9)	-29.3	$-\ 4.6$	82.9		[39]
17 Lipase	Colipase	-36.0 ± 1.7	$-\ 28.7\pm\ 0.6$	$23.5\pm\ 7.7$	-1.31	[40]
18 Lysozyme	N-acetyl-D-Glucosamine	$-\ 9.0\pm0.2$	$-\ 24.3\pm\ 1.0$	$-\ 51\ \pm\ 4$	-0.22 ± 0.16	[41]
	(GlcNAc)$_2$	-21.1 ± 0.2	$-\ 44.3\pm\ 1.5$	$-\ 78\ \pm\ 5$		[41]
	(GlcNAc)$_3$	-28.8 ± 0.7	$-\ 56.8\pm\ 1.0$	$-\ 94\ \pm\ 9$	-0.17 ± 0.15	[41]
	(GlcNAc)$_4$	-30.5 ± 0.3	$-\ 45.8\pm\ 0.5$	$-\ 51\ \pm\ 3$		[41]
19 Malate dehydrogenase (pig heart mitochondria)	NADH	-34.3	$-\ 50.6\pm\ 1.3$			[21]
	NAD$^+$	-15.5 ± 0.4	$-\ 43.5\pm\ 2.9$	$-\ 90.8\pm\ 11.3$		[22]
20 Phosphofructo-kinase (rabbit muscle)	ATP(pH8)		$-\ 24.3$			[42]
	ATP(pH7)		-108.7			[42]
21 Phosphorylase a	FMN	-21.9 ± 0.4	$-\ 37.3\pm\ 2.9$	$-\ 51.6\pm\ 8.3$	-3.01 ± 0.42	[43]
22 Phosphorylase b	AMP	$-74.3\pm3^{\text{a}}$	$173\ \pm12^{\text{a}}$	$-336\ \pm\ 40^{\text{a}}$		[73]
	IMP	$-67.0\pm4^{\text{a}}$	$125\ \pm28^{\text{a}}$	$-195\ \pm\ 95^{\text{a}}$		[73]
23 Phycocyanin	Bu$_4$NBr	$-\ 7.5$	-205.0	-711.3		[44]
24 Ribonuclease A	3'-CMP	-45.2	$-\ 29.7$	51.9	-0.94	[58]
	2'-CMP	-50.2	$-\ 27.6$	75.8	-0.67	[58]
	Cytidine	-21.3	$-\ 25.5$	$-\ 14.1$		[58]
25 Thymidylate synthetase (*L. casei*)	dUMP	-29.7 ± 2.1	$-\ 22.6\pm\ 2.1$	$23.9\pm\ 9.6$		[51]
	FdUMP	-28.5 ± 2.1	$-\ 18.0\pm\ 4.2$	$35.1\pm\ 15.5$		[51]
	dTMP	-27.2 ± 2.1	$2.9\pm\ 2.1$	$101.1\pm\ 8.4$		[51]

Table 1 (continued)

try protein).	Ligand	$\Delta G^{0\prime}$ (kJ mol^{-1})	$\Delta H^{0\prime}$ (kJ mol^{-1})	$\Delta S^{0\prime}$ (J mol^{-1} K^{-1})	$\Delta C_p^{0\prime}$ (kJ mol^{-1} K^{-1})	Refer-ence
Tryptophan synthase (E. coli)						
β_2	PLP		-15.3 ± 0.8[b]		-9.4[b]	[74]
	L-serine		33.5 ± 5.0[b]		-6.5[b]	[75]
$\alpha_2\beta_2$	PLP		-16.5 ± 2.5[b]		-1.7[b]	[74]
	L-serine		105.0 ± 7.1[b]		0[b]	[75]

[a] These are overall thermodynamic parameters and refer to 4 binding sites per active dimer.
[b] These are overall thermodynamic parameters and refer to 2 binding sites per β_2 dimer.

Comments

to dehydrogenases (Entry 2, 11, 15, 19): The majority of the ΔH values has been determined using LKB batch calorimeters. $\Delta G^{0\prime}$ values result from equilibrium constants obtained from equilibrium dialysis, fluorescence, or absorption measurements. All data refer to 25 °C and 1 mole of binding site. Four different buffers were used:
0.2 M potassium phosphate pH = 7   [45–49]
0.1 M sodium phosphate pH = 7.6   [21–23, 30, 35]
0.1 M imidazole pH = 7.6   [22]
0.1 M ionic strength (potassium phosphate) pH = 7   [35]
Directly comparable results such as in entry 2 [34, 35] show reasonably good agreement of data obtained by different groups.

to Entry 1: Warfarin is a coumarin anticoagulant drug; ΔH values are calculated per mole of ligand.

to Entry 3: Experimental conditions: 37 °C, pH 7.5; the ΔH value has been corrected for buffer protonation effects.

to Entry 4: Enthalpy changes are given at 25 °C; 5 mM phosphate buffer, pH 7.4 (0.1 mM DTT, 1 mM EDTA) was used. The changes in heat capacity are independent of temperature between 8° and 30 °C.

to Entry 5: The coenzyme pyridoxal 5′-phosphate (PLP) binds to two different states of the apoenzyme with interaction enthalpies at 25 °C of -8.4 and -33.5 kJ per mole of active site at low and high pH, respectively. Buffer ionization effects have not to be corrected for, since no significant proton evolution occurs over the pH range studied.

to Entry 6: The experimental conditions for the measurements by Shrake et al. [19] are 40 mM potassium phosphate buffer, 30 °C, pH 7. The intrinsic enthalpy of PALA-binding [PALA = bisub-strate analog N-(phosphonacetyl)-L-aspartate] to the R-state of aspartate transcarbamoylase, ATCase (c_6r_6), is listed, the values refer to 1 mole of ATCase. CP = carbonyl phosphate.
   The experiments by Knier and Allewell [20] were performed in different 0.1 M buffers at 25 °C, pH 8.3; 2 mM mercaptoethanol and 0.2 mM EDTA were added to the solutions, ΔH values refer to 1 mole of bound ligand.
   The absence of phosphate which has to be displaced from the enzyme during binding of PALA may influence the thermodynamic binding parameters. On the other hand, the fairly good agreement between the values for PALA-binding reported in the two papers may be due to cancelation of opposing effects or may indicate that the enthalpy of binding of phosphate to ATCase is about zero, as quoted by [19].

to Entry 7: The enthalpy values determined by Shiao and Sturtevant for the interaction of chymotrypsin and indole are pH-dependent varying from -63.6 kJ mol^{-1} at pH 7.8 to -26.8 kJ mol^{-1} at pH 5.6 [26]. Jones and Trowbridge [25], on the other hand, report a pH-independent value of -25.1 kJ mol^{-1} for this reaction between pH 5 and pH 7.

At present it is not possible to rationalize the discrepancy or to discard either value on the basis of obvious mistakes made in handling the compounds or performing the measurements.

Chen and Wadsö [28] determined all three thermodynamic quantities for binding of indole-3-propionic acid simultaneously at pH 5.8 (acetate buffer) with a new syringe titration unit on the LKB batch microcalorimeter.

It should be pointed out that all compounds listed behave as competitive inhibitors to the enzyme except N-acetyl-L-tryptophan which forms an acyl enzyme complex.

ΔH values are given per mole of bound inhibitor.

to Entry 8: Experiments were performed in 0.1 M phosphate buffer pH 7.4 at 25 °C; the ΔH values are not corrected for protonation heats; one proton is transferred from the buffer to the ligand during the binding of methotrexate; however, since the enthalpy of ionization of phosphate buffer at 25 °C is 3.35 kJ mol^{-1} [69], only a slight correction has to be made.

to Entry 9: Thermodynamic parameters for the binding of flavin mononucleotide (FMN) and flavin adenine dinucleotide (FAD) to hen egg white riboflavin binding protein (RBP) are reported. Experiments were performed under a variety of temperature, pH, and buffer conditions. FAD binding to RBP shows a strong pH dependence, anomalous enthalpies in a couple of buffer systems lead to the conclusions that different conformations of the protein are selected by specific protein-buffer interactions. Data in the table: pH 7.4, 26 °C, Pipes-buffer. Data for FMN and FAD binding to hen egg yolk RBP are also reported.

The enthalpy of binding of N^3-carboxymethylriboflavin (3-CM-Rf) to egg white riboflavin binding protein is independent of pH (5.4–8.5) and buffer used. The values in the table are calculated from titration experiments in 50 mM Pipes, 100 mM NaCl at 25 °C, pH 7.4. The thermodynamic binding parameters for a total of nine flavin analogs are reported in the paper.

to Entry 10: The binding of FMN, FAD (see 9 for abbreviations) and 8-carboxylic acid riboflavin (8-COOH-Rf) to the apo flavodoxin from *Azotobacter vinelandii* was studied as a function of temperature and pH in a variety of buffers. No protonation effects have to be accounted for. Equilibrium constants are calculated from titration isotherms. All values in the table result from experiments at 25 °C, pH 7.4 in 50 mM Hepes buffer (20 mM β-mercaptoethanol, 100 mM KCl). The change in heat capacity on binding of FMN to apo flavodoxin is biphasic, the values are -0.9 kJ/(mol · K) below 20 °C and -3.4 kJ/(mol · K) above 20 °C, respectively.

to Entry 11: The experiments were performed in 0.1 M phosphate buffer, pH 7.6, the values are calculated per mole of active site. The authors also report values at 15 °C and thermodynamic quantities for a variety of ternary complexes. According to the authors effects of buffer ionization are insignificant, however, contributions from proton ionization at different pH values are not accounted for.

to Entry 12: Enthalpies of binding of the substrate L-glutamate were measured employing the unadenylylated Mg^{2+}-enzyme in the presence of saturating ADP at 30 °C, pH 7.2; in the case of the feedback inhibitor L-alanine 20 mM phosphate was added to obtain a sufficiently large heat effect.

Enthalpies of binding of the substrate analogs L-methionin-S,R-sulfoximine were determined using the unadenylylated Mg^{2+}-enzyme at 30 °C, pH 7.1; for comparison the parameters for binding L-glutamate to the Mn^{2+}-enzyme are also listed. All values refer to 1 mole of subunit (12 subunits per molecule of enzyme). ΔC$_p$ values were calculated from thermal titration curves.

to Entry 13: Experimental conditions: 25 °C, 0.2 M NaCl, pH 8.5.

to Entry 14: The experiments with the tRNA were performed in 0.05 M potassium phosphate (pH 7), 3 mM MgCl$_2$, 1 mM EDTA, and 1 mM DTE.

to Entry 16: ΔH values were calculated per mole of isopropyl-1-thio-β-D-galactopyranoside (IPTG) and corrected for the buffer ionization effects. For details of the calorimeter see [39].

to Entry 17: The binding of the polypeptide cofactor colipase was measured at 25 °C , pH 7 in various buffers. The change in heat capacity is temperature-dependent between 21° and 37 °C, the ΔH value refers to 1 mole of lipase and a 1:1 binding of lipase and colipase.

to Entry 18: The binding of the saccharide inhibitors was investigated at pH 5 in 0.1 M acetate buffer.

to Entry 20: At pH 8 ATP binds only to the catalytic site of the enzyme; at pH 7 ATP is strongly inhibitory, the ΔH value was calculated by subtracting the value determined at pH 8 from the overall ΔH value. No correction for proton effects ($\Delta n \sim 2$) in the phosphate buffer was made.

to Entry 21: Experiments were performed at pH 6.8 in the presence of 50 mM glucose. The allosteric inhibitor FMN binds to a site located at the surface 10 Å from the catalytic cleft. The unusually strong dependence of ΔC_p on the temperature is expressed by $\Delta C_p = +9519 - 260.9 \cdot t$ J mol^{-1} K^{-1} (t = temperature in °C). K_D was determined by fluorimetric titration.

to Entry 22: Both AMP and IMP bind to four sites per active dimer; the values given in the table are overall thermodynamic parameters at 25 °C, pH 6.9. However, all thermodynamic parameters for the individual binding sites, derived from a Scatchard plot analysis of equilibrium dialysis experiments and thermal titrations, are reported in the paper.

to Entry 23: ΔH values refer to binding of tetrabutylammonium bromide to 1 mole of trimeric phycocyanin. For details of the calorimeter see [44].

to Entry 24: Thermodynamic quantities for 3'-CMP-binding are the average values obtained over the pH range 4–6.5. Measurements with 2'-CMP and cytidine were performed at pH 6.5; for all experiments the temperature is 25 °C and the ionic strength is 0.05.

to Entry 25: The experiments with the substrate 2-deoxyuridylate (dUMP), the substrate analog 5-fluoro-2'-deoxyuridylate (FdUMP) and the product of reductive methylation of dUMP, 2'-deoxythymidylate (dTMP), were performed at 25 °C, pH 7.4 in different buffers with a Tronac titration calorimeter. The ΔH values are corrected for protonation effects and refer to 1 mole of binding site.

to Entry 26: The β_2 subunit of tryptophan synthase from *E. coli* binds 2 moles of the coenzyme pyridoxal 5'-phosphate (PLP). Binding is cooperative in the isolated β_2 dimer and noncooperative in the native $\alpha_2\beta_2$ complex. The values listed in the table are the overall enthalpy values measured at 25 °C, pH 7.5 in 0.1 M sodium pyrophosphate buffer and were calculated on the basis of the β_2 dimer. A complete set of thermodynamic parameters for the individual binding sites as a function of pH and temperature is given in [74].

   The substrate L-serine binds to the β_2 subunit of tryptophan synthase by replacing the covalent Schiff base between PLP and the protein. Binding was studied in 0.1 M sodium pyrophosphate buffer (pH 7.5) as a function of temperature. The listed values refer to 1 mole of β_2 dimer.

   Reference [75] also contains a detailed calorimetric study of indole binding to tryptophan synthase.

Binding of hydrogen ions

try protein).	Ligand	$\Delta G^{0\prime}$ (kJ mol^{-1})	$\Delta H^{0\prime}$ (kJ mol^{-1})	$\Delta S^{0\prime}$ (J mol^{-1} K^{-1})	$\Delta C_p^{0\prime}$ (kJ mol^{-1} K^{-1})	Refer- ence
Lysozyme chymo- trypsinogen A, oxidized cytochrome C						
Carboxyl group	H$^+$		0			[52]
ε-Amino group		-18.4	-43.9	-85.6		[52]
Phenolic group		-18.0	-26.4	-28.0		[52]
α-Amino group		-4.2	-41.8	-126.3		[52]
Imidazole		2.9	-26.4	-98.2		[52]
Ribonuclease A Histidine						
	H$^+$	12	-33.1		19.6	[50]
		105	-37.7	-27.2	35.1	[50]
		119	-28.5		4.2	[50]

Comments

For lysozyme, chymotrypsinogen A, and oxidized cytochrome c the thermal titrations were performed in 0.15 M KCl. Reversibility was established by carrying out the potentiometric titrations both with acid and base. The ΔH values are heats of proton binding per mole of the respective group in globular proteins as calculated from the overall heat of binding and the number of ionizable groups in each protein. An unusually large heat of binding of protons of $125.5 \, kJ \, mol^{-1}$ for chymotrypsinogen A between pH 4.5 and 1.3 leads to the conclusion that the protein undergoes a pH-induced conformational change in this pH region.

The thermal titrations with ribonuclease A were performed at an ionic strength of 0.05 maintained with KCl. Histidine 48 shows a large ionization enthalpy of $100 \, kJ \, mol^{-1}$ ($\Delta G^{0'} = 38.1 \, kJ \, mol^{-1}$; $\Delta S^{0'} = -209 \, J \, mol^{-1} \, K^{-1}$), which again may be assumed to be indicative of a conformational rearrangement of the macromolecule.

b) Binding of inorganic ions

Entry protein No.	Ligand	$\Delta G^{0'}$ (kJ mol^{-1})	$\Delta H^{0'}$ (kJ mol^{-1})	$\Delta S^{0'}$ (J mol^{-1} K^{-1})	$\Delta C_p^{0'}$ (kJ mol^{-1} K^{-1})	Reference
29 (Na$^+$, K$^+$) ATPase	Mg^{2+}	−17.7	−177.8	− 537.9		[53]
	PO$_4^{3-}$	−14.5	−207.1	− 647.2		[53]
30 (Ca^{2+}) ATPase	Mg^{2+}	−12.6	−318.0	−1026.1		[54]
	PO$_4^{3-}$	−11.7	− 96.2	− 283.9		[54]
31 Bovine serum albumin	I$^-$	−22.6	− 72.8±4.2	− 168.3		[55]
32 Troponin C						
(1, 2)	Ca^{2+}	−40.6	− 32.2	28.0		[56]
(3, 4)	Ca^{2+}	−32.2	− 32.2	0		[56]
(1)	Ca^{2+}	−55.9	− 16.6	132.0	0.41	[57]
(2)	Ca^{2+}	−36.5	− 7.3	98.1	0.44	[57]
(3, 4)	Ca^{2+}	−28.5	− 12.9	52.5	−0.18	[57]
33 Ribonuclease	PO$_4^{3-}$	−43.9	− 1.3	143.2		[58]

Comments

Enthalpy changes associated with the binding of Mg^{2+} and phosphate to both the (Na$^+$, K$^+$) and (Ca^{2+}) ATPases were measured at 24.5 °C for the reaction E+L↔EL. In both cases the extraordinarily large enthalpy changes were interpreted as indicating a structural rearrangement of the enzyme.

The data for iodide binding to BSA hold only for the high affinity class of binding sites on the protein and are given per mole of binding site. Details of instrumentation and computation are given in the reference.

The measurements on troponin C by Potter et al. [56] were performed on the Mg^{2+} free protein. From the paper by Yamada and Kometani [57] only the results obtained in 1 mM Mg^{2+} solution (pH 7) are listed, since the discrepancies between the two studies concerning the ΔH values determined with the Mg^{2+} free protein cannot be explained satisfactorily. The changes in heat capacity hold for the temperature range 15° to 25 °C. All molar values refer to 1 mole of binding site.

c) Binding of denaturants

Entry protein No.	Ligand	$\Delta G^{0'}$ (kJ mol^{-1})	$\Delta H^{0'}$ (kJ mol^{-1})	$\Delta S^{0'}$ (J mol^{-1} K^{-1})	$\Delta C_p^{0'}$ (kJ mol^{-1} K^{-1})	Reference
34 Lysozyme	GuHCl		−10.0±0.4			[59]
35 Ribonuclease	Urea		−23			[60]
Trypsin			−23			[60]
β-Lactoglobulin			−23			[60]
Ovalbumin			−23			[60]
Bovine serum albumin			−23			[60]

Comments

The average ΔH for the binding of 1 mole guanidinium hydrochloride (GuHCl) to lysozyme (in 0.1 M NaCl) was taken from the sigmoidal heat vs. GuHCl-concentration curves by extrapolating the postdenatural slope to zero GuHCl-concentration. The enthalpies of interaction of the five globular proteins with urea were determined at an ionic strength of 0.005, pH 7 over a range of urea molality from 0–15 mmol g^{-1}. The value listed is the average binding enthalpy per mole of bound urea after subtraction of the enthalpy of unfolding from the overall value.

Antibody-hapten interactions

try protein).		Ligand	$\Delta G^{0\prime}$ (kJ mol^{-1})	$\Delta H^{0\prime}$ (kJ mol^{-1})	$\Delta S^{0\prime}$ (J mol^{-1} K^{-1})	$\Delta C_p^{0\prime}$ (kJ mol^{-1} K^{-1})	Refer-ence
Anti DNP		DNP-Lys	−45.2	−64	− 63.2		[61]
(rabbit)		TNP-Lys	−38.1	−46.4	− 28.0		[61]
Anti TNP		DNP-Lys	−41.4	−74.5	−110.9	−0.86	[61]
(rabbit)		TNP-Lys	−51.0	−89.1	−127.6	−0.65	[61]
Anti DNP	I	DNP-Lys	−41.8	−58.2±1.5	− 54.8		[62]
fraction	II		−39.5		− 62.8		[62]
(rabbit)	III		−36.1		− 74.1		[62]
	IV		−35.2		− 77.0		[62]
	V		−34.9		− 78.2		[62]
Anti DNP (guinea pig)		DNP-Lys	−50.6	−36.4±4.2	47.7		[63]
MOPC-315		DNP-Lys	−33.9	−69.5	−119.3	−0.75	[64]
		DNP-Gly	−22.6	−83.3	−203.7	−1.22	[64]
		DNP-amino-caproate	−33.1	−60.3	− 91.2	−0.46	[64]
MOPC-460		DNP-Lys	−26.8	−66.1	−131.9	−1.13	[64]
IgG(T) fraction	I	DNP-Gly	−42.7±0.5	−70.4	− 93.1		[65]
(horse)	VI	DNP-Gly	−36.1±0.1	−59.2	− 77.6		[65]

Comments

Most early antibody preparations constitute a heterogeneous population of molecules. However, with respect to thermodynamic quantities, this heterogeneity is only reflected in the affinity of the different fractions towards the hapten. For the five different fractions of rabbit anti-DNP (DNP = dinitrophenol, TNP = trinitrophenol) Halsey and Biltonen [62] have clearly demonstrated that at 24.5 °C the antibody-hapten interactions are entropically controlled.

The change in heat capacity in the study by Barisas et al. [61] is calculated on the basis of a further measurement at 4 °C.

MOPC-315 and -460 are mouse myeloma proteins and therefore comparable to monoclonal antibodies. Only selected ligands are listed, ΔC_p values are calculated on the basis of a further measurement at 4 °C. In contrast, the different affinity classes of equine antibodies of IgG(T) immunoglobulins also exhibit differences in the binding enthalpies. All molar values refer to 1 mole of hapten combining with 1 mole of binding sites; inactive sites were corrected for.

Muscle proteins

ntry protein o.	Ligand	$\Delta G^{0\prime}$ (kJ mol^{-1})	$\Delta H^{0\prime}$ (kJ mol^{-1})	$\Delta S^{0\prime}$ (J mol^{-1} K^{-1})	$\Delta C_p^{0\prime}$ (kJ mol^{-1} K^{-1})	Refer-ence
Myosin	ADP	−33.2	−64.5			[66]
Heavy meromyosin	ADP	−33.3	−70.9			[67]
MS1	ADP	−33.1	−83.0			[68]
MS1	ATPγS		−90.0			[68]

Comments

All experiments with ADP were performed in 0.5 M KCl, 0.01 M $MgCl_2$, 0.02 M Tris buffer (pH 7.8) at 12 °C. The values are corrected for the enthalpy of buffer protonation considering 0.2 moles of protons released during the binding process. A slight dependence of ΔH on the ionic strength is reported.

   The experiments with ATPγS were performed in 0.1 M KCl, 0.01 M $MgCl_2$, 0.02 M TRIS (pH 7.8) at 23 °C in a prototype calorimeter, the details of which can be found in the reference. The ΔH value is corrected for buffer protonation.

   $\Delta G^{0'}$ refers to 1 M ligand concentration in the presence of ATP $= 10^{-3}$ M; ADP $= 10^{-5}$ M; $P_i = 10^{-3}$ M. MS1 = catalytic site contained in each S-1 head of myosin.

Table 2. Binding of NAD^+ analogs to pig heart muscle lactate dehydrogenase in 0.2 M phosphate buffer, pH 7, at 25 °C [a]

Analogs	$\Delta G^{0'}$ (kJ mol^{-1})	$\Delta H^{0'}$ (kJ mol^{-1})	$\Delta S^{0'}$ (J mol^{-1}K^{-1})	$\Delta C_p^{0'}$ (kJ mol^{-1}K^{-1})
Nicotinamide adenine dinucleotide, NAD^+	-19.9	-24.2 ± 1.0 -24.3 ± 1.0 -25.6 ± 1.0 -25.5 ± 1.0	-16.8	-0.350
Thionicotinamide adenine dinucleotide, $TNAD^+$	-19.8	-23.7 ± 1.8	-13.1	-0.014
3-Acetylpyridine adenine dinucleotide, $APAD^+$	-22.1	-33.7 ± 1.7	-38.9	-0.238
2-(3-Acetylpyridino)ethyl adenosine pyrophosphate	-14.7	-7.5 ± 1.7	24.1	
3-(3-Acetylpyridino)propyl adenosine pyrophosphate	-15.9	-11.2 ± 1.6	15.8	-0.047
4-(3-Acetylpyridino)butyl adenosine pyrophosphate	-18.0	-15.6 ± 1.4	8.1	-0.069
5-(3-Acetylpyridino)pentyl adenosine pyrophosphate	-18.8	-17.5 ± 2.0	4.4	-0.091
6-(3-Acetylpyridino)hexyl adenosine pyrophosphate	-20.1	-20.6 ± 2.2	-1.7	-0.110
Nicotinamide 6-mercaptopurine dinucleotide, NMD^+	-20.1	-11.2 ± 2.1	29.9	-0.272
Nicotinamide benzimidazole dinucleotide, NBD^+	-18.4	-11.3 ± 1.2	23.8	-0.133

[a] From reference [76].

References

1. Sturtevant JM (1977) Proc Natl Acad Sci USA 74:2236–2240
2. Hinz H-J (1983) Annu Rev Biophys Bioeng 12:285–317
3. Edsall JT, Gutfreund H (1983) In: Biothermodynamics. Wiley, Chichester
4. Weber G (1975) In: Anfinsen CB, Edsall JT, Richards FM (eds) Advances in protein chemistry, vol 29. Academic Press, New York, pp 1–83
5. Privalov PL (1979) Adv Protein Chem 33:167–241
6. Pfeil W (1981) Mol Cell Biochem 40:3–28
7. Jencks WP (1976) In: Fasman GD (ed) Handbook of biochemistry and molecular biology, 3rd edn. Physical and chemical data, vol 1. CRC Press, Inc, pp 296–299

8. Hinz H-J, Gorbunoff MJ, Price B, Timasheff SN (1979) Biochemistry 18:3084–3089
9. Sturtevant JM (1971) In: Weissberger A, Rossiter BW (eds) Physical methods of chemistry, vol 1, part V. Wiley Interscience, New York, pp 347–425
10. Niekamp CW, Sturtevant JM, Velick SF (1977) Biochemistry 16:436–445
11. Hinz H-J (1983) In: Burgen ASV, Roberts GCK (eds) Topics in molecular pharmacology. Elsevier, Amsterdam, pp 71–122
12. Wadsö I (1975) In: Pain RH, Smith B-J (eds) New techniques in biophysics and cell biology, vol 2. Wiley, London, pp 85–126
13. Grenthe J, Ots H, Ginstrup O (1970) Acta Chem Scand 24:1067–1080
14. Gucker FT, Pickard HB, Planck RW (1939) J Am Chem Soc 61:459–470
15. Ackers GK, Shea MA, Smith FR (1983) J Mol Biol 170:223–242
16. Eftink MR, Anusiem AC, Biltonen RL (1983) Biochemistry 22:3884–3896
17. Crowder III AL, Swenson CA, Barker R (1973) Biochemistry 12:2852–2855
18. Giartosio A, Salerno C, Franchetta F, Turano C (1982) J Biol Chem 257:8163–8170
19. Shrake A, Ginsburg A, Schachmann HK (1981) J Biol Chem 256:5005–5015
20. Knier BL, Allewell NM (1978) Biochemistry 17:784–790
21. Subramanian S, Ross PD (1978) Biochemistry 17:2193–2197
22. Subramanian S, Ross PD (1977) Biochem Biophys Res Commun 78:461–466
23. Fisher HF, Subramanian S, Stickel DC, Colen AH (1980) J Biol Chem 255:2509–2513
24. Fukada H, Sturtevant JM, Quiocho FA (1983) J Biol Chem 258:13193–13198
25. Jones JB, Trowbridge CG (1983) J Biol Chem 258:2135–2142
26. Shiao DDF, Sturtevant JM (1969) Biochemistry 8:4910–4917
27. Shiao DDF (1970) Biochemistry 9:1083–1090
28. Chen A, Wadsö I (1982) J Biochem Biophys Methods 6:307–316
29. Subramanian S, Kaufmann BT (1978) Proc Natl Acad Sci USA 75:3201–3205
30. Subramanian S, Stickel DC, Colen AH, Fisher HF (1978) J Biol Chem 253:8369–8374
31. Shrake A, Park R, Ginsburg A (1978) Biochemistry 17:658–664
32. Gorman EG, Ginsburg A (1982) J Biol Chem 257:8244–8252
33. Takahashi K, Casey JL, Sturtevant JM (1981) Biochemistry 20:4693–4697
34. Schmid F, Hinz H-J, Jaenicke R (1978) FEBS Lett 87:80–82
35. Subramanian S, Ross PD (1979) J Biol Chem 254:7827–7830
36. O'Reilly RA, Ohms JI, Mothley CH (1969) J Biol Chem 244:1303–1305
37. Hinz H-J, Weber K, Floßdorf J, Kula M-R (1976) Eur J Biochem 71:437–442
38. Wiesinger H, Kula M-R, Hinz H-J (1980) Hoppe-Seyler's Z Physiol Chem 361:201–205
39. Donner J, Caruthers MH, Gill SJ (1982) J Biol Chem 257:14826–14829
40. Donner J, Spink CH, Borgström B, Sjöholm I (1976) Biochemistry 15:5413–5417
41. Bjurulf K, Wadsö I (1972) Eur J Biochem 31:95–102
42. Wolfman NM, Hammes GG (1979) J Biol Chem 254:12289–12290
43. Sprang S, Fletterick R, Stern M, Yang D, Madsen N, Sturtevant JM (1982) Biochemistry 21:2036–2048
44. Chen CH, Berns DS (1977) J Phys Chem 81:125–129
45. Hinz H-J, Schmid F (1977) In: Sund H (ed) Pyridine nucleotide dependent dehydrogenases. De Gruyter, Berlin, pp 292–306
46. Hinz H-J, Steininger G, Schmid F, Jaenicke R (1978) FEBS Lett 87:83–86
47. Hinz H-J, Jaenicke R (1973) Biochem Biophys Res Commun 54:1432–1436
48. Hinz H-J, Jaenicke R (1975) Biochemistry 14:24–27
49. Hinz H-J, Schmid F (1979) Hoppe-Seyler's Z Physiol Chem 360:217–219
50. Flogel M, Biltonen RL (1975) Biochemistry 14:2603–2609
51. Beaudette NV, Langerman N, Kisliuk RL (1980) Arch Biochem Biophys 200:410–417
52. Shiao DDF, Sturtevant JM (1976) Biopolymers 15:1201–1211
53. Kuriki Y, Halsey J, Biltonen RL, Racker E (1976) Biochemistry 15:4956–4961
54. Epstein M, Kuriki Y, Biltonen RL, Racker E (1980) Biochemistry 19:5564–5568
55. Lovrien R, Sturtevant JM (1971) Biochemistry 10:3811–3815
56. Potter JD, Hsu F-J, Pownall HJ (1977) J Biol Chem 252:2452–2454
57. Yamada K, Kometani K (1982) J Biochem (Tokyo) 92:1505–1517
58. Flogel M, Albert A, Biltonen RL (1975) Biochemistry 14:2616–2621
59. Pfeil W, Privalov PL (1976) Biophys Chem 4:33–40

60. Paz Andrade MI, Jones MN, Skinner HA (1976) Eur J Biochem 66:127–131
61. Barisas BG, Singer SI, Sturtevant JM (1972) Biochemistry 11:2741–2744
62. Halsey JF, Biltonen RL (1975) Biochemistry 14:800–804
63. Halsey JF, Cebra JJ, Biltonen RL (1975) Biochemistry 14:5221–5224
64. Johnston MFM, Barisas BG, Sturtevant JM (1974) Biochemistry 13:390–396
65. Archer BG, Krakauer H (1977) Biochemistry 16:615–617
66. Kodama T, Woledge RC (1976) J Biol Chem 251:7499–7503
67. KodamaT, Watson ID, Woledge RC (1977) J BiolChem 252:8085–8087
68. Kodama T, Woledge RC (1979) J Biol Chem 254:6382–6386
69. Christensen JJ, Hansen LD, Izatt RM (1976) Handbook of proton ionization heats and re-
    lated thermodynamic quantities. Wiley, New York
70. Nowak HP, Langermann N (1982) Arch Biochem Biophys 214:231–238
71. Mifflin TE, Langermann N (1983) Arch Biochem Biophys 224:319–325
72. Carlson R, Langermann N (1983) Arch Biochem Biophys 229:440–447
73. Mateo PL, Baron C, Lopez-Mayorga O, Jiminez JS, Cortijo M (1984) J Biol Chem
    259:9384–9389
74. Wiesinger H, Hinz H-J (1984) Biochemistry 23:4921–4928
75. Wiesinger H, Hinz H-J (1984) Biochemistry 23:4928–4934
76. Niekamp CW, Hinz H-J, Jaenicke R, Woenckhaus C, Jeck R (1980) Biochemistry 19:3144–
    3152

Chapter 8 **Thermodynamics of Protein-Protein Association**

Philip D. Ross

Symbols

ΔG^0:   standard Gibbs energy change
ΔH^0:   standard enthalpy change
ΔS^0:   standard entropy change
ΔC_p^0:   standard molar heat capacity change

All thermodynamic parameters are expressed per mole of complex formed except for the indefinite association cases of lysozyme, TMV protein and hemoglobin S for which the mole refers to the monomeric protein reacted. Standard states are hypothetical 1 M protein and the pH at which the reaction was measured.

1 Introduction

This chapter is concerned with the thermodynamic parameters associated with protein-protein interactions. Protein-protein interactions may be classified either as homogeneous or heterogeneous, depending upon whether the polypeptide chains involved in association are the same or different. The former class are by definition polymerizations which may result in the formation of oligomers of definite structure and stoichiometry, aggregates composed of a distribution of molecular sizes, or large microtubular structures. Heterogeneous interactions include the association complexes formed between one protein and a different protein or protein fragment and also between nonidentical subunits of the same protein. The examples discussed below fall equally into these two classes.

The aim underlying the selection of results for inclusion in this chapter is to provide, in thermodynamic terms, as broad a picture as possible of protein association processes. Such a description requires information of changes in the Gibbs energy, enthalpy, entropy, and heat capacity. On account of the propagation of error, it is extremely difficult to obtain reliable heat capacity data from Gibbs energy measurements. Therefore the data selection process has been restricted to systems in which the change in heat content has been directly measured by calorimetry. Several examples providing somewhat less thermodynamic information have also been included because these papers provide good descriptions of the various calorimetric techniques that may be used. As a consequence of the application of these selection criteria, the data presented in Table 1 represent a highly selective sample of protein-protein interactions. Further data can be found in [17].

Table 1. Thermodynamics of protein association at 25 °C[a]

Association process	Ref.	ΔG^0 kJ mol^{-1}	ΔH^0 kJ mol^{-1}	ΔS^0 J K^{-1} mol^{-1}	ΔC_p^0 J K^{-1} mol^{-1}	pH
S-peptide + S-protein (ribonuclease)	[1]	−44.3	−166.5	−410	− 6110[b]	7.0
Trypsin (bovine) + inhibitor (soybean)	[2]	−54.8	− 16.7	+128		7.5
Hemoglobin haptoglobin (1:1)	[3]	−38.5	−139	−337	− 3930	5.5
Tryptophan synthase apo-$\beta_2 + 2\alpha = \alpha_2\beta_2$	[4]	−26.4	− 54.6	− 88.5	−15060[b]	7.5
Glucagon trimerization	[5]	−30.6	−131	−337	− 1800	10.6
Lysozyme polymerization	[6]	−16.3	− 26.8	− 35.2		7.0
α-Chymotrypsin dimerization	[7]	−19.6	−148	−431		7.8
β-Lactoglobulin tetramerization	[8]	−57.7	−138	−289		4.6
TMV protein polymerization[c]	[9]		+ 50		− 1440	6.8
Deoxyhemoglobin S gelation	[10]	−14.2	+ 8.4	+ 75.3	− 837[b]	7.2

[a] All thermodynamic parameters are expressed per mole of complex formed except for the indefinite association cases of lysozyme, TMV protein, and hemoglobin S for which the mole refers to the monomeric protein reacted. Standard states are hypothetical 1 M protein and the pH at which the reaction was measured. All data for 25 °C except lysozyme (30 °C), β-lactoglobulin (5 °C) and TMV protein (~ 15 °C). Salt concentrations were typically in the range 0.1–0.3 M NaCl or KCl. When buffers were employed they were usually acetate or phosphate, ionic strength I = 0.1. Consult original papers for these details.
[b] These references report a temperature dependence of ΔC_p^0.
The values reported here refer only to the specific conditions of pH, solvent composition, and temperature at which the reactions were carried out; these values can be quite different when the systems are studied under other conditions. The reader is urged to consult the comments in Section 5 while reading the Table.

2 Explanation of Table 1

The Gibbs energy values, with two well justified exceptions, were obtained by thermodynamic experiments. Therefore all of the entries in the table are true thermodynamic quantities rather than formal thermodynamic parameters derived from nonthermodynamic measurements, such as from the temperature dependence of a spectroscopic signal. The values for the thermodynamic quantities in the table, having been obtained by thermodynamic measurements of equilibrium constants and by calorimetry are inclusive of all of the processes taking place in the system undergoing protein association.

The data listed in Table 1, apart from the conversion from calories to joules, are identically the values reported by the authors in their original publications.

No corrections to the thermodynamic parameters arising from processes involving the hydrogen ion [protonation(s) or ionization(s)] have been made.

No conversion to unitary thermodynamic parameters has been made. Although elimination of trivial statistical effects is useful for comparing the Gibbs energy and entropy changes of protein subunit complexes of differing stoichiometry (as in hemoglobin), the concept appears to lose all meaning in the case of indefinite polymerization.

Results at 25 °C, with exceptions noted above, have been assembled in the table in order to provide an indication of the magnitude of the thermodynamic parameters of protein association reactions at ambient temperature.

3 Commentary on Methods of Gibbs Energy Determination

Ideally, Gibbs energy values are calculated from equilibrium constants measured by direct determination of the concentrations (activities, strictly speaking) of product and reactant species. In the case of extremely tight binding, detection of the concentration of free species is beyond the sensitivity of many physicochemical methods, and equilibrium constant determination is often made by measurement of the forward and reverse rates of the reaction or by assay for enzymatic activity. Such use of kinetic methods requires an understanding of the mechanism of the reaction which, in the last analysis, is a model. This burdens the derived equilibrium constant with an additional assumption (that of the kinetic model) that is not present in the direct determinations.

When enzymatic assay is employed, it is necessary to ensure that the results are not influenced by the concentration of reactants employed [1]. It is highly desirable to extend the range of kinetic measurements to concentration regions that overlap with those from which equilibrium constants have been determined by direct analytical methods.

4 Commentary on Experimental Methodology

Three kinds of calorimetric technique were employed in the determination of the enthalpy changes given in Table 1. These were: isothermal mixing calorimetry – used primarily in the first four papers [1–4], temperature scanning calorimetry – used primarily in the last two papers [9–10], and heat of dilution measurements – used in the middle four papers [5–8]. The techniques of isothermal mixing calorimetry and differential scanning calorimetry are well known and will not be discussed further. The heat of dilution method is specially applicable to associating systems. The method is capable of determining both the equilibrium constant, hence the Gibbs energy change, and the enthalpy change, and thus provides a wholly self-consistent set of thermodynamic data. This technique essentially measures the quantity $K\Delta H$. The values of these parameters are thus coupled. The range of determinable K and ΔH values is discussed by Gill and Farquhar [11], whose paper should be consulted for a critical analysis of this method.

5 Commentary on Specific Systems

S-Peptide + S-Protein of Ribonuclease [1]. This paper is exemplary for the wide range of solution conditions examined and thorough quality of the investigation. The equilibrium constant was determined by enzymatic assay; however, the authors carefully justify this procedure by citing kinetic evidence that the equilibrium expression is correctly formulated and by presenting evidence that the enzyme assay conditions do not perturb the equilibrium. Under various conditions between 0.2 and 0.4 moles of protons are liberated in this reaction, and this would presumably not make a significant contribution to the observed enthalpy change. The energetics of the conformational changes of the reactants and products of this reaction were determined by scanning calorimetry. The very large dependence of ΔC_p upon temperature found for the association reaction is shown to arise from temperature-dependent changes in the conformation of the reactants. After allowance is made for these conformational transitions, it is concluded that ΔH is always negative for S-protein + S-peptide, thus this reaction is enthalpically driven at all accessible temperatures.

Trypsin + Kunitz Soybean Inhibitor [2]. The equilibrium constant for this reaction was determined from potentiometric proton release data. The enthalpy change is negative both at high and at low pH and passes through a maximum in the vicinity of pH 4.2. Therefore, over most of the pH range, this reaction is entropically driven. Despite extensive work on this system by Trowbridge and his collaborators and others [12], the cause of the pH-dependent enthalpy changes has not been specifically identified.

Hemoglobin + Haptoglobin (1:1) [3]. The estimate of the equilibrium constant for this reaction is described by the authors as "crude"; however, an order of magnitude error would only change the free energy by 5.8 kJ mol^{-1} and hardly alter the sense of the results obtained. Between 1.4 and 2 protons are liberated in this reaction and are included in the reported enthalpy values.

Tryptophan Synthase [4]. The equilibrium constant for the association constant was calculated from data given in the text by $K_A = K_{D1}^{-1} K_{D2}^{-1}$. The negative heat capacity change reported for this system is spectacularly large. At 25 °C, 0.75 protons are taken up in this reaction. There is a slight suggestion in the data that the heat capacity change may be temperature-dependent but, as stated by the authors, the experimental uncertainty precludes making a strong statement concerning this point.

Glucagon Trimerization [5]. In this paper, an exponential gradient of glucagon was passed through a flow calorimeter to obtain heat of dilution measurements from which the values of the equilibrium constant and enthalpy change were extracted. Helix formation accompanying trimerization might make a significant contribution to the values of the thermodynamic parameters.

Lysozyme Polymerization [6]. This paper provides an excellent description of the heat of dilution method for obtaining thermodynamic parameters of association using a flow calorimeter. A heat of mixing calorimeter was used to measure the enthalpy change accompanying saccharide binding. The fortuitous circumstance that saccharide binding inhibits polymerization permitted heat of dilution values to be extracted from saccharide binding experiments at different protein concentrations. The pH dependence of the integral heat of dilution of lysozyme was abolished when Glu-35 was esterified, thus implicating that residue in the self-association reaction.

α-Chymotrypsin Dimerization [7]. This work was carried out by flow calorimetry. Although some autolysis might be taking place at the high pH employed, the data appear convincing.

β-Lactoglobulin Tetramerization [8]. A heat of mixing calorimeter was used to make heat of dilution measurements from which self-consistent values of K and ΔH were extracted. This paper also contains some interesting calculations of the effect of the concentration of intermediate species upon the ratio of the van't Hoff to calorimetric enthalpy change.

TMV Protein Polymerization [9]. Scanning calorimetry was used in this investigation. It was found that the kinetics of the polymerization reaction were on roughly the same time-scale as the scan rate. Because this system is so sensitive to variations in pH, ionic strength, and protein concentration, and there is variation in the composition of the reactants and products depending upon these conditions, it is not possible to assign any reliable value of the Gibbs energy change for this reaction. Nevertheless, if one uses any of the Gibbs energy values quoted by the authors in Fig. 8 of their text [9], it is readily apparent that this reaction is strongly entropy driven.

Deoxyhemoglobin S Gelation [10]. Scanning calorimetry was employed in this investigation. The unique lag phase kinetics and enormous concentration dependence of the kinetics of the gelation reaction permitted the temperature of polymerization to be changed both by changing the hemoglobin concentration and the scanning rate. In earlier work, at higher ionic strength, good agreement was observed between the results of scanning experiments and conceptually simpler calorimetric measurements carried out isothermally [13]. A slightly positive temperature coefficient of the heat capacity change was observed in these experiments. The Gibbs energy change was obtained from solubility measurements. At the high hemoglobin concentrations (200 mg ml^{-1}) of the equilibrium solubility, large activity coefficient corrections had to be applied to the solubility concentrations in order to obtain the activity and thus the Gibbs energy change. These corrections were calculated by the hard sphere excluded volume model [14]. It is emphasized that such corrections for nonideality are important in the realm of concentrated protein solutions.

6 Conclusion

Inspection of Table 1 immediately shows that the enthalpy, entropy, and heat capacity changes are large in magnitude. This is a consequence of the extensive area of the contact regions and the multiplicity of interactions occurring in these protein-protein association processes. The signs of the thermodynamic parameters ΔG, ΔH, ΔS, and ΔC_p, are predominantly negative, implying that hydrogen bonding and van der Waals' interactions are predominant for this set of globular proteins [15]. For the most part, the association reactions are enthalpically driven and entropically opposed. It is doubtful that the values for the thermodynamic parameters for association will be substantially altered if it were possible to take account of the heats of ionization of the probable amino acids involved when proton liberation accompanies the association reaction.

A large negative heat capacity change occurs for every system studied. Increasingly exothermic reactions and more negative entropy changes are to be expected at higher temperatures as a consequence of the large negative heat capacity change influencing the enthalpy and entropy changes through the relationships $\partial\Delta H/\partial T = \Delta C_p$ and $\partial\Delta S/\partial T = \Delta C_p/T$.

As more protein association reactions are studied, it is anticipated that a wider spectrum of positive and negative entries will appear in the table of thermodynamic parameters. Association reactions of two very hydrophobic proteins, phycocyanin and tryptophane synthetase, are reported to display positive enthalpy and entropy changes upon association [16]. The trypsin + soybean inhibitor reaction also displays this property at low pH [2].

A quick glance at Table 1 reveals the paucity of extensive thermodynamic data for protein association reactions. Information about heat capacity changes, which may well be the most characteristic feature of these systems, is reported in only about one-half of the cases selected. A great number of calorimetric measurements of the enthalpy change have been carried out and reported in the literature. Such information is useful and adequate if one wants information concerning the variation of an equilibrium constant with temperature. However, if one is interested in the thermodynamic parameters characterizing protein-protein reactions as a class, then more extensive thermodynamic data encompassing determination of ΔG, ΔH, ΔS, ΔC_p, and the temperature dependence of ΔC_p over a wide range of solution conditions (e.g., pH, salt concentration, ionic composition) are required. The work of Hearn et al. [1] provides an excellent model for the scope of a proper characterization. It is this reviewer's opinion that one such thorough investigation which describes the thermodynamic behavior of a protein over a wide range of variables is of much greater value than a large number of fragmentary studies on many different proteins.

References

1. Hearn RP, Richards FM, Sturtevant JM, Watt GD (1971) Biochemistry 10:806–817
2. Yung HYK, Trowbridge CG (1980) J Biol Chem 255:9724–9730
3. Lavialle F, Rogard M, Alfsen A (1974) Biochemistry 13:2231–2234
4. Wiesinger H, Bartholmes P, Hinz H-J (1979) Biochemistry 18:1979–1984

5. Johnson RE, Hruby VJ, Rupley JA (1979) Biochemistry 18:1176–1179
6. Banerjee SK, Pogolotti A Jr, Rupley JA (1975) J Biol Chem 250:8260–8266
7. Shiao DF, Sturtevant JM (1969) Biochemistry 8:4910–4917
8. Atha DH, Ackers GK (1974) Arch Biochem Biophys 164:392–407
9. Sturtevant JM, Velicelebi G, Jaenicke R, Lauffer MA (1981) Biochemistry 20:3792–3800
10. Ross PD, Hofrichter J, Eaton WA (1977) J Mol Biol 115:111–134
11. Gill SJ, Farquhar EL (1968) J Am Chem Soc 90:3039–3041
12. Turner R, Liener IE, Lovrien RE (1975) Biochemistry 14:275–282
13. Ross PD, Hofrichter J, Eaton WA (1975) J Mol Biol 96:239–256
14. Ross PD, Minton AP (1977) J Mol Biol 112:437–452
15. Ross PD, Subramanian S (1981) Biochemistry 20:3096–3102
16. Chen CH, Berns DF (1977) J Phys Chem 81:125–129
17. Hinz H-J (1983) Ann Rev Biophys Bioeng 12:285–317

Chapter 9   **Hemoglobin**

B. GEORGE BARISAS

Symbols

A_i: coefficient used in prediction of Hb osmotic pressure (see Sect. 2)

B_i: coefficient used in prediction of Hb activity coefficient and apparent M_w (see Sect. 2)

c: ratio K_T/K_R of O_2 affinities of hemoglobin T- and R-states

$E_{1/2}$:   midpoint potential defined as the potential of the system when [ox]=[red] (see [26])

F: Faraday (96,487 coulomb mol^{-1})

ΔG_I: Wyman interaction free energy RT ln (K_4/K_1)

ΔG_α: Gibbs energy for O_2-binding to isolated α-chains

$\Delta G_{\alpha 2}$:   Gibbs energy for O_2-binding to α_2-homodimers (average per O_2)

ΔG_β: Gibbs energy for O_2-binding to isolated β-chains

$\Delta G_{\beta 4}$:   Gibbs energy for O_2-binding to β_4-homotetramers (average per O_2)

ΔG_1:   Total Gibbs energy for binding four O_2 to two α-chains and two β-chains

ΔG_2:   Gibbs energy for binding four O_2 to two Hb $\alpha\beta$-heterodimers

ΔG_{2i}:   Gibbs energy for binding ith O_2 to a Hb heterodimer

$\Delta G_{2,4}$:   Total Gibbs energy for binding two O_2 to an α_2-dimer and two O_2 to half a β_4-tetramer

ΔG_4:   Gibbs energy for binding four O_2 to a Hb tetramer

ΔG_{4i}:   Gibbs energy for binding of ith O_2 to a hemoglobin tetramer

$^i\Delta G_{\alpha 2}$:   Gibbs energy of assembly of an α_2-dimer with i-ligands from two α-chains

$^i\Delta G_{\beta 4}$:   Gibbs energy of assembly of a β_4-tetramer with i-ligands from four β-chains

$^i\Delta G_{\alpha\beta}$:   Gibbs energy of assembly of a heterodimer with i-ligands from an α-chain and a β-chain

$^i\Delta G_1$:   Gibbs energy of assembly of a tetramer with i-ligands from two α-chains and two β-chains

$^i\Delta G_2$:   Gibbs energy of assembly of a tetramer with i-ligands from two heterodimers

ΔH_{bc}:   reaction enthalpy corrected for heat of proton uptake or release by buffer

ΔH_{obs}:   uncorrected reaction enthalpy

ΔH_{soln}:   enthalpy of solution of a gas in buffer

i_s: switchover point, namely degree of Hb ligation at which R- and T-state molecules have the same energy

K_i: intrinsic binding constant for ith ligand X to HbX_{i-1} (see [39] for definition)

K_R: ligand binding constant to R-state Hb

K_T: ligand binding constant to T-state Hb

L: equilibrium constant $[T_0]/[R_0]$ between unligated conformational states of Hb

M: molecular weight

M_w: weight-average molecular weight

n: Hill coefficient (see [69] for definition); number of electrons involved in redox reaction (see [26] for definition)

n_{max}:   maximum slope of Hill plot (see [69] for definition)

P: pressure

P_{50}:   ligand partial pressure at which hemoglobin is 50% saturated

P_m: median ligand partial pressure (see [68] for definition)

P_{max}:   ligand partial pressure at maximum Hill plot slope

R: gas constant (8.317 J mol^{-1} K^{-1})

R_i:	R-conformation Hb tetramer binding i-ligands
s:	solubility of a gas in buffer
T:	absolute temperature (K)
T_i:	T-conformation Hb tetramer binding i-ligands
V:	volume; specific volume
W:	Hill plot symmetry index $= K_1K_4/K_2K_3$
X:	ligand activity
$\bar{Y}$:	fractional saturation of protein with ligand
G:	Hb thermodynamic activity coefficient
$\bar{v}$:	number of protons bound per Hb tetramer
$\bar{v}_{Cl,Hb}$:	number of protons linked to chloride binding in deoxy Hb
$\bar{v}_{Cl,HbO_2}$:	number of protons linked to chloride binding in oxy Hb
$\Delta\bar{v}$:	number of protons released upon binding four O_2 to Hb
Δv_i:	number of intrinsic (nonchloride linked) protons released upon binding four O_2 to Hb
$\Delta\bar{v}_2$:	number of protons released per four O_2 bound to $\alpha\beta$-dimers
$\Delta\bar{v}_p$:	number of protons *absorbed* upon organic phosphate binding
$\Delta\bar{v}_{41}$:	number of protons released upon binding of first oxygen to Hb tetramer
$\Delta\bar{v}_{4(2+3)}$:	number of protons released upon binding of second and third oxygens to Hb tetramer
$\Delta\bar{v}_{44}$:	number of protons released upon binding of fourth oxygen to Hb tetramer

Abbreviations

Hb:	hemoglobin;
Hb^+:	methemoglobin;
RBC:	red blood cell;
DPG:	2,3-diphosphoglycerate;
IHP:	inositol hexaphosphate;
IHS:	inositol hexasulfate;
MWC:	Monod-Wyman-Changeux;
R-:	"relaxed" conformation of oxyhemoglobin and methemoglobin;
T-:	"tense" conformation of deoxyhemoglobin.

1 Introduction

Hemoglobin (Hb) is perhaps the most studied protein in the world. There are several reasons why this is so. First, hemoglobins and related proteins are the principal respiratory oxygen carriers in virtually all vertebrates as well as in many invertebrates. Secondly, in humans, virtually every chemical parameter of hemoglobin is optimized for efficient oxygen transport and related processes. This functional efficiency derives in large measure from a positive cooperativity in oxygen binding and such allosteric behavior has been a powerful force motivating studies of the protein. Ever since Adair [1] first determined the tetrameric nature of hemoglobin, scientists have tried to understand in ever greater physical detail how the binding of one oxygen facilitates the binding of subsequent ones. A third factor motivating interest in hemoglobin is its interaction with, and regulation by, a wide range of materials. Besides oxygen, the gases CO_2, CO, and NO all bind to Hb, while hydrogen ions, chloride, and organic phosphate are physiological regulators of Hb function. Fourth, hemoglobin was one of the first proteins to have its three-dimensional structure determined by X-ray crystallography and was the first protein to exhibit ligand-induced structural changes crystallographically. Finally, hundreds of mutant human hemoglobins have been discovered and characterized. These proteins provide molecules altered in single amino acids at known locations. They represent a library of unique tools whereby functional consequences of known protein structural changes may be explored.

Systems Studied. Ackers [2], Barisas [3], Imai [4], Gill [5] and others have pointed out how important are stepwise thermodynamic data on the processes in which hemoglobin participates. Moreover, one needs information as to the thermodynamic linkages between these processes. This chapter tries to summarize the information necessary to predict the Hb species present, and the numbers of various ligands bound to each, over a wide range of temperatures and ligand and protein concentrations. Such a relatively complete set of data can be obtained for hemoglobin A in its reactions with oxygen but for no other hemoglobin or gaseous heme ligand. Thus only human hemoglobin A is discussed, with occasional references being made to hemoglobin S. Whenever the term hemoglobin is used, human hemoglobin A should be understood. Thermodynamic data on human hemoglobin mutants and on hemeproteins from other species range from anecdotal (P_{50} and n) to relatively comprehensive. Readers are referred to books by Lehmann and Huntsman [6] and by Fairbanks [7] where comprehensive references to such studies are collected.

Range of Parameters Studied. Being a protein, Hb retains function only within a relatively narrow range of temperature, pH, and ionic strength. One might expect that the physiological conditions prevailing in the red cell would be the standards to which thermodynamic studies refer. However, this is impractical for at least three reasons. First, the Hb concentration in the RBC is about 5 mM tetramer and the thermodynamic nonideality of the protein at such concentrations is enormous (see Sect. 3). Second, the 5 mM total DPG concentration in the RBC largely saturates deoxy Hb, but does not saturate oxyhemoglobin. Moreover, the thermodynamically important free DPG concentration is inaccessible. Finally, physiological conditions involve a CO_2 partial pressure of 40 torr (5.3 kPa) which is generally impractical to maintain experimentally. The conditions most commonly used in modern Hb studies are therefore different: pH 7.4, Tris or bis-Tris buffer containing 0.1–0.2 M chloride, 25 °C, no organic phosphates present unless stated. Hemoglobin concentrations of 60 µM heme (1 mg ml^{-1}) are common in spectral studies, while 1 mM or greater heme concentrations (16 or more mg ml^{-1}) are often used in calorimetry and in titrations.

Standard State Conditions. Standard states of reactants and products are generally unit activity in aqueous solution. Exceptions are equilibrium constants defined in terms of pressure where the standard state of the gaseous reactant is 1 atm = 101.325 kPa.

2 Molecular Parameters for Human Hemoglobin

The molecular properties of Hb have been under study for over 150 years. In 1825 Englehardt [8] accurately determined the iron content of the protein and this es-

Table 1. Molecular parameters of human hemoglobin A[a]

Parameter	Value	Conditions/method	Reference
Molecular weight[b]	64443	Deoxy, from sequence	[10]
	15740	Deoxy, α-chain	
	16481	Deoxy, β-chain	
Sedimentation coeff $S_{20,w}$	4.45×10^{-13} s	Ultracentrifuge	[11]
Diffusion coeff. $D_{20,w}$	6.40×10^{-7} cm^2 s^{-1}	Ultracentrifuge	[12]
Frictional ratio f/f_0	1.20	Ultracentrifuge	[12]
Partial spec. vol $\bar{V}$	0.750 cm^3 g^{-1}	Dilatometry	[13]
Hydration	0.24 gH$_2$O/gHb	10 GHz dielectric	[14]
Equivalent ellipsoid	$6.4 \times 5.5 \times 5.0$ nm	Deoxy crystallography	[15]
Radius of gyration	2.49 nm (oxy),	Various methods	[16]
	2.42 nm (deoxy)		
Isoelectric point	7.15	4 °C	[17]

[a] Material is oxy Hb unless otherwise specified.
[b] At isoelectric point.

tablished its minimum possible molecular weight as 16 000. The tetrameric structure of the protein became apparent in the 1920's through molecular weight determinations by Adair [1] and by Svedberg [9]. Since that time a considerable volume of data on various molecular parameters of hemoglobin has been amassed. Table 1 summarizes values for those parameters of particular thermodynamic importance.

3 Nonideality and Gelation of Concentrated Hb Solutions

The thermodynamic properties of Hb solutions at concentrations ~ 300 g l^{-1} are important because it is at such concentrations that Hb occurs in the RBC. Moreover at such concentrations in individuals afflicted with sickle cell disease, deoxy HbS can polymerize into oriented arrays of extended helices called tactoids. To this polymerization can be traced the pathology of the disease [18].

Ross and co-workers [19, 20] have shown that the nonideal thermodynamic properties of Hb solutions at concentrations from 1 up to 350 g l^{-1} can be quantitatively predicted by representing the protein as hard, quasispherical noninteracting particles. For such particles equations for the osmotic pressure Π, the apparent weight average molecular weight $M_{w(app)}$, and the activity coefficient Γ are given by

$$\Pi = \frac{RTc}{M} \left(1 + \sum_{i=2}^{7} A_i c^{i-1} \right), \tag{1}$$

$$M_{w(app)} = M / \left(1 + \sum_{i=2}^{7} B_i c^{i-1} \right), \tag{2}$$

$$\ln \Gamma = \sum_{i=2}^{7} i^{-1} B_i c^{i-1}, \tag{3}$$

Table 2. Coefficients for evaluating hemo-globin colligative properties from effective molecular specific volume V [19]

i	A_i	B_i
2	$4.00\,V$	$7.00\,V$
3	$10.00\,V^2$	$22.00\,V^2$
4	$18.36\,V^3$	$43.45\,V^3$
5	$28.24\,V^4$	$67.74\,V^4$
6	$39.53\,V^5$	$95.97\,V^5$
7	$56.52\,V^6$	$158.52\,V^6$

Table 3. Best-fit values of M and V obtained by least-squares analysis of sedimentation equilibrium data at various temperatures[a]

Species	$T\,(°C)$	M/1000	$V\,(ml\,g^{-1})$
CO HbA	2	67.4	0.902
	10	65.5	0.897
	20	64.1	0.890
Deoxy HbA	20	65.0	0.932
	37	63.6	0.950
CO HbS	2	66.9	0.930
	10	63.6	0.902
	20	65.5	0.957
Deoxy HbS	20	65.4	0.930
	37	63.2	0.907
COMBINED		$65.0^b\,(\pm0.4)$	$0.920^b\,(\pm0.010)$

[a] [20]. Values of M and V determined separately for each data set.
[b] Mean of values for individual data sets. Uncertainty is the mean of uncertainties obtained for individual data sets.

where c is the particle concentration in $g\,l^{-1}$. The coefficients A and B are given in Table 2 as functions of an effective molecular specific volume V in $cm^3\,g^{-1}$.

Both the apparent molecular weight M and the specific volume V must be regarded as adjustable parameters to be evaluated experimentally. These authors [20] have examined by sedimentation equilibrium deoxy- and carbonmonoxy HbA and deoxy- and carbonmonoxy HbS at various temperatures. These results are shown in Table 3. It is apparent that neither temperature, ligation state, nor ability to gel (deoxy HbS) has any substantial effect on the thermodynamic properties of the solution-phase monomeric hemoglobin.

Table 4 illustrates the concentration dependences of the apparent activity and activity coefficient of Hb. At concentrations of $360\,g\,l^{-1}$, the thermodynamic activity is over two orders of magnitude larger than the concentration.

Ross et al. [22] have combined sedimentation measurements and differential scanning calorimetry to evaluate the standard state quantities for the gelation of

Table 4. Concentration dependence of hemo-globin activity and activity coefficient in concentrated solution [21]

C (g l⁻¹)	Activity (g l⁻¹)	ln Γ
20	23.1	0.14
40	53.5	0.29
60	94.6	0.46
80	150	0.63
100	226	0.82
120	330	1.01
140	473	1.22
160	679	1.45
180	973	1.69
200	1410	1.95
220	2060	2.24
240	3040	2.54
260	4580	2.87
280	7040	3.22
300	11050	3.61
320	17800	4.02
340	29700	4.47
360	51000	4.95
380	90200	5.47
400	168000	6.04

Table 5. Standard state thermodynamic parameters for gelation of deoxy HbS at 37 °C [a]

Quantity	Value
ΔG^0	$-$ 12.6 kJ mol⁻¹
ΔH^0	0.0 $\pm$ 0.8 kJ mol⁻¹
ΔS^0	41.8 J mol⁻¹ K⁻¹
ΔC_p^0	-824 J mol K⁻¹
$d(\Delta C_p^0)/dT$	35 $\pm$ 10 J mol⁻¹ K⁻²
c_p [b]	0.493 $\pm$ 0.012 g cm⁻³
Kp [c]	0.044 $\pm$ 0.011 kPa⁻¹

[a] Measured in 0.15 M potassium phosphate pH 7.15 from [22].
[b] Concentration of HbS in polymer phase.
[c] O_2 binding constant to polymer. Non-cooperative up to 14% fractional saturation. The free energy change for this process relative to an O_2 standard state of unit molarity can be calculated as $-RT$ ln $(Kp s^{-1})$ where s is the solubility of O_2 in M kPa⁻¹. [25].

HbA from concentrated solutions of monomeric HbS. Their results are presented in Table 5. These scientists, as well as S. J. Gill and coworkers [23], have also studied the properties of oxygen binding to HbS solutions containing polymer. In these and other studies on oxygenation of highly concentrated hemoglobin solutions, the membrane-covered thin-layer cell devised by Gill [24] has been indispensable. Table 5 includes a recent estimate of the oxygen affinity of HbS polymer.

4 Methemoglobin and the Redox Behavior of Hemoglobin

The oxidation of Hb to met Hb is of considerable thermodynamic interest. On the one hand, met Hb possesses a conformation similar to oxy Hb and is thus said to have an R ("relaxed")-conformation [42]. Deoxy Hb possesses a T ("tense")-conformation. The pH dependence of redox potentials for the met Hb/Hb couple can therefore be viewed as resulting from ionizable groups with conformationally-sensitive pK's. This is analogous to the Bohr effect in O_2 binding and has indeed been termed the oxidative Bohr effect. Thus useful structural information derives from the potentiometric titration of hemoglobins. Because oxidation of Hb occurs cooperatively, redox titration curves of Hb exhibit shapes corresponding to variable numbers n of electrons. Both n and the midpoint potential $E_{1/2}$ are therefore necessary to characterize the oxidative Gibbs energy change under

Table 6. Electrochemical parameters for reduction of methemoglobin to deoxyhemoglobin at 30 °C in 0.1–0.3 M ionic strength buffers

pH	$E_{1/2}(v)$ [a]	n [b]
6.0	0.17	1.3
6.4	0.17	1.2
7.0	0.15	1.6
7.7	0.12	2.0
8.2	0.08	2.4
9.2	0.02	2.5

[a] Potential measured vs. standard hydrogen electrode. [26].
[b] Number of electrons. [27].

Table 7. Thermodynamic parameters for the reduction of sperm whale metmyoglobin at 30 °C and pH 7 in 0.1–0.3 M ionic strength buffers [28]

$E_{1/2}(v)$	ΔG^0 (kJ mol^{-1})	ΔH^0 (kJ mol^{-1})	$T\Delta S^0$ (kJ mol^{-1})
+0.050	−4.81	−22.18	−17.37

Table 8. Binding of ligands to methemoglobin [a]

Anion	$-\Delta G^0$	$-\Delta H^0$	ΔS^0	Conditions	Reference
F$^-$ [b]	10.0±0.1	10.0±0.8	0 ±3.0	20°, I=0.05 M, pH 7.0	[32]
SCN$^-$	14.9±0.1	31.3±0.8	− 55.7±3.0	20°, I=0.05 M, pH 7.0	[32]
CN$^-$	50.2±0.1	83.3±0.8	−105.6±3.3	20°, I=0.05 M, pH 7.0	[32]
N$_3^-$	28.8±1.0	74.2±1.2	−151.8±7.4	25.9°, I=0.05 M, pH 7.0	[33]
DPG [c]	23.0±0.4	36.0±1.3	− 43.5	25°, I=0.1 M, pH 6.0	[34]
IHP	27.2	13.0±1.3	47.7	25°, I=0.1 M, pH 7.4	[34]

[a] ΔG^0 and ΔH^0 are given as kJ (mol ligand)$^{-1}$ while ΔS^0 is given as J K^{-1} mol^{-1}.
[b] Data obtained from spectrophotometric titrations.
[c] Data obtained from calorimetric titrations.

given conditions. Table 6 presents these electrochemical data as functions of pH.

The temperature dependences of these redox data permit the Gibbs energies of hemoprotein oxidation to be resolved into enthalpic and entropic components. The former term is particularly important in calorimetry, where heat evolution from spontaneous met Hb formation contributes to instrumental baselines. Data on heats of met Hb formation do not seem to be available but data on myoglobin oxidation are presented in Table 7.

In met Hb the sixth coordination position of each iron is normally occupied by a water molecule. These molecules are moderately acidic, each heme losing a proton with an apparent pK of 8.05 and an apparent ionization enthalpy of 17.6 kJ mol^{-1} [29]. However, this transition is difficult to characterize thermodynamically in the presence of so many other ionizable groups on the protein. Many anions can bind in place of the water and their binding *has* been well characterized thermodynamically. It seems likely that these ligands insert between the iron and the distal histidine, as has been shown to be the case for azide [30, 31]. Met Hb also binds the allosteric affectors DPG and IHP at stoichiometries of one per tetramer. In the thermodynamics of these interactions, met Hb resembles oxy Hb rather than deoxy Hb. Table 8 summarizes the thermodynamics of ligand binding to met Hb.

5 The Bohr Effect

Strictly speaking the Bohr Effect is the reduced oxygen affinity exhibited by Hb at low pH [35], while the Haldane effect refers to the proton release occurring on Hb oxygenation. The terms are frequently used interchangeably and the effects are related by

$$\frac{d\log P_m}{dpH} = -\Delta\bar{v}/4 , \tag{4}$$

where P_m is the median partial pressure of O_2 (or CO) and $\Delta\bar{v}$ is the number of protons released per four oxygens bound. Table 9 summarizes titration data on the number of protons bound per heme to deoxy Hb and oxy Hb molecules at various pH's and temperatures. These specific data have been used by Ackers, Gill, and others to correct calorimetric results for thermal effects of Bohr proton release. Included are proton binding data for met Hb at 30 °C.

It is possible to represent the preceding Bohr effect data as arising from two ionizable groups per subunit, both having different pK values in deoxy and oxy Hb. Both groups are assumed to have characteristic heats of proton ionization. While the Bohr effect is vastly more complicated than this, this formalism provides a convenient method to correct calorimetrically determined enthalpies for heats of Bohr proton release. Table 10 summarizes this representation of the Bohr effect.

The oxygen-linked ionization of hemoglobin has also been examined by Chipperfield et al. [70] and by Ross-Bernardi and Roughton [71] using calorimetric and differential titration methods, respectively. These results are presented in Table 11. They indicate that heat of ionization of the more alkaline Bohr group on deoxy Hb is about 20.9 ± 3.3 kJ mol^{-1} higher that of the corresponding group on oxy Hb. A similar difference appears to exist for the more acid oxygen-linked group but the uncertainties are much larger.

The chloride ion dependence of the Bohr effect has been explored by van Beek et al. [37]. These authors represent the observed Bohr effect $\Delta\bar{v}$ for four oxygens as arising both from intrinsic pK shifts and from differences in hydrogen ion

Table 9. Protons bound per heme by human oxy-, deoxy- and methemoglobin[a]

pH	10 °C		20 °C		30 °C			40 °C	
	HbO$_2$	Hb	HbO$_2$	Hb	HbO$_2$	Hb	Hb$^+$	HbO$_2$	Hb
5.2			+4.07	+3.70	+4.15	+3.89			
5.4			+3.60	+3.26	+3.59	+3.33	.3.64	+3.55	+3.36
5.6	+3.18	+2.93	+3.11	+2.83	+3.02	+2.79	+3.04	+2.93	+2.82
5.8	+2.74	+2.49	+2.60	+2.40	+2.48	+2.36	+2.50	+2.32	+2.29
6.0	+2.31	+2.13	+2.12	+2.02	+1.91	+1.93	+2.00	+1.71	+1.77
6.2	+1.88	+1.80	+1.63	+1.63	+1.39	+1.48	+1.48	+1.10	+1.27
6.4	+1.42	+1.46	+1.12	+1.27	+0.84	+1.03	+0.90	+0.44	+0.71
6.6	+0.93	+1.09	+0.57	+0.81	+0.21	+0.53	+0.26	−0.23	+0.12
6.8	+0.39	+0.69	0	+0.39	−0.42	−0.01	−0.46	−0.89	−0.48
7.0	−0.13	+0.21	−0.57	−0.11	−1.01	−0.53	−1.19	−1.53	−1.10
7.2	−0.73	−0.29	−1.18	−0.67	−1.65	−1.14	−1.88	−2.13	−1.71
7.4	−1.30	−0.77	−1.77	−1.24	−2.22	−1.72	−2.56	−2.66	−2.27
7.6	−1.87	−1.32	−2.33	−1.80	−2.73	−2.27	−3.24	−3.13	−2.80
7.8	−2.40	−1.86	−2.82	−2.33	−3.20	−2.79	−3.83	−3.54	−3.28
8.0	−2.89	−2.37	−3.23	−2.79	−3.60	−3.28	−4.27	−3.88	−3.08
8.2	−3.33	−2.86	−3.62	−3.24	−3.95	−3.73	−4.65	−4.19	−4.04
8.4	−3.68	−3.29	−3.95	−3.63	−4.22	−4.06	−5.07	−4.46	−4.36
8.6	−3.97	−3.67	−4.20	−3.96	−4.45	−4.34	−5.40	−4.72	−4.66
8.8	−4.20	−4.00	−4.42	−4.25	−4.72	−4.63		−5.00	−4.96
9.0	−4.43	−4.27	−4.63	−4.52	−4.98	−4.93		−5.38	−5.36
9.2	−4.62	−4.53	−4.84	−4.79	−5.30	−5.26			
9.4	−4.81	−4.75	−5.08	−5.06					
9.6	−5.05	−5.06	−5.48	−5.47					
9.8	−5.36	−5.36	−6.06	−6.06					
10.0	−5.76	−6.87	−6.87	−6.87					
10.2			−7.91	−7.91					
10.4			−9.36	−9.36					

[a] All measurements made in 0.25 M NaCl. Protein concentration 0.5–1.0% Hb and HbO$_2$ data from [36]. Hb$^+$ data from [26].

binding linked to chloride release upon oxygenation. The intrinsic effect is designated $\Delta\bar{v}_i$ and is defined by

$$\Delta\bar{v} = \Delta\bar{v}_i + \frac{\bar{v}_{Cl,Hb} K_{Cl,Hb}[Cl^-]}{1 + K_{Cl,Hb}[Cl^-]} - \frac{\bar{v}_{Cl,HbO2} K_{Cl,HbO2}[Cl^-]}{1 + K_{Cl,HbO2}[Cl^-]}, \tag{5}$$

where $K_{Cl,Hb}$ and $K_{Cl,HbO2}$ are the binding constants for chloride to the (two) sites on each tetramer and $\bar{v}_{Cl,Hb}$ and $\bar{v}_{Cl,HbO2}$ are the maximum numbers of hydrogen ions linked to such binding. Analysis of the data of Rollema et al. [38] according to this model permitted evaluation of the number of protons representing the intrinsic Bohr effect. These results correspond to a pK shift from 7.8 to 6.9 for two groups (perhaps His-146β) per tetramer and are presented in Table 12. Considerable progress has been made toward identifying the specific residues responsible

Table 10. Proton ionization constants for oxygen-linked acid groups in human hemoglobin [36]. These constants correspond to proton ionization enthalpies of $\Delta H_1 = -6.3\,\text{kJ mol}^{-1}$ for the more acid group and $\Delta H_2 = +37.7\,\text{kJ mol}^{-1}$ for the more basic group

Temperature (°C)	$pK_1(Hb)$	$pK_1(HbO_2)$	$pK_2(Hb)$	$pK_2(HbO_2)$
10	5.42	6.22	8.08	6.68
20	5.46	6.26	7.85	6.45
30	5.50	6.30	7.63	6.23
40	5.54	6.34	7.42	6.02

Table 11. Heats of ionization of oxygen-linked groups on human hemoglobin at 25° [a]

	Deoxy Hb	Oxy Hb	Reference
pK_1	5.13± 0.04	5.60±0.03	[71]
ΔH_1 [b]	−16.3 ±11.3	− 5.9 ±7.9	[71]
pK_2	7.84± 0.01	6.84±0.01	[71]
ΔH_2 [b]	44.8 ± 1.7	23.4 ±2.1	[71]
	46.6 ± 4.2	30.5 ±4.2	[70]

[a] Ionic strength of experimental solutions was about 0.2 M. Measurements cover the pH range 5.5–9.0.
[b] Heats of proton ionization are given in kJ (mol protons)$^{-1}$.

Table 12. Calculated intrinsic component of the alkaline Bohr effect [a]

	Deoxyhemoglobin		Oxyhemoglobin		
	$\bar{v}_{Cl,Hb}$ (protons/$4O_2$)	$K_{Cl,Hb}$ [c] (M^{-1})	$\bar{v}_{Cl,HbO_2}$ (protons/$4O_2$)	K_{Cl,HbO_2} [c] (M^{-1})	$\Delta\bar{v}_i$ (protons/$4O_2$)
5.6	−	−	0.99	30.3	−
6.0	0.51 [b]	−	1.46	10.6	0.20 [b]
6.5	1.23	> 10^3	1.73	4.8	0.43
7.0	1.43	217	1.71	2.7	0.81
7.4	1.45	53.3	1.61	1.2	0.90
8.0	1.25	40.7	−	−	0.56
8.5	0.81	14.3	−	−	0.29

[a] Protein concentration 0.6 mM heme; 25 °C. From [37].
[b] Estimated.
[c] Free energy changes for chloride binding can be calculated as $-RT\ln K$.

for both the intrinsic and chloride-linked Bohr effect. For a detailed discussion of this problem, see Imai [39].

Chu et al. [41] have succeeded in resolving the number of Bohr protons released upon stepwise binding of successive oxygen ligands (Table 13). From pH 7.4 to 8.5 about one-fourth of the total Bohr proton release occurs when the first oxygen binds. The remainder accompanies binding of the second and third

Table 13. Bohr protons released upon stepwise oxygenation of tetramers [41]

pH	$\Delta\bar{v}_{41}$ [a]	$\Delta\bar{v}_{4(2+3)}$ [b]	$\Delta\bar{v}_{44}$ [c]
7.4	0.64 ± 0.07	1.62 ± 0.27	0.05 ± 0.06
8.0	0.38 ± 0.03	1.16 ± 0.12	0.05 ± 0.06
8.5	0.16 ± 0.03	0.78 ± 0.11	0.05 ± 0.06
8.95	-0.03 ± 0.06	0.44 ± 0.23	0.05 ± 0.06

[a] Number of protons released upon binding of first oxygen to Hb tetramer.
[b] Number of protons released upon binding second and third oxygens to Hb tetramer.
[c] Number of protons released upon binding fourth oxygen to Hb tetramer.

Table 14. Dimer Bohr effect contrasted with that in tetramers [a]

pH	$\Delta\bar{v}_2$ (protons released per $4O_2$ bound to dimers)	$\Delta\bar{v}$ (protons released per $4O_2$ bound to tetramers)
7.4	0.22 ± 0.12	2.32 ± 0.19
8.0	0.07 ± 0.05	1.61 ± 0.09
8.5	-0.05 ± 0.04	1.02 ± 0.07
8.95	-0.16 ± 0.09	0.49 ± 0.14
9.5		0

[a] [41]. 1 mM EDTA; total chloride concentration close to 0.18 M; temperature 21.5 °C.

oxygens. Binding of the fourth oxygen does not change the protonation state of the protein. Because of high cooperativity in oxygen binding, the unequal distribution of proton release is still consistent with the observed linearity of plots of proton release vs. oxygen saturation.

It has recently come to light that $\alpha\beta$ dimers also exhibit a Bohr effect [40, 41]. At pH 7.4 oxygenation of these heterodimers releases 0.20 ± 0.08 moles $H^+/4O_2$. The dimer Bohr effect is thus about 20% as large as the effect in tetramers. The authors conclude incidentally that oxygenation of dimers at pH 7.4 involves no change in bound chloride. Table 14 compares the tetramer and dimer Bohr effects estimated under the same conditions.

6 Equilibrium Binding of O_2 to Hemoglobin – Classical Models

It is essentially impossible to discuss the binding of the gaseous ligands O_2, CO, and NO to Hb outside the context of some model or reaction scheme. In the absence of such a model one can present only a binding isotherm and cite the total Gibbs energy of ligating the tetramer. In this section we present hemoglobin O_2-binding data acquired under conditions of general interest and processed in three ways: (1) according to the sequential reaction scheme of Adair [1], (2) by the Hill equation, and (3) according to the two-state concerted model of Monod et al. [42].

The first accurate determinations of hemoglobin O_2-binding isotherms were made by Roughton and coworkers, using extraordinarily careful gasometric procedures [43]. These measurements have the advantage of taking place at high (1–2 mM) heme concentrations where dissociation of the oxy-tetramer into oxy-dimers is minimal. More recently the techniques for automatic spectrophotometric recording of hemoglobin oxygen saturation curves have permitted Imai and his coworkers to achieve great precision in such measurements [39]. However, two points should be borne in mind concerning the accuracy of such measurements. First, such methods depend upon the absolute linearity of Hb spectral shifts with degree of oxygenation from $\bar{Y} = 0$ to $\bar{Y} = 1$. Any nonlinearity will have particularly large effects on the values estimated for the stepwise constants K_2 and K_3. Second, Ackers has pointed out [44] that any appreciable dissociation of oxy Hb nto dimers will likewise affect the reliability of the stepwise constants, especially K_2 and K_3.

The Adair equation [1] can be rewritten [39] to express the reation between the fractional saturation $\bar{Y}$ of a four-site protein and the ligand activity X in terms of the stepwise ligand binding constants K_1–K_4 instead of the experimental constants used originally by Adair [1].

$$\bar{Y} = \frac{K_1X + 3K_1K_2X^2 + 3K_1K_2K_3X^3 + K_1K_2K_3K_4X^4}{1 + 4K_1X + 6K_1K_2X^2 + 4K_1K_2K_3X^3 + K_1K_2K_3K_4X^4} \cdot \qquad (6)$$

The relative values of the K's indicate the cooperativity and the total Gibbs energy change ΔG_4 per 4 moles of ligand bound is given by

$$\Delta G_4 = RT \ln (K_1K_2K_3K_4). \qquad (7)$$

The stepwise reaction parameters for hemoglobin oxygenation are given in Table 15. It has been suggested [39] that, under physiological conditions, values of K_1 and K_4 are reproducible to 20%, while K_2 and K_3 are said to reproduce to ± 30–40%. Much worse reproducibility is often encountered. This table illustrates many of the main features of O_2 binding by hemoglobins. The Bohr effect is clearly evident in the more negative ΔG_4 values observed at pH 9.1 (sets 8–10) relative to pH 7.4. Likewise at pH 6.5 (set 11) oxygenation is less exergonic than at neutral pH. Anion effects are likewise clearly demonstrated. Chloride and phosphate (sets 3 and 5 vs. set 2) decrease overall oxygen affinity mildly while DPG and IHP (sets 6 and 7 vs. set 3) decrease it dramatically. Finally, the preferential binding of CO_2 to deoxy Hb is evident in the reduced O_2 affinity of Hb in the presence of 5% CO_2 (set 4 vs. set 3). Discussion of changes in apparent stepwise O_2 binding constants is beyond the scope of this Review.

The preceding data have also been analyzed by Hill plots [69], $\log \bar{Y}/(1-\bar{Y})$ is plotted vs. $\log P_{O2}$ to obtain the maximum slope or Hill coefficient n_{max} its corresponding oxygen tension P_{max} and the Wyman interaction free energy $\Delta G_I = RT \ln (K_4/K_1)$ [68]. Also included in Table 16 are the oxygen tension at half-saturation P_{50}, the median oxygen tension $P_m = (K_1K_2K_3K_4)^{1/4}$ and the Hill plot symmetry index $W = K_1K_4/K_2K_3$.

Table 15. Four-step analysis of O_2 binding by human hemoglobin[a]

Data set	Conditions[b]	K_1[c]	K_2	K_3	K_4	ΔG_4[d]	Reference
1	pH 7.0, 19 °C, 1.2–2.4 mM heme, 0.6 M phosphate	0.092	0.214	2.49	9.6	−107.9	[43]
2	pH 7.4, 7 mM Cl⁻	0.485	3.6	3.4	31.8	−124.9	[39]
3	pH 7.4, 0.1 M Cl⁻	0.164	0.47	2.3	25.9	−115.7	[4]
4	pH 7.4, 0.1 M Cl⁻ 5% CO_2	0.135	0.15	1.8	27.6	−112.0	[45]
5	pH 7.4, 0.1 M phosphate	0.102	0.23	1.1	25.7	−111.0	[39]
6	pH 7.4, 0.1 M Cl⁻ 2 mM DPG	0.0611	0.025	0.23	28.7	−106.2	[4]
7	pH 7.4, 0.1 M Cl⁻ 2 mM IHP	0.0377	0.10	0.032	6.86	− 94.3	[4]
8	pH 9.1, 19 °C, 1.2–2.4 mM heme, 0.2 M borate	0.431	20.0	7.10	62.1	−129.7	[43]
9	pH 9.1, 2.6 mM Cl⁻	1.28	2.7	26	27.7	−131.3	[4]
10	pH 9.1, 0.1 M Cl⁻	0.446	1.2	11	25.0	−124.4	[4]
11	pH 6.5, 0.1 M Cl⁻	0.89	0.08	0.52	9.3	−103.7	[4]
12	pH 6.5, 0.1 M Cl⁻, 2 mM IHP	0.0302	0.033	0.055	0.25	− 87.0	[4]

[a] Adapted from [39].
[b] Unless otherwise specified, measurements at 25 °C, hemoglobin concentration 60 μM heme, 0.05 M bis-Tris buffer for pH 7.4 and 6.5 and 0.05 M Tris buffer for pH 9.1. The average RMS of residuals of Y for the optical determinations is $\sim 5 \times 10^{-4}$.
[c] Binding constants in kPa^{-1}. Stepwise free energy changes ΔG_{4i} relative to an O_2 standard state of unit molarity can be calculated as $-RT \ln (K_i/s)$ where s is the solubility of O_2 in $M\ kPa^{-1}$.
[d] Total free energy changes given in $kJ\ mol^{-1}$ relative to an oxygen standard state of unit molarity.

Table 16. Hill plot analyses of O_2-binding data by human hemoglobin[a]

Data set	P_{50}[b]	P_m[c]	P_{max}[d]	n_{max}[e]	ΔG_I[f]	W[g]
1	1.225	1.207	1.265	2.97	11.3	1.7
2	0.273	0.271	0.279	2.40	10.38	1.3
3	0.709	0.691	0.785	3.02	12.55	4.1
4	1.040	0.992	1.248	3.15	13.18	14
5	1.164	1.120	1.373	3.14	13.68	11
6	1.87	1.79	2.25	3.08	15.23	31
7	6.51	5.95	9.49	2.53	12.89	84
8	0.124	0.128	0.105	2.47	12.05	0.2
9	0.139	0.143	0.129	2.35	7.61	0.5
10	0.287	0.288	0.285	2.73	9.96	0.9
11	2.51	2.33	3.24	2.88	11.51	20
12	18.1	16.4	27.6	1.77	5.27	4.3

[a] Adapted from [39]. See Table 15 for experimental conditions and references.
[b] O_2 partial pressure at 50% saturation (kPa).
[c] Median O_2 partial pressure (kPa).
[d] O_2 partial pressure at maximum Hill plot slope (kPa).
[e] Hill coefficient.
[f] Interaction free energy in $kJ\ (mol\ heme)^{-1}$.
[g] Hill plot symmetry index. See text.

Finally these data can be analyzed according to the two-state concerted allosteric model of Monod et al. [42]. In this model the fractional saturation Y of the protein with ligand is given by

$$\bar{Y} = \frac{LK_TP(1+K_TP)^3 + K_RP(1+K_RP)^3}{L(1+K_TP)^4 + (1+K_RP)^4}. \tag{8}$$

Here K_T and K_R are the intrinsic association constants of O_2 to the T- and R-states respectively and L is the equilibrium constant between the unligated states, namely $L = [T_0]/[R_0]$. Derived quantities include the ratio of O_2 affinities of the T- and R-states $c = K_T/K_R$, the equilibrium constant between the fully ligated states $Lc^4 = [T_4]/R_4,]$ and the so-called switch-over point i_s, at which degree of ligation R- and T-state molecules possess equal energies. MWC analyses of these data bring out many of the same points as did their analyses by the sequential model. However, two points are made particularly clear concerning the mechanism of anion linkage to O_2 binding. In data sets 2–7 various anions alter both L and c, namely they alter not only the supposed allosteric equilibrium of the unliganded protein, but also the relative O_2 affinities of R- and T-state Hb. This suggests that two-state concerted models are so oversimplified as to need extension even to explain effects of simple allosteric effectors. The second point is the binding affinities of various anions for deoxy- and oxy Hb as evidenced by K_T and K_R. Sets 3, 4, and 5 suggests that 0.1 M chloride and 0.1 M phosphate bind appreciably to deoxy Hb but only slightly to oxy Hb and set 7 shows that 2 mM IHP must bind strongly to both oxy *and* deoxy Hb. Table 22 (later in this chapter) shows the extent to which these inferences are correct.

In summary, hemoglobin oxygenation near physiological conditions is adequately modeled both by a four-step sequential (Table 15) and a two-state concerted model (Table 17). Data to be presented later show that both of these pic-

Table 17. MWC analysis of oxygen binding by human hemoglobin[a]

Data set	$K_T(kPa^{-1})$	$K_R(kPa^{-1})$	c	L	Lc^4	i_s
2	0.697	28.7	0.0243	3.7×10^3	1.3×10^{-3}	2.21
3	0.181	24.8	0.00728	8.7×10^4	2.5×10^{-4}	2.31
4	0.136	29.4	0.00462	7.3×10^5	3.3×10^{-4}	2.51
5	0.110	24.5	0.00451	5.7×10^5	2.4×10^{-4}	2.45
6	0.060	22.5	0.00267	3.0×10^6	1.5×10^{-4}	2.52
7	0.045	6.8	0.00661	2.8×10^6	5.4×10^{-3}	2.96
9	0.111	28.7	0.0387	2.8×10^2	6.2×10^{-4}	1.73
10	0.453	25.1	0.0181	2.7×10^3	2.9×10^{-4}	1.97
11	0.088	12.0	0.00731	6.2×10^5	1.8×10^{-3}	2.71
12	0.0302	0.476	0.0633	4.0×10^3	6.5×10^{-2}	3.01

[a] Adapted from [39]. The meaning of the various quantities is described in the text. See Table 15 for experimental conditions and references. The root mean square of residuals for fractional saturation averages $< 5 \times 10^{-4}$. For K_R and K_T the corresponding free energy changes relative to an O_2 standard state of unit molarity can be calculated as $-RT \ln (K/s)$ where s is the solubility of O_2 in $M \, kPa^{-1}$. For other equilibrium constants the corresponding free energy change is simply $-RT \ln K$.

tures grossly oversimplify the physical situation. Perhaps these models are best regarded as useful means of summarizing gas binding isothermal data obtained under particular conditions.

7 Linked Subunit Assembly and Ligand Binding in Hemoglobin

The most comprehensive study of ligand binding of hemoglobin has been achieved by Ackers' group at the Johns Hopkins University. Over the past decade these workers have examined the energetics of ligand binding to hemoglobin tetramers, dimers, and isolated chains as well as the association of chains and dimers to form higher structures. Full thermodynamic information under consis-

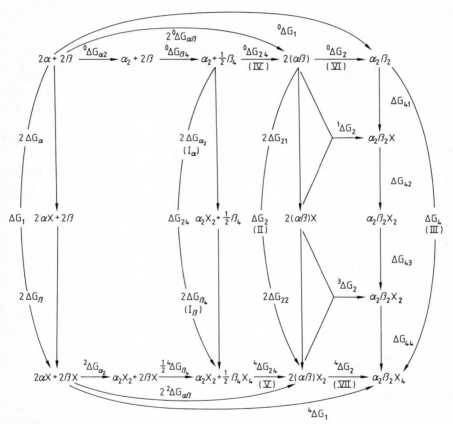

Fig. 1. Subunit assembly and oxygen binding reactions in hemoglobin. *Left superscripts* on ΔG terms specify constant numbers of oxygens bound in protein assembly reactions. In such cases *right subscripts* indicate the particular assembly reaction. In the *absence* of a left superscript, a ΔG value refers to oxygenation of a particular protein. Right subscripts then specify both the protein being considered and the particular oxygenation step. X denotes oxygen or other heme ligand. The same notation is used in Tables 18 and 19 for ΔH and ΔS values

tent conditions is available for this very complicated linkage scheme and this represents an experimental accomplishment of the first magnitude. This linkage scheme is shown in Fig. 1.

The notation for the various steps is somewhat involved and is best understood from reference to Fig. 1. Roman numerals inserted along certain reaction paths are the notation for that reaction used in papers of the Ackers group. The data presented in the following tables derive from four major experimental techniques. First, automatically recorded oxygen saturation curves at various temperatures and protein concentrations have provided the bulk of the ligand binding data. Second, analytical gel chromatography techniques [46] yielded most of the information on association-dissociation equilibria. Third, calorimetric measurements provided valuable constraints on the heats of chain association and of chain and tetramer ligation. Finally, kinetic measurements of the forward and reverse rate constants for the dissociation of the deoxy tetramer into deoxy dimer were critical, since this energetically important equilibrium lies too far towards tetramers for other types of measurement.

All the data obtained in reference to this scheme refer to the following conditions: pH 7.4, 0.1 M Tris buffer, 0.1 M NaCl, 1 mM EDTA, 21.5 °C, total chloride concentration 0.18 M. Standard states of all reactants are one molar. The data are presented in two tables. Table 18 deals with subunit association of various species, while Table 19 presents the stepwise oxygenation thermodynamics of the chains and their associated oligomers.

Table 18. Thermodynamic parameters for ligand-linked hemoglobin subunit association at pH 7.4[a]

Process	ΔG	ΔH	ΔS
(Chains to homodimers or homotetramers)			
$^2\Delta G_{\alpha 2}$	$-22.2 \pm 0.4/2$ [47]	$18.2 \pm 2.1/2$ [47]	$137 \pm 8/2$ [47]
$^0\Delta G_{\alpha 2}$	$-22.2 \pm 0.4/2$ [48]		
$^4\Delta G_{\beta 4}$	$-91.2 \pm 0.4/4$ [47]	$98.3 \pm 4.2/4$ [47]	$642 \pm 17/4$ [47]
	$-94.1 \pm 1.3/4$ [48]		
$^0\Delta G_{\beta 4}$	$-79.9 \pm 1.3/4$ [48]		
(Chains to heterodimers)			
$^0\Delta G_{\alpha \beta}$		$- 68.6 \pm 7.1/2$ [49]	
$^2\Delta G_{\alpha \beta}$		$- 65.7 \pm 2.9/2$ [49]	
(Chains to tetramers)			
$^0\Delta G_1$		$-257.7 \pm 4.6/4$ [49]	
$^4\Delta G_1$		$-115.5 \pm 8.8/4$ [49]	
(Dimers to tetramers)			
$^0\Delta G_2$	$-60.0 \pm 0.4/4$ [50]	$-120.9 \pm 2.1/4$ [51]	$-208 \pm 7/4$ [51]
$^1\Delta G_2$	$-47.2 \pm 0.2/4$ [51]	$- 74.1 \pm 3.3/4$ [51]	$- 90 \pm 11/4$ [51]
$^3\Delta G_2$	$-29.4 \pm 1.4/4$ [51]	$13.0 \pm 15.9/4$ [51]	$146 \pm 52/4$ [51]
$^4\Delta G_2$	$-33.9 \pm 0.3/4$ [50]	$16.3 \pm 6.7/4$ [51]	$169 \pm 24/4$ [51]

[a] Conditions stated in text. ΔG, ΔH given in kJ mol^{-1} and ΔS in J K^{-1} mol^{-1}. The sign "/1", "/2", or "/4" after a thermodynamic quantity indicates whether a mole of monomer, dimer, or tetramer, respectively, is indicated. References are given in brackets.

Table 19. Thermodynamic parameters for hemoglobin oxygenation[a]

Process	ΔG	ΔH	ΔS
(Subunit oxygenation)			
ΔG_α	$-\ 33.9 \pm 0.3/1$ [52]	$-\ 59.4 \pm\ 2.5/1$ [64]	$-\ 86 \pm\ 9/1$ [64]
$\Delta G_{\alpha 2}$	$-\ 33.9 \pm 0.3/1$ [52]	$-\ 59.4 \pm\ 2.5/1$ [64]	$-\ 86 \pm\ 9/1$ [64]
ΔG_β	$-\ 32.3 \pm 1.3/1$ [52]	$-\ 74.1 \pm\ 2.5/1$	$-142 \pm 10/1$
$\Delta G_{\beta 4}$	$-\ 35.8 \pm 0.5/1$ [52]	$-\ 70.7 \pm\ 3.3/1$ [64]	$-121 \pm 12/1$ [64]
ΔG_1	$-132.6 \pm 2.5/4$ [52]	$-266.9 \pm\ 8.4/4$ [49]	$-456 \pm 29/4$
$\Delta G_{2,4}$	$-139.3 \pm 0.8/4$	$-260.2 \pm\ 6.7/4$ [49]	$-411 \pm 23/4$
(Dimer oxygenation)			
ΔG_{21}	$-\ 34.7 \pm 2.9/1$ [51]	$-\ 69.5 \pm\ 2.9/1$ [51]	$-118 \pm 10/1$ [51]
ΔG_{22}	$-\ 34.7 \pm 2.9/1$ [51]	$-\ 69.5 \pm\ 2.9/1$ [51]	$-118 + 10/1$ [51]
ΔG_2	$-138.9 \pm 1.7/4$ [51]	$-277.8 \pm 11.7/4$ [51]	$-472 \pm 39/4$ [51]
(Tetramer oxygenation)			
ΔG_{41}	$-\ 22.6 \pm 0.8/1$ [51]	$-\ 33.9 \pm\ 7.5/1$ [41, 51]	$-\ 39 \pm 26/1$ [41, 51]
ΔG_{42}	$-\ 23.9 \pm 3.9/1$ [51]	$-\ 23.0 \pm\ 9.2/1$ [41, 51]	$3 \pm 34/1$ [41, 51]
ΔG_{43}	$-\ 27.9 \pm 2.2/1$ [51]	$-\ 23.0 \pm\ 9.2/1$ [41, 51]	$14 \pm 32/1$ [41, 51]
ΔG_{44}	$-\ 39.0 \pm 1.5/1$ [51]	$-\ 65.7 \pm 18.0/1$ [41, 51]	$-\ 91 \pm 61/1$ [41, 51]
ΔG_4	$-114.2 \pm 0.3/4$ [51]	$-141.0 \pm\ 2.5/4$ [51]	$-\ 91 \pm\ 9/4$ [51]

[a] Conditions as stated in text. ΔG, ΔH given in $kJ\,mol^{-1}$ and ΔS in $J\,K^{-1}\,mol^{-1}$. The sign "/1", "/2", and "/4" after a thermodynamic quantity indicates the number of O_2 molecules bound. References are given in brackets. The absence of a reference indicates a quantity derived by Imai [39] or by the author.

From these data a number of important conclusions can be drawn. First, the $\alpha\beta$ dimers bind O_2 noncooperatively and with thermodynamic properties similar to α, β, α_2, and β_4. Secondly, assembly of two unliganded $\alpha\beta$ dimers yields the deoxy tetramer, a structure of high stability (26.4 kJ mol^{-1} more than the oxy-tetramer) and low O_2-affinity (12.1 kJ mol $O_2{}^{-1}$ less affinity than the oxy-structure). These effects can be rationalized on the basis of constraining interactions at the $\alpha 1 \beta 2$ interface, which are possible only in the deoxy structure. Third, the 6.3 kcal of linkage free energy is distributed among all four tetramer ligation steps. Ackers [2] points out that, since the transition from the deoxy- to the oxy-structure does not take place at any particular ligation step, concerted models which embody all-or-none transitions between molecular states are conclusively ruled out. Finally, the triliganded molecule has an enhanced affinity for oxygen at its last site relative to that site's affinity in the dimeric species. Ackers terms this "quaternary enhancement" and it occurs as well in β_4, which exhibits enhanced O_2-affinity relative to β.

Recently Chu et al. [40, 41] have examined the pH dependences of the dimer-tetramer assembly and oxygenation processes just described. Analysis of the data yields the changes in Hb protonation at all reaction steps (see Section 4 for data on the dimer Bohr effect which this study has yielded and for the Bohr protons released upon stepwise oxygenation of tetramers). Table 20 lists the resolved pH dependences of stepwise Gibbs energies for oxygenation and subunit association. The protons released upon subunit association at various steps of ligation are given in Table 21. Analysis of these results in combination with the temperature

Table 20. pH Dependence of stepwise Gibbs energies for oxygenation and subunit association[a]

A. Gibbs energy changes for successive tetramer oxygenation steps[b]

pH	$\Delta G'_{41}$ (kJ mol^{-1} O$_2$)	$\Delta G'_{42}$ (kJ mol^{-1} O$_2$)	$\Delta G'_{43}$ (kJ mol^{-1} O$_2$)	ΔG_{44} (kJ mol^{-1} O$_2$)
7.4	-22.7 ± 0.5	-23.2 ± 5.4	-29.1 ± 4.6	-38.3 ± 1.5
8.0	-24.5 ± 0.3	-22.7 ± 5.0	-34.1 ± 5.6	-39.7 ± 2.5
8.5	-25.4 ± 0.5	-27.9 ± 2.5	-32.5 ± 3.8	-39.2 ± 2.2
8.95	-25.3 ± 0.5	-27.7 ± 2.3	-33.0 ± 5.0	-38.7 ± 1.3
9.5	-25.1 ± 0.3	-29.2 ± 3.6	-36.9 ± 3.6	-38.5 ± 0.5

B. Dimer-tetramer association as a function of degree of oxygenation[c]

pH	$^0\Delta G_2$ (kcal mol^{-1})	$^1\Delta G'_2$ (kcal mol^{-1})	$^3\Delta G'_2$ (kcal mol^{-1})	$^4\Delta G_2$ (kcal mol^{-1})
7.4	-60.0 ± 0.4	-48.0 ± 0.8	30.2 ± 1.6	-33.7 ± 0.4
8.0	-58.5 ± 0.4	-47.7 ± 0.4	-34.1 ± 0.8	-37.4 ± 0.4
8.5	-55.5 ± 0.4	-45.4 ± 0.5	-34.9 ± 0.8	-38.7 ± 0.4
8.95	-51.7 ± 0.8	-42.3 ± 0.8	-33.8 ± 1.3	-37.9 ± 0.4
9.5	-48.9 ± 0.4	-39.2 ± 0.3	-32.3 ± 0.4	-35.9 ± 0.4

[a] [41].
[b] Free energy of oxygenation of tetramers stepwise where the reaction step is indicated by the rightmost subscript. The prime affixed to $\Delta G'_{41}$, $\Delta G'_{42}$, and $\Delta G'_{43}$ indicates correction for statistical factors.
[c] Free energy of formation of $\alpha_2\beta_2$-tetramers from $\alpha\beta$-dimers at the degree of tetramer oxygenation indicated by the left superscript. The prime affixed to $^1\Delta G'_2$ and $^3\Delta G'_2$ indicates correction for statistical factors.

Table 21. Protons released upon assembly vs. ligation state of tetramer [41]

pH	$^0\Delta\bar{v}_2$	$^1\Delta\bar{v}_2$	$^3\Delta\bar{v}_2$	$^4\Delta\bar{v}_2$
7.4	-0.62 ± 0.07	-0.06 ± 0.13	$+1.39 \pm 0.22$	$+0.87 \pm 0.11$
8.0	-0.83 ± 0.04	-0.49 ± 0.06	$+0.66 \pm 0.09$	$+0.48 \pm 0.07$
8.5	-1.0 ± 0.06	-0.85 ± 0.05	$+0.05 \pm 0.09$	$+0.15 \pm 0.09$
8.95	-1.16 ± 0.09	-1.17 ± 0.11	-0.49 ± 0.18	-0.14 ± 0.14

[a] Protons released upon assembly of $\alpha\beta$-dimers into $\alpha_2\beta_2$-tetramer at a degree of tetramer oxygenation indicated by the left superscript.

dependency studies of Mills and Ackers [51] confirm a suggestion made in 1974 by Rudolph and Gill [53] that heats of Bohr proton release account for the enthalpies of cooperativity in oxygen binding to tetrameric Hb.

8 Oxygen-Linked Binding of Anions to Hemoglobin

It is well known that both oxy- and deoxyhemoglobin have specific sites for anion binding. This binding is complicated by its linkage both to the oxygenation and

Table 22. Apparent anion binding constants to deoxy- and oxyhemoglobin

Anion	Anion binding constants (M^{-1})		Conditions/method	Reference
	Deoxy Hb[a]	Oxy Hb[a]		
DPG[b]	3×10^4 $(2.4–8.2 \times 10^4)$	8×10^2 $(4–80 \times 10^2)$	20°, pH 7.2, 0.13 M Cl$^-$ gel filtration	[54]
IHP	1.7×10^7 $(0.1–2.4 \times 10^7)$	1.1×10^4 $(1.1–2.5 \times 10^4)$	20°, pH 7.3, 0.1 M Cl$^-$ O_2-equilibrium	[55]
IHS	1.12×10^6	1.27×10^3	25°, pH 7.3, 0.1 M Cl$^-$ O_2-equilibrium	[56]
Chloride[c]	53 (11–453)	1.2 (0.3–10)	25°, pH 7.4 proton titration	[37]
Phosphate	333 (300–4000)	5 (5–10)	25°, pH 7.3, 0.05 M bis-tris O_2-equilibrium	[57]

[a] The numbers in parentheses represent a range of values in the literature. See [39]. Gibbs energy changes corresponding to these binding constants can be evaluated as $-RT \ln K$.
[b] Studies of DPG, IHP, and IHS suggest one binding site per hemoglobin tetramer.
[c] Studies of chloride, phosphate, and other small anions suggest two high affinity sites per hemoglobin tetramer.

Table 23. Buffer-corrected heats of IHP and DPG binding to hemoglobin

Anion	Deoxy Hb			Carbonmonoxy or Oxy Hb			Reference
	ΔH_{obs}[a]	Δv_P[b]	ΔH_{bc}[c]	ΔH_{obs}[a]	Δv_P[b]	ΔH_{bc}[c]	
IHP[d]	-46 ± 4	2.2	-105 ± 13	0 ± 4	1.1	-34 ± 17	[58]
DPG[e]	-46 ± 8	1.1	-75 ± 13	2.9 ± 0.8	1.7	-42 ± 13	[59]

[a] Observed heat of reaction in kJ (mol anion)$^{-1}$.
[b] Protons absorbed per IHP bound [60] or per DPG bound [61].
[c] Buffer-corrected heat of reaction in kJ (mol anion)$^{-1}$.
[d] Performed on carbonmonoxy Hb. pH 7.4, 0.1 M Cl$^-$, 0.2 M bis-Tris, 25 °C.
[e] Performed on oxy Hb. pH 7.2, 0.1 M Cl$^-$, 0.05 M bis-Tris, 25 °C.

the protonation of the protein: deoxy hemoglobin binds anions more strongly than oxyhemoglobin and protons are absorbed when anions bind. The additional complication that the binding of IHP and DPG is frequently studied in buffers containing chloride makes a truly comprehensive analysis of anion binding to Hb difficult in the extreme.

The most model-independent data are the apparent binding constants of anions to oxy- and deoxy Hb under stated conditions. Table 22 summarizes what appear to be the most reliable values for DPG, IHP, inositol hexasulfate, chloride, and phosphate. It is important to note that virtually all studies of DPH, IHP, etc. binding to hemoglobins show a single high affinity site per tetramer. On the other hand, the smaller anions, including chloride, phosphate, sulfate, etc. exhibit approximately two high affinity sites per Hb tetramer.

The heats of anion binding to Hb are difficult to measure and even more difficult to interpret. Given nearly ideal experimental conditions, such as in calorimetric studies of IHP binding by Hb, the linkage of anion to proton binding still gives rise to troublesome complexity. Protons absorbed by the protein during anion binding must be abstracted from some protonated buffer group. This is commonly a highly endothermic process, so that these heats of buffer ionization are large terms. Independent knowledge of the amount of proton release linked to anion binding is necessary even if only correction of apparent heats for buffer effects is desired. It is possible to combine calorimetric anion binding enthalpies with values for anion-linked proton absorption obtained from equilibrium studies. The results are the binding enthalpies of these anions in the absence of thermal contributions from the buffer. Such data for DPG and IHP are presented in Table 23.

9 Miscellaneous Data

Direct calorimetric measurement of heats of ligand binding by hemoglobin has been of great value in establishing the overall thermodynamics of these reactions. Frequently CO has replaced O_2 in such studies of heme ligands. There are two main reasons for this. First, oxidation of Hb to met Hb is particularly troublesome in calorimetry (see Sect. 4) and, with CO, this is greatly reduced. Secondly, the tighter binding of CO relative to O_2 virtually guarantees protein saturation with ligand under calorimetric conditions. Table 24 provides data on the displace-

Table 24. Thermodynamic parameters involving hemoglobin interactions with CO and NO

Reaction	ΔG^{0a}	ΔH^0	ΔS^0	Conditions	Reference
$HbO_2(aq)+CO(g)$	-14.2 ± 0.4	-16.7 ± 2.1	-8.4 ± 7.5	20°, pH 7	[62, 63]
$Hb(aq)+CO(g)$		-65.7 ± 0.8^b		25°, pH 7.0, 0.1 M Cl$^-$	[58]
$1/2\alpha_2(aq)+CO(g)$		-91.2 ± 1.7		25°, pH 7.4, 0.1 M tris Cl$^-$	[64]
$1/4\beta_4(aq)+CO(g)$		-102.1 ± 1.7		25°, pH 7.4, 0.1 M tris Cl$^-$	[64]
$HbCO(aq)+NO(g)$	-16.3			Estimated	[65]

[a] ΔG^0 and ΔH^0 are given in kJ (mol ligand)$^{-1}$ while ΔS^0 is in $J K^{-1} mol^{-1}$.
[b] This is a value corrected for heat of Bohr proton uptake by buffer but *not* for heats of Bohr proton release from the protein.

Table 25. Solubilities of O_2, CO, and NO in water at 20 °C

Gas	s^a	ΔH_{soln}^b	Reference
O_2	13.7	-12.1	[66]
CO	10.2	-12.0	[65]
NO	20.3	-12.6	[65]

[a] Solubilities given as μM concentration per kPa gas partial pressure.
[b] Heats of solution given in kJ mol^{-1}.

ment of O_2 from Hb by CO and on the heats of CO binding to intact Hb and to its α and β chains. Table 25 provides the solubilities and heats of solution of O_2, CO, and NO. These data are useful when, for example, van't Hoff and calorimetric enthalpies are to be compared or when Gibbs energies of gas binding must be converted between pressure and concentration standard states.

10 Conclusion

There exist a number of useful compendia of hemoglobin thermodynamic properties. The book by Antonini and Brunori [16] references many of the important studies done prior to 1971. Imai's recent work [39] focuses on hemoglobin oxygenation reactions and their interpretation in terms of stepwise models. A book by Bunn and his associates [67] contains particularly useful information on thermodynamic aspects of hemoglobin physiologic function. Finally, a recent volume of *Methods in Enzymology* [see for example 65] is devoted to techniques for the study of ligand binding to heme proteins.

This chapter has intentionally been restricted to one central aspect of the field of hemoglobin thermodynamics. Despite the importance and finite scope of this area, many useful data are not available in the literature. In other instances, better data, either more comprehensive, of more recent origin, or obtained under more uniform conditions, are desirable. Thus, in spite of enormous labor already invested in this protein by generations of scientists, there remains much experimentation yet to be done. At least as important is the need for continued creative thought in the interpretation of experimental results.

References

1. Adair GS (1925) Proc R Soc Lond B (Biol Sci] 109:292
2. Ackers GK (1980) Biophys J 32:331
3. Barisas BG, Gill SJ (1979) Biophys Chem 9:235
4. Imai K (1979) J Mol Biol 133:233
5. Gill SJ (1979) Calorimetry Thermal Analysis (Jpn) 6:30
6. Lehmann H, Huntsman RG (1974) Man's hemoglobins, 2nd edn. Lippencott, Philadelphia
7. Fairbanks VF (ed) (1980) Hemoglobinopathies and thalassemias: laboratory methods and clinical cases. Decher Division of Thieme-Stratton, New York
8. Englehardt JF (1825) Doctoral Diss, University of Göttingen, Göttingen, Germany
9. Svedberg T, Fahraeus R (1926) J Am Chem Soc 48:430
10. Braunitzer G, Hilse K, Rudloff V, Hilschmann V (1964) Adv Protein Chem 19:1
11. Rossi-Fanelli A, Antonini E, Caputo A (1961) J Biol Chem 236:391
12. Rossi-Fanelli A, Antonini E, Caputo A (1959) Biochim Biophys Acta 35:93
13. Rossi-Fanelli A, Antonini E, Caputo A (1964) Adv Protein Chem 19:73
14. Schoenborn BP, Featherstone RN, Vogelhut PO, Susskind C (1964) Nature 202:695
15. Muirhead H, Cox JM, Mazzarella L, Perutz MF (1967) J Mol Biol 28:117
16. Antonini E, Brunori M (1971) Hemoglobin and myoglobin in their reactions with ligands. North Holland, Amsterdam
17. Winterhalter KH, Colosimo A (1971) Biochemistry 10:621
18. Serjeant GR (1974) The clinical features of sickle cell diseases. North Holland, Amsterdam
19. Ross PD, Minton AP (1977) J Mol Biol 112:437

20. Ross PD, Briehl RW, Minton AP (1978) Biopolymers 17:2285
21. Minton AP (1977) J Mol Biol 110:89
22. Ross PD, Hofrichter J, Eaton WA (1977) J Mol Biol 115:111
23. Gill SJ, Benedict RC, Fall L, Spokane R, Wyman J (1979) J Mol Biol 130:175
24. Mahoney JD, Ross SC, Gill SJ (1977) Anal Biochem 78:535
25. Sunshine HR, Hofrichter J, Ferrone FA, Eaton WA (1982) J Mol Biol 158:251
26. Brunori M, Wyman J, Antonini E, Rossi-Fanelli A (1965) J Biol Chem 240:3317
27. Brunori M, Taylor JF, Antonini E, Wyman J (1969) Biochemistry 8:2880
28. Brunori M, Saggese U, Rotilio GC, Antonini E, Wyman J (1971) Biochemistry 10:1604–1609
29. Brunori M, Amiconi G, Antonini E, Wyman J, Zito R, Rossi-Fanielli A (1968) Biochim Biophys Acta 154:315
30. Stryer L, Kendrew JC, Watson HC (1964) J Mol Biol 8:96
31. Perutz MF, Mathews FS (1966) J Mol Biol 21:199
32. Anusiem AC, Beetlestone JG, Irvine DH (1968) J Chem Soc (Lond) Sect A 960
33. Anusiem ACI, Lumry R (1973) J Am Chem Soc 95:904
34. Nelson DP, Miller WD, Kiesow LA (1974) J Biol Chem 249:4770
35. Bohr C, Hasselbalch K, Krogh A (1904) Scand Arch Physiol 16:402
36. Antonini E, Wyman J, Brunori M, Fronticelli C, Bucci E, Rossi-Fanelli A (1965) J Biol Chem 240:1096
37. Van Beek GGM, Zuiderweg ERP, DeBruin SH (1979) Eur J Biochem 99:379
38. Rollema HS, DeBruin SH, Janssen LHM, Van Os GAJ (1975) J Biol Chem 250:1333
39. Imai K (1982) Allosteric effects in hemoglobin. Cambridge University Press, Cambridge
40. Chu AH, Ackers GK (1981) J Biol Chem 10:1199
41. Chu AH, Turner BW, Ackers GK (1984) Biochemistry 23:604
42. Monod J, Wyman J, Changeux JP (1965) J Mol Biol 12:88
43. Roughton FJW, Lyster RLJ (1965) Hvalradets Scr 48:185
44. Ackers GK, Johnson ML, Mills FC, Halvorson HR, Shapiro S (1975) Biochemistry 14:5128
45. Imaizumi K, Imai K, Tyuma I (1978) Jpn Soc Biophys Annu Meet Abstracts 728
46. Turner BW, Pettigrew DW, Ackers GK (1981) Methods Enzymol 76:596
47. Valdes R Jr, Ackers GK (1977) J Biol Chem 252:74
48. Valdes R Jr, Ackers GK (1978) Proc Natl Acad Sci USA 75:311
49. Valdes R Jr, Ackers GK (1977) J Biol Chem 252:88
50. Ip SH, Ackers GK (1977) J Biol Chem 252:82
51. Mills FC, Ackers GK (1979) J Biol Chem 254:2881
52. Mills FC, Ackers GK (1979) Proc Natl Acad Sci USA 76:273
53. Rudolph SA, Gill SJ (1974) Biochemistry 13:2451
54. Gustavsson T, deBerdier CH (1973) Acta Biol Med Ger 30:25
55. Benesch RE, Edalji R, Benesch R (1977) Biochemistry 16:2594
56. Ackers GK, Benesch RE, Edalji R (1982) Biochemistry 21:875
57. Nigen AM, Manning JM, Alben JO (1980) J Biol Chem 255:5525
58. Gill SJ, Gaud HT, Barisas BG (1980) J Biol Chem 255:7855
59. Hedlund BE, Lovrien R (1974) Biochem Biophys Res Commun 61:859
60. Brygier J, DeBruin SH, Van Hoff PMKB, Rollema HS (1975) Eur J Biochem 60:379
61. Van Beek GG, DeBruin SH (1979) Eur J Biochem 100:497
62. Antonini E, Wyman J, Brunori M, Bucci E, Fronticelli C, Rossi-Fanelli A (1963) J Biol Chem 238:2950
63. Gaud HT, Barisas BG, Gill SJ (1974) Biochem Biophys Res Commun 59:1389
64. Mills FC, Ackers GK, Gaud HT, Gill SJ (1979) J Biol Chem 254:2875
65. Giardina B, Amiconi G (1981) Methods Enzymol 76:417
66. Wilhelm E, Battino R, Wilcock RJ (1977) Chem Rev 77:219
67. Bunn HF, Forget BG, Ranney HM (1977) Human hemoglobins. Saunders, Philadelphia
68. Wyman J (1964) Adv Protein Chem 19:223
69. Hill AV (1910) J Physiol (Lond) 40:iv.
70. Chipperfield JR, Rossi-Bernardi L, Roughton FJW (1967) J Biol Chem 242:77
71. Rossi-Bernardi L, Roughton FJR (1967) J Biol Chem 242:784

Section IV **Solution Processes**

Chapter 10 Gas-Liquid and Solid-Liquid Phase Equilibria in Binary Aqueous Systems of Nonelectrolytes

SERGIO CABANI and PAOLO GIANNI

Symbols

α:	Bunsen coefficient (see Table 1)
L:	Ostwald coefficient (see Table 1)
L*:	Ostwald coefficient extrapolated to $c \rightarrow 0$
S:	Kuenen coefficient (see Table 1)
k_H:	Henry's law constant (see Table 1)
S_0:	weight solubility (see Table 1)
x:	mole fraction of solute
ΔG_b^*:	standard Gibbs energy change for the transfer of one mole of gas from 1 molar ideal state to ideal aqueous solution of 1 molar concentration
ΔG_h^0:	standard Gibbs energy change for the transfer of one mole of gas from ideal state (1 atm partial pressure) to (hypothetical) ideal aqueous solution of unit mole fraction at 298 K.
ΔH_h^0:	standard enthalpy change for the same process
ΔS_h^0:	standard entropy change for the same process
ΔC_p^0:	standard heat capacity change for the same process
ΔH_s:	molar enthalpies of solution

1 Introduction

Solubility data of nonelectrolytes in water are very useful in many fields from engineering to biology, but until recently critical collections of such data, in summarizing tables, have been neglected. Only in the last few years has the situation noticeably improved, and some excellent reviews containing critical and exhaustive tables on the solubility of gases in water [1] and other solvents [2] have been published. Additionally, a fundamental project has been set up by the IUPAC, *Solubility Data Series* (Pergamon Press/A. S. Kertes), which plans to collect and critically evaluate all existing solubility data covering physical and biological systems. One hundred volumes were planned for publication between 1979 and 1989. By the end of 1982, 17 volumes of the series had already appeared. Eleven of these (from 360 to 650 pages) contain solubility data for gases such as noble gases, hydrogen, nitrogen, oxygen, ozone, oxides of nitrogen, oxides of sulfur, ethane and hydrogen halides in water and nonaqueous solvents. The other six books deal with solubility data of liquid and solid substances. Among the nonelectrolytes, solubility data of alcohols and hydrocarbons in water, and of halogenated benzenes, toluenes, xylenes, and phenols in various solvents have been studied.

The collection of solubility data in pure water for gases and solids is of some interest for biologists. For the sake of an easy consultation these data are presented in short and compact tables. The material is divided in three sections: in the first are solubility data relative to substances (gases) characterized by a normal boiling point lower than 25 °C, in the second partition coefficients between gas phase and water for some liquid and solid compounds having their normal boiling point over 25 °C, and in the third, solubility data in water are reported for some selected categories of solid organic compounds. A final short notice describes the attempts at predicting the solubility in water of nonelectrolytes.

2 Solubility of Gases in Pure Water

The solubility of a gaseous solute in a liquid may be defined in a large variety of ways [2 a, 2 b]. The more common are summarized in Table 1, where the name and the symbol commonly used for the envisaged coefficients are reported together with its operative definition and the formula to be used in order to convert it to mole fraction of ideal gas dissolved in water under a partial gas pressure of 1 atm.

To date, solubility data for gases in water have been reported for ca. 70 compounds. In Table 2 only data for 50 of these are considered, which may be of interest to the biologist either for themselves or because their unique value and the unique way in which they depend on temperature and concentration were used

Table 1. Conversion formulas of various expressions of gas solubility in water to molar fraction (x) of the dissolved gas, under a gas partial pressure of 1 atm

Quantity	Symbol	Definition	Conversion formula[a]
Bunsen coefficient	α	Volume of gas, reduced to 273.15 K and 1 atm, absorbed by the unit volume of the absorbing solvent at the temperature of measurement under a gas partial pressure of 1 atm	$x = \left[1 + \dfrac{1244.1 \cdot \varrho_w}{\alpha} \right]^{-1}$
Ostwald coefficient	L	Ratio of the volume of gas absorbed to the volume of absorbing liquid, both measured at the same temperature	$x = \left[1 + \dfrac{4.5548\,T \cdot \varrho_w}{L} \right]^{-1}$
Kuenen coefficient	S	Volume of gas (cm^3) at a partial pressure of 1 atm, reduced to 273.15 K and 1 atm, dissolved by the quantity of solution containing 1 g of solvent	$x = \dfrac{S}{S + 1244.1}$
Henry's law constant	k_H	Limiting value of the ratio of the gas partial pressure to its mole fraction in solution as the latter tends to zero	$x = \dfrac{1}{k_H}$ [b]
Weight solubility	S_0	Grams of gas dissolved by 100 g of solvent, under a gas partial pressure of 1 atm	$x = \dfrac{S_0 \cdot 18.015}{100\,M_s + S_0 \cdot 18.015}$

[a] The ideal behavior of the solute in the gas phase is assumed. T = temperature (K); ϱ_w = water density (g cm^{-3}); M_s = solute molecular weight.
[b] The Henry's constant is expressed in atmospheres.

Table 2. Mole fraction solubility of gases in water, under a gas partial pressure of 1 atm, as a function of temperature (K): $\ln x = A + B/T + C \ln T + DT + ET^2$ [a]

Gas	T Range	A	B	C	D	E	σ% [b]	n [c]	Reference
Helium	273–348	—	4259.62	14.0094	—	—	0.54	59	[6]
Neon	273–348	—	6104.94	18.9157	—	—	0.47	59	[6]
Argon	273–348	—	7476.27	20.1398	—	—	0.26	42	[7]
Krypton	273–353	—	9101.66	24.2207	—	—	0.32	30	[8]
Xenon	273–348	—	10521.0	27.4664	—	—	0.35	20	[8]
Radon	273–373	—	13002.6	35.0047	—	—	1.02	40	[8]
Hydrogen	274–339	—	6993.54	26.3121	−0.0150432	—	0.44	31	[1]
Deuterium	278–303	—	7309.62	26.1780	−0.0118151	—	0.22	25	[9]
Nitrogen	273–348	—	8632.129	24.79808	—	—	0.72	74	[10]
Oxygen	273–333	1072.48902	27609.2617	191.886028	−0.483090199	2.24445261E-4	0.016	36	[11] [d]
Ozone	277–293	14.9645	1965.31	—	—	—	—	2	[1]
Carbon monoxide	278–323	427.656023	15259.9953	67.8429542	−0.0704595356	—	0.065	14	[12]
Carbon dioxide	273–373	4957.824	105288.4	933.1700	−2.854886	1.480857E-3	0.40	29	[13–16] [d]
Carbon dioxide [e]	273–373	4947.988	105093.7	931.2984	−2.849025	1.477811E-3	0.40	29	[13–16] [d]
Methane	275–328	416.159289	15557.5631	65.2552591	−0.0616975729	—	0.081	16	[23]
Ethane	275–323	−11268.4007	221617.099	2158.42179	−7.18779402	4.05011924E-3	0.102	23	[23]
Ethylene	287–346	176.910	9110.81	24.0436	—	—	0.35	13	[16, 24] [d]
Acetylene	274–343	156.509	8160.17	21.4023	—	—	1.3	8	[1]
Propane	273–347	316.460	15921.2	44.3243	—	—	4.6	25	[1]
Cyclopropane	298–361	326.902	−13526.8	50.9010	—	—	7.4	14	[1]
n-Butane	276–349	290.2380	15055.5	40.1949	—	—	3.4	40	[16–22] [d]
Isobutane	278–343	96.1066	−2472.33	17.3663	—	—	3.1	14	[1]
Neopentane	288–353	437.182	21801.4	61.8894	—	—	6.0	7	[1]
CH_3F	273–353	135.910	7600.23	18.1780	—	—	0.87	12	[1]
CH_3Cl	277–353	172.503	9768.67	23.4241	—	—	6.8	19	[1]
CH_3Br	278–353	163.745	9641.71	22.0397	—	—	4.7	16	[1]
CF_4	275–323	342.437	16250.6	48.3441	—	—	1.0	18	[1]
CH_2FCl	283–352	138.912	8141.19	18.5510	—	—	0.36	4	[1]
CHF_2Cl	297–352	190.691	13083.6	22.7782	0.032357	—	4.1	7	[1]
CHF_3	298–348	19.1037	3214.07	—	—	—	—	3	[1]

Table 2 (continued)

Gas	T Range	A	B	C	D	E	σ%[b]	n[c]	Reference
C_2F_6	278–303	−644.222	29933.6	93.0269	–	–	3.8	7	[26, 27][d]
C_2F_5Cl	298–348	21.3965	2802.35	–	–	–	–	3	[1]
Vinyl chloride[f]	273–358	−2399.543	72028.1	408.576	−0.594578	–	5.6	20	[28–31][d]
C_2F_4	273–343	−184.958	10843.3	23.1458	0.0209611	–	0.77	12	[1]
C_3F_8	278–288	−10180.5	438919	1526.05	–	–	0.70	13	[26]
C_3F_6	278–343	−66.5820	4491.54	6.90981	–	–	1.3	14	[1]
$c\text{-}C_4F_8$	278–303	−759.717	35705.7	110.033	–	–	1.2	7	[26, 27][d]
COS	273–303	−221.211	12025.1	30.3659	–	–	0.80	11	[1]
CH_3NH_2	298–333	−9.1917	2607.10		–	–	–	2	[1]
$(CH_3)_2NH$	298–333	−14.0980	4026.14		–	–	–	2	[1]
$C_2H_5NH_2$	298–333	−12.6231	3628.65		–	–	–	2	[1]
NH_3	273–373	−81.7466	1096.82	16.5603	0.0602469	–	6.0	51	[1]
N_2O	273–313	−158.6208	8882.80	21.2531	–	–	1.2	23	[32]
NO	273–353	−328.097	12541.9	50.7616	−0.0451331	–	0.53	9	[33][d]
H_2S	273–333	−149.537	8226.54	20.2308	0.00129405	–	0.42	20	[1]
SO_2	283–306	−13.0502	2792.62		–	–	2.9	58	[1]
SF_6	275–323	−435.519	20901.8	61.9692	–	–	0.53	16	[34–37][d]
Cl_2	283–313	−108.389	−2428.63	19.1855	0.00892064	–	2.3	13	[25, 27][d]
Cl_2O	273–293	−7.2207	1798.85		–	–	–	3	[1]
ClO_2	283–333	−56.7389	143.179	10.7454	–	–	4.0	9	[1]
Air	273–373	−388.760	14097.6	61.2018	−0.0617537	–	–	11	[33][g]

[a] Experimental data were not corrected for nonideality of the gas phase and chemical reactions with the solvent. The quoted coefficients are valid in the temperature range given in the second column.

[b] Standard deviation of the fit in x as a percentage.

[c] Number of experimental points considered in the fit.

[d] Original data elaborated by the compilers.

[e] Solubility given as the quantity of solute present in solution in nonionized form. Data were corrected using ionization constants given by A.J. Read. J Solution Chem 4, 53 (1975).

[f] Data from [28] were corrected to P=1 atm using vapor pressure data from Dana et al., J Am Chem Soc 49, 2801 (1927).

[g] Calculated data given by the author, chosen at 10K intervals, were elaborated by the compilers.

Table 3. Solubility at 25 °C, and under a gas partial pressure of 1 atm, for selected gases in water[a]

Gas	$x \cdot 10^5$	$L \cdot 10^2$	$S_0 \cdot 10^3$
N_2	1.183	1.601	1.839
O_2	2.301	3.116	4.087
CO	1.774	2.403	2.759
CO_2	60.9	82.4	148.8
CH_4	2.552	3.456	2.273
C_2H_6	3.403	4.608	5.681
C_3H_8	2.70	3.66	6.62
NH_3	18 700	31 100	21 700
N_2O	43.7	59.2	107
NO	3.47	4.70	5.78
H_2S	185.1	251.0	350.7
SO_2	2 510	3 490	9 170
Cl_2	165	223	650

[a] x = solute mole fraction; L = Ostwald coefficient; S_0 = weight solubility. For the definition of these quantities see Table 1. The data were calculated from Eq. (1) using coefficients of Table 2. No correction for ionized forms was made for gases which react with water.

to develop concepts of water and aqueous solutions, largely accepted and used in biophysics [3].

In Table 2 solubility is given as a function of temperature, in the temperature range 0 °–70 °C, or in more limited ranges according to the availability of the data. The temperature dependence of solubility is expressed by a Clarke-Glew-type Equation [4][1]

$$\ln x = A + B/T + C \ln T + DT + ET^2. \tag{1}$$

Parameters A, B, C, D, and E for some gases were taken directly from [1] or from original papers without revision, except for conversion to our units. For other gases, either not treated in [1] or the object of more recent measurements, the above parameters were calculated by a standard linear least-squares procedure. Equation (1) was reduced to the minimum number of parameters necessary to obtain the minimum standard deviation reported in the eighth column. Experimental data showing a deviation in calculated value more than twice the standard deviation of the fit were rejected.

For the sake of convenience, data for some common gases at 25 °C and 1 atm, expressed by the most common solubility coefficients, have been collected in Table 3.

The coefficients A, B, C, D, E of Eq. (1) may be used to obtain thermodynamic quantities such as standard Gibbs energy, enthalpy, entropy, and heat capacity

[1] Benson [5] recently proposed in alternative an equation of the type $\ln x = a + b/T + c/T^2 + d/T^3 + \ldots$ This equation seems to yield a description of the data which is as good as Eq. (1).

Table 4. Standard thermodynamic functions for the transfer of the selected gases from the ideal state of 1 atm partial pressure, into (hypothetical) ideal aqueous solution of unit mole fraction at 298 K [a]

Gas	ΔG_h^0 kJ mol^{-1}	$-\Delta S_h^0$ J mol^{-1}K^{-1}	$-\Delta H_h^0$ kJ mol^{-1}	$\Delta C_{p,h}^0$ J mol^{-1}K^{-1}
N_2	28.121 $\pm$0.005	128.8 $\pm$0.2	10.29 $\pm$0.06	206$\pm$ 7
O_2	26.4739$\pm$0.0003	129.08$\pm$0.02	12.013[b]$\pm$0.006	195$\pm$ 1
CO	27.1183$\pm$0.0013	127.10$\pm$0.16	10.78 $\pm$0.05	215$\pm$ 4
CO_2[c]	18.365 $\pm$0.005	127.2 $\pm$0.5	19.56[d] $\pm$0.15	172$\pm$ 10
CH_4	26.2175$\pm$0.0014	132.37$\pm$0.07	13.25 $\pm$0.02	238$\pm$ 3
C_2H_6	25.5046$\pm$0.0015	150.77$\pm$0.06	19.45 $\pm$0.02	285$\pm$ 3
n-C_4H_{10}	26.593 $\pm$0.034	175 $\pm$2	25.5 $\pm$0.6	334$\pm$ 42
N_2O	19.165 $\pm$0.014	134 $\pm$2	20.7 $\pm$0.7	$-$ 10$\pm$144
NO	25.455 $\pm$0.016	125.0 $\pm$0.7	11.80 $\pm$0.22	198$\pm$ 20
CF_4	30.954 $\pm$0.017	155.0 $\pm$0.9	15.27 $\pm$0.27	402$\pm$ 38
SF_6	30.587 $\pm$0.009	170.2 $\pm$0.5	20.17 $\pm$0.15	515$\pm$ 20

[a] Values here reported were obtained by the compilers using experimental data from references indicated in Table 2. The uncertainties, given as twice the standard deviation, were calculated from the variance and covariance of parameters of Eq. (1). (See e.g. W.E. Wentworth, J Chem Educ 42, 96 (1965), p. 101).
[b] A calorimetric value of 12.06$\pm$0.04 kJ mol^{-1} was reported by Gill et al., J Chem Thermodyn 14, 905 (1982).
[c] Data corrected for ionization.
[d] Calorimetric values were reported of 19.67$\pm$0.10 by Gill et al. (see ref. of [b] and of 19.75$\pm$0.17 kJ mol^{-1} by Berg et al. (1978) J. Chem Thermodyn 10, 1113).

for the transfer of the gas from the ideal state of unit pressure to the hypothetical ideal solution of unit mole fraction according to the following equations:

$$\Delta G_h^0 = -R(AT + B + CT\ln T + DT^2 + ET^3)$$
$$\Delta H_h^0 = R(-B + CT + DT^2 + 2ET^3)$$
$$\Delta S_h^0 = R(A + C\ln T + C + 2DT + 3ET^2)$$
$$\Delta C_{p,h}^0 = R(C + 2DT + 6ET^2).$$

(2)

In Table 4 values for the above-mentioned standard thermodynamic functions ΔX_h^0 (X=G,H,S,C$_p$) have been reported for some selected gases. It should be observed that Eq. (2) produce values of the standard thermodynamic functions of solution, ΔX_h^0, only when the values of x at various temperatures, used to fit Eq. (1), have been obtained in such experimental conditions that their reciprocals represent true Henry's constants. Moreover, significant values of these quantities, and in particular $\Delta C_{p,h}^0$, may be obtained only when the solubility data are of very high precision.

3 Solubility of Gases in Aqueous Solutions of Electrolytes

Extensive bibliographic references on the effect of electrolytes on the solubility of gases in water may be found in [1]. More recently, general equations for calculating gas solubility in sea water of different salinity and temperature have been proposed by Weiss for N_2, O_2 and Ar [38, 39], He and Ne [40], Kr [41], CO_2 [42] and by Wiesenburg [43] for CO, H_2, and CH_4. Solubility data in sea water for SO_2 [37] and H_2S [44] have been reported by Douabul et al.

Finally, Schumpe et al. [45] reported ionic specific constants for the salting out of O_2 in water.

4 Partition Coefficients of Vapors Between Water and the Gas Phase

For volatile liquids and solids soluble in water, an equilibrium is reached, at constant temperature, between the aqueous solution and the gas phase, which may be expressed by a proper partition coefficient. This coefficient may assume different forms, depending on the standard states chosen for the solute in water and in the gas phase. When the unit molar concentration is chosen as standard state in both phases, the partition coefficient is expressed as the ratio of the molar concentration of the solute in the aqueous phase, c_w, and in the vapor phase, c_v, i.e., as an Ostwald coefficient L.

Limiting values of this coefficient, L*, which refer to solutions obeying Henry's law, may be obtained by extrapolating to $c_w = 0$ the L values usually obtained by measurements of partial pressure of the solute over solutions of known concentration using static, dynamic, or transpiration methods. When the solute is very slightly soluble in water, the sole measurement of its solubility and its vapor pressure are sufficient in order to obtain the value of L*.

Values of L* are available [46] for about 350 compounds with values ranging from 10^{-2} to 10^7, i.e., from substances almost insoluble in water to substances whose aqueous solutions display an extraordinary thermodynamic stability.

Table 5 reports the L* values for some nonelectrolytes which may be representative of the most common organic chemicals. Table 5 also gives the values of the related thermodynamic functions of hydration, which refer to the process of transfer of 1 mol of ideal gas at unitary molar concentration into an ideal 1 M aqueous solution. These quantities are very useful for characterizing, from a thermodynamic point of view, the aqueous solutions of nonelectrolytes [3, 47].

A critical review of air-water partition coefficients for chemicals of environmental interest can be found in [48].

5 Solubility of Solid Substances

Solubility data in water for solid organic compounds are very scattered in the chemical literature and difficult to find, especially solubility data of reliable precision. Good quality solubility measurements are certainly not easy to perform. Much care must be paid to factors such as purity of the reagents, temperature control, equilibration time, nature and crystal structure of the solid phase and its possible modification after equilibration with the solvent, etc. All these conditions are rarely accurately controlled, and even more rarely are they well specified in the experimental details, permitting a reasonable estimate of the precision and accuracy of the solubility data.

Table 5. Limiting partition coefficients at 25 °C between the aqueous solution and the gas phase for selected noncharged organic compounds[a]

Compound	$L*$ [b]	ΔG_h^{*} [c] kJ mol^{-1}	$-\Delta H_h^{0}$ [d] kJ mol^{-1}
n-Butane	$2.99\ 10^{-2}$	8.70	25.97
n-Hexane	$1.51\ 10^{-2}$	10.40	31.60
n-Octane	$0.76\ 10^{-2}$	12.10	39.75
Cyclohexane	$1.26\ 10^{-1}$	5.14	33.20
1-Propene	$1.17\ 10^{-1}$	5.31	28.21
1,3-Butadiene	$3.55\ 10^{-1}$	2.57	37.68
1-Propyne	1.68	$-\ 1.28$	15.62
Benzene	4.31	$-\ 3.62$	31.77
Naphthalene	$5.67\ 10^{1}$	-10.01	46.86
Ethanol	$4.74\ 10^{3}$	-20.98	52.40
1-Pentanol	$1.90\ 10^{3}$	-18.72	64.75
1-Octanol	$1.00\ 10^{3}$	-17.13	(75.32)
Cyclohexanol	$1.03\ 10^{4}$	-22.91	70.50
Phenol	$7.07\ 10^{4}$	-27.68	56.94
Dimethyl ether	$2.45\ 10^{1}$	$-\ 7.93$	(40.53)
Diethyl ether	$1.58\ 10^{1}$	$-\ 6.84$	47.01
di-n-Propyl ether	7.02	$-\ 4.83$	(53.49)
Tetrahydrofuran	$3.50\ 10^{2}$	-14.52	47.26
Ethanamine	$2.00\ 10^{3}$	-18.84	54.02
1-Butanamine	$1.41\ 10^{3}$	-17.97	59.04
Diethylamine	$9.59\ 10^{2}$	-17.02	65.12
Triethylamine	$1.65\ 10^{2}$	-12.65	69.68
Aziridine	$9.33\ 10^{3}$	-22.66	49.97
Pyrrolidine	$1.04\ 10^{4}$	-22.94	63.57
Piperidine	$5.57\ 10^{3}$	-21.38	65.41
Pyridine	$2.78\ 10^{3}$	-19.66	49.84
2-Propanone	$6.67\ 10^{2}$	-16.12	40.89
2-Butanone	$4.64\ 10^{2}$	-15.22	45.71
Acetophenone	$2.29\ 10^{3}$	-19.18	–
Acetaldehyde	$3.70\ 10^{2}$	-14.66	–
Butanal	$2.13\ 10^{2}$	-13.29	–
Hexanal	$1.15\ 10^{2}$	-11.76	–
Benzaldehyde	$8.92\ 10^{2}$	-16.84	–
Acetic acid	$8.21\ 10^{4}$	-28.05	52.8
Propanoic acid	$5.57\ 10^{4}$	-27.09	56.5
Acetic acid methyl ester	$2.69\ 10^{2}$	-13.87	42.50
Acetic acid ethyl ester	$1.86\ 10^{2}$	-12.95	45.60
Acetamide	$1.31\ 10^{7}$	-40.63	–
Fluoromethane	1.45	$-\ 0.92$	18.13
Chloromethane	2.56	$-\ 2.33$	23.15
Bromomethane	3.99	$-\ 3.43$	25.53
Iodomethane	4.47	$-\ 3.71$	25.90
Chlorobenzene	6.63	$-\ 4.69$	–
Bromobenzene	$1.18\ 10^{1}$	$-\ 6.11$	–
Ethanethiol	8.90	$-\ 5.42$	–
Dimethylsulfide	$1.35\ 10^{1}$	$-\ 6.45$	–
Acetonitrile	$7.06\ 10^{2}$	-16.26	34.69
Nitroethane	$5.24\ 10^{2}$	-15.52	(38.97)
Nitrobenzene	$1.04\ 10^{3}$	-17.23	–
1,2-Ethanediol	$4.09\ 10^{5}$	-32.03	72.30
1,4-Dioxane	$5.07\ 10^{3}$	-21.15	47.97
1,2-Ethanediamine	$3.70\ 10^{5}$	-31.78	76.10

Table 5 (continued)

Compound	L^{*b}	ΔG_h^{*c} kJ mol^{-1}	$-\Delta H_h^{0d}$ kJ mol^{-1}
Piperazine	2.55 10^5	-30.86	90.37
2-Methoxyethanol	9.11 10^4	-28.31	60.44
Morpholine	1.82 10^5	-30.02	69.45
1,2-Dichloroethane	1.86 10^1	-7.25	–
1,2-Dichlorobenzene	1.00 10^1	-5.71	–
Chloroform	6.04	-4.46	41.17
Tetrafluoromethane	0.52 10^{-2}	13.03	15.06
Tetrachloromethane	8.51 10^{-1}	0.40	–
3-Nitrophenol	1.15 10^7	-40.31	67.68

[a] The values are taken from [46], where a complete list of all the available data may be found. The compounds are listed following the type of functional group, putting at the end polyfunctional molecules.
[b] $L^* = \lim_{c \to 0}(c_w/c_v)$; c_w, c_v = molar concentration of the solute in the aqueous and vapor phase, respectively.
[c] Free energy of the isothermal process: solute (ideal vapor, $c_v = 1$ mol dm^{-3}) → solute (ideal aqueous solution, $c_w = 1$ mol dm^{-3}).
[d] Usually obtained from vaporization (ΔH_v) and solution (ΔH_s) heats: $\Delta H_h^0 = \Delta H_s - \Delta H_v$. Values within parentheses were calculated from group contributions of [46].

Some collections of solubility data of organic compounds are available [49, 50], but contain no recent data. More recent reviews [51, 52] deal mostly with the thermodynamics of the solution process and provide general ideas on correlations between structure and solubility without giving detailed tables of critically revised solubility data.

Compounds of relevance in biochemical research are referred to in some data books [53–56] that report the physicochemical properties of such molecules. However, these sources do not report aqueous solubility or else they limit their information to a fixed temperature, in most cases with generic statements such as "soluble", "very slightly soluble" etc. The solubilities of amino acids in water, collected by Hutchens [53], make the only exception. This is a collection of solubility data in water relative to substances of general biological interest. These data meet the requirement of measurement under carefully controlled experimental conditions, clearly specified in the original papers. Where data from different sources were used, a mean value has been critically assessed. The solubility S is expressed generally as grams of solute per 1000 g of water. Unfortunately, some authors use molar concentration without measuring density, thus making the precise conversion to this unit impossible.

Amino Acids and Related Compounds. The most comprehensive compilation of solubility data of the amino acids in water as a function of temperature is that given by Hutchens [53], in an extension of the original table compiled by Borsook and Huffman [57] and reported later by others [58, 59].

Table 6 is a compilation of data already reported in this tabulation and more recent data. In addition to giving the coefficients of the empirical equation and thus the variation of solubility with temperature, the best value of solubility at

Table 6. Solubility of amino acids in water as a function of temperature: $\log S = a + bt + ct^2$ [a]

Amino acid	T range	a	b × 10²	c × 10⁵	δ% [b]	S [c]		Reference [d]
L-α-Alanine [e]	273–338	2.1048	0.4669	—	0.25	166.9	(166.5)	[68, 60–67]
DL-α-Alanine	273–333	2.0830	0.5608	—	0.66	165.2	(167.2)	[68, 69–71]
β-Alanine	—					891		[72]
L-2-Aminobutanoic acid	285–318	2.3639	−0.1312	4.682	0.41	228.9	(229.3)	[60 [f], 64]
DL-2-Aminobutanoic acid	298–318	2.2466	0.2675	—	—	210.2	(205.8)	[79 [f], 69, 71]
6-Aminohexanoic acid	—					863		[72]
2-Amino-2-methylpropanoic acid	—					117.4		[71]
L-Arginine	293–303	1.8478	1.638	—	—	182.6	(180.9)	[61 [f], 64, 65]
L-Asparagine · H₂O	273–338	0.9289	2.311	−4.981	1.00	25.1 [g]	(29.89)	[74, 61–66, 75, 76]
L-Aspartic acid	273–333	0.3194	1.519	—	2.08	5.04	(5.00)	[68, 61, 64, 65, 70, 76]
DL-Aspartic acid	273–338	0.4181	2.016	−4.999	0.76	7.97	(7.78)	[68, 70]
L-Cystine	273–338	−1.299	1.357	—	1.01	0.107	(0.1096)	[74, 77]
DL-Cystine	273–303	−1.7959	0.8013	27.69	—	0.044	(0.038)	[53, 77]
DL-Cystine	298–323	−2.1087	3.367	−22.56	—	0.044	(0.039)	[53, 77]
meso-Cystine	273–303	−1.7190	0.4514	27.39	—	0.046	(0.037)	[53, 77]
meso-Cystine	298–323	−2.6034	5.890	−49.41	—	0.046	(0.036)	[53, 77]
L-Dibromotyrosine · ½H₂O	273–298	0.0839	1.627	—	1.06		(3.09)	[78]
L-Dibromotyrosine	282–321	0.188	0.9884	—	0.71		(2.72)	[78]
L-Dichlorotyrosine	273–321	0.0065	1.038	4.648	0.62		(1.97)	[78]
L-o-Dihydroxyphenylalanine	—					3.80		[75]
L-Diiodotyrosine	273–320	−0.690	1.92	—	3.15		(0.617)	[68]
DL-Diiodotyrosine	273–321	−0.827	1.43	—	2.54		(0.340)	[78]
D-Glutamic acid	273–333	0.5331	1.613	—	1.36	8.73	(8.64)	[68, 61 [i], 70, 76]
DL-Glutamic acid	273–338	0.9317	1.523	—	3.34	23.5	(20.53)	[68, 70]
L-Glutamine	293–303	1.2900	1.310	—	—	42.5	(41.5)	[61 [f], 62–66, 75, 76 [i]]
Glycine	273–333	2.1516	1.087	−4.114	0.92	250.9	(249.8)	[68, 60–73, 75]
L-Hydroxyproline	273–338	2.4603	0.3891	—	0.61		(361.1)	[80]
L-Hystidine	293–303	1.4433	0.852	—	—	43.5	(45.3)	[61 [f], 62–66, 75]
L-Isoleucine [h]	293–303	1.4685	0.265	—	—	34.24	(34.26)	[61 [f], 64]
D-Isoleucine [h]	273–338	1.5787	0.07862	2.594	1.26		(41.17)	[74]
DL-Isoleucine	273–338	1.2616	0.2512	3.794	1.33	22.09	(22.29)	[68, 70]
L-Leucine	273–338	1.3561	0.02233	3.727	0.79	21.97	(24.26)	[68, 60–67, 71, 75]

	t	a	b	c	δ	S (25°C)	(calc)	References
DL-Leucine	273–343	0.9013	0.2635	4.591	0.70	10.40	(9.91)	[68, 60, 70, 71]
L-n-Leucine	—					16.8		[60, 64, 69]
DL-n-Leucine	273–348	0.9258	0.4524	3.402	0.66	11.44	(11.49)	[68, 60, 69–72]
L-Lysine						5.84[j]		[81]
L-Methionine	293–303	1.5802	0.689	—	—	56.0	(56.5)	[61[f], 62, 63, 65, 66]
DL-Methionine	273–335	1.2597	1.108	− 1.221	0.73		(33.81)	[74]
L-Phenylalanine	273–338	1.2974	0.6982		1.15	27.89	(29.65)	[74, 61–67, 69, 75]
DL-Phenylalanine	273–348	0.9986	0.5252	3.140	1.04	14.15	(14.11)	[68, 70]
L-Proline	283–323	3.1050	0.4206	—	1.04		(1622)	[80]
L-Serine	293–303	2.2704	1.420	—	—		(422)	[61[f]]
DL-Serine	273–335	1.3432	1.520	− 3.548	0.56		(50.23)	[74]
Taurine	273–335	1.5945	1.916	− 8.500	0.86		(104.8)	[74]
L-Threonine	293–303	1.8381	0.593	—	—	98.1	(96.9)	[61[f], 62, 63, 65, 66]
L-Tryptophan	273–338	0.9156	0.4834	2.988	1.44	13.2	(11.35)	[74, 61–64, 66, 67, 75]
L-Tyrosine[k]	273–333	−0.708	1.46	—	1.43	0.464	(0.454)	[68, 61–66, 70]
DL-Tyrosine	273–333	−0.833	1.51	—	1.73		(0.350)	[78]
L-Valine	293–303	1.7117	0.220	—	—	58.7[l]	(58.4)	[61[f], 64, 65, 67]
DL-Valine	273–353	1.7749	0.2389	2.607	0.54	72.1	(70.9)	[68, 70, 71]
L-n-Valine	285–320	1.9870	−0.2252	3.694	—	88.4	(89.9)	[60[f], 64]
DL-n-Valine						85.0		[69]

[a] t = temperature in centigrade degrees. The quoted coefficients are valid in the temperature range given in the second column.

[b] Average deviation $\delta = [\Sigma(S_{exp} - S_{calc})^2/n]^{1/2}$ given as a percentage of S. No value was calculated for systems with less then five experimental points.

[c] S = grams of solute per 1000 grams of water. The quoted figure is the solubility at 25°C, obtained as the average among the data from references of the eighth column. The value within parentheses is calculated through the equation.

[d] References in italics indicate the source of a, b, and c values. All references utilized for calculating the average value at 25°C.

[e] Coefficients a, b, c. [68] refer to a sample of D-alanine.

[f] Values of a, b, and c were calculated by the compilers through a standard least-squares method, using experimental data from the quoted reference.

[g] Solubility of the anhydrous aminoacid.

[h] Even if there is no reason why the L and D isomer should exhibit a different behavior, the relative experimental data were treated separately.

[i] Sample of D-isomer.

[j] Solubility in $g\,dm^{-3}$ at 27°C.

[k] Exactly the same data have also been calculated for the D-isomer between 273 and 323 K [78].

[l] A value of 75.3 has been reported for the D-isomer [71]. However, the solubility of this isomer has been found to be very much dependent on the mode of crystallization: see J Gen Physiol 19, 767 (1936).

Table 7. Solubility of amino acid derivatives in water at 25 °C

Compound	S[b]	Reference
Diglicine	226	[61–64, 66, 67, 73, 75, 76]
Triglycine	64.8	[61–64, 66, 73, 75, 76]
Tetraglycine	11.9	[64]
Diketopiperazine	16.70	[61, 82, 83]
N-Acetylglycine amide	807	[65]
N-Acetylglycilglycine amide	281	[65]
N-Acetyltetraglycine ethyl ester	0.78[c, d]	[84]
N-Benzoylglycine	3.67	[62]
N-Benzoyl-L-phenylalanine	0.85	[62]
N-Benzoyl-L-tryptophan	0.56	[62]
N-Benzoyl-L-tyrosine	3.68	[62]
N-Benzoyldiglycine	3.3	[62]
N-Carbamoylglycine	40.0	[72, 76]
N-Carbamoyl-DL-α-alanine	26.0	[72, 76]
N-Carbamoyl-β-alanine	21.3	[72]
N-Carbamoyl-6-aminohexanoic acid	1.209	[72]
N-Carbamoyl-5-aminopentanoic acid	2.80	[72]
N-Carbamoyl-DL-n-leucine	1.209	[72]
N-Carbamoyldiglycine	22.4	[72]
N-Carbamoyltriglycine	10.44	[72]
N-Carbobenzoxyglycine	4.56	[62, 66]
N-Carbobenzoxy-L-tyrosine	1.53	[62]
N-Carbobenzoxydiglycine	0.70	[62, 66]
N-Carbobenzoxytriglycine	1.1	[62]
N-Carbobenzoxyglycine amide	2.75[c]	[84]
N-Carbobenzoxydiglycine amide	1.18[c]	[84]
N-Carbobenzoxytriglycine amide	0.38[c, e]	[84]
N-Formylglycine	219.8	[72, 76]
N-Formyl-DL-2-aminobutanoic acid	34.6	[72, 76]
N-Formyl-DL-leucine	30.3	[76]
Hydantoin	41.0	[76]
Hydantoin of DL-2-aminobutanoic acid	121.3	[76]
Hydantoin of L-aspartic acid	11.25	[76]
Hydantoin of DL-leucine	1.95	[76]

[a] The solubility S is given as grams per 1000 grams of water.
[b] Average value among those given in the quoted references.
[c] S in $g \, dm^{-3}$.
[d] Values were also measured of $1.14 \, g \, dm^{-3}$ [81] and $0.83 \, g \, dm^{-3}$ [84], both at 27 °C.
[e] In 0.02 M acetate buffer.

25 °C is also given, as the average of all reliable data available at this temperature. In some cases the average value at 25 °C differs from the value calculated through the function $S = f(T)$ more than would be expected on the basis of the average deviation (column 6, Table 6). The larger deviations in data from different authors compared with those for the temperature-fitting function of the single author indicate how critical these measurements are, probably owing to varying control of factors such as purity and crystal form of the solute sample.

Data on the heats of solution of amino acids in water, determined either calorimetrically or through the function $S = f(T)$, may be found in [53, 58, 59, 61, 88].

Table 8. Solubility and enthalpy of solution of purine and pyrimidine bases, and their derivatives, in water at 25 °C[a]

Compound	S^b g dm^{-3}	Reference		ΔH_s^c kJ mol^{-1}	Reference
Adenine	1.04	[89–93]	[94, 98, 99]	33.5±1.0	[100]
Adenosine	5.10	[90–93]	[99]	32.3±0.1	[101]
Caffeine	21.21[d]	[95]	[95]	14.4±0.6	[95, 102, 103]
Cytosine	7.28	[90–92]	[98]	27.2±4.0	[104]
Deoxyadenosine	6.70	[90–92]		–	
Deoxyguanosine	3.45	[93]		–	
Guanine	0.068	[96]		27.2±2.1	[96]
Guanosine	0.502	[91–93]		–	
Hypoxanthine	0.696	[93]		–	
Inosine	e			27.3±0.2	[101]
9-Methyladenine	4.47[d]	[97]	[97]	–	
1-Methylthymine	5.47[d]	[97]	[97]	–	
Theophylline	f			19.7±0.1	[103, 105]
Thymidine	51	[90]		–	
Thymine	3.52	[90–92, 94]		23.4±0.6	[106–108]
Uracil	2.67	[92]	[98, 99]	29.4±1.0	[107, 109]

[a] The quoted values are the average among the data reported in the references given within parentheses.
[b] S = solubility (g dm^{-3}). The uncertainty of S was evaluated only in the case of adenine, adenosine, cytosine and thymine, being ±0.03 g dm^{-3}. References in italics indicate sources of experimental data at temperatures other than 25 °C.
[c] Enthalpy of solution at infinite dilution, obtained calorimetrically. When more than one source was available, the uncertainty was given as the average deviation among the data of different authors.
[d] g kg^{-1}.
[e] A value of 16 g per liter of solvent at 20 °C is reported in [54].
[f] A value of 5.2 g kg^{-1} at 15–20 °C is reported in [49].

Table 7 reports the solubility in water at 25 °C of a number of compounds which may be considered derivatives of the amino acids. The list includes a few peptides, hydantoins, benzoyl, carbamoyl, formyl, and carbobenzoxy derivatives.

The solubility in water of amino acids and related compounds has been studied by many authors also in the presence of other substances. Solubility behavior in the presence of charged molecules (salts, zwitterions) has been reviewed by Greenstein and Winitz [59], Cohn [85], and McMeekin [52]. Recent determinations of the aqueous solubility of amino acids and peptides have been performed in the presence of inorganic salts [61, 86], urea [66,84, 87, 124], guanidinium chloride [61, 63, 84], guanidinium thiocyanate [81] and polyols [62, 65, 67].

The solubility of some amino acids and peptides has also been investigated in heavy water [60, 64, 69].

Purine and Pyrimidine Bases. Some solubility data relative to these bases, their nucleosides, and other common derivatives are collected in Table 8, which also contains reliable recent data of the enthalpy of solution of a few of these compounds.

Other solubility data on related compounds were reported by Albert, who measured the aqueous solubility of a series of purine derivatives [110] and pteridines [111–113]. All these data, also reported in [114], do not meet the requirements of the present compilation and are not included here.

The solubility in water of purine and pyrimidine bases has been also investigated in the presence of salts [89, 99] urea and amides [91, 94], hydroxy compounds [92, 94, 115], amino acids [90, 116], similar bases [94, 97], and other heteroaromatic compounds [93]. Finally, the water solubility of adenine has been studied in the presence of various DNA-denaturing agents [117].

6 Previsions of the Solubility in Water of Nonelectrolytes

To date, no successful theory for predicting the solubility of gases in water has been developed, owing to the complexity of the solvent itself. Some degree of success was obtained by Pierotti [118], applying the scaled particle theory (SPT) in the case of low polarity gases at 298 K. Kim and Brückl [119] have successively demonstrated that SPT previsions of aqueous solubility of inert gases are reasonably quantitative over a broad range of temperature. The solubility behavior of nonpolar gases in water has been correlated and predicted by applying a modified ASOG group contribution method [120]. The use of empirical relationships which make use of group contributions permitted prediction of the water solubility of nonpolar as well as of polar vapors [46, 121].

The solubility in water of liquid and solid compounds has been correlated with the partition coefficient of the solute between water and n-octanol [122, 123]. In particular, a semi-empirical equation has been proposed which allows the evaluation of the solubility in water of either liquid or crystalline organic nonelectrolytes by knowing the melting point, entropy of fusion, and water-octanol partition coefficient [123]. Such an equation was able to reproduce the solubility of a great variety of compounds including some steroid hormones.

References

1. Wilhelm E, Battino R, Wilcock RJ (1977) Chem Rev 77:219
2a. Battino R, Clever HL (1966) Chem Rev 66:395
 b. Wilhelm E, Battino R (1973) Chem Rev 73:1
 c. Clever HL, Battino R (1975) In: Dack MRJ (ed) Solution and solubilities. Techniques of chemistry, vol 8, part 1. Wiley, New York, chapter 7
3a. See for instance Franks F (ed) (1973) Water: a comprehensive treatise. Plenum, New York, vol 2, chapter 5; (1975) ibid, vol 4, chapter 1
 b. Ben-Naim A (1974) Water and aqueous solutions. Plenum, New York
 c. Ben-Naim A (1980) Hydrophobic interactions. Plenum, New York
4. Clarke ECW, Glew DN (1966) Trans Faraday Soc 62:539
5. Benson BB, Krause D Jr (1976) J Chem Phys 64:689
6. Clever HL (ed) (1979) Solubility Data Ser, vol 1. Pergamon, Oxford
7. Clever HL (ed) (1980) Solubility Data Ser, vol 4. Pergamon, Oxford
8. Clever HL (ed) (1979) Solubility Data Ser, vol 2. Pergamon, Oxford
9. Muccitelli JA, Wen WY (1978) J Solution Chem 7:257

10. Clever HL (ed) (1982) Solubility Data Ser, vol 10. Pergamon, Oxford
11. Benson BB, Krause D Jr, Peterson MA (1979) J Solution Chem 8:655
12. Rettich TR, Battino R, Wilhelm E (1982) Ber Bunsen-Ges Phys Chem 86:1128
13. Murray CN, Riley JP (1971) Deep-Sea Res 18:533
14. Yasunishi A, Yoshida F (1979) J Chem Eng Data 24:11
15. Zawisza A, Malesinska B (1981) J Chem Eng Data 26:388
16. Morrison TJ, Billet F (1952) J Chem Soc (Lond) Part III:3819
17. Wetlaufer DB, Malik SK, Stoller L, Coffin RL (1964) J Am Chem Soc 86:508
18. Kresheck GC, Schneider H, Scheraga HA (1965) J Phys Chem 69:3132
19. Claussen WF, Polglase MF (1952) J Am Chem Soc 74:4817
20. Wen WY, Hung JH (1970) J Phys Chem 74:170
21. Ben-Naim A, Wilf J, Yaacobi M (1973) J Phys Chem 77:95
22. Rice PA, Gale RP, Bardhun AJ (1976) J Chem Eng Data 21:204
23. Rettich TR, Handa YP, Battino R, Wilhelm E (1981) J Phys Chem 85:3230
24. Sada E, Kumazawa H, Butt MA (1978) J Chem Eng Data 23:161
25. Ashton JT, Dawe RA, Miller KW, Smith EB, Stickings BJ (1968) J Chem Soc (Lond) Sect A Part II 1793
26. Wen WY, Muccitelli JA (1979) J Solution Chem 8:225
27. Park T, Rettich TR, Battino R, Peterson D, Wilhelm E (1982) J Chem Eng Data 27:324
28. Delassus PT, Schmidt DD (1981) J Chem Eng Data 26:274
29. Nillsson H, Silvegren C, Tornell B (1978) Eur Polymer J 14:737
30. Patel CB, Grandin RE, Gupta R, Phillips EM, Reynolds CE, Chan RKS (1979) Polymer J 11:43
31. Haiduk W, Laudie H (1974) J Chem Eng Data 19:253
32. Clever HL (ed) (1979) Solubility Data Ser, vol 8. Pergamon, Oxford
33. Winkler LW (1901) Ber d Chem Ges 34:1408
34. Hudson JC (1925) J Chem Soc 127:1332
35. Beuschlein WL, Simenson LO (1940) J Am Chem Soc 62:610
36. Tokunaga J (1974) J Chem Eng Data 19:162
37. Douabul A, Riley J (1979) J Chem Eng Data 24:274
38. Weiss RF (1971) J Chem Eng Data 16:235
39. Weiss RF (1970) Deep-Sea Res 17:721
40. Weiss RF (1971) Deep-Sea Res 18:225
41. Weiss RF, Kyser TK (1978) J Chem Eng Data 23:69
42. Weiss RF (1974) Mar Chem 2:203
43. Wiesenburg DA, Guinasso NL Jr (1979) J Chem Eng Data 24:356
44. Douabul A, Riley J (1979) Deep-Sea Res 26:259
45. Schumpe A, Adler I, Deckwer WD (1978) Biotechnol Bioeng 20:145
46. Cabani S, Gianni P, Mollica V, Lepori L (1981) J Solution Chem 10:563
47. Cabani S (1980) Thermodynamics of aqueous dilute solutions of non-charged molecules. In: Bertini I, Lunazzi L, Dei A (eds) Advances in solution chemistry. Plenum, New York
48. Mackay D, Shiu WY (1981) J Phys Chem Ref Data 10:1175
49. Seidell A (1941) Solubilities of organic compounds, vol 2, 3rd ed. Van Nostrand, New York
50. Stephen H, Stephen T (eds) (1963) Solubilities of inorganic and organic compounds. McMillan, New York
51. Getzen FW (1975) Structure of water and aqueous solubility. In: Dack MRJ (ed) Solutions and solubilities, vol 8, part II. In: Weissberger A (ed) Techniques of chemistry. Wiley, New York
52. McMeekin TL (1975) The solubility of biological compounds. In: Dack MRJ (ed) Solutions and solubilities, vol 8, part II. In: Weissberger A (ed) Techniques of chemistry. Wiley, New York
53. Fasman GD (ed) (1976) Handbook of biochemistry and molecular biology, 3rd edn, vol 1. Chemical Rubber, Cleveland
54. Dawson RMC et al. (eds) (1972) Data for biochemical research, 2nd edn. Oxford University Press, Oxford
55. Merck Index (1976) 9th edn. Merck, Rahway NJ (USA)

56. Altman PL, Dittmer DS (eds) (1972) Biology Data Book, vol 1. Soc Exp Biol, Bethesda, Maryland (USA)
57. Borsook H, Huffman HM (1938) In: Schmidt CLA, Thomas CC (eds) The chemistry of the amino acids and proteins. Springfield, p 848
58. Edsall JT, Scatchard G (1943) Solubility of amino acids, peptides and related substances in water and organic solvents. In: Cohn EJ, Edsall JT (eds) Proteins, amino acids and peptides. Hafner, New York
59. Greenstein JP, Winitz M (1961) Chemistry of the amino acids, vol 1, chapter 5. Wiley, New York
60. Klimov AI, Deshcherevskii VI (1971) Biophysics 16:580
61. Bull HB, Breese K, Swenson CA (1978) Biopolymers 17:1091
62. Nozaki Y, Tanford C (1965) J Biol Chem 240:3568
63. Nozaki Y, Tanford C (1970) J Biol Chem 245:1648
64. Bruskov VI, Klimov AI (1973) Biophysics 18:1022
65. Pittz EP, Bello J (1971) Arch Biochem Biophys 146:513
66. Nozaki Y, Tanford C (1963) J Biol Chem 238:4074
67. Gekko K (1981) J Biochem 90:1633
68. Dalton JB, Schmidt CLA (1933) J Biol Chem 103:549
69. Kresheck GC, Schneider H, Scheraga HA (1965) J Phys Chem 69:3132
70. Dunn MS, Ross FJ, Read LS (1933) J Biol Chem 103:579
71. Cohn EJ, McMeekin TL, Edsall JT, Weare JH (1934) J Am Chem Soc 56:2270
72. McMeekin TL, Cohn EJ, Weare JH (1936) J Am Chem Soc 58:2173
73. Conio G, Curletto L, Patrone E (1973) J Biol Chem 248:5448
74. Dalton JB, Schmidt CLA (1935) J Biol Chem 109:241
75. Nozaki Y, Tanford C (1971) J Biol Chem 246:2211
76. McMeekin TL, Cohn EJ, Weare JH (1935) J Am Chem Soc 57:626
77. Loring HS, Du Vigneaud V (1934) J Biol Chem 107:267
78. Winnek PS, Schmidt CLA (1935) J Gen Physiol 18:889
79. Abraham MH, Ah-Sing E, Marks RE, Schulz RA, Stace BC (1977) J Chem Soc Faraday Trans I 73:181
80. Tomiyama T, Schmidt CLA (1936) J Gen Physiol 19:379
81. Dooley KH, Castelino FJ (1972) Biochemistry 11:1870
82. Suzuki K, Tsuchiya M, Kadono H (1970) Bull Chem Soc Jpn 43:3083
83. Gill SJ, Hutson J, Clopton JR, Downing M (1961) J Phys Chem 65:1432
84. Robinson DR, Jencks WP (1965) J Am Chem Soc 87:2462
85. Cohn EJ (1943) In: Cohn EJ, Edsall JT (eds) Protein, amino acids and peptides, chapter 10 and 11. Hafner, New York
86. Robinson DR, Jencks WP (1965) J Am Chem Soc 87:2470
87. Robinson DR, Jencks WP (1963) J Biol Chem 238 PC:1558
88. Spink CH, Auker M (1970) J Phys Chem 74:1742
89. Bruskov VI, Klimov AI (1972) Biophysics 17:158
90. Robinson DR, Grant ME (1966) J Biol Chem 241:4030
91. Herskovits TT, Bowen JJ (1974) Biochemistry 13:5474
92. Herskovits TT, Harrington JP (1972) Biochemistry 11:4800
93. Nakano NI, Igarashi SJ (1970) Biochemistry 9:577
94. Ts'o POP, Melvin IS, Olson AC (1963) J Am Chem Soc 85:1289
95. Cesaro A, Russo E, Crescenzi V (1976) J Phys Chem 80:335
96. Kilday MV (1981) J Res Natl Bur Std 86:367
97. Gill SJ, Martin DB, Downing M (1963) J Am Chem Soc 85:706
98. Scruggs RL, Achter EK, Ross PD (1972) Biopolymers 11:1961
99. Gordon JA (1965) Biopolymers 3:5
100. Kilday MV (1978) J Res Natl Bur Std 83:347
101. Stern JH, Oliver DR (1980) J Chem Eng Data 25:221
102. Stern JH, Devore JA, Hansen SL, Yavuz O (1974) J Phys Chem 78:1922
103. Stern JH, Lowe E (1978) J Chem Eng Data 23:341
104. Kilday MV (1978) J Res Natl Bur Std 83:539
105. Stern JH, Beeninga LR (1975) J Phys Chem 79:582

106. Kilday MV (1978) J Res Natl Bur Std 83:529
107. Ahmed JK, Derwish GAW, Kanbour FI (1981) J Solution Chem 10:343
108. Alvarez T, Biltonen R (1973) Biopolymers 12:1815
109. Kilday MV (1978) J Res Natl Bur Std 83:547
110. Albert A, Brown DJ (1954) J Chem Soc (Lond) 2060
111. Albert A, Brown DJ, Cheeseman G (1952) J Chem Soc (Lond) 4219
112. Albert A, Brown DJ, Wood HCS (1954) J Chem Soc (Lond) 3832
113. Albert A, Lister JH, Pedersen C (1956) J Chem Soc (Lond) 4621
114. Pfleiderer W (1963) The solubility of heterocyclic compounds. In: Katritzky AR (ed) Physical methods in heterocyclic chemistry, vol 1. Academic Press, New York, p 177
115. Anderson JR, Pitman IH (1979) Aust J Pharm Sci 8:117
116. Rohdewald P (1978) Pharm Z 123:371
117. Levine L, Gordon JA, Jencks WP (1963) Biochemistry 2:168
118. Pierotti RA (1976) Chem Rev 76:717
119. Kim JI, Brückl N (1978) Z Phys Chem NF 110:197
120. Tochigi K, Kojima K (1982) Fl Phase Equil 8:221
121. Gianni P, Mollica V, Lepori L (1982) Z Phys Chem NF 131:1
122. Tewari YB, Miller MM, Wasik SP, Martire DE (1982) J Chem Eng Data 27:451 and references cited therein
123. Yalkowsky SH, Valvani SC (1980) J Pharm Sci 69:912; (1979) J Chem Eng Data 24:127
124. Abu-Hamdiyyah M, Shehabuddin A (1982) J Chem Eng Data 27:74

Notes Added in Proof

The present chapter covers the literature up to 1982. We would like to cite some pertinent references that have appeared more recently on this subject.

Solubility of Gases. (a) S.P. Smith and B.M. Kennedy, Geochim. Cosmochim. Acta, 47, 503 (1983) [noble gases]; (b) R. Battino, T.R. Rettich and T. Tominaga, J. Phys. Chem. Ref. Data 12, 163 (1983) [oxygen, ozone]; ibidem 13, 563 (1984) [nitrogen, air]; (c) S.F. Dec and S.J. Gill, J. Solution Chem., 13, 27 (1984) [methane, ethane, propane, butane, 2-methylpropane, 2,2-dimethylpropane, cyclopropane, ethene, propene, 1-butene, ethyne]; (d) T.R. Rettich, R. Battino and E. Wilhelm, J. Solution Chem 13, 335 (1984) [nitrogen]; (e) G. Olofsson, A.A. Oshodi, E. Qvarnström and I. Wadsö, J. Chem. Thermodyn. 16, 1041 (1984) [noble gases, methane, ethane, propane, n-butane, oxygen]; (f) R. Crovetto, R. Fernandez-Prini and M.L. Lapas, Ber. Bunsenges. Phys. Chem 88, 484 (1984) [ethane]; (g) S.F. Dec and S.J. Gill, J. Solution Chem. 14, 417 (1985) [noble gases].

Prevision of solubility: (h) B. Sander, S. Skjold-Jorgensen and P. Rasmussen, Fluid Phase Equilib. 11, 105 (1983); (i) S. Skjold-Jorgensen, Fluid Phase Equilib. 16, 317 (1984).

Solubility of Solids. (j) J.N. Spencer and T.A. Judge, J. Solution Chem. 12, 847 (1983) [adenine]; (k) F.I. Kanbour, J.K. Ahmed and G.A.W. Derwish, J. Solution Chem. 12, 763 (1983) [purine, adenine]; (l) H. DeVoe and S.P. Wasik, J. Solution Chem. 13, 51 (1984) [adenine, guanine]; (m) P.K. Nandi and D.R. Robinson, Biochemistry 23, 6661 (1984) [N-acetyl amino acids ethyl esters]; (n) J.H. Stern and L.P. Swanson, J. Chem. Eng. Data 30, 61 (1985) [uridine, cytidine].

The following books have been furthermore published: (o) J. Wisniak and M. Herskowitz, Solubility of Gases and Solids. A Literature Source Book, Elsevier, Amsterdam (1984). (p) Handbook of Solubility parameters, A.F.M. Barton Ed., C.R.C. Press, Inc., Boca Raton, Florida (1984). (q) Solubility Data Series, H.L. Clever Ed., Pergamon Press, (Oxford): 4 volumes on gases [methane, propane, n-butane, 2-methylpropane, sulfur dioxide, chlorine, fluorine, chlorine oxides, ammonia and amines]; 4 volumes on liquids [hydrocarbons C_5-C_{36}, halobenzenes, halophenols, alcohols]; 1 volume on solids [antibiotics].

Chapter 11 Thermodynamic Parameters of Biopolymer-Water Systems

Madeleine Lüscher-Mattli

Symbols

p/p_0:	equilibrium relative water vapor pressure at a given experimental temperature T and constant (isosteric) water content h (g H_2O/g biopolymer) or n (mol H_2O g^{-1} biopolymer)
h:	water content (g H_2O g^{-1} biopolymer)
h*:	critical water content (g H_2O g^{-1} biopolymer) ("monolayer" content)
v:	experimentally determined specific volume, [cm^3 g^{-1}]; ($v = 1^{-1}$ density)
$\bar{v}_1^0$:	partial specific volume of pure liquid water
$\bar{v}_1$:	partial specific volume of component 1
$\bar{v}_2$:	partial specific volume of component 2
$\Delta\bar{v}_2$:	difference in partial specific volume of component 2, v_2, obtained by extrapolation of v versus W_1 functions to $W_1 = 0$, and v_2 obtained experimentally at $W_1 = 0$
$v_{\phi,i}$:	apparent specific volume of component i
W_i:	weight fraction of component i
$\Delta_a G_i^0, \Delta_a H_i^0, \Delta_a S_i^0$:	standard isosteric molar Gibbs energies, enthalpies and entropies of sorption
$\Delta_a G_m^0, \Delta_a H_m^0, \Delta_a S_m^0$:	standard integral molar Gibbs energies, enthalpies and entropies of sorption
ΔH_v:	enthalpy of vaporization of pure water at 298.15 K
$\Delta_f H$:	enthalpy of water fusion
T_f:	temperature of water fusion
$\Delta_w H^0$:	standard enthalpies of wetting
$\Delta_a H_{cal}^0$:	calorimetrically obtained molar enthalpies of water sorption (or desorption)
$\Delta_a H_{ad}$:	adiabatic molar enthalpies of water sorption
c_p:	experimentally observed specific heat capacity
$\bar{c}_{p,1}$:	partial specific heat capacity of component 1
$\bar{c}_{p,2}$:	partial specific heat capacity of component 2
$\phi c_{p,2}$:	apparent specific heat capacity of the biopolymer
$\bar{c}_{p,1}^0$:	partial specific heat capacity of the pure liquid water

1 Introduction

It has been recognized that biopolymer-associated water plays an important role in the folding and reactivity of native biopolymers. There is, however, a lack of understanding of the basis of this solvent control of biochemical reactivity. Investigation of the energetics of biopolymer hydration processes thus seems to be essential to improve our understanding of the biological (and possibly pathological) functions of water.

A large number of investigations of biopolymer-water systems using a variety of experimental techniques has been reported in the literature. For detailed information we refer to the extensive and comprehensive reviews given by Edsall and Mc Kenzie [1] and Kuntz and Kauzmann [2]. The general picture of biopolymer hydration, which emerged from these studies, can be described as follows:

The primary hydration water (A), interacts directly by H-bond formation with proton donor/proton acceptor groups, or by electrostatic interaction with charged groups of the solvent-accessible biopolymer surface. This water fraction has to be considered as an integral part of the native biopolymer structures. A subfraction of this hydration "shell" ($A' \simeq 1/5$ to $1/6$ of A) corresponds to the very strongly bound, or "monolayer" hydration water. This water fraction appears to play a special role in the stabilization of native biopolymer structures. With increasing water content, the existence of another water phase, *the secondary hydration water* (B), can be observed. The physical properties of this water fraction are still significantly different from those of bulk water. With increasing hydration, i.e., with increasing distance from the biopolymer surface, the properties of this secondary hydration water approach the bulk values (B··· fraction). At water contents above approximately 1.0 g H_2O g^{-1} polymer, the water properties become indistinguishable from those of *bulk water*.

In the present chapter the following thermodynamic parameters of biopolymer-water systems will be considered:

The heats, entropies and free energies of the interaction of the inner-sphere hydration water (A' and A) with biopolymer surface groups. These parameters are mostly obtained by vapor pressure studies on solid biopolymers (lyophilized powders or films). The heats of water sorption or desorption, as well as heats of wetting have also been determined calorimetrically, and will be compared to the van't Hoff data obtained by vapor pressure measurements. A few heats and entropies of activation of biopolymer-bound water have also been obtained by spectroscopic investigations (NMR, IR). These data are, however, scarce and contradictory, and therefore will not be further considered.

Specific heats of biopolymer-water systems have been determined calorimetrically in a large range of water contents. This parameter gives valuable information on hydration events occurring upon successive addition of water to dehydrated biopolymer samples.

The heats and temperatures of water fusion in biopolymer-H_2O systems have been determined calorimetrically (differential scanning calorimetry: DSC) in a variety of biopolymers (mostly proteins) at various degrees of hydration. In this type of study, the amount of unfreezable primary hydration water (A'+A) can be determined. Furthermore the amount, heat, and temperatures of fusion of the secondary hydration water (B'+B) were obtained in detailed DSC studies.

The *specific volumes* of binary biopolymer-water systems of various composition represent another thermodynamic parameter characterizing biopolymer-water interactions. The above-mentioned thermodynamic measurements were mostly obtained on solid biopolymer-water systems. It was shown, however, by

Rupley [3], that the conformation of hydrated biopolymers in the solid state is comparable to that in dilute aqueous solution. The thermodynamic results reported in this chapter are thus also pertinent to native biopolymers in solution. In the following sections, the experimental methods used in the determination of the thermodynamic quantities mentioned above will be briefly described, their sources of error will be discussed, and criteria for selection of best values will be given. Selected values for thermodynamic quantities relevant to biopolymer hydration events will then be presented, compared, and discussed.

2 Experimental Methods, Their Sources of Error, and Criteria for Selection of "Best Values"

2.1 Vapor Pressure Measurements

The usual experiment measures sorption isotherms, i.e., the amount of sorbed water is determined as a function of the relative water vapor pressure at constant temperature and external pressure. There exists a wide variety of gravimetric and volumetric sorption techniques; for detailed information, we refer to a review of water sorption techniques given by Gál [4]. The most convenient technique is the isopiestic or exsiccator method, a discontinuous gravimetric technique.

The relative water vapor pressure is controlled by addition of sulfuric acid or electrolytes to a water reservoir. The amount of sorbed water is determined by weighing after equilibration of the samples in a thermostated chamber for 4–5 days. The dry weight is obtained either at room temperature over phosphorus pentoxide in vacuo (6–7 days), or at elevated temperature (105 °C) in vacuo for approximately 24 h.

The isopiestic method requires relatively simple laboratory equipment, and allows a simultaneous investigation of several samples. The discontinuous registration of the sample weights, however, requires special caution, as outlined below.

More sophisticated sorption techniques allow a continuous registration of the sample weights by gravimetric or volumetric methods. Valuable kinetic information on water sorption or desorption processes is here readily obtained. A detailed description of several types of sorption apparatus is given in [4]. A fully automatized continuous sorption apparatus was developed in our laboratory, and is described by Bolliger et al. [5].

2.1.1 Sources of Error

Errors in Water Content. The determination of the dry weight is of crucial importance to obtain correct water contents. As shown by Rao and Bryan [6] the residual water content of a protein sample after prolonged drying (approx. 20 h) strongly depends on the drying conditions (temperature and residual water vapor pressure). Under the conventional drying methods, the residual water content is approximately 0.1% or smaller, and thus does not measurably affect the experimental water contents, whose accuracy is ± 0.2 to 0.5 w/w.

In discontinuous sorption methods, changes in water content during registration of the sample weight at room temperature and humidity may occur, especially in the low and the high range of relative water vapor pressure. These errors can be largely eliminated by the use of specially constructed sample containers and by a special weighing technique [4].

It is a well-known fact that the water content of a biopolymer sample at a given relative water vapor pressure can vary (by several w/w), depending on the previous history of the material under investigation (hysteresis effect). As discussed in detail by Bryan [7], hysteresis data are not accessible to thermodynamic treatment.

Hysteresis effects can be avoided, if the sorption experiment is started after equilibration of the sample material at high relative humidities. Effects of previous drying (which induces denaturation of the biopolymer) are thereby "cleared", and subsequent desorption isotherms are reversible [8, 9], i.e., accessible to thermodynamic treatment.

Environmental factors, such as sample temperature, purity of sample material and the conformational state of the biopolymer under investigation (native, denatured), also significantly affect the experimental water contents.

Errors in the Relative Water Vapor Pressure (p/p_0). Changes in the temperature of the water reservoir by $+0.1$ °C induce changes the p/p_0 values by $\pm 5 \times 10^{-3}$ units, and changes in pressure of $p \pm 0.013$ atm, induce changes in p/p_0 of 1.4×10^{-2} units. Errors in the concentration of sulfuric acid or electrolyte in the water reservoir may also induce considerable errors in p/p_0. These factors thus have to be carefully controlled.

2.1.2 Criteria for Selection of Best Values

The thermodynamic parameters of biopolymer hydration, derived from vapor pressure experiments at various temperatures (see Sect. 3) can be considered as valid if the following experimental conditions are fulfilled: (1) The dry weight is obtained by an exhaustive and strictly standardized drying procedure. (2) The vapor pressure data refer to desorption isotherms. (3) The sample material is purified (deionized) and corresponds to the biopolymer in its native state. (4) Environmental factors (temperature and pressure) are well controlled.

2.2 Calorimetry

This experimental technique represents a useful tool in thermodynamic and analytical investigations. Spink and Wadsö [10] give a detailed review on calorimetric instrumentation and application of the various calorimetric techniques. In this section we shall restrict ourselves to the calorimetric methods used in determination of thermodynamic quantities of biopolymer-water systems:

Sorption calorimeters are designed to study solid-gas and solid-liquid interactions. Sorption calorimeters of good precision are available commercially (for instance the batch and flow LKB sorption calorimeter of heat conduction type). Heats of water sorption or desorption, as well as heats of wetting, can be determined by this calorimetric technique.

Several authors describe their own constructions of sorption calorimeters, among which the combined gravimetric and calorimetric sorption apparatus, reported by Pineri et al. [11] appears of special interest. In the usual experiment, the sample is dried in the calorimetric cell at constant temperature and in vacuo. Known amounts of water vapor are then added stepwise and the heat evolved (or absorbed) is measured. By this technique differential heats of sorption/desorption are obtained.

Sources of Error. A control of attainment of constant dry sample weight, and constant equilibrium water content is in most cases not possible during the calorimetric experiment. Evaporation or condensation effects in the reaction cell, especially at high or low relative humidities, may introduce serious errors (the heat of evaporation of pure water is 43.93 mJ μmol^{-1} water). Specially designed reaction cells are required to overcome these difficulties.

2.2.1 Heat Capacity Calorimeters

Double drop heat capacity calorimeters, used for precise determination of specific heats of biopolymer-water systems of various composition have been described by Suurkuusk [12] and Yang and Rupley [13]. In these calorimetric measurements, a sample and reference ampule are dropped from a "furnace" of constant temperature T_1 into the twin calorimeter of heat conduction type, kept at a constant temperature T_2. The difference in the heat quantities transferred by the two ampules to the calorimeter are recorded as thermopile vs. time functions. The heat capacities thus obtained refer to the mean of T_1 and T_2 (linearity of the heat capacity in the temperature interval T_1-T_2 assumed). Special constructions of other types of heat capacity calorimeter are reported in the literature [14, 15].

Sources of Error. Control experiments with the double drop calorimeter showed that condensation or evaporation effects associated with the transfer of the ampules from T_1 to T_2 do not introduce significant errors. Changes in the hydration of the sample upon change in temperature were also found to be negligible. Possible changes in the structure of the biopolymer in the experimental temperature range may also introduce errors in the heat capacity terms, and should be controlled by other techniques.

2.2.2 Differential Scanning Calorimetry (DSC)

In a typical DSC measurement, the difference in power necessary to heat the sample and the reference cell at the same rate is measured as a function of temperature. This difference in power compensates the difference in heat capacity of both cells. Commercial DSC instruments of variable sensitivity are available [Perkin-Elmer DSC-2, DuPont 910-DSC are of medium, DASM-1M and DASM-4 (USSR) of very high sensitivity] DSC measurements allow determination of specific heats of BP-H_2O systems over a wide range of temperature and composition.

Furthermore, integral heats and temperatures of phase transitions in biopolymer-water systems can be investigated. In the low temperature range (-70 °

to approx. $+10\,°C$), the heats and temperatures of fusion of biopolymer-associated water have been studied. In the high temperature range ($10\,°$ to $150\,°C$), the thermally induced biopolymer-denaturation processes can be investigated.

Sources of Error. Evaporation or condensation effects in the sample pans. Effect of scanning rate upon the peak maxima. (The temperatures of phase transition T_t may be shifted to higher values at high scanning rates).

In low temperature experiments, the rate and mode of cooling the samples is important. In fast cooling, formation of supercooled water or of vitreous ice may affect the experimental results.

Attention must be paid to correct construction of the base line shifts, due to changes in the heat capacity of the sample during the phase transition. Errors in base line affect the enthalpies of the phase transition investigated.

Criteria for Selection of Best Values. The water contents are properly controlled before and after the calorimetric experiment. The possible artifacts, mentioned above, are controlled for, and eliminated by adequate experimental procedure. The experimental material is well defined with respect to its purity and conformational state.

2.3 Densitometry

The specific volumes (v) of biopolymer-water systems are obtained from density (d) measurements ($v=1/d$).

In investigations of aqueous solutions of biopolymers, digital densitometers of high precision were used (Paar instruments, based on Quarz oscillator principles). With these instruments a precision in v of $\pm 10^{-5}$ 1 kg^{-1} is obtained. In solid biopolymer-water systems the following experimental techniques have been used: Density gradient technique, pycnometry and specific gravimetry. Richards reports a special microbalance [16] and gradient tube technique [17] to study biopolymer crystals.

Sources of Error. In density gradient or pycnometric measurements, absorption of the liquids used (benzene, xylene, bromobenzol etc.) may occur in quasi-dry biopolymer samples. The apparent densities thus obtained are too high (approx. 0.04 to 0.06 kg 1^{-1}), and correspondingly the apparent v values too low. Errors in dry weight determination, i.e., incorrect determination of the mass fractions of biopolymer-water systems, introduce errors in the hydration dependence of v.

Criteria for Selection of Best Values. Solution data of high precision in both densities and mass fractions of biopolymer.

Data on solid biopolymer-water systems are also included in Table 3 because of their general interest for biopolymer-water interactions. The absolute values of v are, however, less accurate in these systems (± 0.003).

3 Tables of Selected Values

The thermodynamic data given in Tables (1) to (3) satisfy, as far as possible, the requirements for selection mentioned in the previous section. A claim for completeness should therefore not be made.

3.1 Thermodynamic Data Obtained from Vapor Pressure Studies

The isosteric and integral Gibbs energies, enthalpies, and entropies of water sorption are given in Table 1 as partial molar excess quantities, referring to the interaction of solid biopolymers with liquid water:

$$(\text{biopolymer})_s + (n_0 H_2O)_i \rightleftharpoons (\text{biopolymer } n_0 H_2O)_s \tag{1}$$
$$\quad p \qquad\qquad p_0 \qquad\qquad\qquad p_0$$

The isosteric thermodynamic parameters corresponding to process (1) are obtained by application of the following equations:

$$\varDelta_a G_i^0 = RT \,(\ln p/p_0)_n, \tag{2}$$

$$\varDelta_a H_i^0 = RT^2 \,(d\ln p/p_0)_n/dT, \tag{3}$$

$$\varDelta_a S_i^0 = RT \,(d \ln p/p_0/dT)_n - R(\ln p/p_0)_n. \tag{4}$$

The integral molar excess quantities are obtained from:

$$\varDelta_a G_m^0 = RT \int_0^{p_0} \ln p/p_0 \, dn, \tag{5}$$

$$\varDelta_a H_m^0 = \int_0^{p_0} \varDelta_a H_i^0 + n_0 H_v, \tag{6}$$

$$\varDelta_a S_m^0 = \frac{\varDelta_a H_m^0 - \varDelta_a G_m^0}{T}, \tag{7}$$

where R is the gas constant, p/p_0 the equilibrium relative water vapor pressure at a given experimental temperature T, and constant (isosteric) water content h or n (given in g or mol H_2O g^{-1} biopolymer).

$\varDelta H_v$ is the enthalpy of vaporization of pure water (43.93 kJ mol^{-1} at 298 K).

As discussed in detail by Bryan [7], Eq. (2) to (7) are valid only for equilibrium conditions. It was shown [8, 9] that desorption isotherms are reversible above a critical water content (h* in Table 1) and are thus accessible to thermodynamic treatment. The data reported in Table 1 were obtained from desorption experiments [8, 19], or from the averages of sorption and desorption data [18].

3.2 Thermodynamic Data Obtained by Calorimetry

1. The calorimetric *enthalpies of water sorption* (given as excess functions $\varDelta_a H_{cal}^0$ in kJ $mol^{-1} H_2O$), represent differential terms, which are related to the isosteric quantities by:

$$\varDelta_a H_d^0 = \varDelta_a H_i^0 - RT. \tag{8}$$

For adiabatic enthalpies the relationship to the isosteric quantities is given by:

$$\Delta_a H_{ad} = \Delta_a H_i^0 + V_g(p/dn_0)_{T,\,nB} + V_g(p/dT)_n(dT/dn_0)_{n,\,B,\,S} \tag{9}$$

2. *The enthalpies of wetting* are defined as the difference in enthalpy of the completely immersed solid, and the enthalpy of the solid and liquid taken separately.

3. *The enthalpies of water fusion* are evaluated from the hydration dependence of the experimentally determined integral heats of water fusion (Q_i).

The slope dQ_i/dh represents the incremental heat of water fusion, from which the molar enthalpy of fusion is readily obtained. The X-intercept of the Q_i vs. h functions represents the amount of unfreezable water (A in Table 2).

4. *The experimental heat capacities* (c_p) are related to the partial specific heat capacities of water ($\bar{c}_{p,\,1}$) and biopolymer ($\bar{c}_{p,\,2}$) by:

$$c_p = \bar{c}_{p,\,1} W_1 + \bar{c}_{p,\,2} W_2. \tag{10}$$

The apparent specific heat of the biopolymer component is defined as:

$$\phi c_{p,\,2} = (c_p - W_1 \bar{c}_{p,\,1}^0)/W_2, \tag{11}$$

where $\bar{c}_{p,\,1}^0$ represents the partial specific heat of pure liquid water, and W_1, W_2 the weight fractions of water and biopolymer, respectively. The partial specific heat capacities of water ($\bar{c}_{p,\,1}$) and biopolymer ($\bar{c}_{p,\,2}$) are related to $\phi c_{p,\,2}$ by:

$$\bar{c}_{p,\,2} = \phi c_{p,\,2} - (d\phi\,C_{p,\,2}/dW_1)W_1, \tag{12}$$

and

$$\bar{c}_{p,\,1} = \bar{c}_{p,\,1}^0 + d\phi\,c_{p,\,2}/dW_1, \tag{13}$$

where $\bar{c}_{p,\,2}$ is obtained from the y intercept, and $\bar{c}_{p,\,1}$ from the slope of $\phi c_{p,\,2}$ vs. W_1 functions.

3.3 Thermodynamic Parameters Obtained by Densitometry

The experimentally determined specific volume (v = 1/density) of a binary biopolymer-water system is related to the partial specific volume of the biopolymer ($\bar{v}_2$) and water ($\bar{v}_1$) by the relationship:

$$v = \bar{v}_1 W_1 + \bar{v}_2 W_2. \tag{14}$$

The apparent specific volume of the biopolymer is defined as:

$$v_{\phi,\,2} = (v - W_1 v_1^0)/W_2. \tag{15}$$

From $v_{\phi,\,2}$ the partial specific volumes of the biopolymer ($\bar{v}_2$) and water ($\bar{v}_1$) in the binary system are obtained from:

$$\bar{v}_2 = v_{\phi,\,2} - \frac{dv_{\phi,\,2}}{dW_1} W_1, \tag{16}$$

and

$$\bar{v}_1 = \bar{v}_1^0 + \frac{dv_{\phi,\,2}}{dW_1}, \tag{17}$$

Table 1. Thermodynamic parameters obtained by vapor pressure studies: Isosteric and integral Gibbs energies, enthalpies and entropies of water sorption

Biopolymer-H_2O system	h	$-\Delta_a G_i^0$ kJ mol⁻¹	$-\Delta_a H_i^0$ kJ mol⁻¹	$-\Delta_a S_i^0$ J mol⁻¹ K⁻¹	$-\Delta_a G_m^0$ kJ mol⁻¹	$-\Delta_a H_m^0$ kJ mol⁻¹	$-\Delta_a S_m^0$ J mol⁻¹ deg⁻¹	Reference
a) Proteins								
Albumin Egg	0.03	6.59	13.85	23.76				[18]
	0.06	3.48	5.69	7.25	3.73	6.57		
	0.12	1.27	3.21	6.34				
	0.18	0.53	1.08	1.81		5.86		
	0.24	0.28	1.33	3.44				
Serum	0.03							[18]
	0.06							
	0.12				4.33	6.07		
	0.18							
	0.24							
Apo-Lactoferrin	0.04	8.53	15.93	30.22				[8]
	0.08	3.72	15.85	39.63				
	0.12	2.05	10.91	29.05	–	–		
	0.18	0.80	8.61	25.59				
	0.22	0.52 (±0.04)	4.42 (±0.3)	15.19 (±0.9)				
α-Chymotrypsin	0.03	7.80	19.63	38.75				[19]
	0.06	4.30	21.00	54.53				
	0.13	1.26	10.10	28.92	–	–		
	0.17	0.64	8.64	26.20				
	0.25	0.20 (±0.01)	6.63 (±0.4)	21.05 (±1.7)				
Collagen	0.03	9.48	37.51	90.99				[18]
	0.09	4.34	12.46	26.89				
	0.12	3.09	5.31	6.99	6.78	15.23		
	0.18	1.67	2.71	3.09		17.36		
	0.24	0.96	1.91	3.00				
Elastin	–	–	–	–	4.184	9.41		[18]
Gelatin	–	–	–	–	6.43	15.90		[18]
						15.50		[18]
Keratin	–	–	–	–	4.08	8.16		[18]
β-Lactoglobulin	0.03	9.58	7.54	91.00	3.84	6.74		[18]
	0.06	6.48	19.91	43.89		6.70		[18]

Compound				0.24	0.96	1.91	2.99	h/h*	Ref
Lysozyme				0.03	7.50	4.74	–	–	[8]
				0.06	4.23	9.71	18.19		
				0.12	1.60	5.19	11.69		
				0.16	0.92	3.35	7.99		
				0.20	0.60	2.08	5.19		
b) Nucleic acids									
Desoxyribonucleic acid	B-form			0.04	8.55	7.81	–		[8]
				0.10	4.08	16.11	39.17		
				0.12	3.30	14.51	34.76		
				0.16	2.40	7.81	17.25		
	A-form			0.18	2.09	13.48	37.33		
				0.24	1.34	7.81	21.49		
				0.34	1.01	4.09	10.09	2.13	
				0.44	0.72	1.65	3.09		
c) Model compounds				h 10²					
Nylon				0.5	7.43	24.88	57.14	–	[18]
				1.0	4.74	11.00	20.50		
				1.5	3.34	3.82	1.26	1.06	
				2.0	2.81	3.41	–		
				5.0	0.87	2.22	2.26		
				10.0	0.10	0.88	1.31		
Amino acids				h/h*					
Glycine				0.5	9.16	16.32	29.83	–	[21]
				0.75	7.24	9.41	12.68		
				1.0	5.86	2.51	(– 7.0)		
Leucine				0.5	8.20	19.87	45.31		
				0.75	6.36	12.22	25.23		
				1.0	4.77	5.02	5.15		
Diketopiperazine				0.5	7.03	19.66	49.50	–	
				0.75	5.52	8.16	13.93		
				1.0	4.31	2.43	(– 1.5)		

h = water content in kg H_2O kg^{-1} polymer.

h* = "monolayer" water content in kg H_2O kg^{-1} polymer.

$\Delta_a G^0$, $\Delta_a H^0$, $\Delta_a S_i^0$ = isosteric Gibbs energies, enthalpies, and entropies of sorption

$\Delta_a G_m^0$, $\Delta_a H_m$, $\Delta_a S_m^0$ = integral molar Gibbs energies, enthalpies, and entropies of sorption

Table 2. Thermodynamic data obtained by calorimetry: Enthalpies of water sorption, and wetting, enthalpies of water fusion, heat capacities

Biopolymer-H_2O system	h $kg\,H_2O\,kg^{-1}\,P$	$\Delta_a H^0_{cal}$ $kJ\,mol^{-1}\,H_2O$		$\Delta_w H^0$ $kJ\,mol^{-1}\,H_2O$	$\Delta_f H^0$ $kJ\,mol^{-1}\,H_2O$	T(K)	Reference
		Sorption	Desorption				
a) Enthalpies of water sorption or desorption: ($\Delta_a H^0_{cal}$)							
Albumin (bovine serum)	0.03	15.5	–	–	–	–	[22]
	0.06	11.7	–				
	0.09	7.5	–				
	0.12	7.5	–				
Albumin (bovine serum)	0.06	–	6.3				[23]
	0.22	–	5.0				
Collagen (calf skin)	0.28	–	6.8				[23]
	0.35	–	10.1				
Collagen (rat tail tendon)	0 –0.12	28.7	35.1				[11]
	0.12–0.24	17.7	8.9				
	0.24–0.38	0	0				
b) Enthalpies of Wetting ($\Delta_w H^0$)							
Collagen (hide)	0	–	–	18.9			[24]
	0.02	–	–	15.9			
	0.04	–	–	13.4			
	0.08	–	–	10.0			
	0.16	–	–	6.0			
				2.8			
Keratin (wool)	0	–	–	29.5	–	–	
Nylon	0	–	–	15.1	–	–	

Table 2 (continued)

c) Enthalpies and temperatures of water fusion ($\Delta_f H^0$, T_f)

Biopolymer-H_2O system	h kg H_2O kg^{-1} P	$\Delta_f H^0$ kJ mol^{-1}	T_f (K)	A kg H_2O kg^{-1} P	Reference
Albumin Egg	—	—	—	0.323 (n) 0.332 (d)	[28]
Serum	—	—	—	0.315 (n) 0.330 (d)	[28]
Casein (native bovine)	0 −0.4 0.4−4.0	0 5.987	— 273 —	0.43	[25]
α-Chymotrypsin	0.3−0.8 0.8−1.4	4.612 ($\Delta_f 5 = 17.55$) 5.184 ($\Delta_f 5 = 21.60$)	260−272 273	0.29	[26]
Collagen (rat tail tendon)	0 −0.3 0.3−0.6 0.6−1.4 >1.4	— 3.239 5.828 5.828	— 251−265 267−270 273	0.26	[27]
Collagen (procollagen)	0.6−2.0	—	269−271	0.465 (n) 0.519 (d)	[28]
Elastin	0 0.3 >0.3	— 5.837	—	0.335	[29]
Keratin (wool)	0 −0.62 0.2−0.34 >0.3	0 4.217 6.012			[14]
β-Lactoglobulin	0 −0.3 0.3−0.7 0.7	0 4.217 5.717			[30]
Nucleic acids	0.75 1.2−2.1		255 263−268	0.61	[31]

Table 2 (continued)

d) Partial specific heat capacities ($\bar{c}p_1$, $\bar{c}p_2$)

Biopolymer-H_2O system	Composition W_2 (weight fraction P)	$\bar{c}_{p,1}$ kJ K^{-1} kg^{-1} H$_2$O	$\bar{c}_{p,2}$ kJ K^{-1} kg^{-1} P	Reference
Albumin (egg)	0 –0.7	4.1778 ± 0.0004	1.534 ± 0.014	[12]
	0.7 –0.95	5.53 ± 0.04	1.231 ± 0.005	
	0.8 –1.0	5.058 ± 0.431	1.021 ± 0.138	[33]
	0 –0.7	4.176	1.657 ± 0.092	[34]
	0.7 ± 0.95	5.217 ± 0.096	1.18 ± 0.255	
Chymotrypsinogen	0 –0.7	4.1808 ± 0.0006	1.529 ± 0.015	[12]
	0.7 –0.95	5.46 ± 0.03	1.223 ± 0.004	
	1.0		1.293	
Keratin (wool)	0.7 –1.0	2.148	1.255	[14]
β-Lactoglobulin	0.83–1.0	3.9622	1.1841	[33]
Lysozyme	0 –0.3	4.1792 ± 0.0004	1.494 ± 0.007	[12]
	0.3 –0.95	5.42 ± 0.03	1.192 ± 0.005	
	0 –0.01	4.1791 ± 0.0001	1.682 ± 0.001	[36]
	0 –0.7	4.183 ± 0.003	1.483 ± 0.009	[13]
	0.7 –0.78	3.35	1.75	
	0.78–0.91	5.8	1.09	
	0.9 –1.0	2.3	1.26 ± 0.01	

Enthalpy of fusion of pure ice: 6.008 kJ mol^{-1} [32] heat capacity ice: 2.092 J mol^{-1}K^{-1}
Entropy of fusion of pure ice: 21.96 J mol^{-1} K^{-1} [32] heat capacity water (0 °C): 4.2258 J mol^{-1}K^{-1}, water (20 °C): 4.1422 J mol^{-1} K^{-1}

Table 3. Thermodynamic parameters obtained by densitometry: specific volumes of biopolymer-water systems

Biopolymer-H_2O system	Composition v W_2=Weight fraction BP	$1\,kg^{-1}$	$\bar{v}_1$ $1\,kg^{-1}\,H_2O$	$\bar{v}_2$ $1\,kg^{-1}\,BP$	Reference
Albumin					
Serum, bovine	0				[37]
	0.00030	1.00289			
	0.00398	1.00192	1.003	0.73604	
	0.03152	0.99453			
	0.1524	0.96217			
	0.21860	0.94427			
	0.28192	0.92704	1.007	0.7234	
		$(\pm 3\text{–}5 \times 10^{-5})$			
Serum, bovine	0–0.2	–	–	0.7348	[38]
Serum, human	0.44	0.882	1.000–0.992[a]	0.737–0.739[a]	[39]
	0.60	0.835			
	0.75	0.796			
	0.80	0.787			
	0.90	0.765			
	0.92	0.800			
	1.0	0.800		0.727[b]	
		(± 0.003)		$\Delta\bar{v}_2 = 0.07$	
Ovalbumin	0–0.31	–	–	0.7463	[38]
	0–0.1			0.7477	[40]
	0.640	0.8865		0.736[b]	[41]
	0.800	0.8057			
	0.822	0.7985		$(\Delta\bar{v} = 0.054)$	
	0.890	0.7807			
	0.895	0.7809			
	0.906	0.7779			
	0.942	0.7779			
	0.984	0.7818			
	0.996	0.7822			
	1.0	0.7902			
	0.146	0.9614	–	0.7358	[42]
	1.0	–	–	0.788	
	>0.8	–	–	0.781	[34]
	<0.8	–	–	0.750	
Hemoglobin (bovine)	0.00354	1.00207	1.003	0.7546[a]	[37]
	0.03053	0.99540		(± 0.00004)	
	0.30708	0.92658			
	0.33260	0.92019			
	0.36785	0.91126			
	0.39016	0.90568			
	0.42362	0.89736			
	0.43888	0.89326	1.0054	0.749	

Table 3 (continued)

Biopolymer – H_2O system	Composition v W_2=Weight fraction BP	$1\,kg^{-1}$	$\bar{v}_1$ $1\,kg^{-1}$	$\bar{v}_2$ $1\,kg^{-1}$	Refer-ence
β-Lactoglobulin (crystals)	0.68	0.8688			[43]
	0.71	0.8562			
	0.76	0.8389			
	0.83	0.8150			
	0.86	0.8058			
	0.89	0.8013			
	0.90	0.8000			
	0.93	0.7974			
	0.95	0.7974			
	0.97	0.7974			
	0.98	0.7987			
	1.00	0.8019	0.983[a]	0.772[b] $\Delta\bar{v}_2=0.04$	
Lysozyme	0.00054	1.00013			[36]
	0.00108	1.00084			
	0.00268	1.00153			
	0.00537	1.00224			
	0.00807	1.00265			
	0.01077	1.00279			
Keratin (wool)	0.752	0.7888	0.988[a]	0.732[b]	[44]
	0.769	0.7834			
	0.800	0.7743	$\Delta\bar{v}_2=0.04$		
	0.833	0.7692			
	0.869	0.7619			
	0.909	0.7605			
	0.935	0.7605			
	0.952	0.7615			
	0.980	0.7637			
	1.0	0.7692			

Symbols used in Table 3: $\bar{v}_1$=partial specific volume of water.
$\bar{v}_2$=partial specific volume of biopolymer.
[a] Values obtained by extrapolation to $W_2=0$.
[b] Values obtained by extrapolation to $W_2=1.0$.

where $\bar{v}_1^0$ is the partial specific volume of pure liquid water, and W_1 and W_2 the weight fractions of biopolymer and water, respectively.

The $\Delta\bar{v}_2$ terms given for some solid biopolymer-water systems in Table 3 correspond to the difference in $\bar{v}_2$ obtained by extrapolation of v vs. W_1 functions to $W_1=0$, and the $\bar{v}_2$ values obtained experimentally at $W_1=0$.

4 Discussion and Correlation of the Thermodynamic Parameters of Biopolymer Hydration

1. The hydration dependence of the thermodynamic data given in Tables 1 to 3 is represented schematically in Fig. 1. All parameters to be discussed below are

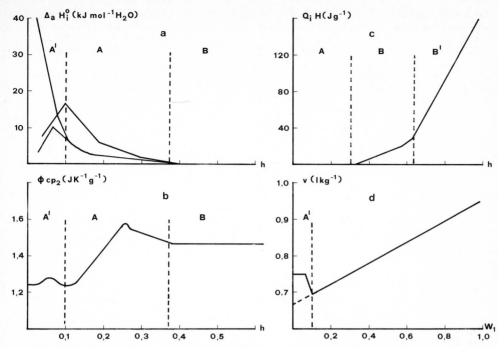

Fig. 1. Schematic representation of the hydration dependence of various thermodynamic functions of biopolymer hydration. **a** The isosteric excess enthalpies of biopolymer hydration. **b** The apparent specific heat capacity of the biopolymer subsystem. **c** The integral heats of water fusion. **d** The specific volume of biopolymer-water systems

measures of the same molecular hydration processes, and thus show parallel changes with hydration.

2. The enthalpies, entropies, and free energies of biopolymer hydration processes (Table 1) have mostly been obtained by vapor pressure studies. On a quantitative level, the situation in this field of research is not satisfactory: reliable and comparable data are scarce, and further work under defined and standardized experimental conditions is required to obtain well-established quantitative results. In a few cases, calorimetrically determined enthalpies of biopolymer hydration are available in the literature (Table 2). In Table 4, some of the selected isosteric and calorimetric enthalpies are compared. The agreement between these enthalpy terms is satisfactory, in view of the experimental accuracy (± 1.5 to 0.5 kJ mol^{-1} H$_2$O) of these parameters. In other cases the agreement between calorimetric and isosteric enthalpies is, however poor, mostly due to different experimental conditions.

On a qualitative level the general features of the thermodynamics of biopolymer hydration processes are, however, well established: in all systems investigated, the enthalpies, entropies, and Gibbs energies of the interaction of water with native biopolymers are of negative sign, the excess values ranging from approximately -5 to -29 kJ mol^{-1} H$_2$O at h*$=0.06$ to 0.1 g H$_2$ g^{-1} polymer. The

Table 4. Comparison of the enthalpies of biopolymer hydration determined by vapor pressure studies and by calorimetry

	h (g H_2O g^{-1} polymer)	$\Delta_a H_i^0$ kJ mol^{-1} H_2O	$\Delta_a H_{cal}^0$ kJ mol^{-1} H_2O	Reference
Albumin	0.03	13.9	15.5	[18, 22]
	0.06	5.7	11.7	
	0.09	–	7.5	
	0.12	3.2	7.5	
Collagen	0 –0.4	17.4	18.9	[18, 24]
	0.28	12.0	6.8	[18, 23]
	0.35	11.2[a]	10.1	
	0 –0.12	28.2[a]	D) 35.1/28.7 (S)	[18, 11]
	0.12–0.24	7.1[a]	D) 8.9/17.7 (S)	
	0.24–0.38	0	0	

[a] Averages of $\Delta_a H_i^0$ for the corresponding range of water content   D=heats of desorption, S=heats of sorption.

hydration dependence of the $\Delta_a H_i^0$ (and $\Delta_a S^0$) terms (Fig. 1a) is identical in all systems investigated: In the low humidity range (A'=0 to approx. 0.06–0.1 g H_2O g^{-1} polymer), pronounced anomalies (maxima or breaks) are observable. Above this critical water content, the thermodynamic parameters of biopolymer hydration decrease monotonically with increasing degree of hydration, and reach the condensation values at h approx. 0.3 to 0.5 g H_2O g^{-1} polymer, i.e., at completion of the primary hydration "shell" (region A in Fig. 1). These common features of the hydration dependence of the enthalpies and entropies of water sorption indicate (a) that only the interaction of the primary hydration water with the biopolymer surface is thermodynamically different from a normal condensation process, and (b) that structural changes occur in the biopolymer subsystem, in a critical, low humidity range (A'), which closely corresponds to the strongly bound "monolayer" hydration water. A further common feature of the thermodynamic parameters of biopolymer hydration is a pronounced, statistically significant enthalpy-entropy compensation pattern, as described previously [8].

   3. The enthalpies and temperatures of fusion of water in binary biopolymer water systems (Table 2, Fig. 1c) show the following general features: In a water content range of 0 to approximately 0.3–0.5 g H_2O g^{-1} polymer (region A), no phase transitions at and below 273 K are observable, i.e., this water fraction is unfreezable until approximately 200 K. This result, which is quite general for all biopolymer-water systems so far investigated, agrees well with the thermodynamic parameters discussed in the preceding section, which indicate that this type of hydration water interacts strongly with polar and charged biopolymer surface groups. At higher degrees of hydration (region B, where h is approx. 0.3 to 0.6–0.8), the water is freezable, but with both enthalpies and temperatures of fusion significantly reduced compared to the bulk values. This water fraction corresponds to the secondary hydration "shell", whose interaction with the biopolymer surface was found to be thermodynamically comparable to a condensation process. At water contents from h=0.6 to approximately 1.0, a third phase

of hydration water is observable (fraction B′). Its enthalpy of fusion corresponds to the bulk value, but the temperatures of fusion are still depressed by 5 to 10 K. At still higher water contents the properties of the biopolymer hydration water become undistinguishable from those of the bulk liquid.

Detailed experimental data in this field of research are rather scarce, and more quantitative detailed studies are required. It appears, however, that both the amount of unfreezable water and the extent of the reduction in $\Delta_f H$ and T_f of the freezable water fractions are dependent on the polarity of the biopolymer surface (elastin, with a very low content in polar side chains does not measurably affect the physical properties of the secondary hydration water, while in collagen and DNA-water systems, the physical properties of the secondary hydration water are strongly influenced by the polar and charged biopolymer surface groups).

4. The heat capacities of a series of biopolymer-water systems (Table 2) have been investigated by several authors over a wide range of composition. The data reported in the literature show in agreement that different hydration regions can be distinguished by characteristic, hydration-dependent heat capacity changes (Fig. 1 b).

In the most detailed study [13], four different hydration ranges are described:

In region A′ (h 0 to 0.1) a heat capacity anomaly at h ∼ 0.05 is observable, which indicates the occurrence of structural changes in the biopolymer subsystem. In region A (h 0.1 to approx. 0.3) the heat capacity increases linearly with water content, indicating that further structural changes are absent in the completion of the primary hydration "shell". In region B the biopolymer surface is fully hydrated. At water contents above h ∼ 0.4, the heat capacity of biopolymer water systems increases linearly with water content, up to infinite dilution. This important result suggests that the biopolymer is approximately the same in the fully hydrated solid state and in dilute solution.

5. The specific volume of biopolymer-water systems has been investigated by several authors over a wide range of composition (Table 3). As shown in Fig. 1 d, the specific volume of the binary system is a linear function of water content above h ≃ 0.1, up to infinite dilution. At the transition from solution to the solid state (approx. at $W_2 = 0.2$), the partial specific volume of the biopolymer slightly decreases (by approx. 0.01 1 kg^{-1}). This decrease is largely compensated for by an increase in the partial specific volume of the water subsystem. In the low humidity range A′ (h ≤ 0.1 g H_2O g^{-1} polymer) the specific volume shows a significant increase in all systems investigated (serum and ovalbumin, lactoglobulin and wool). The increase in the partial specific volume of the biopolymer ($\Delta \bar{v}_2$) amounts to 0.03 to 0.05 1 kg^{-1} upon removal of water fraction A′. It appears justified to correlate this anomaly in the v vs. h functions to the anomalies observed in the $\Delta_a H_i^0$ and cp vs. h functions, and to attribute it to structural changes in the biopolymer subsystem.

The results obtained by entirely different experimental approaches thus show good agreement in their description of biopolymer hydration events. Knowledge of the energetics of biopolymer-water interactions is, however, still in an incomplete stage of development. Future studies in this field of research, which is essential for an improved understanding of the biological functions of water, are therefore highly desirable.

References

1. Edsall IT, Mc Kenzie HA (1983) Adv Biophys 16:53–183
2. Kuntz ID, Kauzmann W (1974) Adv Protein Chem 28:239–345
3. Rupley JA, Yang PH, Gordon T (1980) In: Rowland SP (ed) ACS Symp Ser 127. Water in polymers. Am Chem Soc Wash DC, pp 111–132
4. Gál S (1967) In: Mayer-Kaupp H (ed) Die Methodik der Wasserdampf Sorptionsmessungen. Springer, Berlin Heidelberg New York
5. Bolliger W, Gál S, Signer R (1972) Helv Chim Acta 55:2639–63
6. Rao PB, Bryan WP (1978) Biopolymers 17:1957–1972
7. Bryan WP (1980) J Theor Biol 87:639–661
8. Lüscher-Mattli M, Rüegg M (1982) Biopolymers 21:403–418, 419–429
9. Lüscher-Mattli, unpublished results
10. Spink C, Wadsö I (1976) In: Glick D (ed) Methods of biochemical analysis. Wiley, New York
11. Pineri MH, Escoubes M, Roche G (1978) Biopolymers 17:2799–2815
12. Suurkuusk J (1974) Acta Chem Scand B 28:409–417
13. Yang PH, Rupley JA (1979) Biochemistry 18:2654–2661
14. Haly AR, Snaith JW (1968) Biopolymers 6:1355–1377
15. Picker P, Leduc PA, Philip PR, Desnoyers JE (1971) J Chem Thermodynam 3:631
16. Low BW, Richards FM (1952) Nature 170:412
17. Low BW, Richards FM (1952) J Am Chem Soc 74:1660
18. Bull HB (1944) J Am Chem Soc 66:1499–1507
19. Lüscher M, Rüegg M, Schindler PW (1978) Biochim Biophys Acta 533:428–439
20. Mc Laren AD, Rowen JW (1951) J Polymer Sci 7:289–324
21. Frey HJ, Moore WJ (1948) J Am Chem Soc 70:3644–49
22. Amberg CH (1957) J Am Chem Soc 79:3980–84
23. Berlin E, Kliman PG, Pallansch MJ (1979) J Coll Interface Sci 34:488–494
24. Kanagy JR (1954) J Am Leather Chem Assoc 49:646–658
25. Rüegg M, Lüscher M, Blanc B (1974) J Dairy Sci 57:387–393
26. Lüscher M, Schindler P, Rüegg M (1979) Biopolymers 18:1775–1791
27. Haly AR, Snaith JW (1971) Biopolymers 10:1681–1699
28. Mrevlishvili GM, Privalov PL (1969) In: Kayushin LP (ed) Water in biological systems. Plenum, New York, pp 63–66
29. Ceccorulli G, Scandola M, Pezzin G (1977) Biopolymers 16:1505–1512
30. Rüegg M, Moor U, Blanc B (1975) Biochim Biophys Acta 400:334–342
31. Mrevlishvili GM, Syrnikov YP (1974) Stud Biophys 43:155–170
32. Eisenberg D, Kauzmann W (1969) In: Eisenberg D, Kauzmann W (eds) The structure and properties of water. Oxford University Press, pp 106, 178
33. Berlin E, Kliman PG, Pallansch MJ (1972) Thermochim Acta 4:11–16
34. Bull HB, Breeze K (1968) Arch Biochem Biophys 128:497–502
35. Hutchens JO, Cole AG, Stout JW (1969) J Biol Chem 244:26–32
36. Millero FJ, Ward GK, Chetirkin P (1976) J Biol Chem 251:4001–4004
37. Bernhard J, Pauly H (1975) J Phys Chem 79:584–589
38. Bull HB, Breeze K (1979) Arch Biochem Biophys 197:199–204
39. Low BW, Richards FM (1954) J Am Chem Soc 76:2511–2518
40. Bull HB, Breeze K (1968) J Phys Chem 72:1817–1819
41. Neurath H, Bull HB (1936) J Biol Chem 115:519–528
42. Chick H, Martin CJ (1913) Biochem J 7:92–96
43. McMeekin TL, Groves ML, Hipp NJ (1954) J Polymer Sci 12:309–315
44. King AT (1926) J Textile Ind 12:T53–67

Section V   **Phase Changes**

Chapter 12   The Formation of Micelles

HEINZ HOFFMANN and WERNER ULBRICHT

Symbols

cmc: critical micelle concentration

$\Delta\mu^0$: standard change in chemical potential

μ_1^0: chemical potential of monomers

μ_N^0: chemical potential of aggregates having aggregation number N

K: equilibrium constant

ΔH^0: standard enthalpy change

ΔV^0: standard volume change

ϕ_y: apparent molal quantity

Y: experimental observable Y of solution (volume, heat capacity etc.)

Y_{01}: corresponding molal value of observable Y for pure solvent (water)

Y_2: partial molal quantity of observable Y for component 2

Y_2^0: partial molal quantity of observable Y for component 2 extrapolated to zero concentration

n_i: number of moles of component i

m: molality

E_2^0: expansibility at infinite dilution

K_2^0: compressibility at infinite dilution

ΔG_m^0: standard Gibbs energy of micellization

ΔH_m^0: standard enthalpy of micellization

ΔS_m^0: standard entropy of micellization

τ_1, τ_2: relaxation times

τ_r: residence time of a monomer inside a micelle

k^-: rate constant for a monomer leaving a micelle

k^+: rate constant for a monomer entering a micelle

ΔH: enthalpy of inserting a monomer into a micelle

ΔH_r: enthalpy of formation of micellar nuclei

$\Delta C_{p,m}$: heat capacity change of micelle formation

ΔE_m: expansibility change of micellization

ΔK_m: compressibility change of micellization

ΔV_m: molar volume change of micellization

ΔH_m: enthalpy change of micellization

ΔS_m: entropy change of micellization

n: aggregation number

σ: variance of micellar distribution curve

1 Behavior of Detergents in Water

The dissolution process of substances in a solvent can be described by three steps: In the first step the substances must be molecularly dispersed; in the second, holes have to be formed in the interior of the solvent, and in the third, the dispersed molecules are transferred into the holes. For the first step a positive enthalpy similar to the enthalpy of vaporization is required comprising an entropy increase, in the second step a new interface is formed which requires surface energy, while in the third step the adhesion of the dispersed molecules to the solvent molecules has to be considered. This last process is associated with a negative enthalpy of solvation and a decrease of the entropy. The enthalpy and entropy values for these three steps can be quite different, depending on the nature of the solute and the solvent; this renders it difficult to calculate reliable values for the Gibbs energy of the overall process of dissolution. However, it is possible to derive qualitative rules for the solubility of certain kinds of substance in different solvents. Let us consider, for example, the solution of hydrocarbons in water. The enthalpy of vaporization is positive, but small in comparison to that of salts, due to the small cohesion forces between hydrocarbon molecules. Formation of holes in water requires a large positive Gibbs energy to overcome the strong cohesion forces between the water molecules. In the concave surface around the holes, the water molecules are more strongly ordered than in the bulk water phase, leading to the large entropy decrease of hole formation [1]. The adhesion of the hydrocarbon molecules to the water molecules leads to a gain in enthalpy comparable in magnitude to the positive enthalpy for the other processes since the overall solvation process of short-chain hydrocarbons (up to 6 C-atoms) takes place with a small negative enthalpy (Table 1). The entropy for the adhesion process is always negative. The total enthalpy of solvation is thus usually small; it is slightly negative for short-chain hydrocarbons and becomes progressively less negative with increasing chain length, finally turning positive for hydrocarbons with more than 6 C-atoms in the hydrocarbon chain. The entropy of the solvation process is strongly negative, due to the orientation of the water molecules around the hydrocarbon; in analogy to the solvation of ions this effect is called hydrophobic solva-

Table 1. Values for the Gibbs energy ΔG^0, the enthalpy ΔH^0, the entropy ΔS^0 and the heat capacity ΔC_p for the transfer of hydrocarbons from their pure liquid phase into water at 25 °C and the heat capacity C_p of the pure liquid hydrocarbons. [2]

Hydrocarbon	ΔG^0 (J mol^{-1})	ΔH^0 (J mol^{-1})	ΔS^0 (J/K · mol)	ΔC_p (J/K · mol)	C_p (J/K · mol)
Ethane C_2H_6	16300	−10500	− 88	−	−
Propane C_3H_8	20500	− 7100	− 92	−	−
Butane C_4H_{10}	24700	− 3300	− 96	272	142
Pentane C_5H_{12}	28700	− 2100	−105	402	172
Hexane C_6H_{14}	32400	± 0	−109	440	197
Benzene C_6H_6	19300	+ 2100	− 59	226	134
Toluene $C_6H_5CH_3$	22600	+ 1700	− 71	264	155
Ethylbenzene	26000	+ 2000	− 80	318	184
Propylbenzene	28900	+ 2300	− 88	394	−

tion [2]. As a consequence, the water tends to dispel a dispersed hydrocarbon, which leads to the formation of a separate hydrocarbon phase. However, there are no specific repulsive forces between the water and the hydrocarbon molecules. The poor solubility of hydrocarbons in water does not result from a large positive enthalpy of solution, but from an entropy decrease due to hydrophobic solvation. Table 1 summarizes the values for the enthalpies and entropies of solution for different hydrocarbons in water [3].

The situation is analoguous to that of ionic compounds in water. Indeed the molecular dispersion of salts requires a highly positive enthalpy of vaporization due to the strong electrostatic forces of attraction between the ions, but this enthalpy can be overcompensated by the large negative enthalpy for the solvation of the ions by the dipole molecules of the water. The enthalpy of solvation is so large for a great many ionic compounds that a negative Gibbs energy results for the overall process. Similar arguments hold for nonionic compounds with polar groups. These groups improve solubility of the compound in water as a result of their strong solvation.

Thus for surfactants which consist of a hydrophobic group – normally a hydrocarbon or perfluorhydrocarbon rest – and a hydrophilic group, ionic or nonionic, an unusual solution behavior in water is to be expected. Due to its strong solvation the hydrophilic group is in most cases able to prevent phase separation of detergent and water. The hydrocarbon residue stays in the aqueous medium; however the system tries to minimize the unfavorable water-hydrocarbon interface. There are two possible ways of achieving this objective. First, the surfactant molecules can be brought to the surface of the water: The hydrophilic group remains solvated by water, while the hydrophobic moiety is in contact with the surrounding air.

The surfactant molecules aggregate reversibly to form structures in which the hydrophobic moieties are sequestered from the water, while the hydrophilic groups at the surface of the aggregates are solvated and keep the aggregates in solution. Such aggregates are called micelles. The adsorption process of detergents is described in the literature [1, 4–6]; in the following text the formation of micelles is discussed in detail.

2 Aggregation Behavior of Surfactants

2.1 Determination of the Critical Micelle Concentration cmc

While adsorption of surfactants to the water surface can formally be regarded as a distribution of the surfactant between the bulk water phase and the hypothetical two-dimensional surface phase, and can take place at every surfactant concentration, the aggregation of detergents to micelles was found to show the following features: for concentrations below a characteristic concentration (cmc), the detergent molecules behave as normally dissolved solutes with no detectable association between the molecules [2]; under these conditions the concentration of detergent monomers (c_1) is equal to the total detergent concentration (c_0). Above this characteristic concentration the monomer concentration remains constant to a first approximation and the excess monomers form micelles, whose concentration increases linearly with further surfactant concentration. As no micelles are formed below this characteristic concentration, it is called the critical micelle concentration cmc. In solutions of ionic surfactants the counterion concentration increases with the concentration of monomers up to the cmc; beyond the cmc there is a slower increase with the total surfactant concentration because the major part of the counterions is associated with the highly charged micelles by electrostatic attraction.

Thus the cmc can be measured by methods sensitive either to the monomer or the counterion concentration or to that of micelles. There are consequently many methods available to measure the cmc. As early as 1966, already 71 different methods were cited in the literature [7]; only the most important and common techniques will be described here briefly.

Methods sensitive to the concentrations of the monomers or the counterions are measurements of electric conductivity [8], surface tension [9], EMF using suitable ion-sensitive electrodes [10], transference number [11], colligative properties such as vapor pressure depression [12], freezing point depression [13], or osmotic pressure [14]. The concentration of micelles is detected by density measurements [15], refractive index [16], change of the spectral absorbance [17] in the UV- and IR-region, solubilization of dyes in micelles [18], fluorescence depolarization [19], viscosity [20], light scattering or turbidity [21] and Krafft point solubility [22]. The Krafft point is a characteristic temperature at which the solubility of a detergent seems to increase abruptly. This can be explained as follows: some detergents are so poorly soluble in water that their saturation concentration stays below the cmc; thus no micelles can be formed. As the solubility increases with increasing temperature, the cmc will be reached at a certain temperature and the solid detergent is instantaneously dissolved due to the formation of micelles; this temperature is called the Krafft point.

Figure 1 shows schematically the concentration dependence of some physical properties of surfactant solutions. The cmc can readily be recognized as a break in the curves. For example, from the surface tension measurements it can be seen that the monomer concentration increases only up to the cmc and remains then approximately constant, because the surface tension decreases only until the cmc is reached and remains constant thereafter. This result demonstrates that only monomers, not micelles, are surface active which can be rationalized by the lack of hydrophobic solvation of the micelles.

Conductivity and potentiometric data show a slower increase of the counterion concentration above the cmc since the counterions associated with the micelles do not contribute to the conductivity or the EMF. Light scattering measurements show that below the cmc no strongly scattering micelles exist in the solutions; the micelles are formed just at the cmc and their concentration increases with the surfactant concentration.

Exact measurements of these properties in the cmc region show that the cmc is not represented by a break in the curve but by a transition region. This is demonstrated in Fig. 2 for conductivity measurements [23].

A table with all cmc values which have been published so far is beyond the scope of the present review. Until the end of 1966 thousands of cmc values were

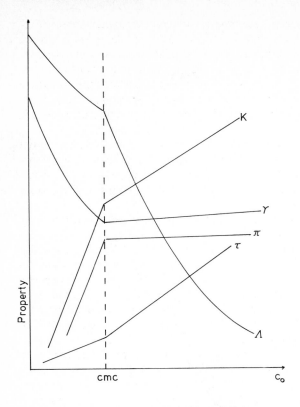

Fig. 1. Schematic plot of some physical properties of surfactant solutions as a function of surfactant concentration c_0; κ specific conductivity; γ surface tension, π osmotic pressure, τ turbidity, Λ equivalent conductivity

summarized for 721 surfactants in a monograph of Mukerjee and Mysels [7]. The monograph, *Critical Micelle Concentrations of Aqueous Surfactant Solutions* was published in 1971 by the National Bureau of Standards (US) in the series *National Standard Reference Data Series* and contains all cmc values available in the literature until the end of 1966. The reliability of the data and the purity of the surfactants investigated is critically reviewed; data of high quality are listed separately. The most important methods for cmc measurements are described, together with their applicability for certain kinds of surfactant. Possible errors involved in the cmc determination by specific methods have been discussed particularly with respect to the cmc as a transition region. Finally, for every entry the reference, together with the method by which the value has been obtained is cited. This work is a very useful data pool for the scientist working in the field of colloid chemistry.

2.2 Influence of the Nature of Surfactants and External Parameters on the cmc

The dependence of the cmc on various parameters has been described in the literature [6]. As general rule, increasing hydrophobicity or decreasing hydrophilicity of the corresponding groups favor formation of micelles, i.e. lower the cmc, and vice versa. This rule can be readily understood, because the hydrophobic

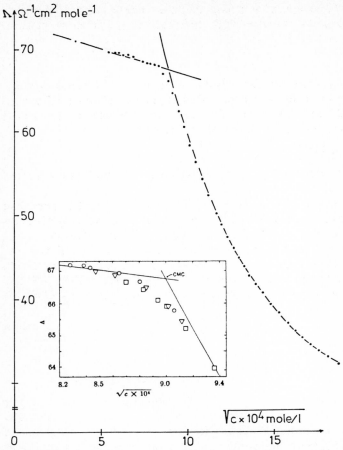

$\Lambda\uparrow\Omega^{-1}cm^2\,mol\,e^{-1}$

$\sqrt{c}\times10^4\,mole/l$

Fig. 2. Plot of the equivalent conductivity Λ as a function of the square root of the total concentration of sodium-n-dodecylsulfate at 25 °C with exact measurements around the cmc

solvation is the main reason for the formation of micelles, while the hydrophilic group prevents the separation of a hydrocarbon phase. An exact theoretical calculation of the cmc is for most of the surfactants not possible at present. But qualitative or semiquantitative relations exist between external parameters or the nature of the detergent and the cmc which permit an adequate estimation of the cmc for almost each detergent.

The most common relation describes the dependence of the cmc on the length of the alkyl chain of the hydrophobic moiety for n-alkyl-detergents. It has a similar form like Traube's rule [24] for the surface activity of n-alkyl-detergents as a function of the chain length and can be written in the form [25]

$$\log(cmc)=A-B\cdot m. \tag{1}$$

Here m is the chain length of the alkyl chain. A and B are empirical constants which are valid for classes of detergents with similar groups and do not possess universal character. The constant A expresses the influence of the hydrophilic group, while B represents the influence of the hydrophobic moiety and can be expressed by the equation

$$B = \omega/2,303 \cdot kT \qquad (2)$$

for nonionic detergents. Here ω represents the energy for the transfer of a CH_2-group from the hydrocarbon environment into the aqueous medium and has been found equal to 2.5 kJ mol^{-1}. For ionic detergents the influence of the electric work has to be taken into account and Eq. (2) becomes

$$B = \omega/(1 + K_g) \cdot 2,303 \cdot kT, \qquad (3)$$

where K_g is the slope of a plot of log (cmc) against the logarithm of the counterion concentration c_c.

Equation (1) can be easily understood if one assumes that with increasing chain length of the hydrophobic residue the unfavorable hydrocarbon-water interface increases and concomitantly the hydrophobic solvation. This favors the tendency to push the hydrocarbon out of the aqueous medium. It follows from Eq. (1) and (2) that each additional CH_2-group in the hydrophobic moiety should lead to a cmc depression of about the factor 3. As can be seen from Fig. 3, this is indeed the case for nonionic detergents and also for ionic detergents, when the

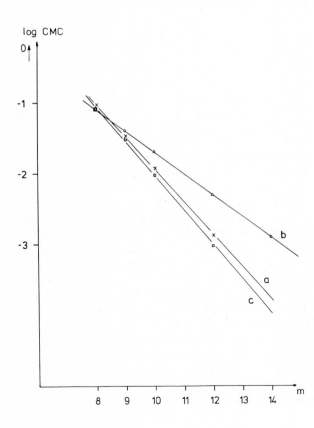

Fig. 3. Logarithmic plot of the cmc against the chain length m of the hydrocarbon tail for some selected surfactant systems at 25 °C; *a* Nonionic detergents (n-alkyldimethylaminoxides); *b* ionic detergents (sodium-n-alkylsulfates); *c* ionic detergents at constant ionic strength I=0.1 M (sodium-n-alkylsulfates with added amounts of NaCl)

influence of the electrostatic interaction is cancelled out by keeping the ionic strength constant by addition of electrolyte. Without additional electrolyte the cmc depression is considerably reduced for ionic detergents (curve b), as can also be seen in Eq. (3). This reduction occurs because the electrostatic repulsion between the charged molecules of ionic detergents impedes formation of micelles and leads to higher cmc values relative to corresponding nonionic surfactants with the same hydrophobic residue. Increase of ionic strength reduces the electrostatic repulsion, thus the cmc values for ionic detergents will approach those of nonionic surfactants. Short-chain detergents with ionic groups have a high cmc value and therefore the ionic strength at the cmc is high; thus the cmc values of such detergents are similar to those of corresponding nonionic surfactants. With increasing chain length the cmc and hence the ionic strength at the cmc decreases; thus the cmc values of long chain detergents without excess electrolyte must be considerably higher than those of nonionics having the same alkyl chain.

A large cmc depression results from replacing alkyl groups by perfluoro-alkyl groups. This behavior is due to the smaller interaction energy between the perfluoro-alkyl groups and water in comparison to the interaction between water and ordinary hydrocarbons. The weaker adhesion forces between water and perfluoro-hydrocarbons can also be seen from the stronger surface activity of the perfluoro-detergents and from the stronger reduction of surface tension of the solutions at the cmc compared with hydrocarbon surfactants. An empirical rule states that a perfluoro-detergent has about the same cmc as a hydrocarbon-detergent with twice the chain length of its alkyl group [26].

For other substituents in the alkyl chain it is found that hydrophobic substituents, for example halides, usually lead to modest cmc reduction, while hydrophilic substituents such as double bonds, hydroxyl groups, etc. result in an increase of cmc [25]. An aromatic group in the hydrophobic moiety has the same effect on the cmc as 3.5 additional CH_2-groups [27]. Branched alkyl chains lead to a smaller interface between water and the hydrocarbon compared to a surfactant having a straight alkyl chain with the same number of C-atoms. Thus the tendency for micelle formation is decreased, which means that for a detergent with a given number of C-atoms in the alkyl chain the cmc depends on the position of the hydrophilic group: the farther the hydrophilic group is located towards the middle of the alkyl chain, the higher is the cmc value [28].

Table 2 lists the cmc values of a series of selected surfactants; the surfactants were selected in such way that the various influences of the hydrophobic groups on the cmc can be seen.

The influence of the hydrophilic group is expressed in Eq. (1) by the parameter A. For ionic detergents this quantity is not strongly dependent on the nature of the hydrophilic group; thus the cmc does not depend greatly on the ionic hydrophilic group.

Nonionic detergents have, in most cases, considerably lower cmc values than ionic ones with the same hydrocarbon moiety. For nonionic surfactants having the common hydrophilic polyglycolether group, the cmc is known to increase somewhat with the increasing number of glycolether groups [29].

Table 3 gives the cmc values of selected surfactants which show the influence of the hydrophilic groups on the cmc.

Table 2. cmc Values for some selected surfactants with different constitution of the hydrophobic chain. [6, 7]

Detergent	T (°C)	cmc (mol l^{-1})
a) Influence of the chain length of the hydrocarbon or perfluorhydrocarbon		
Potassium-n-hexanoate $C_5H_{11}COOK$	25	1.6
Potassium-n-heptanoate $C_6H_{13}COOK$	25	7.5×10^{-1}
Potassium-n-octanoate $C_7H_{15}COOK$	25	4.0×10^{-1}
Potassium-n-nonanoate $C_8H_{17}COOK$	25	2.0×10^{-1}
Potassium-n-decanoate $C_9H_{19}COOK$	25	1.0×10^{-1}
Potassium-n-undecanoate $C_{10}H_{21}COOK$	25	5.0×10^{-2}
Potassium-n-dodecanoate $C_{11}H_{23}COOK$	25	2.5×10^{-2}
Potassium-n-tridecanoate $C_{12}H_{25}COOK$	25	1.3×10^{-2}
Potassium-n-tetradecanoate $C_{13}H_{27}COOK$	25	6.3×10^{-3}
Potassium-n-hexadecanoate $C_{15}H_{31}COOK$	35	1.8×10^{-3}
Potassium-n-octadecanoate $C_{17}H_{35}COOK$	55	4.5×10^{-4}
Potassium-n-perfluorhexanoate $C_5F_{11}COOK$	_a	5.0×10^{-1}
Potassium-n-perfluoroctanoate $C_7F_{15}COOK$	_a	2.7×10^{-2}
Potassium-n-perfluordecanoate $C_9F_{19}COOK$	_a	9.0×10^{-4}
n-Butyl-hexaoxyethylenealcohol $C_4H_9(OC_2H_4)_6OH$	25	7.8×10^{-1}
n-Hexyl-hexaoxyethylenealcohol $C_6H_{11}(OC_2H_4)_6OH$	25	7.0×10^{-2}
n-Octyl-hexaoxyethylenealcohol $C_8H_{17}(OC_2H_4)_6OH$	25	9.9×10^{-3}
n-Decyl-hexaoxyethylenealcohol $C_{10}H_{21}(OC_2H_4)_6OH$	25	9.0×10^{-4}
n-Dodecyl-hexaoxyethylenealcohol $C_{12}H_{25}(OC_2H_4)_6OH$	25	8.7×10^{-5}
n-Tetradecyl-hexaoxyethylenealcohol $C_{14}H_{29}(OC_2H_4)_6OH$	25	1.0×10^{-5}
n-Hexadecyl-hexaoxyethylenealcohol $C_{16}H_{33}(OC_2H_4)_6OH$	25	1.3×10^{-6}
b) Influence of the conformation of the hydrophobic chain (effect of unsaturation or substition)		
Potassium-n-octadecanoate	55	4.5×10^{-4}
Potassium-n-(cis-9)octadecenoate	50	1.2×10^{-3}
Potassium-n-(trans-9)octadecenoate	50	1.5×10^{-3}
Potassium-n-(9,10-dihydroxy)octadecanoate	60	7.5×10^{-3}
Potassium-n-(cis-9)(12-hydroxy)octadecenoate	55	3.6×10^{-3}
Potassium-n-(trans-9)(12-hydroxy)octadecenoate	55	5.5×10^{-3}
c) Influence of branched hydrophobic groups in a detergent molecule		
Sodium-n-tetradecyl-1-sulfate	40	2.2×10^{-3}
Sodium-n-tetradecyl-2-sulfate	40	3.4×10^{-3}
Sodium-n-tetradecyl-3-sulfate	40	4.3×10^{-3}
Sodium-n-tetradecyl-4-sulfate	40	5.2×10^{-3}
Sodium-n-tetradecyl-5-sulfate	40	6.8×10^{-3}
Sodium-n-tetradecyl-6-sulfate	40	8.3×10^{-3}
Sodium-n-tetradecyl-7-sulfate	40	9.7×10^{-3}

[a] Temperature not given in the reference.

The counterions of ionic surfactants do not greatly affect the cmc as long as counterions with a single charge are considered which bind to the micelles only by electrostatic attraction [30]. Bi- and polyvalent counterions lead to an increased ionic strength and hence to a cmc depression [31]. In the same manner the cmc is affected by counterions which are bound to the micelles not only by electrostatic attraction, but also by additional interactions, for example by charge-transfer interaction [32], by hydrophobic interaction [33], by hydrogen

Table 3. cmc Values for some selected n-dodecyl-surfactants and n-decyl-surfactants with different hydrophilic headgroups. [6, 7]

Detergent	$T\,(^\circ C)$	cmc (mol l^{-1})
Sodium-n-tridecanoate	25	1.3×10^{-2}
Sodium-n-dodecyl-1-sulfonate	25	9.8×10^{-3}
Sodium-n-dodecyl-1-sulfate	25	8.2×10^{-3}
n-Dodecylammoniumchloride	25	1.5×10^{-2}
n-Dodecyl-methylammoniumchloride	30	1.5×10^{-2}
n-Dodecyl-dimethylammoniumchloride	30	1.6×10^{-2}
n-Dodecyl-trimethylammoniumchloride	25	2.0×10^{-2}
n-Dodecyl-benzyldimethylammoniumchloride	25	7.8×10^{-3}
n-Dodecyl-pyridiniumchloride	25	1.5×10^{-2}
n-Dodecyl-quinoliniumbromide	25	4.8×10^{-3}
Di-Potassium-n-dodecylmalonate	25	5.0×10^{-2}
n-Decyl-dimethylamineoxide	25	1.9×10^{-2}
n-Decyl-dimethylphosphineoxide	30	4.6×10^{-3}
n-Decyl-β-D-glucoside	25	2.2×10^{-3}
n-Decyl-trioxyethylene-alcohol	25	6.0×10^{-4}
n-Decyl-pentaoxyethylene-alcohol	25	8.0×10^{-4}
n-Decyl-hexaoxyethylene-alcohol	25	9.0×10^{-4}
n-Decyl-nonaoxyethylene-alcohol	25	1.3×10^{-3}

Table 4. cmc Values for some selected ionic detergents with different counterions. [6, 7, 33b, 73a, 184, 185]

a) Anionic: n-Dodecyl-1-sulfate $C_{12}H_{25}SO_4^-$			b) Cationic: n-Dodecyl-N-pyridinium $C_{12}H_{25}NC_5H_5^+$		
Counterion	$T\,(^\circ C)$	cmc (mol/l)	Counterion	$T\,(^\circ C)$	cmc (mol/l)
Li^+	25	8.8×10^{-3}	Cl^-	25	1.5×10^{-2}
Na^+	25	8.2×10^{-3}	Br^-	25	1.2×10^{-2}
K^+	40	7.8×10^{-3}	J^-	25	5.3×10^{-3}
Cs^+	25	6.0×10^{-3}	$CH_3SO_3^-$	25	1.8×10^{-2}
Ag^+	35	7.3×10^{-3}	$C_2H_5SO_3^-$	25	1.6×10^{-2}
NH_4^+	30	6.5×10^{-3}	$C_3H_7SO_3^-$	25	1.3×10^{-2}
$CH_3NH_3^+$	25	5.7×10^{-3}	$C_4H_9SO_3^-$	25	9.4×10^{-3}
$C_2H_5NH_3^+$	25	5.0×10^{-3}	$C_5H_{11}SO_3^-$	25	5.6×10^{-3}
$C_4H_9NH_3^+$	25	2.9×10^{-3}	$C_6H_{13}SO_3^-$	25	3.4×10^{-3}
$C_6H_{13}NH_3^+$	25	1.1×10^{-3}	$C_7H_{15}SO_3^-$	25	2.2×10^{-3}
$C_8H_{17}NH_3^+$	25	2.8×10^{-4}	$C_6H_5SO_3$	25	4.8×10^{-3}
$(CH_3)_4N^+$	25	5.5×10^{-3}	$C_6H_5CO_2^-$	25	5.7×10^{-3}
$(C_2H_5)_4N^+$	25	4.5×10^{-3}	$p\text{-}NH_2\text{-}C_6H_4SO_3^-$	25	9.8×10^{-3}
$(C_4H_9)_4N^+$	25	1.3×10^{-3}	$p\text{-}CH_3\text{-}C_6H_4CO_2^-$	25	3.6×10^{-3}
Mg^{2+}	30	1.3×10^{-3}	$p\text{-}CH_3\text{-}C_6H_4SO_3^-$	25	2.9×10^{-3}
Ni^{2+}	30	1.2×10^{-3}	$o\text{-}HO\text{-}C_6H_4CO_2^-$	25	2.0×10^{-3}
Co^{2+}	30	1.2×10^{-3}	$CF_3CO_2^-$	25	8.0×10^{-3}
Cu^{2+}	30	1.2×10^{-3}	$C_2F_5CO_2^-$	25	4.6×10^{-3}
Mn^{2+}	25	1.1×10^{-3}	$C_3F_7CO_2^-$	25	2.9×10^{-3}
Zn^{2+}	40	1.1×10^{-3}			
Nonionic detergent for comparison			$C_{12}H_{25}(OC_2H_4)_5OH$	25	5.7×10^{-5}

log(cmc) mol/l

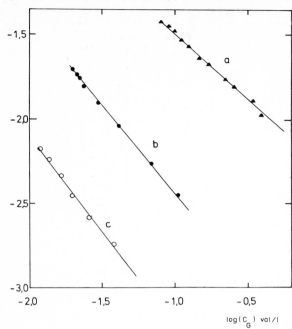

Fig. 4. Double logarithmic plot of the cmc as a function of the concentration c_G of the counterions for some potassium-n-alkylmalonates at 25 °C. *a* potassium-n-dodecylmalonate; *b* potassium-n-tetradecylmalonate; *c* potassium-n-hexadecylmalonate

log(C_G) val/l

bonds [34] or by attraction due to poorly hydrated counterions (Hoffmeister series [35]). These additional bonds between the ionic groups at the micellar surface and the counterions reduce the charge on the micellar surface further and therefore result in further decrease of cmc. Table 4 gives the cmc values of some surfactants, demonstrating the influence of various counterions on the cmc. The influence of electrostatic repulsion is also reduced for ionic detergents, if the ionic strength is increased by addition of an electrolyte. Thus addition of electrolytes leads to a cmc depression for ionic surfactants which increases with increasing ionic strength. This behavior can be quantitatively formulated by the equation [36]

$$\log (cmc) = -K_g \cdot \log (c_c) + const. \tag{4}$$

Here K_g is the constant of Eq. (3). In Fig. 4 the cmc for some detergents is plotted double logarithmically as a function of the counterion concentration; the slope of the plot yields the constant K_g. It can further be concluded from Eq. (4) that for ionic detergents, for which the counterion concentration increases smoothly above the cmc with the surfactant concentration, the cmc, and hence the concentration of monomers, should decrease slightly with increasing surfactant concentration.

The nature of the electrolyte should not affect the cmc depression as long as only counterions are present which bind to the micelles by electrostatic attraction. If various counterions are present in a surfactant solution, micelles are formed with those counterions having the strongest interaction. This behavior can be seen in Fig. 5.

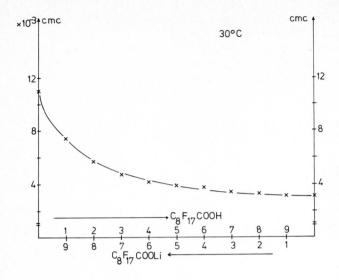

Fig. 5. Plot of the cmc for a binary mixture of surfactants ($C_8F_{17}CO_2H$ and $C_8F_{17}CO_2Li$) as a function of the molar ratio of the two components at 30 °C

The cmc values of nonionic detergents are very little affected by electrolytes, since there are at most only weak repulsive forces between the detergent molecules due to dipole-dipole interactions [37]. Thus the electrolyte can have only secondary effects on the cmc, for example due to the association of water molecules around the ions; i.e., the cmc depression by added electrolytes is generally small for nonionic surfactants and does not increase linearly with the ionic strength [38].

In this chapter the temperature dependence of the cmc is mentioned only briefly, as at is discussed in detail in the following chapter. For many surfactants the cmc values show a minimum at room temperature and are almost independent of temperature in this region [39]. With both increasing and decreasing temperature the cmc values increase slightly. In conclusion, it can be stated that using known cmc values and taking into account the dependence of the cmc on the various parameters discussed above, it is possible to estimate the cmc values for almost every surfactant with rather high accuracy.

3 Thermodynamic Parameters of Micelle Formation

3.1 General Considerations of the Thermodynamics of Micelle Formation

The thermodynamic functions ΔG^0, ΔH^0, and ΔS^0 for micelle formation are of particular interest for a quantitative understanding of the aggregation process. Although we have acquired considerable knowledge of the micellization process, the thermodynamics of micelle formation are far from being completely understood.

A large number of reliable cmc values are available for all kinds of detergent at various temperatures, which permit one to predict some values on the basis of phenomenological rules. However, even the relatively good agreement between experimental and estimated cmc values should not mislead one to believe that the thermodynamics of micelles is understood.

Employing an appropriately defined standard state, the cmc values provide a measure of the change of the chemical potential for the micellization process. Based on the phase model [40] for micelles one obtains

$$\Delta\mu^0 = RT \cdot \ln \text{(cmc)}, \tag{5}$$

and applying the mass action model [41] to the micellar aggregation process the equilibrium constant K for the aggregation step is calculated according to

$$A_{n-1} + A_1 \overset{K}{\rightleftharpoons} A_n$$

and

$$K = c_{A_n}/c_{A_1} \cdot c_{A_{n-1}}. \tag{6}$$

At the maximum of the micellar distribution curve c_{A_n} is equal to $c_{A_{n-1}}$ and one obtains

$$K = 1/\text{cmc}, \tag{7}$$

and furthermore with the use of the relation

$$K = e^{-\Delta\mu/RT} \tag{8}$$

the expression in Eq. (5) is obtained.

The problems begin when we try to split the change of the chemical potential $\Delta\mu_0$ into the heat of micellization ΔH^0 and the entropy of micellization ΔS^0 according to the Helmholtz-Gibbs equation $\Delta\mu^0 = \Delta H^0 - T \cdot \Delta S^0$. In early measurements the heat of micellization was determined in a calorimeter by measuring the integral heat of dilution of the aqueous detergent solution from a concentration above the cmc to a concentration below the cmc [42]. Other authors have obtained ΔH^0 values from the temperature dependence of the cmc using the Van't-Hoff equation

$$d \ln K/dT = \Delta H^0/RT^2, \tag{9}$$

where the equilibrium constant K is equal to the reciprocal cmc according to Eq. (7) [43]. From these enthalpy values the values for the standard entropies of micelle formation were calculated using the Helmholtz-Gibbs equation; the $\Delta\mu^0$ values were obtained using Eq. (5). Such values have been summarized by Shinoda [6] and by Kresheck [44] and are listed in Table 5 for some selected systems. For ionic detergents a factor n $(1 < n < 2)$ has been introduced to the product RT^2 in Eq. (9) correcting for the degree of association of the counterions to the micelles [45].

As can be seen in Table 5, the ΔH^0 values for most detergent systems are rather small, sometimes slightly positive and sometimes slightly negative, but always considerably smaller than $\Delta\mu^0$. It was concluded from these observations that micelle formation in aqueous solution is an entropy-controlled process. An aggregation process is expected to involve a negative entropy. Therefore it is assumed that the increase of the overall entropy results from the loss of the hydrophobic solvation of the hydrocarbon residues during aggregation. At room tem-

Table 5. Standard-values for the enthalpies ΔH^0, the entropies ΔS^0, the heat capacities ΔC_p^0 and the volumes ΔV^0 for the formation of micelles for some selected surfactant systems, summarized from literature data by Shinoda [6] and Kresheck [44]

Surfactant	T (°C)	ΔH^0 (J mol^{-1})	ΔS^0 (J/K·mol)	ΔC_p^0 (J/K·mol)	ΔV^0 (cm^3 mol^{-1})	Reference
n-C$_8$H$_{17}$SO$_4$Na	25	3350	–	–	–	[6]
n-C$_{10}$H$_{21}$SO$_4$Na	25	2090	–	–	–	[6]
n-C$_{12}$H$_{25}$SO$_4$Na	5	3350	–	–	–	[6]
n-C$_{12}$H$_{25}$SO$_4$Na	25	2500	–	–	–	[6]
n-C$_{12}$H$_{25}$SO$_4$Na	25	− 4190	–	–	–	[6]
n-C$_{12}$H$_{25}$SO$_4$Na	40	− 7540	–	–	–	[6]
n-C$_{12}$H$_{25}$SO$_4$Na	50	− 7960	–	–	–	[6]
n-C$_{12}$H$_{25}$SO$_4$Na	70	−20900	–	–	–	[6]
n-C$_{14}$H$_{29}$SO$_4$Na	40	− 6280	–	–	–	[6]
n-C$_{14}$H$_{29}$SO$_4$Na	70	−23000	–	–	–	[6]
n-C$_{11}$H$_{23}$CO$_2$K	25	10900	–	–	–	[6]
	25	− 4190	–	–	–	[6]
n-C$_{11}$H$_{23}$CO$_2$K	40	−10000	–	–	–	[6]
n-C$_{11}$H$_{23}$CO$_2$K	50	0	–	–	–	[6]
n-C$_{11}$H$_{23}$CO$_2$K	70	−20900	–	–	–	[6]
	70	−10000	–	–	–	[6]
n-C$_{12}$H$_{25}$NH$_3$Cl	35	− 5440	–	–	–	[6]
n-C$_8$H$_{17}$SO$_4$Na	25	3350	62	–	–	[44]
	25	6280	72	–	–	[44]
n-C$_{10}$H$_{21}$SO$_4$Na	25	2090	69	–	–	[44]
n-C$_{12}$H$_{25}$SO$_4$Na	25	380	74	−574	–	[44]
	25	2180	80	–	–	[44]
	25	− 1260	69	–	12.1	[44]
n-C$_{14}$H$_{29}$SO$_4$Na	25	–	–	–	15.9	[44]
n-C$_{10}$H$_{21}$N(CH$_3$)$_3$Br	25	420	62	−222	–	[44]
n-C$_{12}$H$_{25}$N(CH$_3$)$_3$Br	25	− 1260	65	−281	–	[44]
	25	− 1380	63	−423	–	[44]
n-C$_{12}$H$_{25}$-NC$_5$H$_5$-Br	25	− 4190	58	−314	–	[44]
n-C$_{12}$H$_{25}$-NC$_5$H$_5$-J	25	−12500	35	−410	–	[44]

perature the entropy increase of the released solvent appears to overcompensate the entropy reduction involved in the aggregation process. Table 5 shows that the ΔH^0 values obtained by different authors disagree in magnitude, sometimes even in sign. Recently Desnoyers et al. [46] carried out detailed thermodynamic measurements on a few systems and succeeded in explaining the reason for the discrepancies of the earlier measurements. These authors showed that the disagreement between the different research groups was based on the methods used to obtain the ΔH^0 values. ΔH^0 values derived from the temperature dependence of the cmc by a van't-Hoff plot can only be reliable if the aggregation number and, in the case of ionic detergents, the degree of dissociation of micelles and counterions do not vary with temperature, a condition which is almost never fulfilled. The same is true if volumes of micellization ΔV are determined from the pressure dependence of the cmc according to

$$d \ln K/dp = \Delta V^0/RT. \tag{10}$$

The calorimetrically measured ΔH^0 values proved to have been strongly affected by a dependence of the relative partial molal enthalpies of the detergents in water on the detergent concentration, as can be seen in Fig. 6. This thermodynamic quantity shows not only a rapid change in the cmc region during the micellization process, but also varies strongly with the concentration of the surfactant above

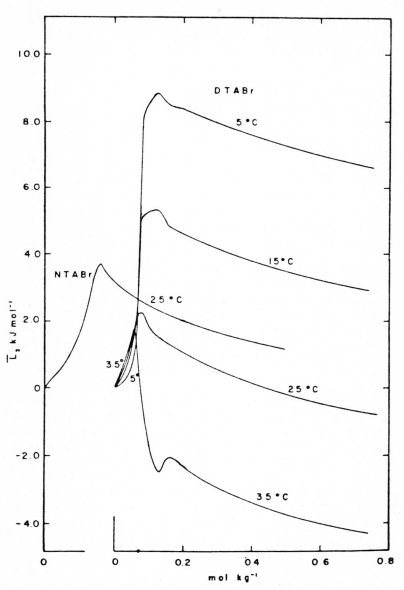

Fig. 6. Plot of the relative partial molal enthalpy L_2 of nonyltrimethylammoniumbromide and decyltrimethylammoniumbromide in water against the concentration at various temperatures

and below the cmc. Thus, if integral heats of dilution are measured by diluting the surfactant from concentrations above the cmc to concentrations below the cmc, totally different values, even different in sign, can be found, depending on the initial and the final concentration of the surfactant. The reason of the large concentration dependence of ΔH on the detergent concentration is not clear at present. Furthermore, as can also be seen in Fig. 6, the relative partial molal enthalpy is strongly temperature-dependent. From these results it can be concluded that it is necessary to measure partial molal quantities of detergents in aqueous solutions over a large concentration and temperature range to obtain reliable thermodynamic state functions for micelle formation.

3.2 Experimental Techniques for the Determination of the Thermodynamic Parameters of Micelle Formation

1. Free Energies. The determination of the nonideal free energy values for aqueous detergent systems requires the measurement of the activity of the solute or of the solvent. As reversible electrodes for the detection of cationic and anionic surfactants are now available [47], the activity of dilute surfactants can be measured directly using the EMF method. The activity of the water can be determined by measuring the vapor pressure of the solutions [48] or the freezing point depression [49]; the activity of the surfactant can be obtained by using the Gibbs-Duhem equation. These techniques are also suitable for ternary systems [50].

2. Enthalpies. Experimentally measurable are enthalpies of dilution or enthalpies of solution for surfactants in water, using microcalorimeters or precision solution calorimeters. Cell-type instruments [51] and continuously operating flow microcalorimeters [52] have been used for this purpose, and many direct calorimetric data for the enthalpy values of micellization have been determined in the past. As already mentioned, only a few data are actually reliable in view of the complexity of the concentration dependence of the excess enthalpies of most ionic detergents below and beyond the cmc.

3. Heat Capacities. Heat capacities can be derived from the variation of the enthalpy of dilution or the enthalpy of solution with temperature [53]. Direct measurements of heat capacities can be carried out with a flow microcalorimeter [52]. This technique yields heat capacities per unit volume; for the conversion of the data to specific heat capacities, the densities of the solutions must be known.

4. Volumes. Volumes are almost always determined from the density values of the solutions, which can be obtained by various techniques, such as pyknometers, float, sinker, or vibrating tube densitometers [54]. A flow version of the latter apparatus was also used very successfully [55] in series with a flow microcalorimeter.

5. Expansibilities. Expansibilities are measured either from the variation of the density with temperature or directly with a dilatometer [56]; for the latter technique a flow type has also been constructed [57].

6. *Compressibilities*. Compressibilities are conveniently obtained by measuring the velocity of sound in the solution [54]. From these measurements the adiabatic compressibility is obtained; the isothermal compressibility can be calculated from these data and the corresponding values for the expansibilities and heat capacities [58].

3.3 Partial Molal Quantities of Surfactants in Aqueous Solutions

The partial molal quantities very often cannot be measured directly. Thus apparent partial molal quantities Φ_y are measured which are defined as follows:

$$\Phi_y = (Y - n_1 \cdot Y_{01})/n_2. \tag{11}$$

Here Y is the experimental observable of the solution, Y_{01} the corresponding molal quantity of the pure solvent water, n_1 the number of moles of water and n_2 the number of moles of the surfactant. The partial molal quantity Y_2 of the surfactant can be obtained from the apparent molal quantity Φ_y and its concentration dependence according to

$$Y_2 = \Phi_y + m \cdot (\delta \Phi_y / \delta m)_{p, T}, \tag{12}$$

with m referring to the molality of the detergent. As these measurements must be carried out in a wide concentration range above and below the cmc, the measurements are limited to surfactants with alkyl chains shorter than 12-C atoms. The surfactants with longer alkyl chains have cmc values too small to render measurements of the apparent molal quantities below the cmc with the necessary accuracy. Under these restrictions the procedure for the measurement of the molal quantities is applicable for the case when Y is the volume V, the heat capacity C_p, the expansibility E, the compressibility K and the enthalpy ΔH; in the latter case only relative values ΔH_2 and no absolute values H_2 are obtained by measuring the heat of dilution of the surfactant solution as a function of detergent concentration. The relative Gibbs energies or the relative chemical potentials $\Delta \mu_2$ cannot be measured directly, as they tend to infinity at infinite dilution; the nonideal contributions μ_2^{ni} can be obtained from the activity coefficient f_2 of the detergent according to

$$\mu_2^{ni} = v \cdot RT \cdot \ln (f_2), \tag{13}$$

where v is the number of ions into which the detergent molecule dissociates in the monomeric state. The activity coefficient f_2 is in most cases obtained from the activity coefficient of the solvent water using the Gibbs-Duhem equation. The ideal contribution to the chemical potential can be calculated according to

$$\Delta \mu^i = v \cdot RT \cdot \ln (m). \tag{14}$$

The partial molal entropies S_2 also cannot be measured directly; they are obtained mainly as nonideal entropies s_2^{ni} from the relative enthalpies and the nonideal chemical potentials using the Helmholtz-Gibbs equation. If the ideal part of the chemical potential is also taken into account, relative partial molal entropies ΔS_2 are obtained.

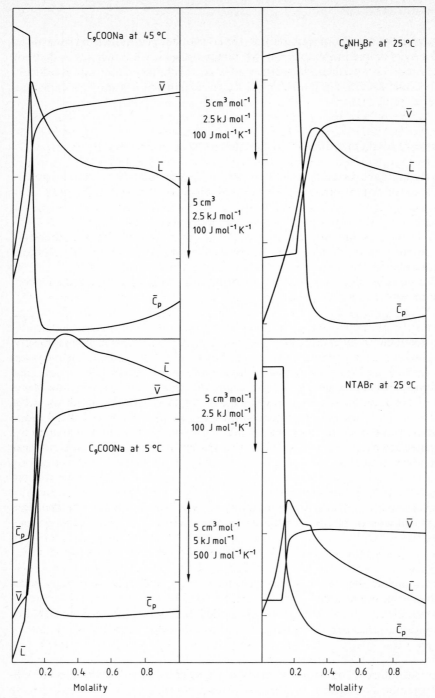

Fig. 7. Plot of the partial molal functions (volume $\bar{v}$, heat capacity $\bar{C}_p$ and relative enthalpy $\bar{L}$) for some selected detergent systems in water against the concentration at various temperatures (sodium-n-decanoate C_9COONa, octylammoniumbromide C_8NH_3Br and nonyltrimethylam-moniumbromide NTABr)

The results obtained show the following general features: the partial molal volumes V_2, the expansibilities E_2, the compressibilities K_2 and the heat capacities C_{p2} are only weakly dependent on the detergent concentration above and below the cmc, but change considerably in the cmc region. This is shown in Fig. 7 for a series of detergents.

As can be seen in Fig. 7, the main change does not occur as a discontinuity at the cmc, but within a small concentration range, which becomes narrower with increasing chain length of the hydrophobic rest. The data for V_2, E_2, K_2, and C_{p2} can be extrapolated for concentrations below the cmc to zero concentration and yield thus standard data V_2^0, E_2^0, K_2^0, and C_{p2}^0. These values reflect the interaction between detergent molecules and water and show an excellent group additivity [59], which means that these parameters increase with a constant increment for each additional CH_2 group. As an example, the increment values for a CH_2-group in aqueous medium are $C_{p2}^0 = 88$ J/K $\cdot$ mol and $V_2^0 = 16.0$ cm^3 mol^{-1}, while the corresponding values in a hydrocarbon medium are 29 J/K $\cdot$ mol and 16.3 cm^3 mol^{-1}.

In Table 6, values for V_2^0, C_{p2}^0, E_2^0, and K_2^0 are listed for some selected detergent systems [46c, 46e]. It can easily be seen from Table 6 that the values for C_{p2}^0 are significantly higher in the aqueous medium in comparison to corresponding values for similar compounds in a nonpolar medium, while the values for the volumes and hence also for the expansibilities and compressibilities are smaller in the aqueous medium. It is generally accepted that this behavior is due to the hydrophobic solvation of the alkyl chains of the surfactant molecules and not to an aggregation of the detergent molecules below the cmc [2]. Especially the lower values of the molal volumes reflect the high order and the tighter packing of the water molecules around the hydrocarbon chains. Most of the partial molal quantities of the surfactants in water show the following concentration behavior: in the premicellar region, i.e., below the cmc, the concentration dependence can be expressed by the equation

$$Y_2 = Y_{02} + A_y \cdot m^{1/2} + B_y \cdot m + \ldots. \tag{15}$$

The constant A_y is determined by the Debye-Hückel theory and reflects the ionic interaction between the detergent ions and the counterions; it is consequently zero for nonelectrolytes. B_y is due to other than electrostatic interactions, mainly to hydrophobic interactions between the hydrocarbon chain and the water molecules; some B_y values are included in Table 7. These values show trends similar to corresponding B_y values of nonelectrolytes with hydrophobic groups in an aqueous medium which do not form micelles [60].

At the cmc the quantities C_{p2}, V_2, E_2, and K_2 change rapidly in a narrow concentration range, as can be seen in Fig. 7. The C_{p2} values decrease significantly, while the other values increase; above the cmc, in the postmicellar region, all quantities have values which are typical for pure liquid hydrocarbons and which are almost independent of concentration. These changes reflect mainly the loss of hydrophobic solvation during the micellization process.

It is more difficult to interpret the ΔH_2 values. At low temperatures they rise steeply at the cmc region, pass over a maximum with increasing concentration and decrease again. At higher temperatures, as shown in Fig. 6, the rise at the cmc

Table 6. Values for the partial molal volumes V_2^0, heat capacities C_{p2}^0, expansibilities E_2^0 and compressibilities K_2^0 at infinite dilution and the corresponding coefficients B_y from Eq. (15) for some surfactants in aqueous solutions at various temperatures. [46c, 46e]

n-Octylammoniumbromide

T (°C)	5	15	25	35	45
V_2^0 $(\mathrm{cm^3\,mol^{-1}})$	169.50	171.62	173.72	175.18	177.54
B_V $(\mathrm{cm^3\,kg\,mol^{-2}})$	− 2.7	− 2.1	− 1.8	− 0.5	− 0.6
C_{p2}^0 $(\mathrm{J\,mol^{-1}\,K^{-1}})$	606	620	622	612	614
B_{C_p} $(\mathrm{J\,kg\,mol^{-2}\,K^{-1}})$	2.5	2.5	− 16	− 45	−130
cmc $(\mathrm{mol\,kg^{-1}})$	0.24	0.22	0.22	0.23	0.26

n-Nonyltrimethylammoniumbromide

T (°C)	5	15	25	35	45
V_2^0 $(\mathrm{cm^3\,mol^{-1}})$	235.52	238.05	240.71	243.45	245.95
B_V $(\mathrm{cm^3\,kg\,mol^{-2}})$	− 7.7	− 3.9	− 3.5	− 3.0	− 1.5
C_{p2}^0 $(\mathrm{J\,mol^{-1}\,K^{-1}})$	745	780	804	804	809
B_{C_p} $(\mathrm{J\,kg\,mol^{-2}\,K^{-1}})$	300	130	0	−140	−260
E_2^0 $(\mathrm{cm^3\,mol^{-1}\,K^{-1}})$	0.25	0.27	0.27	0.25	
cmc $(\mathrm{mol\,kg^{-1}})$	0.150	0.146	0.139	0.138	0.142

For 25 °C: $K_2^0 = -14.2 \times 10^{-4}\,\mathrm{cm^3\,mol^{-1}\,bar^{-1}}$, $B_K = 59\,\mathrm{cm^3\,kg^{-1}\,mol^{-2}\,bar^{-1}}$

n-Decyltrimethylammoniumbromide

T (°C)	5	15	25
V_2^0 $(\mathrm{cm^3\,mol^{-1}})$	250.44	253.62	256.57
B_V $(\mathrm{cm^3\,kg^{-1}\,mol^{-2}})$	− 4.3	− 3.3	− 3.6
C_{p2}^0 $(\mathrm{J\,mol^{-1}\,K^{-1}})$	832	874	873
B_{C_p} $(\mathrm{J\,kg^{-1}\,mol^{-2}\,K^{-1}})$	560	171	217
E_2^0 $(\mathrm{cm^3\,mol^{-1}\,K^{-1}})$	0.32	0.30	
cmc $(\mathrm{mol\,kg^{-1}})$	0.069	0.058	0.057

For 25 °C: $K_2^0 = -16.4 \times 10^{-4}\,\mathrm{cm^3\,mol^{-1}\,bar^{-1}}$, $B_K = 130\,\mathrm{cm^3\,kg\,mol^{-2}\,bar^{-1}}$

Sodium-n-decanoate

T (°C)	5	15	25	35	45
V_2^0 $(\mathrm{cm^3\,mol^{-1}})$	157.36	161.41	164.08	–	168.54
B_V $(\mathrm{cm^3\,kg^{-1}\,mol^{-2}})$	6.4	4.9	3.0	–	16.0
C_{p2}^0 $(\mathrm{J\,mol^{-1}\,K^{-1}})$	720.5	759.8	770	–	762
B_{C_p} $(\mathrm{J\,kg^{-1}\,mol^{-1}\,K^{-1}})$	152	108	− 37.5	–	−114
cmc $(\mathrm{mol\,kg^{-1}})$	0.118	0.098	0.085	–	0.075

becomes smaller and above a certain temperature the ΔH_2 values decrease in the cmc region.

The nonideal Gibbs energy contributions, obtained from the activity coefficients of the solvent as described above, show a discontinuity at the cmc, while the ideal contributions calculated according to Eq. (14), show a steady behavior without anomalies at the cmc. Thus, as the entropies are obtained from the ΔH- and the $\Delta \mu$ values, the entropy values ΔS_2 show the opposite behavior to the ΔH values: at low temperatures they decrease in the cmc region; with increasing temperature the decrease becomes smaller and becomes an increase above a certain temperature.

Table 7. Values for the thermodynamic properties of micellization (heat capacities ΔC_{pm}, volumes ΔV_m, expansibilities ΔE_m, compressibilities ΔK_m, enthalpies ΔH_m and entropies ΔS_m) for some detergents in aqueous solutions at various temperatures. [46c, 46e]

n-Octylammoniumbromide

$T(°C)$		5	15	25	35	45
$\Delta V_m \,(cm^3\,mol^{-1})$		9.8	8.8	8.1	6.9	6.1
$\Delta C_{pm}\,(J\,mol^{-1}\,K^{-1})$		$-$ 349	$-$ 348	$-$ 332	$(-$ 310)	$(-$ 225)
$\Delta H_m\,(J\,mol^{-1})$	a	8700	5280	1975	(-1320)	(-4125)
	b	8860	5375	$-$	(-1235)	(-4025)
$\Delta S_m\,(J\,mol^{-1}\,K^{-1})$	e	31.9	18.7	6.6	$(-$ 4.0)	$(-$ 12.7)

n-Nonyltrimethylammoniumbromide

$T(°C)$		5	15	25	35	45
$\Delta V_m \,(cm^3\,mol^{-1})$		5.44	4.90	4.48	3.80	2.75
$\Delta C_{pm}\,(J\,mol^{-1}\,K^{-1})$		$-$ 390	$-$ 360	$-$ 330	$-$ 300	$-$ 240
$\Delta H_m\,(J\,mol^{-1})$	a	7100	3500	200	-3050	-5700
	b	7400	3650	$-$	-2950	-5650
$\Delta S_m\,(J\,mol^{-1}\,K^{-1})$	c	26	13	2	$-$ 10	-19
	d	27	13.8	$-$	$-$ 8.4	$-$ 17
	e	25.5	12.2	0.7	$-$ 9.9	$-$ 17.9

For 25 °C: $\Delta E_m = -0.05\,cm^3\,mol^{-1}\,K^{-1}$, $\Delta K_m = 77\,cm^3\,mol^{-1}\,bar^{-1}$

n-Decyltrimethylammoniumbromide

$T(°C)$		5	15	25	35	45
$\Delta V_m \,(cm^3\,mol^{-1})$		6.83	6.05	5.05	$-$	$-$
$\Delta C_{pm}\,(J\,mol^{-1}\,K^{-1})$		$-$ 465	$-$ 412	$-$ 390	$(-$ 390)	$(-$ 390)
$\Delta H_m\,(J\,mol^{-1})$	a	8700	4200	250	-3600	-7700
	b	8845	4260	$-$	-3650	-7550
$\Delta S_m\,(J\,mol^{-1}\,K^{-1})$	c	29	14	0	$-$ 12	$-$ 25
	d	29.2	13.7	$-$	$-$ 12.9	$-$ 25.2
	e	30.9	14.5	0.8	$-$ 11.7	$-$ 24.2

For 25 °C: $\Delta E_m = -0.11\,cm^3\,mol^{-1}\,K^{-1}$; $\Delta K_m = 103\,cm^3\,mol^{-1}\,bar^{-1}$

Sodium-n-decanoate

$T(°C)$		5	15	25	35	45
$\Delta V_m \,(cm^3\,mol^{-1})$		10.40	9.72	9.6	$-$	6.4
$\Delta C_{pm}\,(J\,mol^{-1}\,K^{-1})$		$-$ 470	$-$ 460	$-$ 430	$-$	$-$ 370
$\Delta H_m\,(J\,mol^{-1})$	a	17960	13400	8750	$-$	650
	b	17930	13200	$-$	$-$	750
$\Delta S_m\,(J\,mol^{-1}\,K^{-1})$	e	64.6	45.8	29.4	$-$	2.4

a Values obtained graphically from the ΔH_2-values.

b Values calculated according to $\Delta H_m^T = \Delta H_m^{298} + \int_{298}^{T} (\Delta C_{pm})dT$.

c Values obtained graphically from the S_2^{ni}- or from the $(S_2 - S_2^0)$-values.

d Values calculated according to $\Delta S_m^T = \Delta S_m^{298} + \int_{298}^{T} (\Delta C_{pm})dT$.

e Values calculated according to $\Delta S_m = \Delta H_m/T$ using the condition for equilibrium $\Delta G_m = 0$.

3.4 Thermodynamic Functions of Micelle Formation

The thermodynamic functions for the micellization process can now be obtained from the partial molal quantities by the following procedure: The plots of the quantity Y_2 against the concentration above and below the cmc are extrapolated to the concentration where micellization starts; the difference of the two values is the thermodynamic function for micelle formation. As can be seen from the figures, these extrapolations are easy for the C_{p2}-, V_2-, E_2-, and K_2-values, since these parameters are only slightly dependent on the surfactant concentration in these regions; the determination of the enthalpies and entropies of micellization is more problematic, because the ΔH_2- and ΔS_2 values are strongly dependent on the surfactant concentration above as well as below the cmc region.

It seems important to emphasize that the parameters obtained for micelle formation are, according to the method of their determination, no standard values, but values which are obtained at the concentration of the cmc of the detergent system. Again it can be seen from the figures that standard values for the ΔC_{pm}-, ΔV_m-, ΔE_m-, and ΔK_m values could be obtained without great difficulty; the plots of the corresponding partial molal quantities against the surfactant concentration can readily be extrapolated to a defined standard concentration. For the ΔH_m- and ΔS_m values it can be seen that an extrapolation to a defined standard concentration would change these values considerably. Also the extrapolation would not be unambiguous, because the concentration dependence of the ΔH_2- and ΔS_2 values is not linear. For these reasons no attempt was made to determine standard values for the micellization process and only the actual values at the cmc of the surfactants were evaluated from the partial molal quantities. A series of values is listed in Table 7 for some selected systems. The ΔG_m values thus obtained are normally zero, as is required for the equilibrium condition at the cmc. The values for ΔH_m and ΔS_m which are determined directly or from the heat capacities show satisfactory self-consistency.

Recently, an extensive review article containing thermodynamic quantities of micelle formation was published by Stenius et al. [61]. This article contains tables with values for ΔG_m, ΔH_m, ΔS_m, and ΔC_{pm} for a large number of detergent systems from the literature. The authors give brief descriptions of the methods by which these quantities have been determined and of the models for the micellar aggregation used for evaluation of the measurements. They point out that the thermodynamic quantities depend on the association model assumed for calculation, on the actual process taking place in the experiment for the determination of the corresponding quantity, and on the choice of standard states and concentration units. In these tables the original reference is always given, the original paper is recommended in every case, with a careful analysis of the validity of the aggregation model and the arguments for using a certain equation for evaluation of experimental data. This procedure is obviously necessary because the values given by different groups for thermodynamic quantities of the same surfactant vary considerably, as has been pointed out already for an example by Desnoyers [46]. This can also be seen in Table 8, where some selected data for the thermodynamic quantities for aqueous detergent systems taken from the review article are given. This disagreement need not mean that some of the data must be erro-

Table 8. Values for the thermodynamic quantities ΔG_m^0, ΔH_m^0, ΔS_m^0, and ΔC_{pm}^0 for some selected detergent systems in aqueous solution, taken from [61]

Detergent	T K	ΔC_{pm}^0 J/K·mol	ΔG_m^0 kJ mol^{-1}	ΔH_m^0 kJ mol^{-1}	ΔS_m^0 J/K·mol	Reference
Sodiumoctanoate	298	−360	−	−	−	[62]
Sodiumoctylsulfate	298	−325	−	−	−	[55b]
	283	−	−	9.16	−	[63]
	293	−	−	5.82	−	
	303	−	−	− 3.31	−	
	313	−	−	− 4.81	−	
	323	−	−	− 8.21	−	
	294	−	−16.53	5.94	76.43	[53]
	298	−	−16.54	3.77	71.56	
	303	−	−16.53	1.46	66.93	
	308	−	−16.53	− 2.83	58.38	
Sodiumdecylsulfate	298	−425	−	−	−	[55b]
	283	−	−	8.74	−	[63]
	293	−	−	3.18	−	
	303	−	−	− 0.75	−	
	313	−	−	− 5.15	−	
	323	−	−	− 9.87	−	
	294	−	−19.20	4.94	82.06	[53]
	298	−	−19.20	3.14	74.92	
	303	−	−19.20	1.30	67.62	
	308	−	−19.20	0.42	63.67	
Sodiumdodecylsulfate	298	−530	−	−	−	[55b]
	283	−	−	7.94	−	[63]
	293	−	−	3.85	−	
	303	−	−	− 3.31	−	
	313	−	−	− 7.32	−	
	323	−	−	−12.09	−	
	294	−	−23.84	1.09	84.76	[53]
	298	−	−23.84	− 2.13	72.82	
	303	−	−23.85	− 4.02	65.41	
	308	−	−23.85	− 7.07	54.44	
Octyltrimethylammoniumbromide	293	−	−	26.65	136.82	[64]
Nonyltrimethylammonium-bromide	278	−345	−	−	−	[65]
	288	−325	−	−	−	
	298	−295	−	−	−	[55b]
	308	−280	−	−	−	[65]
	318	−250	−	−	−	
Decyltrimethylammonium-bromide	298	−365	−	−	−	[53]
	298	−	−	− 3.89	43.10	[64]
Dodecyltrimethylammonium-bromide	274	−	−	− 0.67	66.11	[64]
	298	−423	−17.9	− 1.39	55.65	[66]
Dodecylpyridiniumchloride	298	−490	−20.6	1.88	75.6	[53]
	303	−	−20.6	− 1.53	62.9	
	308	−	−20.6	− 2.93	57.4	
	298	−	−	−21.4	− 3.35	[64]

Table 8 (continued)

Detergent	T K	ΔC_{pm}^0 J/K·mol	ΔG_m^0 kJ mol^{-1}	ΔH_m^0 kJ mol^{-1}	ΔS_m^0 J/K·mol	Refer- ence
Dodecylpyridiniumbromide	294	–	–	– 2.05	61.7	[53]
	298	– 385	–	– 3.47	56.1	
	303	–	–	– 5.56	48.3	
	308	–	–	– 7.32	41.8	
	298	–	–21.1	– 4.06	56.9	[67]
	303	–	–21.3	– 5.48	52.3	
	308	–	–21.6	– 6.57	48.5	
	313	–	–21.8	– 7.41	46.0	
	318	–	–22.0	– 7.95	44.3	
	323	–	–22.3	– 8.45	42.7	
	328	–	–22.5	– 8.83	41.6	
	333	–	–22.7	– 9.25	40.3	
	338	–	–22.9	– 9.79	38.6	
	343	–	–23.1	–10.63	36.3	
Dodecylpyridiniumiodide	293	–	–	–12.43	39.33	[64]
	298	–	–	–13.5	–	[68]
	303	–	–24.2	–14.5	31.8	
	308	–	–24.3	–16.7	24.7	

neous, because they refer to different experimental conditions or have been calculated on the basis of different standard states or theoretical models. It is therefore possible that the different data coincide if they are recalculated for identical experimental and theoretical conditions; the use of thermodynamic quantities for micelle formation from the literature therefore requires the citation of all conditions and models for which the data were obtained.

For example, the application of the phase separation model to an ionic surfactant with the surfactant ion A and the counterion B forming at the cmc a micelle from n monomers leads to an equilibrium constant or "solubility product" K_s which is given by

$$K_s^{-1} = f_A^m \cdot x_A^n \cdot f_B^m \cdot x_B^n, \tag{16}$$

where f_i and x_i are the activity coefficients and the mole fractions, respectively. Choosing as standard state for the solute ions the hypothetical pure solute in which the environment of each molecule is the same as at infinite dilution, i.e., $f_i = 1$ and $x_i = 1$, the standard Gibbs' energy of the micellization is

$$\Delta G_m^0 = RT \cdot \ln(f_\pm \cdot x_c)^2, \tag{17}$$

where x_c is the mole fraction of the monomers when the micellar phase "precipitates" and $f_\pm$ is the mean acitivity coefficient. Assuming $f_\pm$ equal to 1, one obtains

$$\Delta G_m^0 = 2 RT \cdot \ln(x_c), \tag{18}$$

where x_c is usually identified with the cmc.

The mass action model or multiple equilibrium model which regards micelle formation as an association of p counterions B and n detergent ions A to the micelle A_nB_p leads to Eq.(19) for ΔG_m assuming a hypothetical standard state with $f_i = 1$ and $x_i = 1$ for each species in the equilibrium:

$$\Delta G_m^0 = RT \cdot ((p/n) \cdot \ln(f_B \cdot x_B) + \ln(f_A \cdot x_A) + (1/n) \cdot \ln(f_m \cdot x_m/n)). \tag{19}$$

The index m indicates the quantities for the micellar species and p denotes the number of counterions bound to the micelles; (n–p) counterions which are free in the solution do not contribute to ΔG_m^0.

Under the assumptions $f_A = f_B = f_m = 1$, $(1/n) \cdot \ln(x_m/n) = 0$ and $x_A = x_B = x_c$, one obtains

$$\Delta G_m^0 = (1 + p/n) \cdot RT \cdot \ln(x_c), \tag{20}$$

where x_c is again identified with the cmc. It is evident from Eq.(20) that in the case when all counterions are bound to the micelle (p=n), this equation becomes identical with Eq.(18); when no counterions are bound to the micelles, a condition which is only fulfilled for nonionic detergents (p=0), one obtains

$$\Delta G_m^0 = RT \cdot \ln(x_c). \tag{21}$$

Each ΔG_m value from the literature can therefore only be used for thermodynamic calculations, if it is clearly stated which standard states were defined and how the value was obtained.

Other problems in the determination of thermodynamic quantities for the micellization process can arise from minimal uncertainty of the cmc, which can depend on the method of determination or from neglecting the intermicellar interaction. This is also pointed out in [61] and must also be taken when using the other quantities. ΔH_m values for micelle formation can principally be determined by three methods, (a) temperature dependence of the cmc by the use of the Van't Hoff equation [Eq.(9)], (b) direct calorimetric measurements and (c) kinetic measurements. The first two methods have been already described, the third will be explained in detail in Section 6.

The first method can lead to erroneous results because the degree of association and the aggregation numbers of micelles can vary with temperature. Also the direct calorimetric measurement of the ΔH_m values is not straightforward [46, 50], since the molal heat of dilution as a function of detergent concentration does not only show a jump at the cmc but also a strong concentration dependence above and below the cmc, as can be seen in Figs. 6 and 8. Hence determination of standard values at infinite dilution is often impossible and ΔH_m values at the cmc are given instead. The determination of calorimetric data from measurements of integral dilution heats from a concentration above the cmc to a concentration below the cmc can lead to completely different values, as pointed out previously. Thus ΔH_m values derived from integral heats of dilution can only be used if the concentrations at the beginning and end of the dilution are given.

The entropy values in [61] are obtained from the equation

$$\Delta S_m^0 = (\Delta H_m^0 - \Delta G_m^0)/T. \tag{22}$$

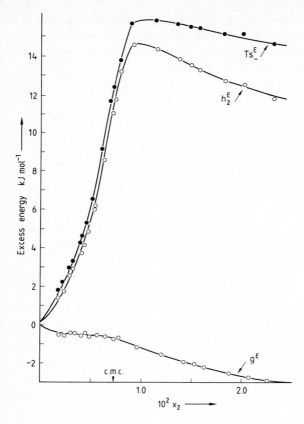

Fig. 8. Plot of the thermodynamic partial molal excess quantities as a function of the molar fraction for sodium octanoate in aqueous solution at 298 K [61]

Therefore for these values the same arguments apply as for the ΔH_m^0- and the ΔG_m^0 values.

The ΔC_{pm} values can only be determined from measurements of the partial molal quantities as a function of the surfactant concentration. The principally possible determination of these quantities from the equation $\delta(\Delta H_m)/\delta T = \Delta C_{pm}$ requires data of the second derivative of the cmc with the temperature which cannot be obtained with the necessary accuracy.

3.5 Molecular Interpretation of Thermodynamic Parameters

The large negative ΔC_{pm}- and the positive ΔV_m-, ΔE_m-, and ΔK_m values are mainly due to the loss of hydrophobic hydration when micelles are formed. The ΔH_m- and ΔS_m values are much more difficult to interpret. As can be seen, the ΔH_m values are positive at low temperatures and become negative at higher temperatures, while the opposite is true for the ΔS_m values. Thus micelle formation is an endothermic entropy-driven process at low temperatures. For high temperatures, micellization becomes increasingly exothermic and the entropy of micellization becomes negative. Therefore the process is no longer entropy- but enthalpy-controlled. As pointed out already, the thermodynamic functions for the transfer of a compound from a hydrocarbon to an aqueous environment can be split into contributions from several processes; all of which are large in size, but of opposite sign, so that the individual contributions can almost compensate for each other. The total transfer functions are therefore the result of differences between large quantities. The same arguments hold for the micellization process, which makes it extremely difficult to account quantitatively for absolute ΔH_m- and ΔS_m values. It can be argued that the contributions from the interface energies are becoming less important with increasing temperature because the interfacial tension decreases also with increasing temperature and would vanish at a critical temperature at which water and the hydrocarbon would become miscible. Consequently the negative entropy contribution resulting from association due to a reduction of the number of particles becomes dominant for the total entropy change.

It is worthwhile to add that heats for micellization have been also determined from kinetic measurements (see Sect. 6). These measurements have shown that the situation is even more complicated than assumed on the basis of the thermodynamic measurements. Micelles do not have one definite size but comprise a large range of aggregation numbers with a concentration maximum around a mean aggregation number. Therefore thermodynamic measurements, for example carried out by calorimetric techniques, give averaged parameters which contain contributions from all species present in the solution. The kinetic methods, on the contrary, allow differentiation between different species, especially between the species around the minimum and the species around the maximum of the distribution curve. Analysis of the kinetic data has shown that the enthalpy for the transfer of a hydrocarbon chain from the bulk water into the micellar aggregate is dependent on the aggregation number of the latter. The enthalpy change is large and positive for aggregates around the distribution minimum which comprises only a few monomers, and it becomes smaller with increasing aggregation number and is finally large and negative for abundant micelles. That means that the ΔH values change continuously from positive to negative with each consecutive association step. In a thermodynamic experiment one observes therefore only the integral heat of micellization over all steps and this integral value can either balance to zero or be positive or negative. As far we know, no attempt has been made at a model in which such data can be explained quantitatively.

4 Models for the Micellar Aggregation Process

4.1 Theoretical Considerations on Micellar Aggregation

Two models have been developed for the theoretical treatment of the micellar association process. In the phase model [40] the micelles are considered as a distinct phase. The chemical potential of a monomer in this phase has a well-defined value μ_M, which is independent of the size of the micelles or of the amount of surfactant material in this phase. The chemical potential of the monomers in the aqueous phase, on the other hand, is given by the sum of a standard value μ_0 and a concentration term $RT \cdot \ln (c_0)$ with the concentration c_0 of the surfactant. Increase of monomer concentration increases the chemical potential of the monomers until it reaches the value of μ_M at the cmc. At this point equilibrium is reached and the micellar phase appears; the cmc is thus understood as a saturation concentration. As μ_M is constant, the chemical potential of the monomers must also be constant above the cmc, a condition which is only fulfilled if the monomer concentration remains constant. The difference between μ_M and μ_0, the Gibbs standard enthalpy for the micellization process, is then given by Eq. (5).

This simple model is a good basis for a qualitative understanding of many properties of micellar solutions. It explains, for instance, why the surface tension of a surfactant solution remains practically constant above the cmc and why the same is true, in a first approximation, for the colligative properties of micellar solutions above the cmc. It follows that the water activity and hence the vapor pressure of water are also insensitive to the detergent concentration. In reality all these properties change somewhat above the cmc, however the changes are usually small enough to be ignored for practical application; they can be measured, however, without too much difficulty. The simple phase model cannot account for these small changes, and cannot provide information on the size of the micelles and their distributory functions.

For this purpose we must use the mass action model [41, 69], in which each micellar aggregate with a different aggregation number is considered as a distinct species, characterized by a chemical standard potential. As each micellar aggregate has to be in equilibrium with the monomers we can therefore write

$$\mu_1^0 + RT \cdot \ln (c_1) = \mu_N^0 + (RT/N) \cdot \ln (c_N/N), \tag{23}$$

where 1 is the index for the monomers and N the index for the aggregate with the aggregation number N. For c_N we can obtain

$$c_N = N \cdot c_1^N \cdot \exp (N \cdot (\mu_1^0 - \mu_N^0)/RT). \tag{24}$$

If we know μ_N^0 as a function of the aggregation number N, we can proceed to calculate c_N for different N values and hence we obtain micellar distribution curves. In order to achieve this objective, Tanford [2] and Ninham et al. [70] split the term μ_N^0 into three contributions, a bulk term g, an interfacial term $\gamma \cdot a$ and an electrostatic term $(2\pi \cdot e_0^2 \cdot d)/(\varepsilon \cdot a)$ and obtained the following:

$$\mu_N^0 = g + \gamma \cdot a + (2\pi \cdot e_0^2 \cdot d)/(\varepsilon \cdot a), \tag{25}$$

where γ is the interfacial tension at the micellar interface, a the surface area per hydrophilic headgroup at the micellar interface, ε the dielectricity constant, d the thickness of the electric double layer and e_0 the elementary charge. In this equation μ_N^0 must go through a minimum for a specific a value because the second term is proportional to a and the third is inverse proportional to a. Inherent in this model is the concept of opposing forces.

In the aggregation process the monomers attempt a state where all the water-hydrocarbon interfaces round the hydrophobic groups are eliminated. This is best realized in a bilayer, where practically all these interfaces have disappeared. On a globular micelle, however, the total micellar surface is more than twice as large as the area occupied by the hydrophilic headgroups, and consequently a large

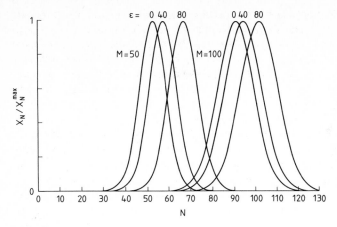

Fig. 9. Plot of the calculated aggregate concentrations X_N normalized to the maximum value X_N^{max} as a function of the aggregation number N for a detergent system assuming different dielectricity constants ε and maximum values M

area with a water-hydrocarbon interface still exists. Thus bilayers could be expected to be the most probable aggregates and spherical micelles should not exist; but, especially for ionic detergents, the closest approach of the headgroups is often prevented by sterical constraints, by electrostatic repulsion between the charged headgroups as expressed by the electrostatic term in Eq. (25) or sometimes, in the case of nonionic detergents, by the bulkiness of the headgroups.

Using these ideas, Ninham and coworkers were able to calculate micellar distribution functions for globular micelles, shown in Fig. 9. These authors used the same principles for the theoretical treatment of transitions in the micellar shape which were found to occur with increasing ionic strength due to addition of electrolytes or to an increase of detergent concentration. According to Ninham, globular micelles should first change to rod-like structures, then to disc-like, and finally to lamellar phases. These transitions should occur at characteristic concentrations which are often called higher cmc's, analogous to the critical micelle concentration.

Similar concepts were proposed by Ruckenstein et al. [71] on the basis of statistical mechanic considerations and later by Jönsson and Wennerström [72]. Ruckenstein et al. [73] succeeded in developing a concept for theoretical calculation of the cmc.

It should be emphasized here that the principles of the phase model and the mass action model are the same, as has been pointed out by Rushbrooke [74]. The simple phase model cannot provide information on the micellar distribution curves, as the chemical potential of a monomer in the micellar phase is assumed to be completely independent of the micellar constitution; it becomes a better approximation for micelles with very large aggregation numbers. There have also been objections to the phase model [75], based on the observation that the cmc is not a sharp transition point but a concentration region, and that the activity and hence the chemical potential of the monomers still changes slightly above the cmc, which should not be the case if the chemical potential of the monomers inside the micelles remained constant. On the other hand, Rusanov [76] has pointed out that these objections are incorrect, as the micelles must be regarded as small particles consisting only of comparatively few monomers; he obtained identical expressions for a theoretical calculation of the cmc using both the mass action model and a refined phase separation model. Many papers have been published dealing with the theoretical calculations of the cmc [77], and many of them have succeeded in predicting the cmc for certain kinds of surfactants.

4.1.1 Size and Shape of Micellar Aggregates

The size and shape of micelles have been discussed extensively. Hartley [40 a] proposed droplet-like spherical micelles with a liquid interior. Debye [21 c] added rod-like micelles and McBain finally postulated micelles which consist of a double membrane [78]. In the meantime, all three proposed models have been confirmed by experiments and it has been shown that a particular detergent can exist in all three types, depending on the conditions of solution. The critical parameter which controls the shape of the micelles is the surface area a which is required by a headgroup of the detergent at the micellar surface [2, 70]; the other two parameters which also influence the aggregation behavior of the surfactants, the volume v of an alkyl chain and the maximum length r of a detergent molecule, are given for a detergent monomer by its configuration, thus the only variable parameter is the surface area a. It follows from simple geometrical considerations [70] that globular micelles are favored if the condition

$$v/a \leqq r/3 \tag{26}$$

is fulfilled. In this case the maximum radius of the spherical micelle and hence its maximum aggregation number is given by the length r of the detergent molecule. For ionic detergents these maximum values are generally not yet reached because electrostatic repulsion between the headgroups prevents the micelles from growing to their full size. On addition of electrolyte or on increasing detergent concentration, the micelles can grow somewhat and therefore it is usually experimentally observed that the aggregation number of spherical micelles increases slightly with salt concentration or with increasing surfactant concentration [79].

   Neglecting this small increase due to the suppression of electrostatic repulsion, the size of spherical micelles does not changes as a function of detergent concentration in first approximation and hence the micellar concentration is found to increase linearly with increasing detergent concentration. Most normal ionic detergents, where the counterions are bound to the micellar interface only by electrostatic forces, show this aggregation behavior. When excess electrolyte is added to the detergent solutions or counterions are used which bind more strongly to the micelles due to effects already discussed in Section 2.2, the electrostatic repulsion between the headgroups is reduced and thus the area a which is required by the headgroups is also reduced. This can also occur simply by an increase of the surfactant concentration. Under these circumstances the headgroups on the micellar surface can move together more closely and hence the unfavorable water-hydrocarbon interface decreases. When the required area becomes smaller than a minimum value a_{min}, which is determined by the geometry of the detergent molecule, spherical micelles become thermodynamically unstable and rod-like micelles are formed which permit closer packing of the headgroups. This transition occurs at a well-defined electrolyte concentration or also at a certain detergent concentration, which is usually referred to as the cmc_{II} in analogy to the critical concentration cmc for the micelle formation. For rod-like micelles the geometrical constraint is

$$v/a \leqq r/2. \tag{27}$$

This implies the following rules: the small semi-axis is equal to the length of an extended detergent molecule and more or less independent of the detergent concentration, which is also the case for the radius of a spherical micelle, while the large semi-axis of the micelle and hence its aggregation number is increasing with increasing detergent concentration above the cmc_{II}. Such behavior has been confirmed for a number of detergent systems forming rod-like micelles [80]. From the theoretical treatment of the aggregation process, it is expected that the aggregation number n and hence the length L of the rods increases with the square root of the total detergent concentration above the cmc_{II} [70]. For a number of systems it was shown that L seems to increase even faster than expected in theory [81].

Even for rods, the hydrocarbon-water interface has not yet been eliminated completely. Therefore rods share a tendency to reduce this interface further. If by addition of electrolyte or by increasing detergent concentration electrostatic repulsion of the headgroups can be further reduced, the value of the required area a again becomes smaller than a critical value and rod-like structures become thermodynamically unstable. The stable structures under these conditions are disc-like micelles or lamellar bilayers. In this case the geometrical constraint is

$$v/a \leqq r. \tag{28}$$

The thickness of the discs or of the double layers is again determined by the length of the detergent molecules and independent of the concentration, while their large semi-axis increases with increasing detergent concentration. Systems under conditions for which the double layer is the thermodynamically stable state can very often be transformed into vesicles [82]. At present it does not seem to be clear, however, whether vesicles are the thermodynamically stable species or whether they transform back into lamellar layers.

4.1.2 Determination of Higher cmc's and Their Dependence on the Structure of the Detergent and External Parameters

In principle, higher cmc's can be determined using the same methods described for determination of cmc_I. However, methods which are only sensitive to the concentration of monomers are usually not suitable, because the monomer concentration does not change at the cmc_{II} in most cases. Therefore surface tension measurements are mostly not sensitive to higher cmc's with a few exceptions [83]. On the other hand, conductivity measurements are well suited for the determination of higher cmc's, because this method is sensitive to counterion concentration. The degree of dissociation of rod-like micelles is, however, usually smaller than that of spherical micelles. This leads to a second break in the plot of the conductivity against the detergent concentration [81, 84]; an example is shown in Fig. 10.

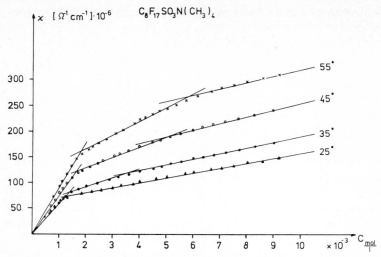

Fig. 10. Plot of the specific conductivity κ as a function of the total concentration for $C_8F_{17}SO_3N(CH_3)_4$ at various temperatures

Well suited for determination of higher cmc's are also methods which are sensitive to the concentration, the molecular weight or the assymetry of micelles. Such methods are rheological [85], light scattering [86], or fluorescence depolarization measurements [87]. In principle, also methods like dynamic light scattering, electric or streaming birefringence measurements (see later) are suitable to determine the higher cmc's; magnetic birefringence measurements have also been carried out to detect the cmc_{II} [88]. However, these transitions in the micellar shape occur very often at rather high detergent concentrations, where the micelles are subject to considerable intermicellar forces. These interactions can considerably influence the experimental results and it is sometimes difficult to distinguish between shape and interaction effects. For instance, light scattering intensity can be reduced by intermicellar repulsion to such an extent that it becomes smaller than in the case of spherical micelles. This can render it very difficult to recognize a sphere-rod transition. To some extent, the same applies to methods which are based on the determination of the asymmetry of micellar aggregates, for example by measuring the electric, magnetic or streaming birefringence of solutions. These methods only show sufficiently large results, if the axis ratio is larger than 3. Slightly above the cmc_{II}, however, this ratio is only a little larger than one which makes it extremely difficult to determine the transition concentration with accuracy. For all these reasons, the higher cmc's of the detergents are far less well known than the cmc_I. At present, no summary with data concerning higher cmc's is available. Only a few systems have been studied in detail with different techniques; a selection of reliable data is given in Table 9 [81, 85].

The dependence of the higher cmc's on detergent structure and on other parameters such as electrolyte concentration or nature of the counterions follows the same rules as the dependence of the cmc_I on these parameters, because basically also the transitions are still determined by hydrophobic solvation. Increas-

Table 9. Values for the cmc_{II} for some detergent systems in aqueous solutions at different temperatures [73a, 73b, 77a, 79b, 186]

Detergent	T (°C)	cmc_I (mol l^{-1})	cmc_{II} (mol l^{-1})	Reference
Sodium-n-dodecyl-1-sulfate	25	8.2×10^{-3}	6.5×10^{-2}	[79b]
Potassium-n-decanoate	20	1.0×10^{-1}	1.1	[77a]
Potassium-n-dodecanoate	20	2.4×10^{-2}	5.0×10^{-1}	[77a]
Potassium-n-tetradecanoate	20	6.3×10^{-3}	2.7×10^{-1}	[77a]
n-Tetradecylpyridinium-n-heptanesulfonate	25	5.2×10^{-4}	1.9×10^{-3}	[73a]
	40	5.5×10^{-4}	2.8×10^{-3}	
n-Tetradecylpyridinium-n-octanesulfonate	30^a	2.4×10^{-4}	2.4×10^{-4}	[73a]
	40	3.2×10^{-4}	4.5×10^{-4}	
n-Tetradecylpyridinium-perfluorobutyrate	25	6.7×10^{-4}	2.0×10^{-3}	[186]
n-Hexadecylpyridinium-o-salicylate	25	1.5×10^{-4}	4.5×10^{-4}	[73b]
n-Octadecylpyridinium-o-salicylate	35	3.1×10^{-5}	6.0×10^{-5}	[186]
Tetraethylammonium-perfluorooctanesulfonate	25^a	9.4×10^{-4}	9.4×10^{-4}	[186]
	45	1.2×10^{-3}	2.6×10^{-3}	
	50	1.1×10^{-3}	5.5×10^{-3}	
Tetramethylammonium-perfluorooctanesulfonate	25^a	1.1×10^{-3}	1.1×10^{-3}	[186]
	30^a	1.2×10^{-3}	1.2×10^{-3}	
	35	1.3×10^{-3}	2.8×10^{-3}	
	42	1.5×10^{-3}	4.5×10^{-3}	
	45	1.6×10^{-3}	4.8×10^{-3}	
	48	1.65×10^{-3}	5.2×10^{-3}	
	51	1.73×10^{-3}	5.5×10^{-3}	
	56	1.75×10^{-3}	6.0×10^{-3}	

^{a} Two separated cmc's could not be detected at these temperatures; above the measurable cmc rod-like micelles are formed; there is no concentration region at these temperatures where spherical micelles exist.

ing the hydrophobicity of the tail or lowering of the hydrophilicity of the head-group favors the formation of rods or discs, and hence leads to lower values for the higher cmc's. Low cmc_{II} values, for example, are observed with detergents with reduced micellar charge by using weakly dissociating counterions [81, 89] or pronouncedly with nonionic detergents [90]. Moreover, double-tail detergents of perfluoro-detergents [80c, 80d, 89, 91] usually have a strong tendency to form anisometric micelles. In extreme cases these detergents form anisometric micelles from the first detectable cmc and there is no concentration range for the existence of globular micelles. This seems very often to be the case for double-tail detergents when each chain consists of more than 12 CH_2 groups [92].

Apart from the cmc_{II} values, very little is known about the thermodynamic parameters associated with these transitions. The determination of such parameters is likely to be difficult due to interference from concentration-dependent interaction effects. It is interesting to note that the cmc_{II} for all systems known is always strongly dependent on temperature. It is shifted to higher values with increasing temperature as shown in Fig. 10; this means that increase of the temperature strongly favors spherical micelles. When the temperature is raised to about 100 °C, practically all detergent systems which are known to form rod-like micelles at room temperature form again globular micelles [81, 84]. Judging by these observations, one can assume that the transition from spheres to anisometric micelles seems to be strongly exothermic.

We have to be aware, however, that the sequence of transitions need not always be the same in each system. As already mentioned, there are systems which do not form globular micelles, but start the aggregation with rod-like micelles or bilayers; other systems can show a transition directly from spherical micelles to bilayers [80c, 80d, 2, 91, 93]. In particular the double-tail detergents seem to favor this sequence.

4.1.3 Methods for the Determination of Size and Shape of Micelles

Without going into the theories of the methods, we give a brief summary of the methods which are usually used for the characterization of the micellar aggregates. These methods can be ordered into several groups:

a) Scattering Methods. This group comprises the light scattering technique [21, 94], the small angle X-ray scattering method (SAXS [95]) and the small angle neutron scattering method (SANS [96]). The last two methods permit determination of molecular weight, the radius of gyration and usually also the shape of the micelles; the first method would in principle give the same information, but only for particles whose dimensions are not significantly smaller than the wavelength of the scattered light, which means in the case of micellar aggregates that usually only the molecular weight can be determined from the light scattering measurements. The aggregation number can easily be derived from the molecular weight. SANS experiments are especially well suited for the study of the structure of micelles because of the possibility of varying the contrast ratio of the scatterers with respect to the solvent by using mixtures of H_2O and D_2O as solvent [97]. The scattering length of H-atoms for neutrons has a positive sing, while that of D-atoms is negative; it is thus possible to adjust any desired scattering length of the solvent mixture between these two extremes and so inspect details of the micellar structure by the use of specifically deuterated or fluorinated surfactant compounds.

b) Methods for the Measurement of the Translational or Rotational Diffusion Coefficient of Micelles. This group comprises the dynamic light scattering technique [98] which measures the temporal fluctuations of the intensity of the scattered light as a function of the scattering angle, the NMR pulsed gradient spin echo method [99] for determining the translational diffusion coefficient, the measurement of the polarization ratio of the fluorescence light of a suitable dye indicator which is solubilized by the micelles [100] and the measurements of the electric [101] and the streaming [102] birefringence. These last two techniques monitor build-up, decay and magnitude of optical birefringence of the solutions in an electric field or a shear field; birefringence is usually characterized by a sign, an amplitude, and one or several time constants. Also rheological measurements [103] can be included in this group, as the viscosity of the solutions is affected by the presence of the aggregates. The hindrance of the flow by such aggregates depends on the size and shape of the aggregates; measurements of the viscosity as a function of detergent concentration can give information about micellar parameters. The flow resistance of the solutions is strongly influenced by the rotational diffusion coefficient of anisometric aggregates; the presence of stiff overlapping rods always leads to highly viscous solutions, because the free rotation of the rods in such a situation is restricted [104].

c) Direct Observations of the Aggregates. Recent successful attempts have made micelles visible under the electron microscope [105] using freeze-etching technique. The solutions are first dispersed into small droplets and these droplets are cooled and frozen rapidly. Freezing of the droplets must be so fast that the system has no time to adjust equilibrium. Therefore cooling rates of 10^4 K s^{-1} or faster have to be employed for studies of detergents with C_{16}-hydrocarbon chains (see Sect. 6). Such cooling rates can be obtained by dispersing the droplets in liquid nitrogen. After freezing, the samples are broken at low temperature, several layers of water molecules are taken off under high vacuum and the protruding micelles are stained from one direction with a metal. Finally the metal print of the micelles can be seen under the electron microscope. Using this complicated technique, Bachmann finally succeeded in obtaining good photographs of micelles. From

Table 10. Values for the aggregation numbers n, the radii r and the lengths l of micellar aggregates for some selected detergent systems in aqueous solutions at 25 °C at total concentrations c_0 close to the cmc

Detergents forming spherical aggregates

tergent	n	r (Å)	Reference
dium-n-octyl-1-sulfate	20	–	[6]
dium-n-nonyl-1-sulfate	31	–	[98a]
dium-n-decyl-1-sulfate	43[a]	–	[98a]
dium-n-dodecyl-1-sulfate	60[a]	25[b]	[94a, 72d]
		16.0	
dium-n-dodecyl-1-sulfate + 0.02 M NaCl	66	–	[6]
dium-n-dodecyl-1-sulfate + 0.03 M NaCl	72	–	[6]
hium-n-dodecyl-1-sulfate	63	–	[6]
tramethylammonium-n-dodecyl-1-sulfate	76	–	[6]
)ctyltrimethylammoniumbromide	23	–	[98a]
Nonyltrimethylammoniumbromide	30	–	[98a]
Decyltrimethylammoniumbromide	40[a]	–	[98a]
Dodecyltrimethylammoniumbromide	50[a]	–	[98a]
Tetradecyltrimethylammoniumbromide	84[a]	19.5	[98a, 72d]
Hexadecyltrimethylammoniumbromide	88[a]	–	[98a]
Dodecyltrimethylammoniumchloride	65	–	[78b]
Dodecyltrimethylammoniumchloride + 0.01 M NaCl	67	–	[78b]
Dodecyltrimethylammoniumchloride + 0.05 M NaCl	74	–	[78b]
Dodecyltrimethylammoniumchloride + 0.1 M NaCl	81	–	[78b]
Dodecyltrimethylammoniumchloride + 0.2 M NaCl	91	–	[78b]
dium-n-octyl-p-benzenesulfonate	–	12.8[c]	[98b]
		19.5	
hium-perfluorooctanesulfonate	37	15.5	[125]

Table 10 (continued)

b) Detergents forming rod-like aggregates

Detergent	r (Å)	l (Å)	Referen
n-Dodecylammonium-trifluoroacetate $c_0 = 2 \times 10^{-2}$ M	20 [d]	340	[85a]
n-Dodecylammonium-trifluoroacetate $c_0 = 3 \times 10^{-2}$ M	20 [d]	410	[85a]
n-Dodecylammonium-trifluoroacetate $c_0 = 4 \times 10^{-2}$ M	20 [d]	450	[85a]
n-Dodecylammonium-trifluoroacetate $c_0 = 5 \times 10^{-2}$ M	20 [d]	475	[85a]
Tetramethylammoniumperfluorooctanesulfonate (T=15 °C) $c_0 = 3 \times 10^{-3}$ M	22 [d]	175	[81]
Tetramethylammoniumperfluorooctanesulfonate (T=15 °C) $c_0 = 4 \times 10^{-3}$ M	22 [d]	235	[81]
Tetramethylammoniumperfluorooctanesulfonate (T=15 °C) $c_0 = 5 \times 10^{-3}$ M	22 [d]	260	[81]
Tetramethylammoniumperfluorooctanesulfonate (T=15 °C) $c_0 = 6 \times 10^{-3}$ M	22 [d]	320	[81]
Tetramethylammoniumperfluorooctanesulfonate (T=15 °C) $c_0 = 7 \times 10^{-3}$ M	22 [d]	355	[81]
Tetramethylammoniumperfluorooctanesulfonate (T=15 °C) $c_0 = 8 \times 10^{-3}$ M	22 [d]	375	[81]
Tetramethylammoniumperfluorooctanesulfonate (T=15 °C) $c_0 = 9 \times 10^{-3}$ M	22 [d]	400	[81]
Tetramethylammoniumperfluorooctanesulfonate (T=15 °C) $c_0 = 1 \times 10^{-2}$ M	22 [d]	430	[81]
Tetramethylammoniumperfluorooctanesulfonate (T=25 °C) $c_0 = 1 \times 10^{-2}$ M	22	206	[72d]
Tetramethylammoniumperfluorooctanesulfonate (T=25 °C) $c_0 = 2 \times 10^{-2}$ M	22	276	[72d]
n-Hexadecylpyridinium-o-salicylate $c_0 = 1 \times 10^{-3}$ M	21.5	100 [a]	[73b]
n-Hexadecylpyridinium-o-salicylate $c_0 = 2 \times 10^{-3}$ M	21.5	220 [a]	[73b]
n-Hexadecylpyridinium-o-salicylate $c_0 = 3 \times 10^{-3}$ M	21.5	335 [a]	[72c, 73
n-Hexadecylpyridinium-o-salicylate $c_0 = 4 \times 10^{-3}$ M	21.5	450 [a]	[73b]
n-Hexadecylpyridinium-o-salicylate $c_0 = 5 \times 10^{-3}$ M	21.5	580 [a]	[72c, 73
n-Hexadecylpyridinium-o-salicylate $c_0 = 1 \times 10^{-2}$ M	21.5	760 [a]	[72c, 73
n-Hexadecylpyridinium-o-salicylate $c_0 = 1 \times 10^{-3}$ M $+ 1 \times 10^{-4}$ M NaSal	21.5	370 [a]	[187]
n-Hexadecylpyridinium-o-salicylate $c_0 = 1 \times 10^{-3}$ M $+ 2 \times 10^{-4}$ M NaSal	21.5	670 [a]	[187]
n-Hexadecylpyridinium-o-salicylate $c_0 = 1 \times 10^{-3}$ M $+ 5 \times 10^{-4}$ M NaSal	21.5	2500 [a]	[187]
n-Hexadecylpyridinium-o-salicylate $c_0 = 1 \times 10^{-3}$ M $+ 1 \times 10^{-3}$ M NaSal	21.5	3900 [a]	[187]
Diethylammonium-perfluorononanoate $c_0 = 5 \times 10^{-3}$ M [e]	21	274	[72d]
Diethylammonium-perfluorononanoate $c_0 = 1.5 \times 10^{-3}$ M [e]	21	306	[72d]

[a] Average values from different measurements.
[b] Value for the hydrodynamic radius.
[c] Value for the radius of the inner core of the micelles.
[d] r-Values estimated from the length of a detergent molecule [2].
[e] Detergent forms disclike aggregates.

these pictures he could obtain the dimensions, shapes, and even number density of the micellar aggregates without complications due to micellar interactions.

Table 10 summarizes results which have been obtained on various micellar systems with different techniques [6, 80c, 80d, 81, 89, 93a, 106]. Globular micelles show increasing radii and aggregation numbers with increasing chain length of the hydrophobic moiety, while the values for a given system increase only slightly with detergent concentration. For rod-like aggregates the data show that the short radii of the rods are independent of the detergent concentration and also given by the length of a detergent molecule as in the case of spheres, while the lengths of the rods increase with total detergent concentration. Here we would like to emphasize that determining the absolute sizes of the micelles is not easy (see Sect. 4.4). Very often equations have been used for evaluation of experimental results derived under the assumption that no interaction is present between the aggregates. But all micellar aggregates show some degree of mutual interaction, mostly due to electrostatic repulsion between them, and this interaction increases with increasing concentration of the micelles. In contrast to polymer solutions, in almost all cases it is impossible to eliminate this intermicellar interaction either by dilution or by addition of electrolyte, because the size of the micelles is partly determined by the intermicellar interaction energy, which means that the system would be changed by dilution or addition of electrolytes. In the last few years, however, methods have been developed which can correct micellar interaction, so that the true size of micelles can be evaluated from experimental data (see references in section 4.4). Here it must be pointed out that data from the literature on micellar sizes and shapes should be viewed with caution, if they date from before the availability of these theories.

4.3 Configuration of the Monomers Inside Micelles

Recently, the question of the configuration of the alkyl chains inside the micelles was raised again by different groups. In earlier investigations it was usually assumed that the alkyl chains in the micellar interior are disordered, as in a liquid hydrocarbon [40 a]. This conclusion was mainly based on the observation that the detergent solution above the cmc could be cooled considerably below the Krafft point without crystallization of the surfactant. If the chain were in an ordered state comparable to that in a crystal, undercooling would be difficult. This view is supported by values for the partial molal volumes and heat capacities of the surfactants above the cmc which are in the same range as for normal liquid hydrocarbons (see Sect. 3.3).

Contrary to this view, Fromhertz [107] has recently proposed a detergent block model characterized by a considerable amount of order of the alkyl chains. According to this model, these alkyl chains are mainly in the all-trans configuration with a few gauche configurations permitted. Hence the order inside the micelles should be similar to that in lipid bilayers.

Considerably less order is assumed by the models recently proposed by Flory [108] and Gruen [109]. In these models the order parameter for a CH_2-group along the chain varies continuously from the α-position to the ω-position. In particular the model of Gruen seems to be very well suited to account for the order parameters which have been experimentally determined from NMR [110] and SANS measurements [111]. The electric birefringence measurements from which the mean order parameter of all chains in the rod-like micelles can be determined [112] also clearly show that the degree of order is very low, in clear contradiction to the Fromhertz model. The situation could be different in perfluoro-detergent micelles. The chains of these compounds are considerably stiffer than those of hydrocarbon compounds and this would also favor more ordered states inside the micelles.

4.4 Interaction Between Micellar Aggregates

In this section the influence of the intermicellar interaction on experiments for determination of micellar sizes and shapes is briefly discussed. The theories for the
evaluation of these experiments have usually been developed by regarding the
micelles as noninteracting particles. This is obviously only a rough approximation, because micelles are always highly charged species. This applies especially
to micelles formed by ionic detergents, even if one takes into account that about
70–80% of the counterions are normally bound to the micellar surface [4, 113].
The intermicellar interaction is mainly due to electrostatic repulsion between the
evenly charged aggregates and to mutual steric hindrance; it increases with increasing surfactant concentration and with transition to unisometric aggregates.
The electrostatic repulsion gives rise to a nearest-neighbor order between the particles [114]. The interaction therefore affects all kinds of experiments used for the
determination of size and shape of micelles; for example, the nearest-neighbor
order gives rise to maxima in the scattering curves of X-ray and SANS measurements [80c, 80d, 95, 115] or to a decrease of the scattered intensity of visible light
[94a, 116]. Further, interaction leads to nonexponential correlation functions in
dynamic light scattering measurements [117], to nonexponential decay functions
in electric birefringence measurements [117], and to an increased viscosity or also
viscoelasticity [118, 119] of the surfactant solutions. The evaluation of such measurements without taking into account the intermicellar interaction can therefore
lead to completely erroneous results and it is necessary to inspect data from the
relevant literature carefully. Such corrections are necessary in any case, because
it is not possible to suppress interaction by the addition of an electrolyte or by
dilution of the solutions, since such procedures would change the aggregates and
hence affect the results.

Several theories have been developed to calculate the influence of the intermicellar interaction [117, 118, 120–122], most of them valid only for spherical aggregates. For experimental compensation of the interaction it is necessary to combine as many different experimental techniques as possible; each technique is specifically affected by interaction, and it is thus often possible to eliminate its influence.

In a recent paper [123], it was found that the intermicellar interaction can even
affect the sizes of unisometric aggregates; the aggregates grow with increasing surfactant concentration, according to the theory, until their rotational volumes
start to overlap at a characteristic concentration c*. Above this concentration the
interaction prevents an increase of the overlap and consequently the number of
the aggregates has to increase again with increasing surfactant concentration,
while their lengths start to decrease again slightly (see Fig. 11).

It should be briefly mentioned that at higher surfactant concentrations also
attractive interactions between the aggregates can take place which can be explained by the DLVO theory [124]. These forces lead to the formation of coacervates and of lyotropic liquid crystalline mesophases. Normally micelles in surfactant solutions can be compared with gas molecules which can move more or less
freely; the coacervate then represents the liquid state of this micellar "gas", while
the lyotropic mesophase can be regarded as its solid state. Many monographs

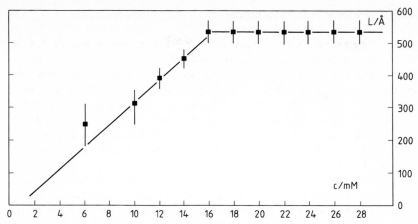

Fig. 11. Plot of the lengths L of rod-like aggregates as a function of the detergent concentration for $C_{14}H_{29}$-NC_5H_5-$C_7H_{15}SO_3$ solutions at 30 °C

have been published on coacervates [125] and lyotropic liquid crystals [126]; this subject will not be discussed here in detail.

5 The Behavior of Surfactants in Nonpolar Solvents; Formation of Reversed Micelles

In nonpolar solvents the surfactants have a lyophilic component, which is now the hydrocarbon group, and a lyophobic one which is the polar head group. This polar group can be only poorly solvated by the nonpolar solvent and would be insoluble without the lipophilic hydrocarbon residues. In numerous investigations which have been summarized recently in a review article by Eicke [127], it has been well established that surfactants in nonpolar solvents form so-called reversed micelles, where the polar part is located in the micellar interior without contact with the solvent, while the nonpolar hydrocarbon groups are surrounding the polar core; they are in contact with the solvent and are responsible for the solubility of the aggregates [127].

It has been known for some time that strong solvent effects are involved in the formation of reversed micelles; such micelles are formed in some nonaqueous solvents, but not in all solvents [128]. From these observations it has been established that the aggregation tendency for reversed micelles is more pronounced in more apolar solvents. However, polarity of the solvent cannot be the only important parameter for detergent association; it has been realized that in solvents which are hydrogen bond donors or acceptors no reversed micelles can be detected, even if these solvents can be regarded as nonpolar solvents, for example ethylacetate or dioxane [129].

There is also another problem connected with the investigations on reversed micelles. One characteristic property of micelles is the ability to solubilize compounds which are themselves insoluble in the solvent. Micelles in aqueous solution incorporate hydrophobic compounds into their hydrophobic interior, making them soluble in water. Similarly, reversed micelles have the ability to incorporate polar compounds, especially water, into their polar core and to solubilize them in the nonpolar solvent. Water, however, is always present to a certain amount in nonpolar solvents and cannot be excluded completely. These traces of water can be incorporated into the reversed micelles and can considerably affect their properties [130]. Thus the systems consisting of detergent and nonpolar solvent can also be regarded as ternary systems, the third component being water. It has been shown that at least traces of water, which forms hydrogen bonds between the ions, are required for the formation of reversed micelles in most, if not in all cases [131].

A critical concentration for the formation of reversed micelles has also been found. As micellar aggregation numbers for reversed micelles are usually rather small, an exact determination of the cmc is difficult, because the cmc is no longer well defined and must be replaced by a transition region [132]. Principally, this transition region can be measured by the same techniques as are suitable for aqueous micelles with some exceptions. Electrochemical methods and conductivity measurements cannot be applied in most cases, due to the low conductivity of the solution. Surface tension measurements are not suitable either because the surfactants in nonpolar media are not surface-active, with the exception of perfluorodetergents in hydrocarbon solvents [6].

The other methods, for example measurements of colligative properties [133], light scattering [134], fluorescence spectroscopy of solubilized dyes [135], Krafft point solubility [136], change of NMR signals [137], dielectric increment [138] or positron annihilation [139] are mainly sensitive to the concentration of the aggregates and hence principally suitable for the determination of the cmc. However, these techniques are also usually rather insensitive to the cmc of surfactants in apolar media due to the low cmc values and the small aggregation numbers. Values of the cmc of reversed micelles have been summarized by Shinoda [6] and more recently by Eicke [127]. In Table 11, a series of cmc values for some selected detergent systems in various solvents is given. From the table it can be seen that, contrary to aqueous micelles, the cmc is not strongly dependent on the length of the alkyl chain.

This behavior can be readily understood, when recalling that the driving force for the aggregation does not come from an unfavorable hydrocarbon-solvent interface, but is due to an electrostatic attraction between the ionic groups and the counterions. The observed small cmc dependence on the alkyl chain length can result from interactions of the hydrocarbon chain with the solvent leading to a decreased solubility of the surfactant with increasing chain length [140].

The failure to understand the thermodynamics of micelle formation exists also for reversed micelles. Calorimetric measurements of ΔH_m measurements are likely to be very difficult because of the small cmc values and aggregation numbers. This may be are reason why we did not find references in the literature. There are some ΔH_m values derived from the temperature dependence of the cmc [141]. This method has many disadvantages because of the ill-defined cmc for the reversed micelles and thus there are serious disagreements between the data obtained by the different research groups.

Table 11. Values for the cmc and the aggregation numbers n for some selected detergents in nonpolar solvents. [127]

Detergent	Solvent	T(°C)	cmc (mol/l)	n
Aerosol OT	Cyclohexane	28	1.3×10^{-3}	56
Aerosol OT	Cyclohexane	37	3.9×10^{-4}	17
Aerosol OT	Cyclohexane	25	–	39
Aerosol OT	Benzene	28	2.7×10^{-3}	24
Aerosol OT	Benzene	37	3.5×10^{-4}	13
Aerosol OT	Benzene	25	–	15
Aerosol OT	CCl_4	25	1.6×10^{-4}	17
Aerosol OT	CCl_4	37	4.0×10^{-4}	17
Aerosol OT	2,2,4-Trimethylpentane	25	4.9×10^{-4}	15
Aerosol OT	n-Octane	25	–	30
Aerosol OT	n-Decane	25	–	37
Aerosol OT	n-Dodecane	25	–	28
$n\text{-}C_4H_9NH_3\text{-}C_2H_5CO_2$	Benzene	33	$(5 \pm 0.5) \times 10^{-2}$	4
$n\text{-}C_6H_{13}NH_3\text{-}C_2H_5CO_2$	Benzene	33	$(2.7 \pm 0.5) \times 10^{-2}$	7
$n\text{-}C_8H_{17}NH_3\text{-}C_2H_5CO_2$	Benzene	33	$(1.5\text{–}1.7) \times 10^{-2}$	5
$n\text{-}C_{10}H_{21}NH_3\text{-}C_2H_5CO_2$	Benzene	33	$(8\text{–}10) \times 10^{-3}$	–
$n\text{-}C_{12}H_{25}NH_3\text{-}C_2H_5CO_2$	Benzene	33	$(3\text{–}7) \times 10^{-3}$	–
$n\text{-}C_{12}H_{25}NH_3\text{-}n\text{-}C_3H_7CO_2$	Benzene	26	1.8×10^{-3}	–
$n\text{-}C_{12}H_{25}NH_3\text{-}n\text{-}C_5H_{11}CO_2$	Benzene	26	2.0×10^{-2}	–
$n\text{-}C_{12}H_{25}NH_3\text{-}n\text{-}C_7H_{15}CO_2$	Benzene	26	2.5×10^{-2}	–
$n\text{-}C_{12}H_{25}NH_3\text{-}n\text{-}C_9H_{19}CO_2$	Benzene	26	3.5×10^{-2}	–
$n\text{-}C_4H_9NH_3\text{-}C_2H_5CO_2$	CCl_4	33	$(2.3\text{–}2.6) \times 10^{-2}$	3
$n\text{-}C_6H_{13}NH_3\text{-}C_2H_5CO_2$	CCl_4	33	$(2.1\text{–}2.4) \times 10^{-2}$	7
$n\text{-}C_8H_{17}NH_3\text{-}C_2H_5CO_2$	CCl_4	33	$(2.6\text{–}3.1) \times 10^{-2}$	5
$n\text{-}C_{10}H_{21}NH_3\text{-}C_2H_5CO_2$	CCl_4	33	$(2.2\text{–}2.7) \times 10^{-2}$	5
$n\text{-}C_{12}H_{25}NH_3\text{-}C_2H_5CO_2$	CCl_4	33	$(2.1\text{–}2.5) \times 10^{-2}$	4
$n\text{-}C_{12}H_{25}NH_3\text{-}C_3H_7CO_2$	CCl_4	26	1.6×10^{-2}	–
$n\text{-}C_{12}H_{25}NH_3\text{-}C_7H_{15}CO_2$	CCl_4	20	1.3×10^{-2}	–

Nevertheless, numerous theoretical papers have been published about the association process of reversed micelles. The thermodynamic models used in these considerations are the phase separation model [132, 142] and the mass action model [136 a, 143], because these models are not restricted to the solvent water. Based on the mass action model, the attempt was made to calculate the free energy for the transfer of a monomer from the solution into the micelle as a function of the aggregation number, as was done in the case of aqueous micelles. In these calculations different contributions to the Gibbs energy due to molecular parameters were introduced, for example an interfacial term for the solute molecules, an electrostatic term for the charged headgroups, and a term for the hydrogen bond interaction [144]. In other concepts additional entropic contributions were introduced [145] and these models could qualitatively account for the aggregation behavior of surfactants in nonpolar solvents.

The experimental determination of the size and shape of reversed micelles can be carried out using the same techniques as for aqueous micelles. Thus measurements of colligative properties [133], light scattering [21 b, 94 b, 134, 146], viscosity [147], fluorescence spectrum of a solubilized indicator [135], sedimentation

rate in an ultracentrifuge [148], dynamic light scattering [98] or dielectric dispersion [149] have been carried out not only to determine the cmc, but also to obtain
the aggregation numbers of reversed micelles. Some n values have been included
in Table 10. As can be seen from these values, there is no strong influence of the
length of the hydrocarbon residues on micellar size, which was also found for the
cmc, but there is a great influence of the solvent on the aggregation numbers [150]
which is not only given by the polarity of the solvent, but by a solubility parameter defined by Hildebrand [151]; the larger this parameter is, the smaller become
the micelles which are formed in this solvent.

Reversed micelles usually assume rod-like structures [127]. Spheres seem to
occur seldom or not at all, because there is no charge on the micellar surface
which would favor a spherical shape by the repulsive forces in the surface.

6 Dynamics of Micelle Formation

Re-adjustment of equilibrium after a perturbation is too fast to be resolved with
conventional techniques. With the advent of relaxation techniques it became possible to study the kinetics of micelle formation and many investigations have been
carried out especially on aqueous systems, but also on detergent systems in nonpolar media [152]. Only aqueous detergent solutions will be discussed in detail.

Pressure jump [153], temperature jump [154] and ultrasound absorption [155]
techniques, but also flow methods such as stopped flow [156] have been used to
study micellar kinetics. Different results were obtained using different techniques
until it was realized, using the pressure jump and the shock wave method [158],
that in micellar equilibria there are two well-separated relaxation processes. For
the investigated systems the faster process was observed in the microsecond range
and the slower one in the millisecond range; the slower process was shown to be
strongly affected by impurities of the surfactants. At the same time, but independently of these measurements, Aniansson and Wall [159] showed on the basis of
theoretical considerations that the adjustment of equilibria in micellar systems
should proceed in two separate steps. The faster step was associated with the exchange of monomers between the micelles and the bulk phase, while the slower
one is due to the formation and dissolution of whole micelles. This process proceeds via the micellar nucleus; the nucleus represents an oligomer at the minimum
of the micellar distribution curve which is schematically shown in Fig. 12. As the
equilibrium concentration of the nuclei is many orders of magnitude smaller than
the concentration of aggregates, the relatively long lifetime of the micelles can
thus be readily understood without the need of a single slow step.

Aniansson and Wall derived for the two relaxation times τ_1 and τ_2 the following expressions:

$$1/\tau_1 = (k^-/\sigma^2) + (k^-/n) \cdot (c_0 - cmc)/cmc, \tag{29}$$

and

$$1/\tau_2 = (k^- \cdot c_i/c_m) \cdot (cmc + n^2 \cdot c_m)/(cmc + \sigma^2 \cdot c_m), \tag{30}$$

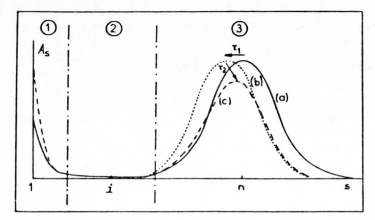

Fig. 12. Scheme of the shift of micellar equilibria in a relaxation experiment in two separated steps, according to the theory of Aniansson and Wall. A_s concentration of the species with the aggregation number s; i aggregation number of the micellar nuclei, i.e. the species with the lowest concentrations and n the mean aggregation number of the micelles. **1** range of monomers and oligomers with s < i; **2** range of micellar nuclei; **3** range of micelles

with

$$c_m = (c_0-cmc)/n. \tag{31}$$

Here, k^- means the rate constant for a monomer leaving the micelle, k^+ the rate constant for a monomer entering the micelle, σ the variance of the micellar distribution curve, c_i the concentration of micellar nuclei, c_m the micellar concentration, c_0 the total surfactant concentration and n the mean aggregation number of micelles. For spherical micelles, an excellent agreement between the theory and the experimental data was always observed [33 b, 160].

Assuming a Gauss function for the micellar distribution curve, the k^--, k^+-, n-, σ-, c_i values and also the i values for the aggregation number of the nucleus can be obtained from the two relaxation times, because there is proportionality between c_i and cmc^i, and i can hence be derived from the variation of the τ_2- and the cmc values with the ionic strength [33 b]. From the temperature dependence of the two relaxation times the enthalpy values for the incorporation of a monomer into the micelle and for the formation of the nucleus can be further evaluated. These values are summarized in Table 12 for some selected surfactant systems. From these values it can be seen that the insertion of monomers into micelles takes place in diffusion-controlled steps; the small changes of the k^+ values with the chain length of the detergents are probably caused by electrostatic effects, different in magnitude for the detergent systems due to the different ionic strengths of the systems. The stability of the micelles is determined by the k^- values for the monomers leaving the micelles; these values decrease significantly with increasing chain length of the hydrophobic rests. The n values are in good agreement with the n values obtained from other techniques. Information is also available now for σ, $\Delta H_{nucleus}$, ΔH^-, c_i, and i; these parameters cannot be ob-

H. Hoffmann and W. Ulbricht

Table 12. Values for the kinetic parameters k^-/n, k^-/σ^2, the rate constant k^+ for the insertion of a monomer into the micelle, the rate constant k^- for the dissociation of a monomer leaving a micelle, the mean aggregation number n of the micelles, the variance σ of the micellar distribution curve, the aggregation number i of the micellar nucleus, the concentration c_i of the micellar nuclei close to the cmc, the enthalpy ΔH for the insertion of a monomer into the micelle and the heat ΔH_i for the formation of micellar nuclei for some detergent systems, obtained only from the two relaxation times τ_1 and τ_2. [33b, 182, 75a]

Detergent	T (°C)	k^-/n (s⁻¹)	k^-/σ^2 (s⁻¹)	k^- (s⁻¹)	k^+ (l mol⁻¹ s⁻¹)	n	σ	i	c_i (mol l⁻¹)	ΔH (kJ mol⁻¹)	ΔH_i (kJ mol⁻¹)
$C_6H_{13}SO_4Na$ [a]	25	7.8×10^7	3.8×10^7	1.3×10^9	3.2×10^9	17	6	–	–	–	–
$C_7H_{15}SO_4Na$ [a]	25	3.3×10^7	6.9×10^6	7.3×10^8	3.3×10^9	22	10	–	–	–	–
$C_8H_{17}SO_4Na$ [a]	25	3.7×10^6	–	1.0×10^8	7.7×10^8	27	–	–	–	–	–
$C_9H_{19}SO_4Na$ [a]	25	4.2×10^6	–	1.4×10^8	2.3×10^9	33	–	–	–	–	–
$C_{10}H_{21}SO_4Na$ [a]	40	2.2×10^6	–	9.0×10^7	2.7×10^9	41	–	–	–	–	–
$C_{11}H_{23}SO_4Na$ [a]	25	8.2×10^5	–	4.0×10^7	2.6×10^9	52	–	–	–	–	–
$C_{12}H_{25}SO_4Na$	25	1.7×10^5	6.0×10^4	1.0×10^7	1.2×10^9	59	13	7	2.7×10^{-10}	−29	140
$C_{14}H_{29}SO_4Na$	25	1.2×10^4	2.8×10^3	1.2×10^6	5.7×10^8	97	20	7	3.6×10^{-12}	−29	140
$C_{14}H_{29}C_5H_5NCl$	5	7.0×10^4	2.8×10^4	4.5×10^6	1.0×10^9	64	13	7	6.1×10^{-12}	−24	110
$C_{14}H_{29}C_5H_5NCl$	15	1.1×10^5	5.4×10^4	5.4×10^6	1.3×10^9	49	10	7	3.9×10^{-11}	–	–
$C_{14}H_{29}C_5H_5NBr$	5	1.8×10^4	1.2×10^4	1.1×10^6	4.1×10^8	60	10	7	8.4×10^{-13}	−40	170
$C_{14}H_{29}C_5H_5NBr$	15	3.7×10^4	2.6×10^4	1.9×10^6	7.7×10^8	52	9	7	6.1×10^{-12}	–	–
$C_{14}H_{29}C_5H_5NBr$	25	7.8×10^4	5.8×10^4	3.4×10^6	1.3×10^9	44	8	7	–	–	–
$C_{14}H_{29}C_5H_5N\text{-}CH_3SO_3$	15	1.7×10^5	7.8×10^4	8.3×10^6	1.8×10^9	49	10	7	7.4×10^{-10}	−24	126
$C_{14}H_{29}C_5H_5N\text{-}C_2H_5SO_3$	15	1.3×10^5	6.0×10^4	6.5×10^6	1.6×10^9	50	10	7	1.8×10^{-10}	−29	120
$C_{14}H_{29}C_5H_5N\text{-}C_3H_7SO_3$	15	7.6×10^4	4.6×10^4	4.0×10^6	1.3×10^9	53	9	7	1.0×10^{-11}	−34	120
$C_{14}H_{29}C_5H_5N\text{-}C_4H_9SO_3$	15	4.4×10^4	3.0×10^4	2.4×10^6	1.1×10^9	55	9	7	3.2×10^{-13}	−38	120
$C_{14}H_{29}NH_3Cl$	25	1.3×10^4	6.5×10^3	9.5×10^5	2.6×10^8	73	12	7	1.2×10^{-13}	−54	210
	35	3.3×10^4	1.8×10^4	4.0×10^6	1.0×10^9	79	14	7	4.7×10^{-13}	–	
$6\text{-}C_{12}H_{25}\text{-}C_6H_4\text{-}SO_3Na$	25	1.5×10^4	3.1×10^4	3.0×10^5	1.3×10^8	20	3	7	3.5×10^{-8}	−22	65
$6\text{-}C_{12}H_{25}\text{-}C_6H_4\text{-}SO_3NH_2(CH_3)_2$	25	1.5×10^4	1.8×10^4	4.0×10^5	2.3×10^8	26	5	7	1.4×10^{-9}	−43	110
$6\text{-}C_{12}H_{25}\text{-}C_6H_4\text{-}SO_3NH_2(C_2H_5)_2$	25	1.2×10^4	1.5×10^4	3.5×10^5	2.3×10^8	29	5	7	1.7×10^{-10}	−33	96
$6\text{-}C_{12}H_{25}\text{-}C_6H_4\text{-}SO_3NH_2(C_3H_7)_2$	40	3.2×10^3	3.9×10^3	7.7×10^4	6.6×10^7	24	4	7	1.1×10^{-9}	–	–

[a] For these systems only the fast relaxation process could be measured; the given data were obtained by using n-values from the literature.

Table 13. Values for the residence time $\tau_r = n/k^-$ of a monomer inside the micellar aggregate for various detergent systems with different types of aggregates at 25 °C. [72d]

Detergent	cmc (mol l^{-1})	Type of aggregate	τ_r (µs)
$C_{12}H_{25}SO_4Na$	8.3×10^{-3}	Sphere	6
$C_{12}H_{25}NH_3J$ (20 °C)	1.1×10^{-2}	Sphere	9
$C_{12}H_{25}NH_2CF_3CO_2$	6.8×10^{-3}	Rod	15
$C_{14}H_{29}SO_4Na$	2.1×10^{-3}	Sphere	83
$C_{14}H_{29}C_5H_5NCl$ (15 °C)	4.1×10^{-3}	Sphere	9
$C_{14}H_{29}C_5H_5NBr$	2.6×10^{-3}	Sphere	14
$C_{14}H_{29}NH_2Cl$	3.7×10^{-3}	Sphere	77
$C_{16}H_{33}C_5H_5NCl$	9.9×10^{-4}	Sphere	170
$C_{16}H_{33}C_5H_5NBr$	6.8×10^{-4}	Sphere	300
$C_{16}H_{33}C_5H_5N$-o-salicylate	1.5×10^{-4}	Sphere	1160
$C_{16}H_{33}C_5H_5N$-o-salicylate	4.5×10^{-4}	Rod	2860
$C_8F_{17}SO_3Li$	6.4×10^{-3}	Sphere	< 10
$C_8F_{17}CO_2NH_4$	7.9×10^{-3}	Sphere	11
$C_8F_{17}CO_2N(CH_3)_4$	3.1×10^{-3}	Sphere	13
$C_8F_{17}SO_3N(CH_3)_4$	1.3×10^{-3}	Rod	83
$C_8F_{17}SO_3N(C_2H_5)_4$	8.8×10^{-4}	Rod	160
$C_8F_{17}CO_2NH(CH_3)_3$	3.8×10^{-3}	Rod	38
$C_8F_{17}CO_2NH_2(C_2H_5)_2$	3.1×10^{-3}	Disc	182
$C_8F_{17}CO_2NH_3CH_3$	5.9×10^{-3}	Disc (probably)	550

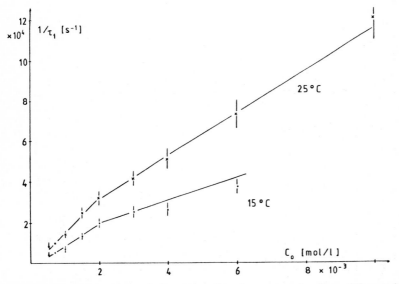

Fig. 13. Plot of the reciprocal short relaxation time τ_1 as a function of the total concentration c_0 for $C_{14}H_{29}$-NC_5H_5-$C_7H_{15}SO_3$ at various temperatures

tained in most cases by other techniques. As already mentioned in Section 3.4, the micellar aggregation process starts with large positive enthalpies, which become smaller with increasing aggregation number and turn to negative values in the range of complete micelles.

From the concentration dependence of τ_1, the values for k^-/n, that is the reciprocal lifetime of a monomer inside the micelle, can be evaluated according to Eq. (29). Such values are summarized in Table 13 for different detergent systems forming various kinds of aggregates. As can be seen from Table 13 this lifetime is not strongly affected by the structure of the surfactant, but depends significantly on the shape of the aggregates [80 d, 81 c, 93 a, 119 c, 160 b]. The shortest lifetimes are observed for spherical aggregates, the longest for discs. Thus the cmc_{II} can also be recognized as a break in the plots of $1/\tau_1$ as a function of the total detergent concentration c_0, demonstrated, for example, in Fig. 13.

Finally, it should be noted that information on the micellization process can also be obtained from the evaluation of the amplitudes of the two relaxation processes, in particular shown by Kahlweit et al. [161].

References

1. Adamson AW (1976) In: Physical chemistry of surfactants. Wiley, New York
2. Tanford C (1980) The hydrophobic effect. Wiley, New York
3. Gill SJ, Nichols NF, Wadsö I (1976) J Chem Thermodynam 8:445
4. Hiemenz PC (1977) Principles of colloid and surface chemistry. Dekker, New York
5. Matijević E (1969) Surface and colloid science, vol 1 and 2. Plenum, New York
6. Shinoda K, Tamamushi B, Nagakawa T, Isemura T (1963) Colloidal surfactants. Academic Press, New York
7. Mukerjee P, Mysels KJ (1971) Critical micelle concentrations of aqueous surfactant systems. NSRDS-NBS 36
8. a. Malsh J, Hartley GS (1934) Z Phys Chem (Leipzig) A 170:321
   b. Wright KA, Abbott AD, Sievertz V, Tartar HV (1939) J Am Chem Soc 61:549
9. a. Powney J, Addison CC (1937) Trans Faraday Soc 33:1242
   b. Addison CC, Elliott TA (1950) J Chem Soc (Lond) 3103 Part I
10. a. McBain JW, Williams RC (1933) J Am Chem Soc (Lond) 55:2250
    b. Hutchinson E, Winslow L (1957) Z Phys Chem (Frankf) 11:165
11. Hartley GS, Callie B, Sanis CS (1936) Trans Faraday Soc 32:795
12. McBain JW, Salmon CS (1920) J Am Chem Soc 43:426
13. Gonick E, McBain JW (1947) J Am Chem Soc 69:334
14. Hess L, Suranyi LA (1939) Z Phys Chem (Leipzig) A 184:321
15. a. Bury CR, Parry GA (1935) J Chem Soc (Lond) 626 Part I
    b. Harkins WD, Mattoon RW, Corrin ML (1946) J Coll Sci 1:105
16. a. Hess K, Philippoff W, Kiessig JH (1939) Koll Z 88:40
    b. Klevens HB (1948) J Phys Coll Chem 52:130
17. Harkins WD, Krizek H, Corrin ML (1951) J Coll Sci 6:576
18. a. Hartley GS (1934) Trans Faraday Soc 30:444
    b. Sheppard SE, Geddes AL (1945) J Chem Phys 13:63
    c. Corrin ML, Harkins WD (1947) J Am Chem Soc 69:679
19. Arkin L, Singleterry CR (1948) J Chem Soc 70:3965
20. a. Hess K, Kiessig JH, Philippoff W (1941) Fette und Seifen 48:377
    b. Wright KA, Tartar HV (1939) J Am Chem Soc 61:544
21. a. Debye P (1949) J Phys Coll Chem 53:1
    b. Debye P (1949) Ann NY Acad Sci 51:575
    c. Debye P, Anacker EW (1951) J Phys Coll Chem 55:644

22. Krafft F, Wiglow H (1895) Ber Dtsch Chem Ges 28:2566
23. Mukerjee P, Mysels KJ, Dulcin CI (1958) J Phys Chem 62:1390
24. Traube Liebigs I (1891) Ann Chem 265:27
25. Klevens HB (1953) J Am Oil Chem Soc 30:74
26. Klevens HB, Raison M (1954) J Chim Phys 51:1
27. Paquette RG, Lingafelter EC, Tartar HV (1943) J Am Chem Soc 65:686
28. a. Winsor PA (1948) Trans Faraday Soc 44:463
    b. Evans HC (1956) J Chem Soc (Lond) 579 Part I
29. Hsiao L, Dunning HN, Lorenz PB (1956) J Phys Chem 60:657
30. Goddard ED, Harva O, Jones TG (1953) Trans Faraday Soc 49:980
31. a. Lange H (1951) Koll Z 121:66
    b. Lelong ALM, Tartar HV, Lingafelter EC, O'loane JK, Cadle RD (1951) J Am Chem
    Soc 73:5411
32. Mukerjee P, Ray A (1966) J Phys Chem 70:2144, 2150
33. a. Lange H, Schwuger MJ (1971) Koll Z; Z Polymere 243:120
    b. Hoffmann H, Nüßlein H, Ulbricht W (1977) In: Mittal KL (ed) Micellization, solubi-
    lization and microemulsions. Plenum, New York, p 263
34. Lindman B, Ekwall P (1969) Koll Z 234:1115
35. Anacker EW, Ghose HM (1968) J Am Chem Soc 90:3161
36. a. Shinoda K (1955) J Phys Chem 59:432
    b. Shick MJ, Fowkes FM (1957) J Phys Chem 61:1062
37. Shinoda K, Yamaguchi T, Hori R (1961) Bull Chem Soc Jpn 34:237
38. Shinoda K, Yamaguchi T, Kinoshita K (1959) J Phys Chem 63:648
39. a. Ginn ME, Kinney FB, Harris HC (1960) J Am Oil Chem Soc 37:183
    b. Adderson JE, Taylor H (1964) J Coll Sci 19:495
40. a. Hartley GS (1936) In: Aqueous solutions of paraffin chain salts. Hermann and Cie, Pa-
    ris
    b. Hutchinson E, Inabe A, Bailey LG (1955) Z Phys Chem (Frankf) 5:344
41. a. Davies DG, Bury CR (1930) J Chem Soc (Lond) 2263 Part II
    b. Grindley J, Bury CR (1929) J Chem Soc (Lond) 679 Part I
42. a. Hutchinson E, Manchester KE, Winslow L (1954) J Phys Chem 58:1124
    b. Goddard ED, Hoeve CAJ, Benson GC (1957) J Phys Chem 61:593
    c. White P, Benson GC (1958) J Coll Sci 13:584
    d. White P, Benson GC (1959) Trans Faraday Soc 55:1025
43. Stainsby G, Alexander AE (1950) Trans Faraday Soc 46:587
44. Kresheck GC (1975) In: Franks F (ed) Water: a comprehensive treatise, vol 4, chapter 2.
    Plenum, New York
45. Matijević E, Pethica BA (1958) Trans Faraday Soc 54:587
46. a. Desnoyers JE, De Lisi R, Ostiguy C, Perron G (1979) In: Mittal KL (ed) Solution
    chemistry of surfactants, vol 1. Plenum, New York, p 221
    b. Desnoyers JE, De Lisi R, Perron G (1980) Pure Appl Chem 52:433
    c. De Lisi R, Perron G, Desnoyers JE (1980) Can J Chem 58:959
    d. De Lisi R, Perron G, Paquette J, Desnoyers JE (1981) Can J Chem 59:1865
    e. De Lisi R, Ostiguy C, Perron G, Desnoyers JE (1979) J Coll Interface Sci 71:147
47. Feinstein HE, Rosano HL (1967) J Coll Interface Sci 24:73
48. a. Treiner C, Le Besnerais A (1977) J Chem Soc Faraday Trans I 73:44
    b. Elworthy PH, Florence AT (1966) Koll Z; Z Polymere 208:157
49. Desnoyers JE, Ostiguy C, Perron G (1978) J Chem Educ 55:137
50. Danielsson I, Rosenholm JB, Stenius P, Backlund S (1976) Prog Coll Polymer Sci 61:1
51. Benjamin L (1966) J Coll Interface Sci 22:386
52. Leduc PA, Fortier JL, Desnoyers JE (1974) J Phys Chem 78:1217
53. Kresheck GC, Hargraves WA (1967) J Coll Interface Sci 48:481
54. a. Brun TS, Høiland H, Vikingstad E (1978) J Coll Interface Sci 63:89
    b. Desnoyers JE, Arel M (1967) Can J Chem 45:359
    c. Shinoda K, Soda T (1963) J Phys Chem 67:2072
55. a. Picker P, Tremblay E, Jolicoeur C (1974) J Solution Chem 3:377
    b. Musbally GM, Perron G, Desnoyers JE (1974) J Coll Interface Sci 48:494

56. Cabani S, Conti G, Matteoli E (1976) J Solution Chem 5:751
57. Fortier JL, Simard MA, Picker P, Jolicoeur C (1979) Rev Sci Inst 50:1474
58. Desnoyers JE, Philip PR (1972) Can J Chem 50:1094
59. Perron G, Desnoyers JE (1979) Fl Phase Equil 2:239
60. a. Franks F, Reid DS (1973) In: Franks F (ed) Water: a comprehensive treatise, vol 2, chapter 5. Plenum, New York
   b. Roux G, Perron G, Desnoyers JE (1978) Can J Chem 52:2808
   c. Jolicoeur C, Lacroix F (1976) Can J Chem 54:624
61. Stenius P, Backlund S, Ekwall P (1980) In: Franzosini P, Sanesi M (eds) Thermodynamic and transport properties of organics salts. IUPAC Chemi Data Ser 28. Pergamon, Oxford, p 295
62. Leduc PA, Desnoyers JE (1973) Can J Chem 51:2993
63. Moroi Y, Nishikodo N, Uehara H, Matuura R (1975) J Coll Interface Sci 50:254
64. Jolicoeur C, Philip PR (1974) Can J Chem 52:1834
65. Musbally GM, Perron G, Desnoyers JE (1976) J Coll Interface Sci 54:80
66. Jones MN, Pilcher G, Espada L (1970) J Chem Thermodynam 2:333
67. Anacker EW (1970) In: Jungermann E (ed) Cationic surfactants. Dekker, New York, p 203
68. Jones MN, Piercy J (1972) J Chem Soc Faraday Trans I 68:1839
69. a. Hall DG, Pethica BA (1967) In: Schick MJ (ed) Nonionic surfactants. Dekker, New York
   b. Hill TL (1963) Thermodynamics of small systems. Benjamin, New York
70. Israelachvili JN, Mitchell DJ, Ninham BW (1976) J Chem Soc Faraday Trans II 72:1525
71. Nagarajan R, Ruckenstein E (1979) J Coll Interface Sci 71:580
72. Jönsson B, Wennerström H (1981) J Coll Interface Sci 80:482
73. a. Ruckenstein E, Nagarajan R (1981) J Phys Chem 85:3014
   b. Nagarajan R, Ruckenstein E (1977) J Coll Interface Sci 60:221
74. Rushbrooke GS (1949) In: Introduction to statistical mechanics, chapter 12. Oxford University Press, Clarendon, London
75. Ben Naim A (1979) Hydrophobic interaction. Plenum, New York
76. Rusanov AI (1982) J Coll Interface Sci 85:157
77. a. Stigter D (1954) Rec Trav Chim 73:593
   b. Lange H (1950) Koll Z 117:48
   c. Stauff J (1952) Koll Z 125:79
   d. Overbeek JTG, Stigter D (1956) Rec Trav Chim 75:1263
   e. Shinoda K (1953) Bull Chem Soc Jpn 26:101
   f. Hoeve CAJ, Benson GC (1957) J Phys Chem 61:1149
   g. Stauff J, Rasper J (1957) Koll Z 151:148
78. a. McBain JW (1950) Coll science. Heath, Boston
   b. McBain JW (1942) In: Kramer EO (ed) Advances in colloid science, vol 1. Wiley Interscience, New York, p 124
79. a. Stigter D, Mysels KJ (1955) J Phys Chem 59:45
   b. Mysels KJ, Princen LH (1959) J Phys Chem 63:1696
80. a. Tanford C (1972) J Phys Chem 76:3020
   b. Tartar HV (1955) J Phys Chem 59:1195
   c. Kalus J, Hoffmann H, Reizlein K, Ulbricht W, Ibel K (1982) Ber Bunsen-Ges Phys Chem 86:37
   d. Hoffmann H, Kalus J, Reizlein K, Ulbricht W, Ibel K (1982) Coll Polymer Sci 260:435
81. a. Hoffmann H, Rehage H, Platz G, Schorr W, Thurn H, Ulbricht W (1982) Coll Polymer Sci 260:1042
   b. Hoffmann H, Platz G, Rehage H, Schorr W, Thurn H, Ulbricht W (1981) Ber Bunsen-Ges Phys Chem 85:255
82. a. Huang C, Charlton JP (1972) Biochim Biophys Acta 46:1660
   b. Rosemann M, Litman BJ, Thompson TE (1975) Biochemistry 14:4826
   c. Rothmann JE, Dawidowicz EA (1975) Biochemistry 14:2809
   d. Hauser H, Barratt MD (1973) Biochem Biophys Res Commun 53:399

    e. Thompson TE, Huang C, Litman BJ (1974) In: Moscona AA (ed) The cell surface in development. Wiley, New York

83. a. Bauernschmitt D, Hoffmann H, Platz G (1981) Ber Bunsen-Ges Phys Chem 85:203
    b. Hoffmann H, Platz G, Rehage H, Schorr W (1982) Adv Coll Interface Sci 17:275

84. Miura M, Kodama M (1972) Bull Chem Soc Jpn 45:428

85. a. Sata N, Tyuzyo K (1953) Bull Chem Soc Jpn 26:177
    b. Kodama M, Miura M (1972) Bull Chem Soc Jpn 45:2265

86. a. Kodama M, Kubota Y, Miura M (1972) Bull Chem Soc Jpn 45:2953
    b. Ikeda S, Ozeki S, Tsunoda M (1980) J Coll Interface Sci 73:27

87. a. Kim J, Kim C, Song P, Lee K J Coll Interface Sci 80:294
    b. Kubota Y, Kodama M, Miura M (1973) Bull Chem Soc Jpn 46:100

88. Porte G, Poggi Y (1978) Phys Rev Lett 41:1481

89. Schorr W, Hoffmann H (1981) J Phys Chem 85:3160

90. Becher P, Arai H (1968) J Coll Interface Sci 27:634

91. a. Brockerhoff H (1977) In: van Tamelen EE (ed) Bioorganic chemistry, vol 3. Academic Press, New York
    b. Kunitake T, Okahata Y (1977) J Am Chem Soc 99:3860
    c. Kunitake T, Okahata Y (1978) Bull Chem Soc Jpn 51:1877

92. Nagamura T, Mihara S, Okahata Y, Kunitake T, Matsuo T (1978) Ber Bunsen-Ges Phys Chem 82:1093

93. a. Hoffmann H, Platz G, Ulbricht W (1981) J Phys Chem 85:1418
    b. Kunitake T, Okahata Y (1977) Chem Lett 1337
    c. Deguchi K, Mino J (1978) J Coll Interface Sci 65:155

94. a. Kerker M (1969) The scattering of light and other electromagnetic radiation. Academic Press, New York
    b. Zimm BH (1948) J Phys Chem 16:1093, 1099

95. a. Reiss-Husson F, Luzzati V (1964) J Phys Chem 68:3504
    b. Reiss-Husson F, Luzzati V (1966) J Coll Interface Sci 21:534

96. Jacrot B (1976) Rep Prog Phys 39:911

97. Stuhrmann H (1979) Chem Unserer Zeit 13:11

98. Berne B, Pecora R (1976) Dynamic light scattering. Wiley, New York

99. a. Stejakal EO, Tanner JE (1965) J Chem Phys 49:288
    b. Bull T, Lindman B (1974) Mol Cryst Liq Cryst 28:155

100. a. Almgren M, Swarup S (1982) J Phys Chem 86:4212
    b. Almgren M, Löfroth JE (1981) J Coll Interface Sci 81:486
    c. Lianos P, Zana R (1980) J Phys Chem 84:3339

101. a. Fréder'cq E, Houssier C (1973) Electric birefringence and electric dichroism. Clarendon, Oxford
    b. O'Konski CT (1976) Molecular electrooptics. Dekker, New York
    c. Jennings BR (1979) Electrooptics and dielectrics of macromolecules and colloids. Plenum, New York

102. a. Tsetkov V (1962) Vysokomol Soedin 4:894
    b. Janeschitz-Kriegl H, Burchard W (1968) J Polymer Sci 6:1953
    c. Peterlin A, Stuart HA (1943) Doppelbrechung, insbesondere künstliche Doppelbrechung. In: Hand- und Jahrbuch der Chemischen Physik, Leipzig

103. a. Einstein A (1906) Ann Phys 9:301; (1911) ibid 34:592
    b. Smoluchowski M v (1916) Koll Z 18:190
    c. Ostwald W (1924) Z Phys Chem 111:62
    d. Staudinger H (1930) Ber Dtsch Chem Ges 63:222
    e. Simha R (1940) J Phys Chem 44:25
    f. Simha R (1945) J Chem Phys 13:188
    g. Schurz J (1974) Einführung in die Strukturrheologie. Kohlhammer, Stuttgart
    h. Eirich FR (1956) Rheology. Academic Press, New York

104. Edwards SF, Evans KE (1982) J Chem Soc Faraday Trans II 78:113

105. a. Bachmann L, Schmitt-Fumian W (1973) In: Benedetti EL, Favard P (eds) Proc Soc Franc Microsc Electr. Paris, pp 63–72
    b. Bachmann L, Schmitt-Fumian W (1971) Proc Natl Acad Sci USA 68:2149

    c. Kutter P, Schmitt-Fumian W, Bachmann L (1976) Freeze-etching of micellar solutions. 6th Eur Congr Electron Microsc Proc, Jerusalem
106. a. Leibner JE, Jacobus J (1977) J Phys Chem 81:130
     b. Cabos C, Delord P (1978) J Phys 39:432
107. Fromhertz P (1981) Chem Phys Lett 77:460
108. Dill KA, Flory PJ (1981) Proc Natl Acad Sci USA 78:676
109. Gruen DWR (1981) J Coll Interface Sci 84:281
110. a. Tiddy GJT (1981) Chem Soc SPR Ser NMR, vol 10, chapter 12
     b. Cabane B (1981) J Phys 42:847
     c. Lindman B, Wennerström H (1980) In: Topics in current chemistry, vol 87. Springer, Berlin Heidelberg New York
     d. Persson BO, Drakenburg T, Lindman B (1976) J Phys Chem 80:2124
111. Bendedouch D, Chen S, Koehler WC (1983) J Phys Chem 87:153
112. a. Schorr W (1982) PhD Thesis, University Bayreuth
     b. Hoffmann H, Schorr W (1985) Ber Bunsen Ges Phys Chem 89:538
113. a. Overbeek JTG (1950) In: Mark H, Verwey EJW (eds) Advances in colloid chemistry, vol 3. Wiley Interscience, New York
     b. Overbeek JTG, Wiersma P (1967) In: Bier M (ed) Electrophoresis, vol 2. Academic Press, New York
114. a. Onsager L (1949) Ann NY Acad Sci 51:627
     b. De Gennes PG, Pincus P, Velasco RM, Brochard F (1967) J Phys 37:1461
     c. Ise N, Okubo T (1980) Accounts Chem Res 13:303
115. a. Brown JC, Pusey PN, Goodwin JW, Ottewill RH (1975) J Phys A Math Gen 8:664
     b. Cebula D, Thomas RK, Harris NM, Tabony J, White JW (1978) Faraday Discuss 65/6
     c. Magid LJ, Triolo R, Johnson JS Jr, Koehler WC (1982) J Phys Chem 86:164
116. a. Huglin MB (1972) Light scattering from polymer solutions. Academic Press, New York
     b. McIntyre D, Gormick F (1964) Light scattering from dilute polymer solutions. Gordon and Breach, New York
117. Doi M, Edwards SF (1978) J Chem Soc Faraday Trans II 74:560
118. a. Doi M, Edwards SF (1978) J Chem Soc Faraday Trans II 74:918
     b. Doi M (1975) J Phys 36:607
     c. Doi M (1980) J Polymer Sci 18:1991
119. a. Ulmius J, Wennerström H, Johansson LBA, Lindblom G, Gravsholt S (1979) J Phys Chem 83:2232
     b. Gravsholt S (1979) Naturwissenschaften 66:263
     c. Hoffmann H, Platz G, Rehage H, Reizlein K, Ulbricht W (1981) Makromol Chem 182:451
     d. Ulmius J, Lindman B, Lindblom G, Drakenburg T (1974) J Coll Interface Sci 65:88
     e. Barker CA, Saul D, Tiddy GJT, Wheeler BA, Willis E (1974) J Chem Soc Faraday Trans I 70:154
     f. Saul D, Tiddy GJT, Wheeler BA, Wheeler PA, Willis E (1974) J Chem Soc Faraday Trans I 70:163
120. Hayter JB, Penfold J (1981) Mol Phys 42:109
121. Brown JC, Pusey PN, Goodwin JW, Ottewill RH (1977) J Coll Interface Sci 58:357
122. a. Pusey PN, Tough RJA (1981) In: Pecora R (ed) Dynamic light scattering and velocimetry: applications of photon correlation spectroscopy. Plenum, New York
     b. Hess W (1980) In: Degiorgio V, Corti M, Giglio M (eds) Light scattering in liquids and macromolecular solutions. Plenum, New York
     c. Ackerson BJ (1976) J Chem Phys 64:242
123. a. Hoffmann H, Rehage H, Schorr W, Thurn H (1984) In: Mittal KL, Lindman B (eds) Surfactants in solution, vol 1. Plenum, New York, p 425
     b. Hoffmann H, Kalus J, Thurn H, Ibel K (1983) Ber Bunsen-Ges Phys Chem 87:1120
124. Verwey EJ, Overbeek JTG (1948) Theory of stability of lyophobic colloids. Elsevier, Amsterdam
125. Vassiliades AE (1970) In: Cationic surfactants. Surfactant Sci Ser, vol 4. Dekker, New York

126. a. Brown GH, Wolken JJ (1979) Liquid crystals and biological structures. Academic Press, New York
    b. Gray PW, Winsor PA (1974) Liquid crystals and plastic crystals. Wiley, New York
    c. Friberg S (1976) Lyotropic liquid crystals. In: Advances in chemistry series, vol 152. Am Chem Soc Wash
    d. De Gennes PG (1974) The physics of liquid crystals. Clarendon, Oxford
    e. Kelker M, Hatz R (1980) Handbook of liquid crystals. Chemie, Weinheim
    f. Helfrich W, Heppke G (1980) Liquid crystals of one- and two-dimensional order. Springer series in chemical physics, vol 11. Springer, Berlin Heidelberg New York
    g. Ekwall P (1975) In: Brown G (ed) Advances in liquid crystals. Academic Press, New York
    h. Luckhurst GR, Gray GW (1979) The molecular physics of liquid crystals. Academic Press, New York
    i. Tiddy GJT (1980) Phys Rep 57:1
127. Eicke HF (1980) In: Topics in current chemistry 87:85. Springer, Berlin Heidelberg New York
128. a. Ray A (1969) J Am Chem Soc 11:6511
    b. Ray A (1971) Nature 231:313
129. Eicke HF, Christen H (1974) J Coll Interface Sci 46:417
130. a. Eicke HF, Rehak J (1976) Helv Chim Acta 59:2883
    b. Debye P, Prins WJ (1958) J Coll Soc 13:86
    c. Honig JG, Singleterry CR (1954) J Phys Chem 58:201
131. a. Soldatov V, Oedberg L, Högfeldt E (1975) In: Ion exchange and membranes, vol 2. Gordon and Breach, p 83
    b. Zundel G (1969) In: Hydration and intermolecular interaction. Academic Press, New York
    c. Eicke HF, Christen H (1976) Helv Chim Acta 61:2883
132. Singleterry CR (1955) J Am Oil Chem Soc 32:446
133. Tavernier S (1977) In: Inverted micellar structure of anionic surfactants in apolar medium. Proefschrift, Universiteit Antwerpen
134. Anacker EW (1970) In: Jungermann E (1970) Cationic surfactants. Dekker, New York
135. Kaufmann S, Singleterry CR (1955) J Coll Sci 10:139; (1952) ibid 7:453
136. a. Kertes AS, Gutmann H (1976) Surfactants in organic solvents. In: Matijević E (ed) Surface and colloid science, vol 8. Wiley, New York, p 193
    b. Stauff J (1960) Kolloidchemie. Springer, Berlin Göttingen Heidelberg
137. a. Fendler EJ et al. (1973) J Phys Chem 77:1432
    b. Fendler EJ et al. (1973) J Chem Soc Faraday Trans I 68:280
138. Eicke HF, Christen H (1974) J Coll Interface Sci 48:281
139. a. Madia WJ, Nichols AL, Ache HJ (1975) J Am Chem Soc 97:5041
    b. Jean Y, Ache HJ (1978) J Am Chem Soc 100:984
140. El Seoud OA et al. (1973) J Phys Chem 77:1876
141. a. Muller N (1977) In: Mittal KL (ed) Micellization, solubilization and microemulsions, vol 1. Plenum, New York, p 229
    b. Kitahara A (1970) In: Jungermann E (ed) Cationic surfactants. Dekker, New York
142. a. Shinoda K (1953) Bull Chem Soc Jpn 26:101
    b. Hutchinson E, Inaba A, Bailey LG (1955) Z Phys Chem 5:344
    c. Fowkes FM (1962) J Phys Chem 66:1843
143. a. Mukerjee P (1975) In: Olphen Van, Mysels KJ (eds) Enriching topics from colloid and surface science. La Jolla
    b. Eicke HF, Denss A (1978) J Coll Interface Sci 64:386
144. a. Pilpel N (1960) Trans Faraday Soc 56:893; (1961) ibid 57:1426
    b. Kitahara A, Kon-no K (1975) In: Colloidal dispersions and micellar behavior. ACS Symp Ser 9:225
145. Eicke HF (1977) In: Mittal KL (ed) Micellization, solubilization and microemulsions, vol 1. Plenum, New York, p 429
146. Huglin MB (1978) In: Determination of molecular weights by light scattering. Topics Curr Chem 77:141

147. a. Tanford C (1961) In: Physical chemistry of macromolecules. Wiley, New York
     b. Einstein A (1956) Investigations on the theory of brownian movement. Daver Publ, New York
     c. Frisch WL, Simha R (1956) In: Eirich F (ed) Rheology: theory and applications, vol 1. Academic Press, New York, p 525
148. a. Mathews MB, Hirschhorn E (1953) J Coll Sci 8:86
     b. Zulauf M, Eicke HF (1979) J Phys Chem 83:480
149. a. Bergmann K, Eigen M, Maeyer De L (1963) Ber Bunsen-Ges Phys Chem 67:819
     b. Eicke HF, Shepherd JWC, Steinemann A (1974) Helv Chim Acta 57:1951
150. Little RC, Singleterry CR (1964) J Phys Chem 68:3453
151. a. Hildebrand JH, Scott RL (1964) The solubility of nonelectrolytes. Dover Publ, New York
     b. Hildebrand JH, Prausnitz JM, Scott RL (1970) Regular and related solutions. Van Nostrand Reinhold, New York
152. a. Eicke HF, Christen H, Hopmann RFW (1975) In: Wolfram E (ed) Int Conf Coll Surface Science, vol 1. Budapest Akademiai Kiado, p 489
     b. Zana R (1978) In: Proc Sect Solution Chem Surfactants, 52nd Coll Surface Sci Symp, Knoxville
153. a. Mijnlieff PF, Dithmarsch R (1965) Nature 208:889
     b. Takeda K, Yasunaga T (1972) J Coll Interface Sci 40:127
     c. Herrmann U, Kahlweit M (1973) Ber Bunsen-Ges Phys Chem 77:1119
     d. Tanjić T, Hoffmann H (1973) Z Phys Chem NF 86:322
154. a. Kresheck GC et al. (1966) J Am Chem Soc 88:246
     b. Bennion BC et al. (1969) J Phys Chem 73:3288
     c. Takeda K, Yasunaga T (1973) J Coll Interface Sci 45:406
     d. Tondre C, Lang J, Zana R (1975) J Coll Interface Sci 52:372
155. a. Yasunaga T, Fujii S, Miura M (1969) Coll Interface Sci 30:399
     b. Zana R, Lang L (1968) Comp Rend Acad Sci Paris 266:1347
     c. Rassing J, Sams P, Wyn-Jones E (1974) J Chem Soc Faraday Trans II 70:1247
156. a. Lang J, Auborn JJ, Eyring EM (1972) J Coll Interface Sci 41:484
     b. Yasunaga T, Takeda K, Harada S (1973) J Coll Interface Sci 42:457
157. a. Inoue H, Nakagawa T (1966) J Phys Chem 70:108
     b. Muller N, Platko E (1971) J Phys Chem 75:547
     c. Muller N (1972) J Phys Chem 76:3017
158. a. Folger R, Hoffmann H, Ulbricht W (1974) Ber Bunsen-Ges Phys Chem 78:986
     b. Bauer R, Bösl O, Hoffmann H, Ulbricht W (1974) Ber Bunsen-Ges Phys Chem 78:1271
     c. Lang J, Tondre C, Zana R, Bauer R, Hoffmann H, Ulbricht W (1975) J Phys Chem 79:276
159. Aniansson EAG, Wall SN (1974) J Phys Chem 78:1024; (1975) ibid 79:857
160. a. Aniansson EAG, Almgren M, Wall SN, Hoffmann H, Kielmann I, Ulbricht W, Zana R, Lang J, Tondre C (1976) J Phys Chem 80:905
     b. Hoffmann H, Nagel HR, Platz G, Ulbricht W (1976) Coll Polymer Sci 254:812
     c. Hoffmann H, Tagesson B (1978) Z Phys Chem NF 110:113
     d. Hoffmann H, Lang R, Pavlović D, Ulbricht W (1979) Croat Chim Acta 52:87
161. Teubner M, Dieckmann S, Kahlweit M (1974) Ber Bunsen-Ges Phys Chem 78:986

Chapter 13 Unfolding of Proteins

WOLFGANG PFEIL

Symbols

K_{unf}	equilibrium constant for protein unfolding
K_{unf}^0	equilibrium constant for protein unfolding at standard conditions (zero denaturant concentration)
k	denaturant binding constant
a	denaturant activity
Δn	preferential interaction parameter
ΔG_{unf}	Gibbs energy change for protein unfolding
ΔG_{unf}^0	Gibbs energy change for protein unfolding at standard conditions (zero denaturant concentration)
ΔG_{unf}^{res}	Gibbs energy change for protein unfolding per mole of residues
ΔH_{unf}	enthalpy change for protein unfolding
ΔH_{unf}^{cal}	enthalpy change for protein unfolding obtained by calorimetry
ΔH_{unf}^{vH}	enthalpy change for protein unfolding obtained by van't Hoff treatment
$\Delta C_{p, unf}$	heat capacity change for protein unfolding
T	temperature
T_{trs}	transition temperature (referring to $\Delta G_{unf} = 0$)
N, X, U	native, intermediate and unfolded states of protein
k_i	rate constant
vH	equilibrium treatment by means of the van't Hoff equation
*	the value was calculated using other thermodynamic quantities in the original paper

Abbreviations

DSC	differential scanning calorimetry
GuHCl	guanidine hydrochloride

1 Introduction

This chapter contains tabulated data of Gibbs energy changes (ΔG_{unf}) enthalpy changes (ΔH_{unf}), and heat capacity changes ($\Delta C_{p, unf}$) associated with protein unfolding. The data are relevant for various purposes, ranging from practical applications to theoretical considerations on structure stabilizing forces. For the diversity of these aspects the reader is referred to reviews and monographs [1–17].

The tabulated data reflect the experimental accuracy in determination of biothermodynamic quantities. Various approaches for the determination of the Gibbs energy changes of protein unfolding are commonly used. Some of the approaches include a priori assumptions or suffer from experimental limitations. Therefore, the approaches will be briefly introduced below in order to provide the user with additional information which enables him to assess critically whether or not the presented data meet the accuracy required for the intended purpose.

The negative Gibbs energy change for protein unfolding can be considered as a measure of the net stabilization energy of the protein structure. As in the case of former data compilations [11, 16], ΔG_{unf} values are often required for comparisons between mutants, proteins in liganded and unliganded form etc. The comparison is, of course, only valid for identical experimental conditions and data treatment. Therefore, data obtained by various approaches (or extrapolation procedures) are included in Table 1 without selection, in order to offer as many ΔG_{unf} values as possible for the purpose of comparison.

In Table 3, which contains values of enthalpy changes and heat capacity changes of protein unfolding, preference was given to calorimetric measurements. In particular, differential scanning calorimetry offers the possibility to determine $\Delta C_{p,\,unf}$ both from single calorimetric recordings and from the temperature dependence of the (calorimetric) heat, ΔH_{unf}^{cal}, according to Kirchhoff's law $[d(\Delta H_{unf}^{cal})/dT]_p = \Delta C_{p,\,unf}$. Indirect determinations of the heat capacity change associated with protein unfolding (making use of the second derivative of the logarithm of the equilibrium constant as a function of temperature) are rather sensitive to error propagation [14, 18]. In comparison to direct calorimetric [20] measurement, all procedures which use urea or guanidine hydrochloride as denaturants are rather difficult to evaluate quantitatively due to the unavoidable assumptions of the heat involved in protein-denaturant interaction (see, e.g., [19]). Table 3 contains values of heat capacity changes, enthalpy changes, and transition temperatures. The enthalpy values refer, where possible, to the observed transition temperatures.

Tables 1–3 contain, in addition to the thermodynamic quantities, information concerning the experimental conditions, the approaches used, and references. For further details the reader is referred to the original papers. Most of the data taken from the original sources were calculated in Joules. Some few data were taken from figures or were calculated based on other thermodynamic quantities given in the original papers (the latter values have been marked by *). The values of heat capacity changes in Table 3 are in most cases assumed to be temperature- and pH-invariant within the range of experimental conditions [21]. Gibbs energy values refer, if calculation was possible, to standard conditions: $T^0 = 25\ ^\circ C$, $pH^0 = 7.0$ and denaturant concentration $c = 0$. In view of the common interest in having ΔG_{unf} over a wide range of pH and temperature, we refer to phase diagrams for combined pH and temperature dependence for α-chymotrypsin, cytochrome b_5, cytochrome c, lysozyme, myoglobin, parvalbumin and ribonuclease (see Table 1).

2 The Approaches in Determining ΔG_{unf}

The following section contains information concerning the approaches used to determine Gibbs energy changes associated with protein unfolding as well as abbreviations used in Tables 1–3. The most frequently used means for unfolding proteins are heat, acid, urea, and guanidine hydrochloride (GuHCl). Since

Table 1. Gibbs energy change for protein unfolding. Units: Gibbs energy change (ΔG_{unf}) is given in kJ mol^{-1}; temperature is given in degrees Celsius (°C). *Abbreviations:* DSC = differential scanning calorimetry; GuHCl = guanidine hydrochloride; Eq. (No) = equation, see text; k = proposed denaturant binding constant in Eq. (1); * = calculated values based on thermodynamic quantities given in the original paper

otein	pH	T °C	ΔG_{unf} kJ/mol	Approach	Remarks	Reference
tin						
actin	7.5	25	27	Heat	Optical methods	[40]
kaline phosphatase						
ɔoalkaline phosphatase	7.5	30	83.1	DSC		[41]
ɔophosphoryl alkaline phosphatase	7.5	30	23.4	DSC		[41]
ɔomonomer	7.5	30	31.8	DSC	In the presence of formamide	[42]
olipoprotein A1						
ɔolipoprotein A1	9.2	25	11.3	Urea	Eq. (3)	[43]
ɔolipoprotein A1	9.2	37	10.0	DSC		[43]
ɔolipoprotein A1	8	23	15.5	GuHCl	Eq. (1) k = 0.8	[44]
ɔolipoprotein A1	9	37	43.9	DSC	Reconstituted with dimyristoyl-lecithin	[45]
olipoprotein C II	7.4	25	11.7	GuHCl	Eqs. (1) and (2)	[46]
Arabinose binding protein						
Arabinose-binding protein (E. coli)	7.4	25	38.5*	DSC		[163]
partate aminotransferase						
ɔoenzyme	7.1	25	251 ±13	Urea	Eq. (3)	[47]
ɔoenzyme, multiple subform α	8.2	30	126	DSC		[48]
ridoxal form, multiple subform α	6.0	30	226	DSC		[48]
ridoxal phosphate form	7.1	25	331 ± 8	Urea	Eq. (3)	[47]
ridoxal phosphate form multiple subform α	8.2	30	238	DSC		[48]
ridoxal phosphate form multiple subform α	8.2	30	222	DSC		[48]
ridoxamine form multiple subform α	8.2	30	159	DSC		[48]
ridoxamine phosphate form	7.1	25	243 ±13	Urea	Eq. (3)	[47]
ɪrious liganded forms						[47, 48]
ɪrbonic anhydrase						
ɪrbonic anhydrase, bovine	–	–	45	Hydrogen exchange		[49]
ɪrbonic anhydrase, bovine guadinated	–	–	51	Hydrogen exchange		[49]
Chymotrypsin						
Chymotrypsin	3	25	30.1	Heat	Optical method	[50]
Chymotrypsin	4	25	48.5 ± 2.1	DSC		[21]
Chymotrypsin	4.0	25	45	GuHCl	Eq. (2)	[188]

Table 1 (continued)

Protein	pH	T °C	ΔG_{unf} kJ/mol	Approach	Remarks	Referen
α-Chymotrypsin	4.0	25	49	GuHCl	Eq. (3)	[188]
α-Chymotrypsin	4.3	25	57.3	GuHCl	Eq. (1), k=1.2	[51]
α-Chymotrypsin	4.3	25	47.7	GuHCl	Eq. (1), k=0.8	[51]
α-Chymotrypsin	4.3	25	49.8	GuHCl	Eq. (1), k=0.6	[51]
α-Chymotrypsin	4.3	25	32.6	GuHCl	Eq. (2)	[24]
α-Chymotrypsin	4.3	25	35.1	Urea	Eq. (2)	[24]
α-Chymotrypsin	4.3	25	43.5	GuHCl	Eq. (3)	[24]
α-Chymotrypsin	4.3	25	36.8	Urea	Eq. (3)	[24]
α-Chymotrypsin	7	25	58.6	Heat	Optical method	[7]
α-Chymotrypsin	–	–	51.5	GuHCl	Eq. (1)	[53]
α-Chymotrypsin phase diagram	2 – 4	0–100		DSC		[14]
Chymotrypsinogen						
Chymotrypsinogen	3	25	31	Heat	Optical method	[54, 55]
α-Chymotrypsinogen	4.0	25	40.6*	DSC	Extrapolated to zero protein concentration	[182]
Chymotrypsinogen	7	25	63 ±21	Heat	Optical method	[56]
Chymotrypsinogen	–	–	52	Hydrogen exchange		[49]
Chymotrypsinogen, guadinated	–	–	59	Hydrogen exchange		[49]
Collagen					Thermostability data see [17] Table 2	
Cytochrome b₅						
Cytochrome b₅	7.4	25	21	DSC		[57]
Cytochrome b₅	8.0	25	16	DSC	Reconstituted with dimyristoyl-lecithin	[57]
Cytochrome b₅ lower temperature range	7.4	25	13.6	Heat	Optical methods	[58]
Cytochrome b₅ higher temperature range	7.4	25	28.4	Heat	Optical methods	[58]
Cytochrome b₅ fragment 12–97	7	25	27.6 ± 2.0	DSC		[59]
Cytochrome b₅ fragment 1–90	7	25	25 ± 2	DSC		[60]
Cytochrome b₅ fragment 1–90, phase diagram	7 –10	0–100		DSC		[60]
Cytochrome c						
Bovine cytochrome c, ferric form	4.8	25	37.7 ± 2.5	DSC		[21]
Bovine cytochrome c, ferric form, phase diagram	2.5– 5	0–100		DSC		[14]
Bovine cytochrome c, ferric form	6.5	25	64.4	GuHCl	Eq. (1)	[52]

Table 1 (continued)

otein	pH	T °C	ΔG_{unf} kJ/mol	Approach	Remarks	Reference
ovine cytochrome c, ferric form	6.5	25	35.1	GuHCl	Eq. (2)	[52]
ovine cytochrome c, ferric form	7	25	32.1	GuHCl	Eq. (2)	[61]
ovine cytochrome c, ferrous form	7	25	34.6	GuHCl	Eq. (2)	[61]
andida krusei cyt. c, ferric form	7	25	58.6	GuHCl	Eq. (1)	[52]
andida krusei cyt. c, ferric form	7	25	30.3	GuHCl	Eq. (2)	[52]
aicken cyt. c, ferric form	7	25	32.1	GuHCl	Eq. (2)	[61]
og cyt. c, ferric form	7	25	30.4	GuHCl	Eq. (2)	[61]
onkey cyt. c, ferric form	7	25	32.1	GuHCl	Eq. (2)	[61]
og cyt. c, ferric form	7	25	47	GuHCl	See [62]	[62]
orse cyt. c, ferric form	4.0	16± 1	41.9 ± 5.9	Urea	Eq. (2) electrophoresis	[63]
orse cyt. c, ferric form succinylated	4.0	16± 1	38.9 ± 8.4	Urea	Eq. (2) electrophoresis	[63]
orse cyt. c, ferric form	5.6	23	17	Urea	Eq. (2)	[64]
orse cyt. c, ferric form	–	–	39	Hydrogen exchange		[49]
orse cyt. c, ferric form guanidinated	–	–	44	Hydrogen exchange		[49]
orse cyt. c, ferric form	6.5	25	53.1	GuHCl	Eq. (1)	[52]
orse cyt. c, ferric form	6.5	25	30.4	GuHCl	Eq. (2)	[52]
orse cyt. c, ferric form	7	25	24.9	GuHCl+heat		[65]
orse cyt. c, ferric form	7	25	30.9	GuHCl	Eq. (2)	[61]
orse cyt. c, ferrous form	7	25	33.4	GuHCl	Eq. (2)	[61]
orse cyt. c, ferric form	7.6	25	43.9	GuHCl	Eq. (2)	[164]
orse cyt. c, ferric form	7.6	25	80.3	GuHCl	Eq. (1)	[164]
abbit cyt. c, ferric form	7	25	32.1	GuHCl	Eq. (2)	[61]
ana cyt. c, ferric form	7	25	32.7	GuHCl	Eq. (2)	[61]
ytochrome c-552						
ytochrome c-552	7.5	25	119.2 ± 0.6	GuHCl	Eq. (1), k = 1.2	[66]
ytochrome c-552	7.5	25	121	GuHCl	Eq. (1)	[67]
o-2 *Cytochrome c,* yeast	7.2	20	13	GuHCl	Eq. (2)	[68]
o-1 Cytocrome c, yeast iodacetamide blocked at Cys-102	7	20	8.8	GuHCl	Eq. (2)	[69]
hydrofolate reductase						
ihydrofolate reductase, E. coli, mutants	–	15	See [180]	Urea		[180]
NAse I						
NAse I	7	25	39	GuHCl	Eq. (2), 1 mM Ca^{2+}	[70]
NAse I	7	25	37	GuHCl	Eq. (2), no Ca^{2+}	[70]
brinogen					Thermostability data see [71]	

Table 1 (continued)

Protein	pH	T °C	ΔG_{unf} kJ/mol	Approach	Remarks	Reference
Flagellin						
Flagellin, monomeric, normal type	7	25	72*	DSC		[165]
Galactoside-binding protein in the presence of Ca^{2+}	7.8	25	11*	DSC		[145]
Glucose dehydrogenase						
Glucose dehydrogenase	–	–	3.9	Enzyme activity		[166]
Glucose dehydrogenase	–	–	6.9	GuHCl	Eq. (2)	[166]
Glycinin (11 S globulin from Soybean)						
Glycinin	7.6	25	64*	DSC	In the absence of NaCl, calculated per protomer	[190]
Growth factor, epidermal						
Growth factor, epidermal	7.5	25	67 ±29	GuHCl	Mean of various extrapolation procedures	[72]
Growth factor, epidermal cyanogen bromide derivative	7.5	25	18	GuHCl	Mean of various extrapolation procedures	[72]
Growth hormone						
Bovine growth hormone	7.9– 8.5	25	58.6	GuHCl	Eq. (3)	[73]
Ovine growth hormone	7.9– 8.5	25	40.6	GuHCl	Eq. (3)	[73]
Rat growth hormone	7.9– 8.5	25	33.5	GuHCl	Eq. (3)	[73]
Hemolysin						
Hemolysin of vibrio parahemolyticus	8.15	25	32	DSC		[86]
Histones H1 and H5					Thermostability data see [74]	
Immunoglobulin						
Wes L chain	7	20	23.0	GuHCl	Eq. (3)	[75]
Fab fragment, C region	–	–	62.8	GuHCl	Eq. (3), $16.7\,kJ\,mol^{-1}$ domain-domain interaction	[76]
Fab fragment, V region	–	–	50.2	GuHCl	Eq. (3), $5.4\,kJ\,mol^{-1}$ domain-domain interaction	
Human immunoglobulin G domain R.A.Fc (t)	7.5	25	24	GuHCl	Eq. (1), k = 0.6	[77]
Human immunoglobulin G domain pFc (c_{H3})	7.5	25	23	GuHCl	Eq. (1), k = 0.6	[77]
Human immunoglobulin G domain (c_{H2})	7.5	25	24	GuHCl	Eq. (1), k = 0.6	[77]
Fragment c_L intact	7.5	25	24	GuHCl	Eq. (1), k = 0.6	[78]
Fragment c_L intact	8.0	25	22	GuHCl	Eq. (1), k = 0.6	[189]
Fragment c_L intact	10.3	25	22	GuHCl	Eq. (1), k = 0.6	[189]
Fragment c_L intact	7.5	25	26	GuHCl	Eq. (1), k = 1.2	[78]
Fragment c_L reduced	7.5	25	7	GuHCl	Eq. (1), k = 0.6	[78]
Fragment c_L reduced	7.5	25	8	GuHCl	Eq. (1), k = 1.2	[78]

Table 1 (continued)

otein	pH	T °C	ΔG_{unf} kJ/mol	Approach	Remarks	Reference
ulin						
sulin	3	20	27	Hydrogen exchange		[35]
c-Repressor, headpiece						
c-Repressor, headpiece	7.25	20	10.0	Urea	Eq. (2)	[79]
c-Repressor, headpiece	–	20	8.4 ± 3.3	Heat		[79]
c-Repressor, headpiece	7.25	20	12.1	Urea	Eq. (3)	[79]
c-Repressor, headpiece	8.0	25	10.8	DSC		[80]
Lactalbumin						
vine α-lactalbumin	5.5	25	15.9	GuHCl	Eq. (1), pH-function	[81]
vine α-lactalbumin	5.5	25	17.6	GuHCl	Eq. (2)	[82]
vine α-lactalbumin	6.3	25	21.9	DSC		[83]
vine α-lactalbumin	6.65	25	27.2	GuHCl	Three-state model	[84]
vine α-lactalbumin	7	25	18.4	GuHCl	Eq. (1), pH-function	[81]
vine α-lactalbumin	7	25	28.5	GuHCl	Eq. (1)	[85]
vine α-lactalbumin	7	25	17.6	GuHCl	Eq. (2)	[87]
vine α-lactalbumin	7	25	18.4	Urea	Eq. (2)	[87]
vine α-lactalbumin	7	25	17.6	GuHSCN	Eq. (2)	[87]
vine α-lactalbumin	7.3	25	26.4	GuHSCN	Eq. (1)	[88]
vine α-lactalbumin	7.3	25	18.0	GuHSCN	Eq. (11) in [23]	[88]
vine α-lactalbumin	7.3	25	23.8	GuHSCN	Anion binding	[88]
vine α-lactalbumin	7.3	25	17.6	GuHSCN	Eq. (3), $\alpha = 0.14$	[88]
vine α-lactalbumin	7.3	25	22.2	GuHSCN	Eq. (3), $\alpha = 0.16$	[88]
vine α-lactalbumin	–	–	18	Hydrogen exchange		[49]
vine α-lactalbumin calcium free form	8.0	25	12.5 ± 0.3	DSC	Ionic strength 0.1	[184]
vine α-lactalbumin guanidinated	–	–	26	Hydrogen exchange		[49]
man α-lactalbumin	6 – 7	25	26.4	GuHCl	Eq. (1)	[89]
man α-lactalbumin calcium free form	8.0	25	6.1 ± 0.1	DSC	Ionic strength 0.1	[184]
ctic dehydrogenase						
ctic dehydrogenase	7.6	20	110	Pressure denaturation and dissociation		[90]
Lactamase						
Lactamase	7	25	25.0	GuHCl	Eq. (3)	[112]
Lactamase	7	20	8.5 ± 0.9	GuHCl	Eq. (2)	[112, 113]
Lactamase	6.6	24	9.1 ± 0.8	GuHCl	Eq. (2)	[113]
Lactamase	6.6	24	8.8 ± 0.8	Biguanide HCl	Eq. (2)	[113]
Lactamase	6.6	24	8.8 ± 0.7	Propyl-biguanide HCl	Eq. (2)	[113]
Lactamase	6.6	24	19.2 ± 1.3	Hexyl-biguanide HCl	Eq. (2)	[113]

Table 1 (continued)

Protein	pH	T °C	ΔG_{unf} kJ/mol	Approach	Remarks	Referen-
β-Lactamase	6.6	24	19.2	n-Decyl biguanide HCl	Eq. (2)	[113]
β-Lactamase	7.0	20	8.52± 0.9	GuHCl	Transition N→H	[114]
β-Lactamase	7.0	20	11.19± 0.74	GuHCl	Transition H→U	[114]
β-Lactamase	7.0	20	10.15± 1.50	Urea	Transition N→H	[114]
β-Lactamase	7.0	20	10.09± 1.61	Urea	Transition H→U	[114]
β-Lactoglobulin						
Bovine β-lactoglobulin A	3.15	25	42.7	Urea	Eq. (2)	[11]
Bovine β-lactoglobulin A	3.15	25	55.2	GuHCl	Eq. (2)	[11]
Bovine β-lactoglobulin B	3.15	25	49.0	Urea	Eq. (2)	[11]
Bovine β-lactoglobulin	5.5	0	> 33	Hydrogen exchange		[35]
Bovine β-lactoglobulin A and B	7.2	16±1	31.4 ± 3.3	Urea	Electrophoresis	[63]
Modified β-lactoglobulin B: Mixed disulfide derivatives						
Native	2.83	25	46.4	Urea	Eq. (2)	[160]
Propyl-	2.83	25	16.3	Urea	Eq. (2), apparent ΔG_{unf}, see [160]	[160]
Aminoethyl-	2.83	25	3.0	Urea	Eq. (2), apparent ΔG_{unf}, see [160]	[160]
Carboxyethyl	2.83	25	27.4	Urea	Eq. (2), apparent ΔG_{unf}, see [160]	[160]
Hydroxyethyl	2.83	25	22.2	Urea	Eq. (2), apparent ΔG_{unf}, see [160]	[160]
Goat β-lactoglobulin	–	25	60.7	Estimation, minimal value		[26]
Goat β-lactoglobulin	3.2	25	109	GuHCl	Eq. (1), k=1.2	[51]
Goat β-lactoglobulin	3.2	25	93	GuHCl	Eq. (1), k=0.8	[51]
Goat β-lactoglobulin	3.2	25	52.3	GuHCl	Eq. (2)	[24]
Goat β-lactoglobulin	3.2	25	43.9	Urea	Eq. (2)	[24]
Goat β-lactoglobulin	3.2	25	77.0	GuHCl	Eq. (3)	[24]
Goat β-lactoglobulin	3.2	25	50.6	Urea	Eq. (3)	[24]
Lipase						
Lipase from aspergillus	7	–	46	Heat	Optical method (CD)	[91]
Lipase from rhizopus	7	–	16.8	Heat	Optical method (CD)	[91]
Lysozyme						
Hen egg white lysozyme	7	25	44.4	GuHCl	Eq. (11) in 23, k=1.2	[23]
Hen egg white lysozyme	7	25	48.1	GuHCl	Eq. (13) in 23, k=1.2	[23]
Hen egg white lysozyme	7	25	59.4	GuHCl	Eq. (15) in 23, k=1.2	[23]
Hen egg white lysozyme	7	25	73.6	GuHCl	Eq. (16) in 23, k=1.2	[23]
Hen egg white lysozyme	7.3	25	52.7	GuHCl	Eq. (3)	[88]

Table 1 (continued)

otein	pH	T °C	ΔG_{unf} kJ/mol	Approach	Remarks	Reference
en egg white lysozyme	7	25	60.7 ± 3.3	DSC		[21, 92]
phase diagram	1.5–7.0	0–100		DSC		[14, 16, 92]
en egg white lysozyme	5.5	25	37.6	GuHCl	Eq. (2)	[167]
en egg white lysozyme	5.5	25	37.2	GuHCl	Eq. (2)	[87]
en egg white lysozyme	5.5	25	38.1	GuHSCN	Eq. (2)	[87]
en egg white lysozyme	5.5	25	36.8	Urea	Eq. (2)	[87]
en egg white lysozyme	2.9	25	29.7	GuHCl	Eq. (1), k = 0.1	[11]
en egg white lysozyme	2.9	25	38.9	GuHCl	Eq. (1), k = 0.6	[51]
en egg white lysozyme	2.9	25	44.8	GuHCl	Eq. (1), k = 1.2	[51]
en egg white lysozyme	2.9	25	24.3	GuHCl	Eq. (2)	[24]
en egg white lysozyme	2.9	25	38.1	GuHCl	Eq. (3)	[11, 24]
en egg white lysozyme	2.9	25	28.9	Urea	Eq. (1), k = 0.1	[11]
en egg white lysozyme	2.9	25	50.6	Urea	Eq. (1), k = 1.0	[11]
en egg white lysozyme	2.9	25	24.3	Urea	Eq. (2)	[24]
en egg white lysozyme	2.9	25	27.6	Urea	Eq. (3)	[24]
en egg white lysozyme	5.4	20	31.0	Hydrogen exchange		[93]
en egg white lysozyme	4.5–8.7	0	≧ 42	Hydrogen exchange		[35, 94]
en egg white lysozyme	9	25	54.4	Hydrogen exchange		[95]
en egg white lysozyme	–	–	45.2	Hydrogen exchange		[49]
en egg white lysozyme, guanidinated	–	–	45.2	Hydrogen exchange		[49]
en egg white lysozyme	–	25	40.6	Heat	Viscosity	[96]
en egg white lysozyme	4	60	17.6	DSC		[31]
en egg white lysozyme	4	60	18.0	Acid denaturation		[31]
en egg white lysozyme	2	25	24.3	DSC	Extrapol. to zero alcohol conc.	[97]
len egg white lysozyme	< 1	25	37.7	GuHCl	Eq. (1), k = 1.2	[23]
ysozyme, phage T4						
hage T4 lysozyme	3	49	8.8	Heat		[99]
hage T4 lysozyme	3	49	6.7	GuHCl	Eq. (2)	[99]
hage T4 lysozyme wild type	2	0	30	Heat	Optical method	[98]
hage T4 lysozyme mutant e R1	2	0	18	Heat	Optical method	[98]
hage T4 lysozyme mutant e RRR	2	0	18	Heat	Optical method	[98]
•hage T4 lysozyme wild type	–	25	25	Heat		[99]
•hage T4 lysozyme mutant (Arg 96→His)	–	25	8	Heat		[99]
•hage T4 lysozyme mutant (Glu 128→Lys or Ala)	–	25	17	Heat		[99]
•hage T4 lysozyme mutant N14	–	25	10	Heat		[99]
•hage T4 lysozyme mutant N19	–	25	7.5	Heat		[99]

Table 1 (continued)

Protein	pH	T °C	ΔG_{unf} kJ/mol	Approach	Remarks	Reference
Phage T4 lysozyme wild type	3	25	15.0*	Heat	Optical method	[168]
Phage T4 lysozyme mutant (Glu 128→Lys)	3	25	15.2*	Heat	Optical method	[168]
Phage T4 lysozyme mutant (Arg 96→His)	3	25	12.7*	Heat	Optical method	[168]
Phage T4 lysozyme mutant (Met 102→Val)	3	25	11.2*	Heat	Optical method	[168]
Phage T4 lysozyme mutant (Ala 146→Thr)	3	25	8.2*	Heat	Optical method	[168]
Myoglobin						
Cow metmyoglobin	7	25	35.1	GuHCl	Eq. (1)	[100]
Cow metmyoglobin	7	25	47.3	GuHCl	Eq. (3)	[101]
Dog metmyoglobin	7	25	26.4	GuHCl	Eq. (1)	[100]
Horse metmyoglobin	6	25	44.8	GuHCl	Eq. (1), k=1.2	[51]
Horse metmyoglobin	6	25	36.8	GuHCl	Eq. (1), k=0.8	[51]
Horse metmyoglobin	6	25	39.7	GuHCl	Eq. (1), k=0.6	[51]
Horse metmyoglobin	6	25	40.6	GuHCl	Eq. (3)	[51]
Horse metmyoglobin	6.6	25	31.8 ± 0.8	GuHCl	Eq. (2)	[87]
Horse metmyoglobin	6.6	25	33.9 ± 0.8	Urea	Eq. (2)	[87]
Horse metmyoglobin	7	25	49.4	GuHCl	Eq. (1)	[102]
Horse metmyoglobin	7	25	43.5	GuHCl	Eq. (3)	[102]
Horse metmyoglobin	7	25	46.0	GuHCl	Eq. (3)	[101]
Horse metmyoglobin	7	25	41.3	pH	See [102]	[102]
Horse metmyoglobin	7	25	49.0	pH	See [102]	[102]
Horse metmyoglobin	7	25	40.6	GuHCl	Eq. (1), k=0.6	[169]
Horse metmyoglobin	7	25	38.5	GuHCl	Eq. (3), α=0.24	[169]
Horse metmyoglobin	7	25	77.4	GuHCl	Eq. (3), α=0.4	[169]
Horse metmyoglobin	7	25	50.2	pH		[169]
Horse metmyoglobin	7	25	66.9	GuHCl	Three-state model	[169]
Horse metmyoglobin	7.1	25	31.8	GuHCl	Eq. (1)	[100]
Horse metmyoglobin	7.6	25	62.8	Urea	Eq. (3)	[103]
Horse metmyoglobin	4.8	25	10.5	Heat	Optical method	[104]
Horse metmyoglobin, azide	4.8	25	13	Heat	Optical method	[104]
Horse metmyoglobin, cyanide	4.8	25	16	Heat	Optical method	[104]
Horse metmyoglobin fluoride	4.8	25	11	Heat	Optical method	[104]
Horse metmyoglobin	11	25	21	Heat	Optical method	[104]
Horse metmyoglobin, azide	11	25	30	Heat	Optical method	[104]
Horse metmyoglobin, cyanide	11	25	33	Heat	Optical method	[104]
Horse metmyoglobin, fluoride	11	25	24	Heat	Optical method	[104]
Human metmyoglobin	7	25	36.8	GuHCl	Eq. (3)	[101]
Porpoise metmyoglobin	7	25	34.7	pH		[101]
Seal metmyoglobin	7	25	56.9	pH		[101]
Turtle metmyoglobin	7	25	30.1	GuHCl	Eq. (3)	[101]
Sperm whale apomyoglobin	8.1–9.8	25	37.7	Hydrogen exchange		[105, 106]

Table 1 (continued)

otein	pH	T °C	ΔG_{unf} kJ/mol	Approach	Remarks	Reference
erm whale metmyoglobin	6.2– 9.4	21	50.2	Hydrogen exchange		[105, 107]
erm whale metmyoglobin	7	25	48.7	pH	See [102]	[102]
erm whale metmyoglobin	7	25	58.6	pH	See [102]	[102]
erm whale metmyoglobin	7	25	55.3	GuHCl	Eq. (1)	[102]
erm whale metmyoglobin	7	25	59.8	GuHCl	Eq. (3)	[102]
erm whale metmyoglobin	7	25	56.9	GuHCl	Eq. (3)	[101]
erm whale metmyoglobin	7.6	25	62.8	Urea	Eq. (3)	[103]
erm whale metmyoglobin	9	25	58.6	pH	Eq. (3)	[30]
erm whale metmyoglobin	7.1	25	44.4	GuHCl	Eq. (1)	[100]
erm whale metmyoglobin	7.1	25	56.5	GuHCl	Eq. (1)	[108]
erm whale metmyoglobin	7.1	25	43.1	GuHCl	Eq. (2)	[108]
erm whale metmyoglobin, cyanide	7.1	25	61.9	GuHCl	Eq. (1)	[108]
erm whale metmyoglobin, cyanide	7.1	25	47.3	GuHCl	Eq. (2)	[108]
erm whale metmyoglobin, cyanide	10	25	50.2 ± 3.3	DSC		[21]
erm whale metmyoglobin, cyanide, phase diagram	10.5–12.5	0–100		DSC		[14]
erm whale metmyoglobin, azide	7.1	25	59.0	GuHCl	Eq. (1)	[108]
erm whale metmyoglobin, azide	7.1	25	44.8	GuHCl	Eq. (2)	[108]
yosin fragments					Thermostability data see Table 2	
valbumin						
valbumin	7	25	24.8	GuHCl	Eq. (1), k = 1.2	[109]
valbumin	–	–	56.5	Hydrogen exchange		[105]
pain						
pain	3.8	25	76*	DSC		[153]
aramyosin					Thermostability data see Table 2	
arvalbumin (carp, component III)						
arvalbumin	7	25	46.0	DSC	0.1 mM CaCl$_2$	[111]
arvalbumin	7	25	83.7	DSC	33 mM CaCl$_2$	[111]
arvalbumin, phase diagram	2–7	20–70		DSC	0.1 mM and 33 mM CaCl$_2$	[16]

Table 1 (continued)

Protein	pH	T °C	ΔG_{unf} kJ/mol	Approach	Remarks	Reference
Penicillinase see *β-Lactamase*						
Pepsin						
Pepsin	6.5	25	45.2	DSC	7.9 and 12.5 kJ/mol for the N-terminal lobe, 16.3 and 8.4 kJ/mol for the C-terminal lobe, 26 kJ/mol for the N-terminal lobe in the presence of pepstatin	[17, 115]
Pepsinogen						
Pepsinogen	6–8	25	31.8	Urea	Eq. (1), k=0.1	[116]
Pepsinogen	6–8	25	49.0	Urea	Eq. (1), k=1.0	[116]
Pepsinogen	6–8	25	27.2	Urea	Eq. (2)	[116]
Pepsinogen	6.5	25	65.7	DSC	25.9 kJ/mol and 39.7 kJ/mol for the two cooperative blocks of pepsinogen	[17, 115]
Phosphoglycerate kinase						
Phosphoglycerate kinase from Thermus thermophilus	7.5	25	49.7 ± 0.9	GuHCl	Eq. (1)	[117]
Phosphoglycerate kinase from yeast	7.5	25	22.3 ± 0.5	GuHCl	Eq. (1)	[117]
Phosphoglycerate kinase from yeast	7	25	17.9	GuHCl	Eq. (2)	[118]
Phosphoglycerate kinase from yeast	7	25	20.6	GuHCl	Eq. (3)	[118]
Phosphoglycerate kinase from horse muscle	7.4	23	12	GuHCl	Eq. (2) optical method	[170]
Phosphoglycerate kinase from horse muscle	7.4	23	26	GuHCl	Eq. (2) enzyme activity	[170]
Phycocyanin						
Phycocyanin	6	25	18.4	Urea	Eq. (2)	[119]
Pyruvate kinase						
Pyruvate kinase from rabbit muscle	6.5	5	6.9	GuHCl	Eq. (2), transition $A_4 \rightarrow A_4{}^*$	[120]
Pyruvate kinase from rabbit muscle	6.5	5	66.5	GuHCl	Eq. (2), transition $A_4{}^* \rightarrow 2A_2$	[120]
Pyruvate kinase from rabbit muscle	6.5	5	65.3	GuHCl	Eq. (2), transition $A_2 \rightarrow 2A_1$	[120]
Ribonuclease, bovine pancreatic						
Ribonuclease, bovine pancreatic	2.50	30	3.8	Heat		[8, 121]

Table 1 (continued)

otein	pH	T °C	ΔG_{unf} kJ/mol	Approach	Remarks	Reference
bonuclease, bovine pancreatic	2.77	30	8.6	Heat		[8, 121]
bonuclease, bovine pancreatic	3.0	25	19.7 ± 0.4	GuHCl	Eq. (2)	[171]
bonuclease, bovine pancreatic	3.0	25	15.9± 0.8	LiClO$_4$	Eq. (2)	[171]
bonuclease, bovine pancreatic	3.0	25	53.1 ± 0.8	LiCl, LiBr	Eq. (2)	[171]
bonuclease, bovine pancreatic	3.15	30	12.9	Heat		[8, 121]
bonuclease, bovine pancreatic	3.28	25	21.9*	DSC		[122]
bonuclease, bovine pancreatic	4.04	25	31.0*	DSC		[122]
bonuclease, bovine pancreatic	4	16±1	67	Urea	Electrophoresis	[63]
bonuclease, bovine pancreatic	4.7	38	27.2	Hydrogen exchange		[35]
bonuclease, bovine pancreatic	4.7	38	27.6	DSC		[21]
bonuclease, bovine pancreatic	4.8	25	31.4 ± 0.4	GuHCl+LiCl LiBr, NaBr	Eq. (2)	[172]
bonuclease, bovine pancreatic	5	25	44.4*	DSC		[122]
bonuclease, bovine pancreatic	5.0	25	39.6*	DSC	Extrapolated to zero protein concentration	[182]
bonuclease, bovine pancreatic	5	25	44	DSC	Extrapolated to zero dimethyl-sulfoxide concentration	[123]
bonuclease, bovine pancreatic	5.5	25	45.2 ± 2.5	DSC		[21]
bonuclease, bovine pancreatic	5.8	25	36	DSC		[181]
bonuclease, bovine pancreatic	6	25	54.5	GuHCl	Eq. (1)	[124]
bonuclease, bovine pancreatic	6	25	37.2	GuHCl	Eq. (11) in [23]	[124]
bonuclease, bovine pancreatic	6.5	25	59.4	GuHCl	Eq. (1), k = 1.0	[125]
bonuclease, bovine pancreatic	6.5	25	61.5	GuHCl	Eq. (1), k = 1.3	[125]
bonuclease, bovine pancreatic	6.5	25	59.8	GuHCl	Eq. (3)	[125]
bonuclease, bovine pancreatic	6.6	25	38.9	GuHCl	Eq. (2)	[24]
bonuclease, bovine pancreatic	6.6	25	32.2	Urea	Eq. (2)	[24]
bonuclease, bovine pancreatic	6.6	25	61.9	GuHCl	Eq. (3)	[24]

Table 1 (continued)

Protein	pH	T °C	ΔG_{unf} kJ/mol	Approach	Remarks	Referen
Ribonuclease, bovine pancreatic	6.6	25	50.6	Urea	Eq. (3)	[24]
Ribonuclease, bovine pancreatic	6.6	25	79.1	GuHCl	Eq. (1), k = 1.2	[51]
Ribonuclease, bovine pancreatic	6.6	25	67.4	GuHCl	Eq. (1), k = 0.6	[51]
Ribonuclease, bovine pancreatic	6.6	25	68.2	GuHCl	Eq. (1), k = 0.8	[51]
Ribonuclease, bovine pancreatic	7	25	58.6*	DSC		[122]
Ribonuclease, bovine pancreatic	7	25	31.4	GuHCl	Eq. (2)	[87]
Ribonuclease, bovine pancreatic	7	25	31.0	Urea	Eq. (2)	[87]
Ribonuclease, bovine pancreatic	7	25	29.7	GuHSCN	Eq. (2)	[87]
Ribonuclease, bovine pancreatic	7.35	25	41.8 ± 2.1	Urea	Eq. (2)	[126]
Ribonuclease, bovine pancreatic	7.6	25	43.1	GuHCl	Eq. (2)	[164]
Ribonuclease, bovine pancreatic	7.6	25	78.2	GuHCl	Eq. (1)	[164]
Ribonuclease, bovine pancreatic	–	21	75.7	Heat		[99]
Ribonuclease, bovine pancreatic	–	21	78.7	GuHCl	Eq. (2)	[99]
Ribonuclease, bovine pancreatic	–	43	37.7	Heat		[99]
Ribonuclease, bovine pancreatic	–	43	38.5	GuHCl	Eq. (2)	[99]
Ribonuclease, bovine pancreatic	–	–	48	Hydrogen exchange		[49]
Ribonuclease, bovine pancreatic	Neutral	25	47–102	Prediction		[26]
Ribonuclease, bovine pancreatic, phase diagram	2.5–7.0	0–100		DSC		[14]
Ribonuclease (RNase), modified:						
RNase des-(1–20)	7	25	7.8*	DSC		[122]
RNase des-(1–20) mixed with the peptide (1–20)	7	25	24.5*	DSC		[122]
RNase des-(121–124)	7	26	42.3 ± 2.9	GuHCl	Mean of various approaches	[125]
RNase des-(120–124)	7	26	3.6	GuHCl	Eq. (3)	[127]
RNase des-(119–124)	7	26	3.3	GuHCl	Eq. (3)	[127]
RNase glycosylated	7	26	63.2 ± 0.8	GuHCl	Mean of various approaches	[128]
RNase succinylated	4	16 ± 1	49.6	Urea	Electrophoresis	[63]
RNase guanidinated	–	–	54	Hydrogen exchange		[49]

Table 1 (continued)

otein	pH	T °C	ΔG_{unf} kJ/mol	Approach	Remarks	Reference
um albumin, bovine						
um albumin, bovine	–	–	32.6	Hydrogen exchange		[105]
um albumin, bovine	7.15	25	> 30	DSC		[110]
um albumin, bovine	–	–	39	Hydrogen exchange		[49]
um albumin, guanidinated	–	–	47	Hydrogen exchange		[49]
btilisin inhibitor						
otilisin inhibitor	7	25	59*	DSC (see also Table 3)		[161]
ermolysin fragment (206–316)						
ermolysin fragment	7.5	37	13	Heat	Optical method	[129]
ermolysin fragment	7.5	25	23	GuHCl	Eq. (2)	[129]
ermolysin fragment	7.5	25	31	GuHCl	Eq. (3)	[129]
ermolysin fragment	7.5	25	23	Urea	Eq. (2)	[129]
ermolysin fragment	7.5	25	27	Urea	Eq. (3)	[129]
ioredoxin						
ioredoxin, E. coli	7.0	25	30.5	GuHCl	Eq. (2)	[185]
xins						
totoxin I	4.5	25	24*	DSC		[186]
urotoxin I	5.7	25	24*	DSC		[186]
urotoxin II	5.0	25	36*	DSC		[186]
opocollagen					Thermostability data see [17]	
opomysin					See Table 2	
oponin C					Thermostability data see [17]	
opsin	–	25	67	DSC		[15]
ypsin inhibitor, ovine pancreatic						
ypsin inhibitor ovine pancreatic	3	25	51.7	DSC		[173]
ypsin inhibitor ovine pancreatic	–	25	46	DSC		[15]
ypsin inhibitor ovine pancreatic	7	45	48	Hydrogen exchange		[130]
yptophan synthase, α-subunit (Trp-S, s)						
o-S, s	7.8	25	49.0	Urea	Eq. (1), three-state fit (UV)	[132]
o-S, s	7.8	25	43.1	Urea	Eq. (1), three-state fit (CD)	[132]
o-S, s	7.2	25	56.9	Urea	Eq. (1), three-state fit (UV)	[132]
o-S, s	7.2	25	50.2	Urea	Eq. (1), three-state fit (CD)	[132]
o-S, s	6.5	25	52.7	Urea	Eq. (1), three-state fit (UV)	[132]
o-S, s	6.5	25	47.7	Urea	Eq. (1), three-state fit (CD)	[132]

Table 1 (continued)

Protein	pH	T °C	ΔG_{unf} kJ/mol	Approach	Remarks	Referen
Trp-S, s, wild type (Gly 211)	7.8	25	24.1	DSC		[131]
Trp-S, s, mutant (Arg 211)	7.8	25	22.2	DSC		[131]
Trp-S, s, mutant (Glu 211)	7.8	25	26.4	DSC		[131]
Trp-S, s, wild type (Glu 49)	7	25.8	36.8	GuHCl	Eq. (1), CD	[133]
Trp-S, s, wild type (Glu 49)	7	25.8	42.7	GuHCl	Eq. (1), fluorescence	[133]
Trp-S, s, wild type (Glu 49)	7	25	36.8	GuHCl	See [134]	[134]
Trp-S, s, wild type (Glu 49)	7	25	35.1	DSC		[135]
Trp-S, s. wild type (Glu 49]	7	25	28	Heat		[177]
Trp-S, s, mutant (Gln 49)	7	25.8	26.4	GuHCl	Eq. (1), CD	[133]
Trp-S, s, mutant (Gln 49)	7	25	26.4	GuHCl	See [134]	[134]
Trp-S, s, mutant (Gln 49)	7	25	31.0	DSC		[135]
Trp-S, s, mutant (Gln 49)	7.2	25	25	Heat		[177]
Trp-S, s, mutant (Ser 49)	7	25	31.0	GuHCl	See [134]	[134]
Trp-S, s, mutant (Ser 49)	7	25	32.2	DSC		[135]
Trp-S, s, mutant (Met 49)	7	25.8	56.1	GuHCl	Eq. (1), CD	[133]
Trp-S, s, mutant (Met 49)	7	25	55.6	GuHCl	See [134]	[134]
Trp-S, s, mutant (Val 49)	7	25	50.2	GuHCl	See [134]	[134]
Trp-S, s, mutant (Tyr 49)	7	25	36.8	GuHCl	See [134]	[134]
Trp-S, s, mutant (Leu 49)	7	25	63	GuHCl	Eq. (1), CD	[178]
Trp-S, s				Data for further pH values see [134, 177–179]		
Trp-S, s, from E. coli	7	25	27.6	Heat		[179]
Trp-S, s, from Salmonella typhimurium	7	25	29.3	Heat		[179]
Trp-S, s, hybrid	7	25	27.2	Heat		[179]

Table 2. Thermodynamic parameters for unfolding of coiled coil structures. ΔG_{unf} is given in Joule per mol of residues (ΔG_{unf}^{res}) at 25 °C according to [17]. ΔH_{unf} is given in kJ per mol of residues (ΔH_{unf}^{res}) at 25 °C

Protein	T_{trs} °C	ΔG_{unf}^{res} J mol^{-1}	$\Delta C_{p, unf}$ J (K · g)	ΔH_{unf}^{res} kJ mol^{-1}	Reference
Helical parts of collagens (source):					
Collagen of cod skin	15	−183		4.58	[17, 136]
Collagen of halibut	18	−120		4.63	[17, 136]
Collagen of frog skin (*Rana temp.*)	25	0		5.20	[17, 136]
Collagen of pike skin	27	35		5.30	[17, 136]
Collagen of carp swim bladder	30	90		5.40	[17, 136]
Collagen of rat skin	37	257		6.55	[17, 136]
Collagen of sheep skin	37	230		6.03	[17, 136]
Myosin, fragment LMM	50	170	0.25	1.80	[17, 137]
Myosin, fragment LF-3	53	200	0.25	1.90	[17, 137]
Paramyosin, fragment TRC-1	77	310	0.29	1.30	[17, 137]
Paramyosin, fragment TRC-2	83	260	0.42	0.30	[17, 137]
Tropomyosin, fragment (11–189)	58	240	0.27	2.00	[17, 138]
Tropomyosin, fragment (190–284)	46	180	0.38	2.40	[17, 138]
Tropomyosin, fragment CNIA (11–127)	57	260	0.27	2.20	[17, 139]
Tropomyosin, fragment Cy2 (190–284)	42	150	0.30	2.50	[17, 139]

unfolding proceeds either at elevated temperature or in the presence of a denaturant, this necessitates extrapolation procedures in order to obtain ΔG_{unf} at standard conditions or other desired conditions. In most cases, reversible equilibrium is assumed between the states "native" (N) and "unfolded" (U). A variety of experimental methods is available for the determination of the degree of conversion (α) which permits calculation of an equilibrium constant $K = \alpha/(1-\alpha)$, for a reversible N↔U transition. Such equilibrium treatment, however, includes inevitably the two-state assumption which is one of the most uncertain factors in determining thermodynamic quantities of protein unfolding (even when assuming consecutive two-state transitions, e.g., N↔X↔U). As the only exception, scanning calorimetry (DSC) does not require a priori assumptions, and provides, therefore, an independent criterion for checking the validity of the two-state assumption [15, 21, 22].

a) Unfolding by Denaturants. Studies using urea or GuHCl offer advantages which make this approach the most popular one. In fact, no other denaturation procedure gives a greater extent of unfolding than can be achieved using, in particular, GuHCl [5, 6, 13]. Furthermore, experimental measurements can be performed with good accuracy directly at the desired temperature and pH, using commonly available optical instruments without requiring special experimental equipment. A weak point of the approach is, however, the extrapolation of the apparent equilibrium constant (K_{unf}) determined at a given GuHCl concentration to zero denaturant concentration (K_{unf}^0).

Table 3. Heat capacity changes $\Delta C_{p,\text{unf}}$ for protein unfolding. Abbreviations: DSC = differential scanning calorimetry; vH = equilibrium treatment by means of the van't Hoff equation; GuHCl = guanidine hydrochloride

Protein	$\Delta C_{p,\text{unf}}$ J/(K·g)	$\Delta C_{p,\text{unf}}$ kJ/(K·mol)	pH	T_{trs} °C	ΔH_{unf} J/g	ΔH_{unf} kJ mol^{-1}	Approach	Reference
Actin (G-actin)			7.5	57.1		197 (25 °C)	vH	[40]
Alkaline phosphatase	0.59	5.4	7.5	52.0	17.6		DSC	[42]
L-Arabinosa-binding protein (E. coli)	0.40 ±0.01	13.2 ±0.3	7.4	53.5	19.2 ±0.2	635 ±5	DSC	[163]
Aspartate aminotransferase, multiple subform α:								
Apoenzyme	0.49		8.2	67.0	22.8		DSC	[48]
Pyridoxal form	0.30		6.0	78.8	25.3		DSC	[48]
Pyridoxal phosphate form	0.36		8.2	78.5	26.9		DSC	[48]
Pyridoxamine form	0.44		8.2	74.0	24.1		DSC	[48]
Carbonic anhydrase B	0.58			57		227	DSC	[15]
α-Chymotrypsin	0.50 ±0.08	12.6	3.8	57		685	DSC	[15, 21, 140]
Chymotrypsinogen		10–18 temperature-dependent from 20° to 60 °C					DSC	[141]
Chymotrypsinogen		15.9	2.0	42		418	DSC	[141]
Chymotrypsinogen		11.7	3.0	54		586	DSC	[141]
Chymotrypsinogen		8.4	2.0	50		515	DSC	[174]
Chymotrypsinogen		14.0	4.0	61.9		637	DSC	[182]
						Dependence on protein concentration		
Chymotrypsinogen	≦ 0.2	15.2	5.0	62	see Table 2	620	DSC	[152, 175]
Collagen								[136]
Cytochrome b_5								
Cytochrome b_5 in aqueous solution		3.6 ±0.7	7.4	59.1 ±0.1		271 ±36	DSC	[57]
Cytochrome b_5 incorporated in dimyristoyllecithin		6.2 ±1.0	7.4	63.1 ±0.5	267 ±25		DSC	[57]
Cytochrome b_5 tryptic fragment (1–90)	0.55 ±0.08	6.0 ±0.9	7.4	70.7 ±0.3		336 ±12	DSC	[57, 60]

							Method	Reference
Cytochrome c, bovine								
Ferric form	0.58	6.61					DSC	[142]
Ferric form	0.61 ±0.02	7.3					DSC	[15, 21]
Ferric form		7.5 ±0.2					DSC	[143]
Cytochrome c-552		5.2 ±7.0	4.8	78.0 ±0.3		444 ±12	vH, GuHCl	[66]
Fibrinogen, D fragment		7.1	3.5	74.0		350		[144]
Flagellin, monomeric			7.0	47.5 –48.0		1340 ±40	DSC	[165]
Galactoside-binding protein in the presence of Ca^{2+}	0.50	28 ±2	7.8	61	11.2		DSC	[145]
Glycinin (11 S globulin from soybean)	0.18 ±0.02		7.6	80	21	108	DSC	[190]
Glycinin		0 ±2	7.5	40			vH, GuHCl	[72]
Growth factor, epidermal	0.13		8.15	51.4	23.4	182	DSC	[86]
Hemolysin	0.54		7.0	59.5		213	DSC	[74]
Histone GH1 fragment	0.50		7.0	63.0			DSC	[74]
Histone GH5 fragment								
Insulin	0.35 ±0.21	2.1 ±1.2	9.6	25		90.7	Flow calorimetry, insulin reduction	[146]
lac-Repressor, headpiece	0.21	1.255	8.0	65		118 ±13	DSC	[80]
α-Lactalbumin, bovine	0.28 ±0.06	4.0 ±0.8	6.3	62.0 ±0.2		276 ±9	DSC	[83]
Apo α-lactalbumin (α-LA)								
Apo α-LA, bovine		4.95 ±0.14	8.0			187 ±18 (25 °C)	vH	[183]
Apo-α-LA, bovine		5.2 ±0.6	8.0			175 ±15 (25 °C)	DSC	[184]
Apo α-LA, human		5.2 ±0.6	8.0			175 ±15 (25 °C)	DSC	[184)
β-Lactoglobulin		10.80 ±0.36					Flow calorimetry, urea	[147]
Lipoprotein, in dimyristoyllecithin		10.5				134 (37 °C)	DSC	[45]
Lysozyme (hen egg white)		5.75				560	vH, GuHCl	[148]
Lysozyme	0.46 ±0.04	6.7 ±0.4	4.0	77.0			DSC	[21, 149]
Lysozyme	0.46 ±0.02	6.57 ±0.29	4.5	78.5		585 ± 6	DSC	[31]
Lysozyme		6.7 ±0.8					DSC in the presence of GuHCl	[20]
Lysozyme		6.3 ±0.8	1.9	25.0		226 ±17	Isothermal GuHCl titrations	[20]
Lysozyme		6.49 ±0.50	2.0	53.03 ±0.52		382 ± 8	DSC	[97]

Table 3 (continued)

Protein	$\Delta C_{p,unf}$ J/(K·g)	$\Delta C_{p,unf}$ kJ/(K·mol)	pH	T_{trs} °C	ΔH_{unf} J/g	ΔH_{unf} kJ mol^{-1}	Approach	Reference
Lysozyme	0.46	5.6	5.0	75.5		509	DSC	[150]
Lysozyme, phage T4	0.46	8.56					DSC	[15]
Lysozyme, phage T4, wild type		8.8	3.0	54.5		544	vH	[98]
Lysozyme, phage T4, wild type	0.46	8.8	2.2	42.4		448	vH	[98]
Lysozyme, phage T4, mutant eR1	0.46	8.8	2.2	36.1		351	vH	[98]
Lysozyme, phage T4, mutant eRRR	0.46	8.8	2.2	35.8		357	vH	[98]
Lysozyme, phage T4, wild type		9.46	3.0	56.6		326 (320 K)	vH	[168]
Lysozyme, phage T4, mutant Glu 128→Lys		9.41	3.0	51.3		328 (320 K)	vH	[168]
Lysozyme, phage T4, mutant Arg 96→His		10.0	3.0	43.0		299 (320 K)	vH	[168]
Lysozyme, phage T4, mutant Met 102→Val		9.75	3.0	43.9		273 (320 K)	vH	[168]
Lysozyme, phage T4, mutant Ala 146→Thr		10.5	3.0	47.2		338 (320 K)	vH	[168]
Myoglobin								
Sperm whale, cyanomet form	0.63	11.3	11.0	72.0		561	DSC	[152]
Sperm whale, cyanomet form	0.65	11.6	10.5	78.5		628	DSC	[15, 21]
Myosin								
Myosin fragment LMM	0.25				see Table 2		DSC	[17, 137]
Myosin fragment LF-3	0.25				see Table 2		DSC	[17, 137]
Ovalbumin		11.3 ±1.7	7.0	25		216	vH, GuHCl	[109]
Papain		13.7*	3.8	83.8		889*	DSC	[153]
Paramyosin								
Paramyosin, fragment TRC-1	0.29				see Table 2		DSC	[17, 137]
Paramyosin, fragment TRC-2	0.42				see Table 2		DSC	[17, 137]

Parvalbumin							
Parvalbumin in the presence of Ca²⁺	0.40 ±0.04	4.6 ±0.5	7.0	90	500±30	DSC	[111]
Parvalbumin in the absence of Ca²⁺	0.49 ±0.04	5.6 ±0.5	7.0	35	168±12	DSC	[111]
Pepsin in the presence of NaCl	0.544±0.042		7.1	60.5	406	DSC	[115]
Pepsinogen		25.5 ±1.3	6.0	66.0	1134	DSC	[154]
Pepsinogen	0.607 ±0.042				132	DSC	[115]
Pepsinogen		21.8	6–8	25		vH, Urea	[116]
Plasminogen human, various fragments	0.35 ±0.45					DSC	[187]
Ribonuclease, bovine		2.8 ±0.3				DSC	[155]
Ribonuclease, bovine		9.6 ±1.7				DSC	[156]
Ribonuclease, bovine		8.42	2.5	30	283	DSC	[121]
Ribonuclease, bovine		5.02	3.0	53.5 ±0.1	370± 9	DSC	[181]
Ribonuclease, bovine		4.57	5.0	61.0	453	DSC	[182]
						Dependence on protein concentration	
Ribonuclease, bovine		5.02	5.8	61.5 ±0.1	428±17	DSC	[181]
Ribonuclease, bovine		6.3				DSC	[157]
Ribonuclease, bovine		9.2 ±2.1	7.0	61.3	703	DSC	[124]
Ribonuclease, bovine		8.66±0.29				DSC	[122]
Ribonuclease, bovine		4.2				DSC	[158]
Ribonuclease, bovine	0.33	5.12	5.5	64	486	DSC	[159]
Ribonuclease, bovine	0.38	4.2 ±1.1	4.1	63.9	382	DSC	[15, 21]
Ribonuclease, bovine	0.38	8.70	7.0	47.7	448	DSC	[123]
Ribonuclease, S		5.94	7.0	37.6	230	DSC	[122]
Ribonuclease, S-protein			7.15	62	750	DSC	[122]
Serum albumin, bovine	0.38 ±0.08	25 ±5				DSC	[110]
Subtilsin inhibitor	0.20 ±0.01		7.0	80.1*(c=0)	511±11	DSC	[161]
Thermolysin, fragment (206–316)		7.9	7.5	66	272 (37°C)	vH	[129]
Toxins							
Cytotxin I	0.26 ±0.02		4.5	75*	32*	DSC	[186]
Neurotoxin I	0.32		5.7	75*	29.5*	DSC	[186]
Neurotoxin II	0.26 ±0.02		5.0	96*	37.5*	DSC	[186]

Table 3 (continued)

Protein	$\Delta C_{p,\,unf}$ J/(K·g)	$\Delta C_{p,\,unf}$ kJ/(K·mol)	pH	T_{trs} °C	ΔH_{unf} J/g	ΔH_{unf} kJ mol^{-1}	Approach	Reference
Tropomyosin, fragment (11–189)	0.27				see Table 2		DSC	[17, 138]
Tropomyosin, fragment (190–284)	0.38				see Table 2		DSC	[17, 138]
Tropomyosin, fragment CNIA (11–127)	0.27				see Table 2		DSC	[17, 139]
Tropomyosin, fragment Cy2 (190–284)	0.30						DSC	[17, 139]
Troponin C	0.25 ±0.08		7.2	37 / 58 / 84		146 / 264 / 251	DSC	[162]
βTrypsin	0.5	11.9	5.0	70	10.9 (25 °C)		DSC	[15]
Trypsin inhibitor bovine pancreatic	0.46	3.0	4.0	100		430	DSC	[15]
Trypsin inhibitor bovine pancreatic	0.015	0.096	7.0	102.9		293	DSC	[173]
Tryptophan synthase, α-subunit								
Wild type (Gly-211)		8.8 ±1.7	7.8	57.8 ±0.5		392 ±25	DSC	[131]
Mutant (Arg-211)		11.7 ±3.8	7.8	58.0 ±0.2		464 ±17	DSC	[131]
Mutant (Glu-211)		11.3 ±2.9	7.8	59.6 ±0.6		456 ±25	DSC	[131]
Wild type (Glu 49)	0.42 ±0.04		7.0	62.4		97 ± 5 (25 °C)	DSC	[135]
Wild type (Glu 49)			7.0	61.2		82.4 (25 °C)	vH	[177]
Mutant (Ser-49)	0.42 ±0.04	10.3	7.0	59.4		97 ± 5 (25 °C)	DSC	[135]
Mutant (Gln-49)	0.42 ±0.04		7.0	60.5		97 ± 5 (25 °C)	DSC	[135]
Mutant (Gln 49)		11.2	7.2	59.6		59.8 (25 °C)	vH	[177]
From E. coli		9.8	7.0	62*		84 (25 °C)	vH	[179]
From Salmonella typhimurium		9.8	7.0	55*		167 (25 °C)	vH	[179]
Hybrid		9.8	7.0	55*		155 (25 °C)	vH	[179]

Various methods have been proposed:
1. Equations derived from denaturant binding models [23]. The most commonly used expression is Eq. (1):

$$K_{unf} = K^0_{unf} (1 + k a_{\pm})^{\Delta n},$$ (1)

assuming identical, noninteracting denaturant binding sites (k = binding constant, Δn = preferential interaction parameter, $a_{\pm}$ = mean ion activity of GuHCl).
2. Linear extrapolation according to Eq. (2) [11, 24, 25]:

$$\ln K^0_{unf} = \ln K_{unf} + m \text{ (denaturant)}$$ (2)

(m = molarity).
3. Equation (3), which is based on studies on model compounds [5, 6, 11, 26]:

$$\Delta G_{unf} = \Delta G^0_{unf} + \alpha \Sigma n^0_i \delta g_{tr,i},$$ (3)

where α = average degree of exposure of group of i-th type, $\delta g_{tr,i}$ = Gibbs energy change during transfer of i-th side chain from water to denaturant.

Thus, through the extrapolation procedure to zero denaturant concentration, additional complications are introduced into the determination of the Gibbs energy change. For a critical comparison of the extrapolation procedures the reader is referred to [51].

b) *Unfolding by Protonation.* Acid (or alkaline) denaturation is associated with the existence of two conformational states (of charge Z_1 and Z_2) with different degrees of protonation stages:

$$(d \ln K_{unf}/d \ln a_{H^+})_T = Z_1 - Z_2 = \Delta v_{unf}.$$ (4)

The Gibbs energy change can be obtained as a pH-dependent function:

$$\Delta G_{unf}(\ln a_{H^+}) = \Delta G^0_{unf} + RT \int_{(\ln a_{H^+})_0}^{\ln a_{H^+}} \Delta v_{unf}(\ln a_{H^+}) d \ln a_{H^+}$$ (5)

Experimental determination of the key parameter Δv_{unf}, which is pH-dependent itself, is possible by methods including and excluding the two-state assumption [14, 27–31].

c) *Determination of the Gibbs Energy Change from Melting Curves.* Melting curves can be measured with high accuracy [8]. However, a transition usually proceeds in a rather small range at elevated temperature. Therefore, extrapolation to the desired reference temperature (e.g., standard temperature $T^0 = 25$ °C) is difficult and sensitive to error propagation [14]. The temperature dependence of the equilibrium constant K_{unf} for protein unfolding is given by Eq. (6):

$$(d \ln K_{unf}/dT)_{pH} = \frac{\Delta H^0_{unf}}{RT^2} + (1/RT^2) \int_{T^0}^{T} \Delta C_{p,unf} dT.$$ (6)

Equation (6) contains the enthalpy change ΔH^0_{unf} at T^0 and the heat capacity change $\Delta G_{p,unf}$. Equation (6) must be integrated a second time to obtain ΔG_{unf} values (see next section).

d) Thermal Denaturation Monitored by Scanning Calorimetry (DSC). Scanning calorimetry is at present a well-developed, highly sensitive technique of great value in studying thermal transitions of biomacromolecules [32, 33]. The advantages of scanning calorimetry for determination of protein stability are evident [21]:

1. determination of (calorimetric) enthalpy change (ΔH_{unf}^{cal}) without any assumption (from the area under the peak);
2. simultaneous determination of van't Hoff heat (ΔH_{unf}^{vH}) (which includes the two-state assumption) from the same calorimetric recording when analysing the peak height;
3. direct determination of the heat capacity change $\Delta C_{p, unf}$ from single calorimetric recordings.

Thus scanning microcalorimetry enables precise determination of the important thermodynamic variables for protein unfolding. Moreover, it provides the possibility of checking the correctness of the two-state assumption. The complete temperature function ΔG_{unf} can be obtained from the experimental data according to Eq. (7):

$$\Delta G_{unf}(T) = \Delta H_{unf}^{T_{trs}} \cdot (T_{trs} - T)/T_{trs} - \int_{T}^{T_{trs}} \Delta C_{p, unf} dT$$

$$+ \int_{T}^{T_{trs}} \Delta C_{p, unf} d\ln T. \tag{7a}$$

In Eq. (7 a) and (7 b), ΔH_{unf} is the enthalpy change at the transition temperature T_{trs}. T_{trs} refers to the temperature when $\Delta G_{unf} = 0$. Since $\Delta C_{p, unf}$ can be regarded as being temperature-independent based on calorimetric investigations [15, 21] the integration indicated in Eq. (7 a) gives:

$$\Delta G_{unf}(T) = \Delta H_{unf}^{T_{trs}} \cdot (1 - T/T_{trs}) - \Delta C_{p, unf}[(T_{trs} - T) + T\ln(T/T_{trs})]. \tag{7b}$$

The key values in Eq. (7 a) and (7 b), ΔH_{unf} and $\Delta C_{p, unf}$, are according to calorimetric measurements identical for thermal and GuHCl denaturation (e.g., lysozyme [2]). Thus, there is no reason to assume substantial differences between ΔG_{unf} values obtained by the two methods. In other words, the effect of the often discussed residual structure remaining after thermal denaturation should not be overestimated (for examples, see [16]).

e) Determination of Protein Stability from Peptide Hydrogen Exchange in the Native Protein. Hydrogen exchange [34–37] provides an interesting possibility of determining ΔG_{unf} while avoiding complete unfolding of the protein structure. If it is assumed that the exchange of peptide hydrogen from the internal part of the native protein with hydrogen from the surrounding water takes place as a result of local unfolding (unfolded state U), it follows:

$$N \underset{k_2}{\overset{k_1}{\longleftrightarrow}} U \overset{k_{ex}}{\longrightarrow} U_{ex}. \tag{8}$$

The corresponding kinetic equation can be reduced under certain conditions to an expression which gives a relation between the experimentally determinable rate constant k_{obs} and the kinetic constants of Eq. (8):

$$k_{obs} = (k_1/k_2) \, k_{ex}. \tag{9}$$

Since k_{ex} can be obtained from independent experiments, the k_1/k_2 ratio representing the equilibrium constant of microunfolding becomes available. Although the procedure is not straightforward [11], it has long been considered as a valuable complement to the approaches given above. However, the assumption made in Eq. (8) must be scrutinized more critically in view of recent results [38, 39].

References

1. Anson ML (1945) Adv Protein Chem 2:361–386
2. Putnam FW (1953) In: Neurath H, Bailey K (eds) The proteins, vol 1 B. Academic Press, New York, pp 807–892
3. Kauzmann W (1959) Adv Protein Chem 14:1–63
4. Joly M (1965) A physico-chemical approach to the denaturation of proteins. Academic Press, London
5. Tanford C (1968) Adv Protein Chem 23:121–282
6. Tanford C (1970) Adv Protein Chem 24:1–95
7. Lumry R, Biltonen R (1969) In: Timasheff SN, Fasman G (eds) Structure and stability of biological macromolecules. Dekker, New York, pp 65–212
8. Brandts JF (1969) In: Timasheff SN, Fasman G (eds) Structure and stability of biological macromolecules. Dekker, New York, pp 213–290
9. Kuntz ID, Kauzmann W (1974) Adv Protein Chem 28:239–345
10. Franks F, Eagland D (1975) Crit Rev Biochem 3:165–219
11. Pace CN (1975) Crit Rev Biochem 3:1–43
12. Edelhoch H, Osborne JC Jr (1976) Adv Protein Chem 30:183–250
13. Lapanje S (1978) Physicochemical aspects of protein denaturation. Wiley, New York
14. Pfeil W, Privalov PL (1979) In: Jones MN (ed) Biochemical thermodynamics, studies in modern thermodynamics, vol 1. Elsevier Scientific, Amsterdam, pp 75–115
15. Privalov PL (1979) Adv Protein Chem 33:167–241
16. Pfeil W (1981) Mol Cell Biochem 40:3–28
17. Privalov PL (1982) Adv Protein Chem 35:1–104
18. Larson JW, Hepler LG (1969) In: Coetze JF, Ritchie CD (eds) Solute-solvent interactions. Dekker, New York, pp 1–44
19. Tanford C, Aune KC (1970) Biochemistry 9:206–211
20. Pfeil W, Privalov PL (1976) Biophys Chem 4:33–40
21. Privalov P, Khechinashvili NN (1974) J Mol Biol 86:665–684
22. Lumry R, Biltonen R, Brandts JF (1966) Biopolymers 4:917–944
23. Aune KC, Tanford C (1969) Biochemistry 8:4586–4590
24. Greene RF Jr, Pace CN (1974) J Biol Chem 249:5388–5393
25. Schellman JA (1978) Biopolymers 17:1305–1322
26. Tanford C (1964) J Am Chem Soc 86:2050–2059
27. Hermans J Jr, Scheraga HA (1961) J Am Chem Soc 83:3283–3292
28. Hermans J Jr, Scheraga HA (1961) J Am Chem Soc 83:3293–3300
29. Acampora G, Hermans J Jr (1967) J Am Chem Soc 89:1543–1547
30. Hermans J Jr, Acampora G (1967) J Am Chem Soc 89:1547–1552
31. Pfeil W, Privalov PL (1976) Biophys Chem 4:23–32
32. Sturtevant JM (1974) Annu Rev Biophys Bioeng 3:35–51
33. Privalov PL (1980) Pure Appl Chem 52:479–497
34. Linderstrøm-Lang KU, Schellman JA (1959) In: Boyer PD, Lardy H, Myrbäck K (eds) The enzymes, 2nd edn, vol 1. Academic Press, New York, pp 443–510
35. Hvidt A, Nielsen SO (1966) Adv Protein Chem 21:287–386
36. Willumsen L (1971) CR Trav Lab Carlsberg 38:223–295

37. Hvidt A, Wallevik K (1972) J Biol Chem 247:1530–1535
38. Gregory RB, Knox DG, Percy AJ, Rosenberg A (1982) Biochemistry 24:6523–6530
39. Woodward C, Simon I, Tüchsen E (1982) Moll Cell Biochem 48:135–160
40. Contaxis CC, Bigelow CC, Zarkadas CG (1977) Can J Biochem 55:325–331
41. Chlebowsi JF, Mabrey S, Falk MC (1977) J Biol Chem 252:7042–7052
42. Chlebowsi JF, Mabrey S, Falk MC (1979) J Biol Chem 254:5745–5753
43. Tall AR, Shipley GG, Small DM (1976) J Biol Chem 251:3749–3755
44. Edelstein C, Scann AM (1980) J Biol Chem 255:5747–5754
45. Tall AR, Small DM, Deckelbaum RJ, Shipley GG (1977) J Biol Chem 252:4701–4711
46. Mantulin WW, Rohde MF, Gotto AM Jr; Pownall HJ (1980) J Biol Chem 255:8185–8191
47. Ivanov VI, Bocharov AL, Volkenstein MV, Karpeisky MY, Mora S, Okina EI, Yudina LV (1973) Eur J Biochem 40:519–526
48. Relimpio A, Iriarte A, Chlebowski JF, Martinez-Carrion M (1981) J Biol Chem 256:4478–4488
49. Cupo P, El-Deiry W, Whitney PL, Awad WM Jr (1980) J Biol Chem 255:10828–10833
50. Biltonen RL, Lumry R (1969) J Chem Soc 91:4256–4264
51. Pace CN, Vanderburgh KE (1979) Biochemistry 18:288–292
52. Knapp JA, Pace CN (1974) Biochemistry 13:1289–1294
53. Greene RF, personal communication to Pace CN cited in Ref 52
54. Brandts JF (1964) J Am Chem Soc 86:4291–4301
55. Brandts JF (1964) J Am Chem Soc 86:4302–4313
56. Brandts JF, personal communication to Pace CN cited in Ref 11
57. Bendzko P, Pfeil W (1983) Biochim Biophys Acta 742:669–676
58. Sugiyama T, Miki N, Miura R, Miyake Y, Yamano T (1982) Biochim Biophys Acta 706:42–49
59. Bendzko P, Pfeil W (1980) Acta Biol Med Ger 39:47–53
60. Pfeil W, Bendzko P (1980) Biochim Biophys Acta 626:73–78
61. Mc Lendon G, Smith M (1978) J Biol Chem 253:4004–4008
62. Brems DN, Cass R, Stellwagen E (1982) Biochemistry 21:1488–1493
63. Hollecker M, Creighton TE (1982) Biochim Biophys Acta 701:395–404
64. Harrington JP (1981) Biochim Biophys Acta 671:85–92
65. Kawaguchi H, Noda H (1977) J Biochem (Tokyo) 81:1307–1317
66. Nojima H, Hon-nami K, Oshima T, Noda H (1978) J Mol Biol 122:33–42
67. Hon-nami K, Oshima T (1979) Biochemistry 18:5693–5697
68. Nall BT, Landers TA (1981) Biochemistry 20:5403–5411
69. Zuniga EH, Nall BT (1983) Biochemistry 22:1430–1437
70. Okabe N, Fujita E, Tomita K-I (1982) Biochim Biophys Acta 700:165–170
71. Privalov PL, Medved LV (1982) J Mol Biol 159:665–683
72. Holladay LA, Savage CR Jr, Cohen S, Puett D (1976) Biochemistry 15:2624–2633
73. Holladay LA, Hammonds RG JR, Puett D (1974) Biochemistry 13:1653–1661
74. Tiktopulo EI, Privalov PL, Odintsova TI, Ermokhina TM, Krasheninnikov IA, Aviles FX, Cary PD, Crane-Robinson C (1982) Eur J Biochem 122:327–331
75. Rowe ES, Tanford C (1973) Biochemistry 12:4822–4827
76. Rowe ES (1976) Biochemistry 15:905–916
77. Sumi A, Hamaguchi K (1982) J Biochem (Tokyo) 92:823–833
78. Goto Y, Hamaguchi K (1979) J Biochem (Tokyo) 86:1433–1441
79. Schnarr M, Maurizot J-C (1981) Biochemistry 21:6164–6169
80. Hinz H-J, Cossmann M, Beyreuther K (1981) FEBS Lett 129:246–248
81. Kita N, Kuwajima K, Nitta K, Sugai S (1976) Biochim Biophys Acta 427:350–358
82. Contaxis CC, Bigelow CC (1981) Biochemistry 20:1618–1622
83. Pfeil W (1981) Biophys Chem 13:181–186
84. Kuwajima K, Nitta K, Yoneyama M, Sugai S (1976) J Mol Biol 106:359–373
85. Sugai S, Yashiro H, Nitta K (1973) Biochim Biophys Acta 328:35–41
86. Uedaira H, Honda T, Takeda Y, Miwatani T, Uedaira H, Ohsaka A (1983) Int Symp Thermodynam Proteins Biol Membranes, Granada, Spain, Abstract p–II–20
87. Ahmad F, Bigelow CC (1982) J Biol Chem 257:12935–12938
88. Takasa K, Nitta K, Sugai S (1974) Biochim Biophys Acta 371:352–359

89. Nozaka M, Kuwajima K, Nitta K, Sugai S (1978) Biochemistry 17:3753–3758
90. Müller K, Lüdemann H-D, Jaenicke R (1982) Biophys Chem 16:1–7
91. Tombs MP, Blake G (1982) Biochim Biophys Acta 700:81–89
92. Pfeil W, Privalov PL (1976) Biophys Chem 4:41–50
93. Nakanishi M, Tsuboi M, Ikegami A (1972) J Mol Biol 70:351–361
94. Hvidt A (1964) CR Trav Carlsberg 34:299–317
95. Knox DG, Rosenberg A (1980) Biopolymers 19:1049–1063
96. Sophianopoulos AJ, Weiss BJ (1964) Biochemistry 3:1920–1928
97. Velicelebi G, Sturtevant JM (1979) Biochemistry 18:1180–1186
98. Elwell ML, Schellman JA (1977) Biochim Biophys Acta 494:367–383
99. Schellman JA, Hawkes RB (1980) In: Jaenicke R (ed) Protein folding. Elsevier, North-Holland, Biomedica, Amsterdam, pp 331–343
100. McLendon G (1977) Biochem Biophys Res Commun 77:959–966
101. Puett D, Friebele E, Hammonds RG Jr (1973) Biochim Biophys Acta 328:261–277
102. Puett D (1973) J Biol Chem 248:4623–4634
103. Schechter AN, Epstein CJ (1968) J Mol Biol 35:567–589
104. Cho KC, Poon HT, Choy CL (1982) Biochim Biophys Acta 701:206–215
105. Ottesen M (1971) Methods Biochem Anal 20:135–168
106. Abrashi H (1970) Trav Lab Carlsberg 37:129–144
107. Abrashi H (1970) Trav Lab Carlsberg 37:107–128
108. McLendon G, Sandberg K (1978) J Biol Chem 253:3913–3917
109. Ahmad F, Salahuddin A (1976) Biochemistry 15:5168–5175
110. Leibman DY, Tiktopulo EI, Privalov PL (1975) Biofizika 20:376–379
111. Filimonov VV, Pfeil W, Tsalkova TN, Privalov PL (1978) Biophys Chem 8:117–122
112. Robson B, Pain RH (1976) Biochem J 155:331–344
113. Mitchinson C, Pain RH, Vinson JR, Walker T (1983) Biochem Biophys Acta 743:31–36
114. Adams B, Burgess RJ, Carrey EA, Mackintosh IR, Mitchinson S, Thomas RM, Pain RH (1980) In: Jaenicke R (ed) Protein folding. Elsevier-North-Holland Biomedical, Amsterdam, pp 447–467
115. Privalov PL, Mateo PL, Khechinashvili NN, Stepanov VM, Revina LP (1981) J Mol Biol 152:445–464
116. Ahmad F, McPhie P (1978) Biochemistry 17:241–246
117. Nojima H, Ikai A, Oshima T, Noda H (1977) J Mol Biol 116:429–442
118. Burgess RJ, Pain RH (1977) Biochem Soc Trans 5:692–694
119. Chen C-H, Kao OHW, Berns DS (1977) Biophys Chem 7:81–86
120. Doster W, Hess B (1981) Biochemistry 20:772–780
121. Brandts JF, Hunt L (1967) J Am Chem Soc 89:4826–4838
122. Tsong TY, Hearn RP, Wrathall DP, Sturtevant JM (1970) Biochemistry 9:2666–2677
123. Jacobson AL, Turner CL (1980) Biochemistry 19:4534–4538
124. Salahuddin A, Tanford C (1970) Biochemistry 9:1342–1347
125. Puett D (1972) Biochemistry 11:1980–1990
126. Barnard EA (1964) J Mol Biol 10:235–262
127. Puett D (1972) Biochemistry 11:4304–4307
128. Puett D (1973) J Biol Chem 248:3566–3572
129. Vita C, Fontana A (1982) Biochemistry 21:5196–5202
130. Richarz R, Sehr P, Wagner G, Wüthrich K (1979) J Mol Biol 130:19–30
131. Matthews CR, Crisanti MM, Gepner GL, Velicelebi G, Sturtevant JM (1980) Biochemistry 19:1290–1293
132. Matthews CR, Crisanti MM (1981) Biochemistry 20:784–792
133. Yutani K, Ogasahara K, Suzuki M, Sugino Y (1979) J Biochem (Tokyo) 85:915–921
134. Yutani K, Ogasahara K, Kimura A, Sugino Y (1982) J Mol Biol 160:387–390
135. Yutani K, Khechinashvili NN, Lapshina EA, Privalov PL, Sugino Y (1982) Int J Pept Protein Res 20:331–336
136. Privalov PL, Tiktopulo EI, Tischenko VM (1979) J Mol Biol 127:203–216
137. Potekhin SA, Privalov PL (1979) Mol Biol (Mosc) 13:666–671
138. Potekhin SA, Privalov PL (1978) Biofizika 23:219–223
139. Potekhin SA, Privalov PL (1982) J Mol Biol 159:519–535
140. Tischenko VM, Tiktopulo EI, Privalov PL (1974) Biofizika 19:400–404

141. Jackson WM, Brandts JF (1970) Biochemistry 9:2294–2301
142. Privalov PL, Khechinashvili NN (1974) Biofizika 19:14–18
143. Pfeil W, unpublished results
144. Medved LV, Privalov PL, Ugarova TP (1982) FEBS Lett 146:339–342
145. Strickland DK, Andersen TL, Hill RL, Castellino FJ (1981) Biochemistry 20:5294–5297
146. Fukada H, Takahashi K (1982) Biochemistry 21:1570–1574
147. DiPaola G, Belleau B (1978) Can J Chem 56:848–850
148. Tanford C, Aune KC (1970) Biochemistry 9:206–211
149. Khechinashvili NN, Privalov PL, Tiktopulo EI (1973) FEBS Lett 30:57–60
150. Fujita Y, Iwasa Y, Noda Y (1982) Bull Chem Soc Jpn 55:1896–1900
151. Atanasov BP, Privalov PL, Khechinashvili NN (1972) Mol Biol (Mosc) 6:33–41
152. Privalov PL, Khechinashvili NN, Atanasov BP (1971) Biopolymers 10:1865–1890
153. Tiktopulo EI, Privalov PL (1978) FEBS Lett 91:57–58
154. Mateo PL, Privalov PL (1981) FEBS Lett 123:189–192
155. Beck K, Gill SJ, Downing M (1965) J Am Chem Soc 87:901–904
156. Danford R, Krakauer H, Sturtevant JH (1967) Rev Sci Instrum 38:484–487
157. Reeg C (1969) Diss Boulder, Colorado, see Ref 122
158. Privalov PL, Tiktopulo EI, Khechnashvili NN (1973) Int J Pept Protein Res 5:229–237
159. Tiktopulo EI, Privalov PL (1974) Biophys Chem 1:349–357
160. Cupo FF, Pace CN (1983) Biochemistry 22:2654–2658
161. Takahashi K, Sturtevant JM (1981) Biochemistry 20:6185–6190
162. Tsalkova TN, Privalov PL (1980) Biochim Biophys Acta 624:196–204
163. Fukada H, Sturtevant JM, Quiocho FA (1983) J Biol Chem 258:13193–13198
164. Saito Y, Wada A (1983) Biopolymers 22:2105–2122
165. Fedorov OV, Khechinashvili NN, Kamiya R, Asakura S (1984) J Mol Biol 175:83–87
166. Thompson RE, Morrical SW, Campell DP, Carper WR (1983) Biochim Biophys Acta 745:279–284
167. Ahmad F, Contaxis CC, Bigelow CC (1983) J Biol Chem 258:7960–7965
168. Hawkes R, Grutter MG, Schellman J (1984) J Mol Biol 175:195–212
169. Bismuto E, Colonna G, Irace G (1983) Biochemistry 22:4165–4170
170. Desmadril M, Mitraki A, Betton JM, Yon JM (1984) Biochem Biophys Res Commun 118:416–422
171. Ahmad F (1983) J Biol Chem 258:11143–11146
172. Ahmad F (1982) J Biol Chem 259:4183–4186
173. Moses E, Hinz H-J (1983) J Mol Biol 170:765–776
174. Biltonen R, personal communication to Brandts JF cited in Ref 141
175. Privalov PL, Khechinashvili NN (1971) Mol Biol (Mosc) 5:718–725
176. Betton J-M, Desmadril M, Mitraki A, Yon JM (1984) Biochemistry 23:6654–6661
177. Ogasahara K, Yutani K, Suzuki M, Sugino Y (1984) Int J Peptide Protein Res 24:147–154
178. Yutani K, Ogasahara K, Aoki K, Kakuno T, Sugino Y (1984) J Biol Chem 259:14076–14081
179. Yutani K, Sato T, Ogasahara K, Miles EW (1984) Arch Biochem Biophys 229:448–454
180. Matthews CR, Perry KM, Touchette NA (1984) 188th Natl Meeting ACS, see Biochemistry 23:3558
181. Fujita Y, Noda Y (1984) Bull Chem Soc Jpn 57:1891–1896
182. Fujita Y, Noda Y (1984) Bull Chem Soc Jpn 57:2177–2183
183. Hiraoka Y, Sugai S (1984) Int J Pept Protein Res 23:535–542
184. Pfeil W, Sadowski M (1985) Studia biophys. 109:163–170
185. Kelley RF, Stellwagen E (1984) Biochemistry 23:5095–5102
186. Khechinashvili NN, Tsetlin VI (1984) Molekularnaya Biol 18:786–791
187. Novokhatny VV, Kudinov SA, Privalov PL (1984) J Mol Biol 179:215–232
188. Santoro MM, Miller JF, Bolen DW (1984) ASBC/AAI Annual Meeting, see Fed Proc 43:1838
189. Ashikari Y, Arata Y, Hamaguchi K (1985) J Biochem (Tokyo) 97:517–528
190. Danilenko AN, Grozav EK, Bibkov TM, Grinberg VY, Tolstoguzov VB (1985) Int J Biol Macromol 7:109–112

Chapter 14 The Thermodynamics of Conformation Transitions in Polynucleotides

Vladimir V. Filimonov

Symbols

poly (A): polyriboadenylic acid
poly (C): polyribocytidylic acid
poly (G): polyriboguanidylic acid
poly (U): polyribouridylic acid
s: equilibrium constant for unstacking a base pair on top of a stacked sequence
θ: degree of unstacked bases
σ: cooperativity parameter [Eq. (6)]
T_{tr}: transition temperature
ΔG_{tr}: Gibbs energy change of the transition
ΔS_{tr}: transition entropy
ΔH^0: standard enthalpy change
$\Delta H_{tr}(T)$: temperature dependent transition enthalpy
ΔH_{unst}: unstacking enthalpy
Δc_p: apparent change in isobaric heat capacity
$\Delta c_{p,tr}$: heat capacity difference of helix and coil form
Δc_p^{unf}: heat capacity change of helix unfolding
δC_p^{exp}: temperature dependence of excess heat capacity
δC_p^{int}: intrinsic heat capacity change
$c_{p,293}$: partial specific heat capacity of native structure at 293 K
$c_{p,373}$: partial specific heat capacity of unfolded structure at 373 K
$\Delta c_{p,unf}^{app}$: apparent specific heat capacity change of unfolding obtained from linear extrapolations of pre- and posttransitional baselines into the transition region
ΔQ: observed heat effect [kJ kg^{-1}]
$\Delta X/T$): change in a temperature dependent physical observable (absorption, enthalpy, etc.)
ψ_h: parameter reflecting counterion density of helix form
ψ_c: parameter reflecting counterion density of coil form
sc: scanning calorimetry
mc: isothermal mixing calorimetry
hcm: heat capacity measurements
UV: ultraviolet absorption
ORD: optical rotatory dispersion

1 Introduction

The thermodynamics of conformational transitions of nucleic acids represents a broad field of modern physical chemistry. As it is almost impossible to give a full and critical account of the vast amount of experimental data and theoretical considerations in the literature, we indicate here only generally adopted facts and relations and illustrate the latest state of those problems which are not yet solved.

We will, when possible, restrict ourselves to experimental data obtained by calorimetric techniques, for the following reasons: indirect approaches to determine changes of the thermodynamic functions, particularly the enthalpy changes in the case of polynucleotides, are ineffective due to the heterogeneity in molecular weight (complexes of synthetic polynucleotides) or the block organization and sequence heterogeneity (natural RNA and DNA). In all these cases, analyses of melting curves on the basis of a two-state model, or even more sophisticated models, are not useful for the determination of fundamental thermodynamic parameters.

It is assumed that the reader is basically acquainted with the two most frequently used calorimetric techniques. The first, isothermal mixing calorimetry, allows determination of the heat effect of the reaction at constant temperature by mixing two solutions in the calorimeter cell. The second, differential scanning calorimetry, allows measurement of the difference between the heat capacities of solution and pure solvent as a function of temperature. In the best case this technique gives the temperature dependence of partial heat capacity of polymers for a given composition of solution in the temperature range of 0 ° to 110 °C [44]. For details of calorimetric techniques the reader is referred to original papers or review articles [73, 74].

Calorimetric experiments still use high amounts of polymeric materials. The low accessibility of biopolymers or synthetic compounds, explains why no data were obtained for most of them and why the data which have been published are very often fragmentary and insufficiently precise for practical application. Another typical shortcoming is the lack of exact structural and thermodynamic definitions of conformational states between which the transitions occur. This means that in many cases the exact structural interpretation of thermodynamic function changes and the meaning of their comparison is not immediately obvious.

Different aspects of problems which are omitted in this compilation are discussed in detail in a number of review articles or books [3, 8, 19, 20, 23, 41, 42, 51, 65]. All abbreviations symbols, and units are used in accordance with recommendations of IUB-IUPAB.

2 The Thermodynamics of Conformational Transitions in Single-Stranded-Polynucleotides

It is generally assumed that for the majority of polynucleotides conditions can be chosen in such a way that the polymers exist in single-stranded form. It has also been experimentally shown that when double helices or structures of a higher complexity are melted or cannot be formed, single-stranded polynucleotides can still possess some ordered structure due to stacking interactions between bases. The degree of order changes very slowly with temperature and is rather insensitive to the ionic strength. The thermodynamics of unstacking was studied on different polynucleotides using various, mainly spectroscopic, techniques. We will focus our attention on the data obtained for long homopolynucleotides, since these polymers have been studied most intensively and the results published for them are quite representative. Other physicochemical properties of polynucleotides in a single-stranded form have been described in a number of review articles [20, 41, 42, 65].

Most of the published data refer to poly(A) and poly(C), since these polynucleotides do not form double helical or more complex structures in aqueous solutions at neutral pH and, at the same time, have a rather stable residual structure. Double helical structures which poly(U) can form at high ionic strength or in the presence of other stabilizing agents are very unstable, and it is generally assumed that at moderate ionic strength poly(U) has no residual structure at temperatures higher that 20 °C [29, 55, 67, 70]. On the other hand, it is known that poly(G) and poly(I) in neutral salt solutions form multistrand complexes which, in the case of poly(G), are very stable [20, 24].

From the first physicochemical studies on poly(A) and poly(C), it was found that their optical properties at neutral pH change gradually in the temperature range from zero to 100 °C. Some other physical parameters change in parallel, proving the existence of a conformational transition which proceeds "noncooperatively". Therefore many authors used the simplest two-state model to estimate the changes of thermodynamic functions from the temperature dependence of optical parameters.

If it is assumed for simplicity that the two nearest bases can be either in a stacked or in an unstacked conformation and that the thermodynamic parameters of the transition do not depend upon the state of the nearest neighbors, one can introduce an equilibrium constant s for the transition between the two forms

$$s = \exp(-\Delta H°/RT + \Delta S°/R),\tag{1}$$

where $\Delta H°$ and $\Delta S°$ are standard enthalpy and entropy changes of unstacking. Since the amount of stacks in long polymers is nearly one per base, all the thermodynamic functions are expressed per mole of bases. The degree of unstacked bases can be expressed as

$$\Theta = s/(1+s),\tag{2}$$

and the temperature dependence of any physical parameter X which is different in the stacked (X_{st}) and the unstacked form (X_{unst}) by

$$\Delta X(T) = \Theta(T) \cdot \Delta X_{tot}(T),\tag{3}$$

where $X_{tot}(T)$ is the total change of the parameter

$$\Delta X_{tot}(T) = X_{st}(T) - X_{unst}(T).\tag{4}$$

Usually $\Delta X(T)$ is determined from the experimental melting curve together with ΔX_{tot} if X_{st}, and X_{unst} are known or can be estimated by reasonable extrapolation over the temperature range of the transition. Then one can easily find s(T), and from the van't Hoff relation:

$$d(\ln s)/dt = \Delta H°/RT^2,\tag{5}$$

the standard enthalpy change of transition.

A more sophisticated approach using a one-dimensional Ising model was suggested by many authors [1, 13, 48, 64, 68]. According to this model, two equilibrium constants are introduced. The first, s, is for the unstacking of one base on the border between stretches of stacked and unstacked conformations. The other, σs, is an equilibrium constant for the unstacking of two bases within a fully stacked region. From theoretic considerations it is usually assumed that the co-operativity factor σ does not depend on temperature, i.e., it is entropy-determined [1].

Within the framework of this model Θ is defined as

$$\Theta(T) = 0.5\{1 + (1\text{-}s) \cdot [(s\text{-}1)^2 + 4\sigma s]^{-1/2}\}. \tag{6}$$

It is seen that at $\sigma = 1$ (all the stacks are equivalent) the expression (6) turns into (2). As long as σ is constant the Eq. (3) yields for the enthalpy change

$$\Delta H(T) = \Theta \cdot \Delta H_{tr}(T). \tag{7}$$

The apparent change in heat capacity, ΔC_p, can be obtained from

$$\Delta C_p(T) = \frac{d\Delta H}{dT} = \frac{d\Theta}{dT} \Delta H_{tr} + \Theta \Delta C_{p,tr} = \delta C_p^{exc} + \delta C_p^{int}, \tag{8}$$

where $\Delta C_p(T)$ is the temperature dependence of the heat absorption of the system minus the heat absorption of the fully stacked form, $\Delta C_{p,tr}$ is the difference in the heat capacities of two forms; δC_p^{exc} and δC_p^{int} are temperature dependences of the excess heat capacity and the intrinsic heat capacity change, correspondingly. Since ΔH is usually assumed to be temperature-independent [64] for transitions in polynucleotides

$$\Delta C_p(T) = \delta C_p^{exc} = \sigma s(1+s) \frac{(\Delta H^0)^2}{RT^2} [(1-s)^2 + 4\sigma s]^{-3/2}, \tag{9}$$

which at $\sigma = 1$ is equivalent to

$$\Delta C_p(T) = \frac{s}{(1+s)^2} \cdot \frac{(\Delta H^0)^2}{RT^2}. \tag{10}$$

More sophisticated models for the analysis of polynucleotide melting curves have not been used.

The main results can be roughly divided into two groups: the data obtained by fitting procedures based on formulas (1)–(10) (upper part of Table 1) or by direct evaluation of the heat effect from calorimetric data. Special attention should be paid to the data of Freier et al. [13] which were obtained by a combination of different approaches.

From an inspection of Table 1, it is seen that there is remarkable disagreement even between the enthalpy values calculated from optical curves for one type of polymer using a two-state approximation. This disagreement is associated with the very large width of the melting process which makes extrapolations of physical parameters of both states over the transition temperature range rather uncertain. This uncertainty leads to large systematic errors in $\Theta(T)$. In this situation, as was convincingly shown by Freier et al. [13], the UV melting curves of polymers can be equally well fitted when either the two-state or the general Ising model is assumed. Moreover, good fits for poly(A) and poly(C) were obtained for all

Table 1. Thermodynamic parameters of transition between ordered (all bases are stacked) and coiled conformations of single-stranded homo-polynucleotides in neutral water solutions containing salts of alkali metals

Polymer		T_{tr}(K)	ΔH_{tr} (kJ mol^{-1})	ΔS_{tr} (J K^{-1} mol^{-1})	σ	ΔG_{tr} (kJ mol^{-1})	Method	Reference	
(A)$_n$	n=∞	318	56.5	167	1[a]	8.0(273 K)	UV	[32]	
(A)$_n$	n=2–∞	315	33.5	117	1[a]	4.6(273 K)	ORD	[5]	
(A)$_n$	n=∞	320	54.4	170	1[a]	—	UV	[31]	
(A)$_n$	n=2,3,5,9,∞	320	39.3	123	0.6	—	UV	[1]	Fitting of data in
(A)$_n$	n=2,4,6,∞	304	27.2	30	0.71	—	ORD	[48]	accordance with
(A)$_n$	n=∞	318	35.6	112	0.57	—	sc, hcm	[64]	the adopted model
(A)$_n$	n=∞	324	48.5	150	1[a]	—	UV, ORD	[61]	
(dA)$_n$	n=∞	328	52.3	153	1[a]	—	UV	[52]	
(dA)$_n$	n=∞	–	54.4	–	1[a]	–	UV	[6]	
(C)$_n$	n=∞	336	35.6	106	1[a]	–	UV	[47]	
(C)$_n$	n=∞	318	40.2	125	1[a]	5.6(273 K)	UV	[32]	
(C)$_n$	n=∞	327	38.5	118	1[a]	–	CD	[47]	
(C)$_n$	n=∞	327	37.7	115	0.9–1	–	UV, sc	[13]	
(A)$_n$	n=∞	–	37.7	–	–	–	mc, hcm	[9]	
(A)$_n$	n=∞	–	11.3 (298 K)	–	–	–	mc	[49]	
(A)$_n$	n=∞	–	(12.6±2.3)	(41±8)	–	–	sc	[12]	
(A)$_n$	n=7	313	14.2	–	1	–	sc	[69]	
(A)$_n$	n=∞	–	(17.6±1)[b]	–	–	–	sc	[43]	

T_{tr} is the midpoint of the heat-induced transition in polymer; ΔH_{tr} is the enthalpy change expressed in joules per mole of bases or stacks, it is generally assumed that ΔH_{tr} does not depend on temperature and ΔC_p is equal to zero; ΔS_{tr} is the entropy change at $T=T_{tr}$, if $\Delta C_p=0$ it is also independent of temperature at given conditions; σ is the cooperativity parameter in one-dimensional Ising model[1] (see also the text); $\Delta G_{tr}(T)$ is the Gibbs energy change for most cases calculated as $\Delta G_{tr}(T) = \Delta H_{tr} - T\Delta S_{tr}$.

Methods: UV – data calculated from temperature dependence of optical density; ORD – from temperature dependence of optical rotatory dispersion; sc – from data of scanning calorimetry; mc – isothermal mixing calorimetry; hcm – heat capacity measurements. n=∞ means that n>50.

[a] Data were fitted assuming a two-state model of transition.

[b] Calculated from the data on (A)$_n$ · (U)$_n$ melting under assumption that at 95 °C the transition in (A)$_n$ is complete.

values of σ between 0.05 and 1.0. In parallel, ΔH and ΔS change so that $\Delta H_{tr} \simeq$ 9 kJ/mol at $\sigma = 0.05$ and $\Delta H_{tr} \simeq 48$ kJ mol^{-1} at $\sigma \simeq 1$ in the case of poly(A). This ambiguity is not very surprising if one takes into account that in adopting the Ising model we introduce one additional variable parameter. Attempts were made to deduce σ from the dependence of optical parameters on the chain length [1, 48]. The results do not seem to be convincing, since the differences in the melting temperature or optical parameters for σ-values of 0.05–1.0 become significant only at a chain length of about 2–10 where end effects are likely to be important. For this reason or far another, the enthalpy values reported for practically the same σ differ substantially [1, 48]. It seems therefore that optical data alone can give only a rough estimate of the thermodynamic parameters and do not allow to choose the appropriate model of the transition. (Only two parameters could be determined relatively well, $\Theta(T)$ and T_{tr}.)

Scanning calorimetry data in principle are more informative since they allow either a direct determination of the heat effect by integration of δC_p^{exc} or an evaluation of the "effective" enthalpy change which can be found by analysis of the melting curve shape according to formulas (9) and (10). However, δC_p^{exc} must first be found from the calorimetric records which requires knowledge of $C_p^{int}(T)$. When the peak of excess heat capacity is sharp, the c_p^{int} can be easily found by extrapolation. For single-stranded polynucleotides such extrapolations are impossible since, for example, the fully stacked form of poly(A) exists in neutral solutions only at subzero temperatures, while in the best case C_p^p (partial heat capacity) was measured only from 0° to 95 °C [12]. That is why, to estimate δC_p^{exc}, Filimonov and Privalov made two assumptions, the first that δC_p^{int} for poly(A) is close to that of tRNA and the second that $\Delta C_{p, tot}$ is zero. The latter assumption is usually taken for granted. The assumption that δC_p^{int} equals the value observed for tRNA is supported by the following considerations. The partial specific heat capacities of various tRNA's coincide within a 10% accuracy irrespective of their base content [45]. Almost the same values of C_p^p were observed by these authors for $(A)_n \cdot (U)_n$ (12). From these assumptions, the C_p^{int} for poly(A) was approximated by a linear function $C_p^{int} \simeq (0.0075 \cdot T - 1.0225) \pm 0.12$ (kJ K^{-1} kg^{-1}). After subtraction of C_p^{int} from C_p^{app}, the resulting excess heat capacity function was integrated from 263 K to 363 K, giving the value of 12.6 kJ mol^{-1} for ΔH_{tr}. It is worthwhile to mention that δC_p^{exc} calculated in this manner can be relatively well described by formula (9) at $\sigma = 0.05$ and $\Delta H = 13$ kJ mol^{-1}. Although these values do not contradict the results of optical melting curve analysis [13], the question arises how to explain such a low value of σ, which means a rather high cooperativity. In particular, it means that $(A)_2$ must be unstacked at temperatures near 0 °C, which is not really the case. One can argue, of course, that σ could be length- and temperature-dependent, but then it is necessary to suggest some physical grounds for such a strong dependence.

Suurkuusk et al. [64] have used another approach. They tried to fit the temperature dependence of C_p^{app} in the framework of the Ising model using different linear approximations of C_p^{int}. This approach can principally solve the task, but the reliability of the results will depend on the quality of the calorimetric data. Two observations made by these authors are striking. Firstly, ΔH_{tr} and σ were practically independent on large variations of slopes and intercepts of C_p^{int} simulated. This insensitivity of fitting could be explained only by the narrow temperature range in which C_p^{app} was known (20–70) °C and the low accuracy of calorimetric data. Even more striking is that the best fit was achieved at $C_p^{int}(T) = (1.67 \cdot T–420)$ J K^{-1} mol^{-1}. This means that near 250 K the heat capacity of poly(A) becomes negative and that at 273 K it is equal only to 0.1 kJ K^{-1} kg^{-1}. This value is approximately ten times lower than the heat capacity of tRNA and seven times lower than the value reported by the same authors for double-stranded $(A)_n \cdot (U)_n$. All these discrepancies are very difficult to explain if ΔH_{tr} is really as high as 36 kJ mol^{-1}.

The consideration of both most recent calorimetric investigations of poly(A) shows that the main source of errors is the inaccurate approximation of C_p^{int}. Coming back to the results of Filimonov and Privalov, one can try to estimate the possible error in ΔH_{tr} arising from the uncertainty in absolute values of C_p^{int}. If one assumes a 10–15% error in C_p^{int} to be realistic this will lead to an error of about 4 kJ mol^{-1} in the determination of the area under δC_p^{exc} between 260 and 360 K [12]. In fact, since the transition in poly(A) and poly(C) according to optical data must be broader [13], the error could be even higher and reach 6–7 kJ mol^{-1}. For similar reasons, all attempts to determine ΔH_{tr} by direct integration of δC_p^{exc} will lead to its underestimation [12, 69]. Therefore it seems that most of the "direct" calorimetric data give the lower limit of ΔH_{tr} which in the case of poly(A), can be estimated as 16–18 kJ mol^{-1}.

Other results based on calorimetric data alone are also not free of various assumptions and long extrapolations. Most of them were obtained by application of the Hess law to multistep reaction, like $(A)_n$ unfolding [43], poly(A) transfer into solutions with acidic pH [49] or dissolving at different temperatures [9]. Perhaps the most promising approach to the practical solution of this problem was suggested by Freier et al. [13]. They employed a combination of UV and scanning calorimetry data. The temperature dependences of poly(C) hypochromicity in water solutions of different contents were used to find $\Theta(T)$. After functions $\Theta(T)$ had been established with a high accuracy, they were used for the treatment of calorimetric data. Unfortunately, the calorimetric data themselves were not of the best possible quality, which leaves their analysis somewhat ambiguous. Nevertheless, the conclusions made in this work seem to be convincing and the thermodynamic parameters reliable. These authors made an attempt to apply their method for treatment of the calorimetric data for poly(A) [12, 64]. It was found that the data of Suurkuusk et al. correspond to $\sigma = 0.6$, $\Delta H_{tr} \simeq 33$ kJ mol^{-1}, and those of Filimonov and Privalov to $\sigma = 0.3$, $\Delta H_{tr} \simeq 23$ kJ mol^{-1}. This uncertainty is likely to reflect the disagreement between experimental calorimetric data obtained with instruments of different sensitivity.

In summary, this means that the true values of thermodynamic parameters for the transitions in single-stranded polynucleotides still remain rather uncertain. Only a thorough reinvestigation of polynucleotides by different approaches, including calorimetric studies with a higher accuracy, can possibly help to solve this problem.

3 Thermodynamic Parameters of the Polynucleotide Complexes

Here we will consider mainly the direct calorimetric data on the formation or breakdown of complexes between synthetic and natural (DNA) polynucleotides. The thermodynamics of complex formation between oligonucleotides is described in another chapter of this book. Other important aspects of helix-coil transitions are described in special literature.

We will restrict ourselves to data illustrating well-established properties of polynucleotide complexes at neutral pH. Some results concerning special questions or those which are not yet confirmed will not be treated.

3.1 Complexes of Polyribonucleotides

Synthetic homo- and heteropolynucleotides can form complexes of various types [41, 42]. Of the large set of possible structures, the complexes with the Watson-Crick type of base pairing have attracted the greatest attention. Unfortunately if turned out that helices consisting only of $G \cdot C$ pairs are so stable that they cannot be melted in the scanning calorimeter. On the other hand, the rather stable self-structure of poly(G) makes interpretation of mixing experiments very problematic. Since heteropolynucleotides with a repeating sequence were very expensive, previous efforts were concentrated on the poly(a)-poly(U) system. It has long been known that at neutral pH poly(A) and poly(U) can form either the double helical complex $poly(A) \cdot poly(U)$ or the triple helix $poly(A) \cdot 2\,poly(U)$, depending on the ionic strength, temperature, and polymer concentrations. The "phase" diagram for this system is well established [42, 48]. When concentrations of both bases are equal and the salt concentration is lower than 0.2 M, the transition

$$poly(A) \cdot poly(U) \leftrightarrow poly(A) + poly(U) \tag{11}$$

can be induced by a temperature increase. At higher salt concentrations the triple helical complex is formed from the double-stranded one at heating

$$poly(A) \cdot poly(U) \leftrightarrow 1/2\,poly(A) \cdot 2\,poly(U) + 1/2\,poly(A), \tag{12}$$

which at further heating dissociates completely

$$1/2\,poly(A) \cdot 2\,poly(U) + 1/2\,poly(A) \leftrightarrow poly(A) + poly(U). \tag{13}$$

All three transitions are highly cooperative, but at least one of the products of the reactions (11)–(13), namely poly(A) has some residual structure which breaks down noncooperatively according to reaction

$$poly(A)_{st} \leftrightarrow poly(A)_{unst}. \tag{14}$$

Since this transition is accompanied by a rather high enthalpy change, it must be taken into account. Generally speaking, the products of unfolding reactions in biopolymers often possess some residual structure which is very seldom characterized structurally or thermodynamically. Neglecting this fact can lead to incorrect conclusions.

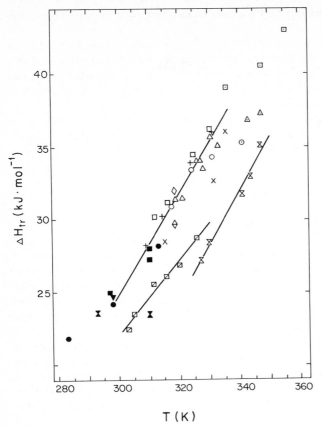

Fig. 1. Temperature dependence of the heat effects for the dissociation of polyribonucleotide double helical complexes. T corresponds to T_{tr} for scanning calorimetry data or to the temperature at which the heat of the reaction was measured by mixing calorimetry (*solid symbols*). *Open symbols with dots* inside correspond to the data calculated for the reaction poly(A)·poly(U)↔ poly(A)+poly(U) (11) from the experimental data for the reactions poly(A)·poly(U)↔ 1/2[poly(A)·2poly(U)]+1/2poly(A) and 1/2[poly(A)·2poly(U)]+1/2poly(A)↔poly(A)+ poly(U). All heat effects are calculated per mole of base pairs. The following symbols correspond to the reaction (11): ○, ⊙ [28]; △, ▲[43]; □, ▫[63]; + [12]; x [64]; ■ [53]; ● [49]; ▼[62]. For comparison data for (A_7U_7) melting are also plotted: ⬦ heat effect calculated per mole of base pairs; ◇ the same calculated per mole of base pair stacks [4]. The symbols ⨉ [21] and 𝕀 [54] correspond to the heats of the reaction poly(I)·poly(C)↔poly(I)+poly(C) and the symbol ⬚ [63] to the reaction poly(A,G)·poly(U)↔poly(A,G)+poly(U). All data were obtained at neutral pH and different concentrations of NaCl and KCl. The parameters of linear approximations shown in the figure are specified in Table 2. The absolute errors of ΔH_{tr} determination are not always given in the original papers. It can be assumed that the average error does not exceed 3%

The experimentally observed heats of reaction (11) are plotted in Fig. 1 as function of temperature. The correctness of such plotting of the results obtained by different techniques and at different salt concentrations is supported by one of the well-established facts: chlorides of alkali metals, such as NaCl or KCl, which have strong influence on the stability of polynucleotide complex formation, scarcely change the enthalpy of that process. This conclusion is drawn from the coincidence (within the limits of experimental error) of the enthalpy values obtained at one temperature, but at different salt concentrations. *Therefore at neutral pH and in presence of some neutral salts, the ΔH of secondary structure unfolding depends only on the temperature and the conformational state of the products.* This means in particular that the slopes of $\Delta H_{tr} = f(T_{tr})$ must correspond to ΔC_p^{app} of the transition. Taking into account that the transition [14] is very broad and therefore its ΔH changes almost linearly around T_{tr} [which for poly(A) is about 318 K for all salt concentrations], one can anticipate a linear dependence of $\Delta H_{tr, A \cdot U}$ from the temperature in the vicinity of 318 K. Such a linearity is really observed within the limits of experimental error. The least-squares linear regression of all data obtained for the temperature range (300-335) K is shown in Fig. 1 (see also Table 2). The slope of the line corresponds to $\Delta C_p^{app} = (350 \pm 30)$ J K^{-1} mol^{-1}. This value is good agreement with direct estimates of ΔC_p^{app} from calorimetric measurements [12, 64].

Assuming that ΔC_p^{app} for the reaction (11) arises only from melting of the residual structure in poly(A), one can roughly estimate either the enthalpy of reaction (14) or the maximal value for ΔH of the reaction (11) at temperatures where poly(A) is completely unstacked. For a two-state model, $T_{tr, poly(A)} = 318$ K and $\delta C_p^{exc}(T_{tr}) = 350$ J K^{-1} mol^{-1} we obtain, using Eq. (10), about 34 kJ mol^{-1} for ΔH_{unst} and about 48 kJ mol^{-1} for $\Delta H_{A \cdot U}^{max}$. The value of ΔH_{unst} is in a rather good agreement with the enthalpies listed in Table 1 for a two-state approximation.

At present it is impossible to point out how large is the contribution of poly(A) to $\Delta C_p^{app}(T_{tr})$ of the complex and, therefore, to determine precisely ΔC_p for the helix-coil transition. The latter could be estimated from the temperature dependence of ΔH_{tr} on polynucleotide complexes whose constituents do not have a residual structure in a single-stranded form. For example, it is very probable that alternating polymers like poly(A–U) or poly d(A–T) exist in a fully unstacked form. However, until now no data for poly(A–U) have appeared, and those for poly d(A–T) are rather contradictory (see below). From the data obtained for the natural DNA and RNA it seems that the ΔC_p (helix-coil) is not large and in most cases does not exceed 110 J K^{-1} mol^{-1} (0.24 kJ K^{-1} kg^{-1}) [45, 46, 60].

The slopes of $\Delta H_{tr} = f(T_{tr})$ are much higher both for poly(I)·poly(C) and poly(A, G)·poly(U), as one can judge from the scanning calorimetry data [21, 63].

The data of Hinz et al. [21] can be approximated by a linear function with a slope $\simeq 370$ kJ K^{-1} mol^{-1}, which is very close to that of poly(A)·poly(U). If again the main part of this slope can be explained by the residual structure of poly(C), this coincidence need not be accidental, since poly(C) has almost the same enthalpy of unstacking as poly(A) and its residual structure is nearly 10 K more stable. Inasmuch as the existence of the residual structure in poly(C) is well established, it is very difficult to explain the practical independence of the heat liberated at mixing of poly(I) and poly(C) both on the ionic strength and temperature [54]. This observation seems even more striking in the light of a subsequent publication of the same authors in which they report the remarkable dependence on temperature of the heat of the reaction between poly(BrC) and poly(I) at almost the same salt conditions [56], although it is known that poly(BrC) has a more stable structure and therefore is more stacked in the (298–310) K temperature range [32]. (See also Table 2).

Recently Stulz measured the heat effect of the reaction

$$\text{poly}(A, G) \cdot \text{poly}(U) \leftrightarrow \text{poly}(A, G) + \text{poly}(U),$$

which is very important for estimating the enthalpy change of the $G \cdot U$ pair formation [63]. From the results plotted in Fig. 1, it follows that the enthalpy change for $G \cdot U$ is almost the same as for $A \cdot U$, but $G \cdot U$ opposition definitely destabilizes the helix. Unfortunately no data are available for other complementary heteroribopolynucleotides.

A strong temperature dependence is also observed for the heat effect of reaction (13) (Fig. 2, Table 2). Although the scattering of experimental data is much higher in this case, one can try to estimate the slope of $\Delta H_{tr} = f(T_{tr})$. It turned out to be very close to ΔC_p^{app} of poly(A) · poly(U) melting which is not very surprising if the main part of the slope is again associated with the residual structure in single-stranded poly(A). It is also remarkable that, according to the latest data of Stulz, the enthalpy of the reaction (13) exhibits no tendency to level off at high temperatures, as was assumed by Neumann and Ackermann [43]. If the observation by Stulz is true (it is supported by unpublished results of Filimonov and Privalov), this could mean both a ΔH_{unst} in poly(A) [43] higher than $17.6 \, \text{kJ mol}^{-1}$ and a nonzero ΔC_p for the helix-coil transition. In any case, the data on triple helix melting should be checked once more before some definite conclusions about the real dependence of ΔH_{tr} on temperature can be drawn.

It is noteworthy that the heat effect of formation of the triple helical complexes between poly(U) and adenosine monophosphates is very close to the heat of poly(A) · 2 poly(U) formation. The possible structural meaning of this correlation is discussed in the original paper [18].

Thus, at moderate temperatures the enthalpies of formation/breakdown of the double and triple helical complexes of homopolyribonucleotides as a rule exhibit a significant temperature dependence which is most likely connected mainly with the temperature dependence of the degree of stacking in single stranded polymers. The enthalpy values for the formation of the complexes studied so far, are not very different.

3.2 Complexes of Polydesoxyribonucleotides

Historically, calorimetric studies of natural DNA's began earlier and were carried out more systematically than investigations of their synthetic analogs. Only recently, due to efforts of a group of scientists from the University of New Jersey, have the data for a number of polymeric and oligomeric synthetic DNA's appeared. When considering the data for natural DNA, it is necessary to remember that they are heteropolynucleotides of very large length consisting, in many cases, of a number of blocks with different stabilities and $A \cdot T/G \cdot C$ ratios. Natural DNA's, as a rule, melt in a wider temperature range than synthetic analogs, and their melting curves exhibit a fine structure. Such average parameters as T_{tr} and ΔH_{tr} are rather artificial and characterize the system only roughly. This circum-

Table 2. Thermodynamic parameters of dissociation of double and triple-helical complexes of synthetic polynucleotides in water solutions containing salts of monovalent cations at neutral pH

Reaction	Solvent	T_{ex} (K)	ΔH_{tr} (kJ K^{-1} mol^{-1})	ΔS_{tr} (kJ K^{-1} mol^{-1})	ΔC_p^{app} (kJ K^{-1} mol^{-1})	Method	Reference
poly(A)·poly(U)⇌poly(A)+poly(U)	pH 7, (0.01÷0.2) M NaCl	(300÷335)	(0.35·T−80.0) ± 1.0	(0.35 −80.0/T_{tr}) ± 0.003	(0.35 ±0.05)^e	sc, mc	a
poly(A)·2poly(U)⇌poly(A)+2poly(U)	pH 7, (0.2÷1.0) M NaCl	(300÷360)	(0.33·T−56.0) ± 3.0	(0.33 −56.0/T_{tr}) ± 0.01	(0.33 ±0.05)^e	sc, mc	b
poly(I)·poly(C)⇌poly(I)+poly(C)	pH 6.9, 001 M citrate (0.06÷1.0) M NaCl	(328÷348)	(0.375·T_{tr}−95.5) ± 1.0	(0.375−95.5/T_{tr}) ± 0.005	(0.375±0.005)^e	sc	[21]
poly(I)·poly(C)⇌poly(I)+poly(C)	pH 8.0, 0.005 M EDTA (0.02÷0.4) M NaCl	(293 ÷310)	23.4	—	0^e	mc	[54]
poly(I)·poly(BrC)⇌poly(I)+poly(BrC)	pH 8.0, 0.005 M EDTA (0.02÷0.1) M NaCl	298 310	30.2 32.5	—	~0.19e	mc	[56]
poly(I)·poly(MeC)⇌poly(I)+poly(MeC)	pH 8.0, 0.005 M EDTA 0.1 M NaCl	(298÷310)	26.3	—	0^e	mc	[56]
poly[d(A-T)]·poly[d(A-T)] ⇌poly[d(A-T)]+poly[d(A-T)]	pH 7, 0.01 M cacodylate, 0.001 M citrate, 0.005 M NaClO$_4$	314	33.1±0.6	105±2	(0.096±0.067)^d	sc	[59]
poly[d(A-T)]·poly[d(A-T)]⇌poly[d(A-T)]+poly[d(A-T)]	pH 6.8, 0.005 M cacodylate, (0.01÷0.6) mol kg^{-1} Na$_2$SO$_4$	(311÷344)	(0.2·T_{tr}−32.0) ± 1.0	(0.2−32.0/T_{tr}) ± 0.003	—	sc	[17]^c
poly[d(A-T)]·poly[d(A-T)] ⇌poly[d(A-T)]+poly[d(A-T)]	1.0 mol kg^{-1} Na$_2$SO$_4$	348.5	30.8÷1.3	88.3	—	sc	[17]
poly[d(A-T)]·poly[d(A-T)] ⇌poly[d(A-T)]+poly[d(A-T)]	1.8 mol kg^{-1} Na$_2$So$_4$	352.3	25±0.9	71.0	—	sc	[17]
poly[d(A-T)]·poly[d(A-T)] ⇌poly[d(A-T)]+poly[d(A-T)]	3.3 mol kg^{-1} Na$_2$SO$_4$	355.5	25	71.0	—	sc	[17]
poly[d(A-T)]·poly[d(A-T)] ⇌poly[d(A-T)]+poly[d(A-T)]	pH 6.85, 0.01 M phosphate, 1.0 M NaCl or 1.0 M Me$_4$NCl	348.6 354.2	28.5 29.3	82 83	— —	sc sc	[38] [38]
poly[d(A-T)]·poly[d(A-T)] ⇌poly[d(A-T)]+poly[d(A-T)]	pH 7.0, 0.01 M sodium phosphate, 0.001 M EDTA	316.7	31.8	100	—	sc	[36]
poly[d(A-T)]·poly[d(A-T)] ⇌poly[d(A-T)]+poly[d(A-T)]	The above plus (0.01÷1.0) M NaCl	(321.7÷347.7)	29.7±3	—	0^e	sc	[37]

	Conditions	T_{tr}	ΔH_{tr}		ΔC_p^{app}	Method	Ref.
poly(dA)·poly(dT) ⇌poly(dA)+poly(dT)	pH 7.0, 0.01 M sodium phosphate, 0.001 M EDTA 0.01 M NaCl	329.2	36.0	109	0^e	sc	[37]
poly[d(A-C)]·poly[d(T-G)] ⇌poly[d(A-C)]+poly[d(T-G)]	The same as above	350.2	23.4	67	–	sc	[37]
poly(A,G)·poly(U) ⇌poly(A,G)+poly(U) $nA_x·2(U)_n ⇌ nA_x + 2(U)_n$	pH 7.5, 0.001 M Tris HCl, (0.01÷1.0) M NaCl pH 7.2, 0.1 cacodylate 1.0 M NaCl	(303÷326)	$(0.26·T_{tr}−55.0)$ ± 1.0	$(0.26−55.0/T_{tr})$ 0.003 ±	$(0.26±0.03)^e$	sc	[63]
A_j = adenosine		298	50.5	170	–	sc	[18]
A_x = 2′,3′-cAMP		(292÷304)	45.0	151	0^e	sc	[18]
A_x = 2′(3′)-AMP		(301÷304)	42.9	141	–	sc	[18]
A_x = 5′-AMP		(280÷294)	28.6	99	0^e	sc	[18]
A_x = 5′-ADP		(284÷293)	26.1	90	0^e	sc	[18]
A_x = 5′-ATP		282	25.6	91	–	sc	[18]
A_x = 2-aminoadenosine	pH 7, 0.01 M cacodylate, 0.6 M NaCl	322.7	66.0	205	–	sc	[58]
A_x = adenosine		305	53.6	175	–	sc	[58]
A_x = desoxyadenosine		293	53.6	–	–	mc	[58]
A_x = adenine		293	53.6	–	–	mc	[58]
A_x = 2,6-diaminopurine		293	65.0	–	–	mc	[58]

T_{ex} is the temperature (or temperature range) at which the data were obtained (or in which corresponding empirical relations are valid); T_{tr} is the transition temperature usually determined from scanning calorimetry data as temperature of half-conversion; it coincides with T_{ex} for scanning calorimetry data; $\Delta G(T_{tr}) = \Delta H_{tr}(T_{tr}) - T_{tr}\Delta S_{tr}(T_{tr}) = 0$; ΔC_p^{app} is the apparent ΔH_{tr} is the enthalpy change for corresponding reaction at $T = T_{ex}$ and given salt conditions; ΔC_p^{app} is the apparent change of partial heat capacity determined directly from calorimetric data (d) or estimated from the slopes of $\Delta H_{tr} = f(T_{tr}) - (e)$.

All thermodynamic parameters are expressed per mole of base pairs or base triplets depending on stoichiometry of the reaction.

Methods: sc – scanning calorimetry; mc – isothermal mixing calorimetry.

[a] Empiric relations based on a linear least-square fitting of data from references [12, 28, 43, 49, 53, 62, 63] for temperature range (300÷335) K. ΔC_p^{app} was estimated from the slope of $\Delta H_{tr}(T)$. For details see Fig. 1 and the text. As within the limits of error enthalpy does not depend on ionic strength, ΔC_p^{app} could not be used for the estimation of ΔC_p^{app}.

[b] Empiric relations based on a linear least-square fitting of data shown in Fig. 2 (references in the legend to the figure). ΔC_p^{app} is estimated from the slope of $\Delta H = f(T_{tr})$.

[c] Empiric relations based on a linear least-squares fitting of data from [17] for the temperature range (311÷344). As it was not shown independently that the heat effect does not depend on Na_2SO_4 concentration, the slope of $\Delta H = f(T_{tr})$ could not be used for the estimation of ΔC_p^{app}.

[d] Apparent ΔC_p has been evaluated from the slope of Δh vs. transition temperature curve.

[e] Apparent ΔC_p was measured directly.

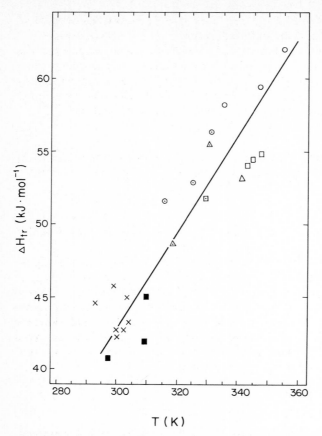

Fig. 2. Dependence of the heat of the reaction $poly(A) \cdot 2\,poly(U) + poly(A) \leftrightarrow 2\,poly(A) + 2\,poly(U)$ (13) on the transition temperature T_{tr} (scanning calorimetry) or on the temperature of the experiment (mixing calorimetry, *solid symbols*). *Open symbols with dots* inside correspond to the data calculated for the reaction (13) from the experimental data obtained for the reactions $poly(A) \cdot poly(U) \leftrightarrow poly(A) + poly(U)$ and $2[poly(A) \cdot poly(U)] \leftrightarrow poly(A) \cdot 2\,poly(U) + poly(A)$. All heat effects are calculated per mole of base triplets $(A \cdot 2\,U)$ and are represented as follows: $\triangle$, $\triangle$ [28]; $\square$, $\square$ [43]; $\circ$, $\odot$ [63]; ■ [53]. For comparison the data of Gukowskii et al. are plotted for the reaction $nA_x \cdot 2\,poly(U) \leftrightarrow nA_x + 2\,poly(U)$, where A_x corresponds to 3′, or 5′, or 2′, 3′-phosphates of adenosine (see also Table 2). All data were obtained at neutral pH and different concentrations of NaCl and KCl. The parameters of linear approximation shown in figure are specified in Table 2. The absolute errors of ΔH_{tr} determination are unknown

stance diminishes the value of information which is obtained from calorimetric studies of natural DNA's. Therefore we have a limited compilation of the data available for natural DNA's to illustrate only the general thermodynamic properties of helix-coil transitions in these molecules.

In parallel with experimental studies (most of them were based on other than calorimetry approaches) the theory of helix-coil transitions in heteropolynucleotides has been developed. Modern theories allow sufficiently accurate prediction of the melting profiles of DNA's [30, 33, 66]. The desirable parameters were found empirically from studies of natural and model complexes of polynucleotides.

The first and most accurately documented observation concerns the linear dependence of T_{tr} (which in case of long DNA is determined as the temperature of half conversion) on the $G \cdot C$ content. Different authors, however, report slightly different coefficients. Mandel et al. have deduced the following empiric relation from an analysis of melting curves of DNA's from 33 types of bacteria

$$T_{tr} (0.1 \text{ M NaCl, } G \cdot C) = [50.2 \cdot (G \cdot C) + 322.9] \cdot K, \tag{15}$$

where T_{tr} is the melting temperature of DNA with a given $G \cdot C$ content at 0.1 M NaCl in solution with neutral pH [34]. Lyubchenko et al. [33] have used another relation

$$T_{tr} (0.1 \text{ M NaCl, } G \cdot C) = [42.4 \cdot (G \cdot C) + 325.6] \cdot K. \tag{16}$$

Taking into account that T_{tr} is an average parameter, it is difficult to tell which relation is more correct.

It is known also from studies of various polynucleotide complexes that T_{tr} increases approximately linearly with the logarithm of monovalent ion activity in solution if the concentration of salt is not very high. The slope of the $T_{tr} = f(\log a)$ depends on the chemical composition of salt, the chemical structure of polynucleotides, their length, etc. For most of the natural DNA's in NaCl solutions the slope is approximately equal to (19 ± 2) K. Tsong and Battersby have analyzed a large body of data on DNA melting and found that in standard solutions containing a 1 mM citric acid buffer the following relation is valid [71]

$$T_{tr}(a_{Na}, G \cdot C) =$$
$$-5.2 \log a_{Na}(G \cdot C) + 18.4 \log a_{Na} + 41.0(G \cdot C) + 342.5. \tag{17}$$

Application of the polyelectrolyte theory to helix-coil transitions in polynucleotide complexes gives the following expression [28, 35, 46, 50]

$$\frac{dT_{tr}}{d \log a} = \frac{2.303 \cdot RT_{tr}^2}{\Delta H_{tr}(T_{tr})} \cdot (\psi_h - \psi_c), \tag{18}$$

where a is the activity of monovalent cations, ΔH_{tr} is the enthalpy change per mole of base pairs, T_{tr} is the temperature of transition, ψ_h and ψ_c are the parameters reflecting the density of counterion shell screening phosphate groups for helical and coiled forms, respectively. The average number of Na^+ ions liberated at melting of double helices according to calorimetric data is about 0.34 per two phosphates [16, 28, 46], which is in good qualitative agreement with the structural parameters of A- and B forms [50].

Inspection of Fig. 3 shows that the average heat of DNA melting increases approximately linearly with T_{tr} at moderate ionic strength. (When the Na_2SO_4 concentration is higher than 0.5 M or the concentration of tetraalkylammonium chlorides is higher than 1 M, the enthalpy of thymus DNA and poly d(A–T) melting starts to drop) [16, 17, 25]. Since it has been shown independently that the stabilizing effect of NaCl is mainly of entropic nature, the data of Privalov et al. [46] for DNA of the T7 phage and the data of Shiao and Sturtevant for thymus DNA

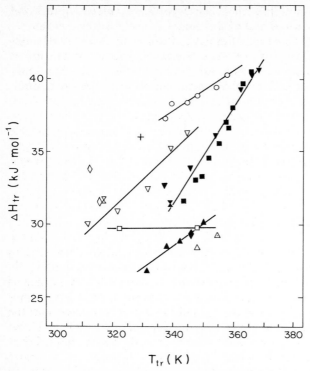

Fig. 3. Dependences of the heat effects of DNA melting on transition temperature according to scanning calorimetry data. The ΔH_{tr} values are expressed per mole of base pairs. T_{tr} is the transition temperature which for natural DNA's corresponds to the point of half-conversion. The data for calf thymus DNA are taken from the following references: Shiao and Sturtevant [60], (neutral pH, NaCl up to 0.05 M) — ▲; Rüterjans [57] — ◆; Klump and Burkart [27] (neutral pH, 0.01 M NaCl) — ✗; Grünwedel [16] (neutral pH, Na_2SO_4 up to 0.5 M) — ▼; Klump [25] (neutral pH, tetramethyl or tetraethyl ammonium chlorides up to 1 M) — ■. The last data were taken from figures and are approximate. *Open circles* data of Privalov et al. [46] for DNA from phage T7 (neutral pH, NaCl up to 0.2 M). *Other symbols* correspond to the reactions poly-[d(A–T)]·poly[d(A–T)]↔2 poly[d(A–T)] (19) and poly(dA)·poly(dT)↔poly(dA)+poly(dT) (20). The data for reaction (19) are depicted by: ▽ [17]; □ [37]; ◇ [59]; ✗ [36]; △ [38] (1 M NaCl); [38] (1 M tetramethylammonium chloride) and for reaction (20) by: + [37]. For salt conditions and other details consult Table 2 and the text. The slopes of linear approximations are: for phage T7 DNA and calf thymus DNA in NaCl about 160 J K^{-1} mol^{-1}, for calf thymus DNA in Na_2SO_4 and tetraalkylammonium chloride about 330 J K^{-1} mol^{-1}

can serve as basis for the estimation of $\Delta C_{p,tr}$ in DNA. The slopes of the respective lines in Fig. 3 correspond to about 160 J K^{-1} mol^{-1}. The same slight dependence of ΔH_{tr} on T_{tr} follows from the data of Klump and Ackermann [26], and was observed by Baba and Kagemoto for DNA from *Micrococcus lysodeikticus* also at low concentrations of $MgCl_2$ in solution [2].

It seems, however, that the stabilizing effect of some salts on the DNA structure could not be of a purely entropic nature. As seen from Fig. 3, the data obtained for calf thymus DNA by Gruenwedel at various Na_2SO_4 concentrations and by Klump in the presence of tetraalkylammonium salts agree very well and follow the same quasilinear dependence of ΔH_{tr} upon T_{tr}, but the slope of this dependence is higher than for the data of Shiao and Sturtevant [16, 25]. The observed difference, particularly at high temperatures, exceeds experimental errors and indicates that different ions have a qualitatively different effect on the parameters of helix-coil transition in natural DNA's. It is known that tetraalkylammonium salts have a different effect on the stability of A–T and G·C rich helices resulting in a decrease of the T_{tr} dependence on the G·C content and in the narrowing of the temperature range of the transition (7). Recently, Marky et al. [38] have shown that replacement of NaCl by tetramethylammonium chloride does not substantially alter the enthalpy of melting of $d(A–T)·d(A–T)$ (see Table 2) which makes the observed disagreement of data for calf thymus DNA difficult to explain. It could be directly connected with the structural heterogeneity of long natural DNA's and the right answer could be found only by investigating more simple model polynucleotides.

Unfortunately, a similar disagreement is observed for published results on the melting of synthetic polynucleotide complexes, namely for relatively well studied poly d(A–T) (see Table 2 and Fig. 3). The data of Gruenwedel for melting of poly d(A–T) at different Na_2SO_4 concentrations exhibit a remarkable dependence of ΔH_{tr} both on the salt content and T_{tr}. The apparent increase of ΔH_{tr} with temperature corresponds to $\Delta C_p^{app} \simeq 200$ kJ mol^{-1}, which is somewhat higher than the ΔC_p of DNA in NaCl solutions, but lower than the slope of $\Delta H_{tr} = f(T_{tr})$ for thymus DNA in Na_2SO_4. On the other hand, Marky and Breslauer claimed that within a 5–10% accuracy the enthalpy of d(A–T) melting does not depend either on the NaCl concentration or T_{tr} and is equal to 29.7 kJ mol^{-1} [37]. Later the same authors obtained 31.8 kJ mol^{-1} for polymer at a low temperature [36]. From an inspection of all the data available for the poly (d(A–T) double helix, it appears that the effects of salts and temperature on ΔH_{tr} could not be resolved against the background of large experimental errors and one can assume that the enthalpy of poly d(A–T)·poly d(A–T) melting is equal to 32 ± 4 kJ mol^{-1} irrespective of the salt concentration and temperature.

Recently, heat effects for some polynucleotides have been reported [37]. These data are very important for elucidation of the base sequence effect on the thermodynamic parameters of helix-coil transitions. However, some observations made by Marky and Breslauer cannot be rationalized. In particular, they have pointed out that the enthalpy of poly(dA)·poly(dT) melting does not depend on temperature. This conclusion disagrees with well-established facts for the poly(A)·poly(U) system. The origin of disagreement is not clear, since poly(dA) is known to have a residual stacking with almost the same thermodynamic properties as poly(A). The other very important observation of Marky and Breslauer concerns the rather strong dependence of ΔH_{tr} and the stability of A–T opposition upon the concrete sequence of base pairs in DNA. Further investigations should be carried out to establish more precisely other details and thermodynamic parameters of helix-coil transitions in DNA.

A number of data concerning heat denaturation of DNA were not considered here, since some of them seem to be qualitative and others are difficult to interpret. For example, it was found that at high concentrations of Na_2SO_4 and

Me$_4$NCl the enthalpy of DNA melting starts to drop [17, 25]. Qualitatively, urea gives a similar effect [27]. The temperature of the transition decreases when the pH is shifted in both directions from neutrality. The determination of true enthalpy values in this case depends upon knowledge of the heats of base protonation in the polymeric form, otherwise the data obtained in acidic or alkaline regions are hardly comparable with those measured at neutral pH [46, 60].

4 Thermodynamics of RNA Unfolding

It is well known that natural RNA molecules consist, as a rule, of a single polynucleotide chain which at normal conditions is folded into a three-dimensional structure. The following three levels of RNA structure organization are usually considered: the nearest-neighbor interactions in the single-stranded regions, the secondary structure which consists of relatively short double-helical regions and the tertiary structure which is maintained by hydrogen bonds, stacking interactions and salt bridges between phosphate groups. An example of such a structural hierarchy is the well-known model of tRNA structure. The general principles of RNA folding are similar to those of globular proteins; however, the thermodynamic properties of unfolding reactions are somewhat different. There are many small globular proteins whose melting can be considered as a two-state process, but even small RNA's, like tRNA, exhibit a multistep temperature denaturation. The sequential, step-by-step unfolding of RNA is particularly well expressed in the absence of divalent cations or other specific stabilizing agents and is well documented by various experimental techniques. Most of the known facts, together with theoretical considerations of the secondary structure unfolding, are summarized in a number of review articles [3, 20, 23, 48, 51].

Unfortunately the applicability of the various physicochemical approaches to the investigations of thermodynamics of RNA unfolding is very limited due to the extreme complexity of this process. In fact, the overall heat effect of the unfolding can be measured only by direct calorimetric techniques. This, however, is not easy for the process, which covers, in most cases, more than 60 K. Nevertheless, the overall heat of unfolding was measured for a number of RNA's. These data are listed in Table 3 and need only some general comments.

The overall stability of RNA structure depends on the ionic strength or divalent ion activity in almost the same manner as the stability of long helices. There is one exception, however. It seems that tertiary structure elements and defective elements of secondary structure are more sensitive to cation concentration, which usually results in a narrowing of the temperature range of the transition at an increase of salt (especially MgCl$_2$) concentration in the solution [51]. Although magnesium ions remarkably increase the cooperativity of unfolding of various RNA's, none of the molecules studied so far has exhibited a two-state unfolding at millimolar concentrations of MgCl$_2$ in solution.

Melting enthalpies of various natural RNA's are rather similar within the limits of experimental error, irrespective of either the NaCl and MgCl$_2$ concentration in solution. It follows from this observation that the stabilizing action of NaCl and MgCl$_2$ is mainly of an entropic nature and therefore, ΔC_p^{unf} must be relatively small [32, 45].

Table 3. Integral thermodynamic parameters of melting of the structure of naturally occurring RNA's

RNA	Solvent	ΔQ (kJ kg^{-1})	ΔH_{unf} (kJ mol^{-1})	$C_{p,293}$ (kJ K^{-1} kg^{-1})	$C_{p,373}$ (kJ K^{-1} kg^{-1})	$C_{p,unf}^{app}$ (kJ K^{-1} kg^{-1})	Method	Reference
tRNA^{Phe} (yeast)	pH 7, 0.01 M cacodylate or 0.005 M sodium phosphate, $(0 \div 0.15)$ M NaCl and/or $(0 \div 0.001)$ M MgCl$_2$	50.0 ± 2.0	1250 ± 50	1.1 ± 0.12	1.67 ± 0.12	0.24	sc	[22, 45]
tRNA$_1^{Val}$ (yeast)	The same as above	54.4 ± 2.0	1350 ± 50	1.1 ± 0.12	1.67 ± 0.12	0.24	sc	[45]
tRNA$_1^{Ile}$ (yeast)	The same as above	52.3 ± 1.0	1300 ± 25	1.1 ± 0.12	1.67 ± 0.12	0.24	sc	[10, 45]
tRNA$_f^{Met}$ (E. coli)	The same as above	55.2 ± 2.0	1380 ± 50	1.1 ± 0.12	1.67 ± 0.12	0.24	sc	[45]
tRNA$_{II}^{Ser}$ (yeast)	The same as above	48.7 ± 1.0	1350 ± 50	1.1 ± 0.12	1.67 ± 0.12	0.24	sc	[10, 45]
tRNA^{Asp} (yeast)	The same as above	54.7 ± 2.0	1330 ± 50	1.1 ± 0.12	1.67 ± 0.12	0.24	sc	[11, 45]
5S RNA, ribosomal (E. coli), A-form	pH 7, 0.01 M cacodylate or 0.005 M sodium citrate or 0.005 M sodium phosphate $(0 \div 1.0)$ M NaCl and/or $(0 \div 0.001)$ M MgCl$_2$	54.0 ± 4.0	2100 ± 160	1.0 ± 0.15	1.6 ± 0.15	0.20	sc	[39]
5S RNA, ribosomal (E. coli), B-form	The same as above	47.0 ± 4.0	1840 ± 160	1.0 ± 0.15	1.6 ± 0.15	0.20	sc	[39]
16S RNA, ribosomal (F. coli)	The same as above	38.9 ± 3.6	~ 21700	1.17 ± 0.15	–	–	sc	[40]

ΔQ is the complete specific heat effect obtained by integration of the temperature dependence of the excess heat capacity; ΔH_{unf} is the enthalpy calculated per mole of RNA; C_p is the value of partial specific heat capacity at the given temperature for the compact $(C_{p,293})$ and unfolded $(C_{p,373})$ states of RNA; $\Delta C_{p,unf}^{app}$ is the maximal estimate of the heat capacity change at unfolding of a molecule obtained from linear extrapolation of the temperature dependence of heat capacity of the initial state to the whole region of the transition.

This conclusion agrees fairly well with the data available for other nucleic acids. It is confirmed also by direct estimates of ΔC_p^{unf} from calorimetric measurements which give an average value of about 0.24 kJ K^{-1} kg^{-1}, although the absolute difference between the heat capacities of the folded conformer at elevated temperature is much higher. From the scanning calorimetry data obtained at different salt conditions it follows that the large overall change of the heat capacity is caused mainly by its temperature dependence. The high values of ΔC_p^{unf} reported by many authors were observed as a rule in the presence of MgCl$_2$ and are most probably connected with the polynucleotide chain hydrolysis catalyzed by Mg^{2+} at high temperatures [45].

The overall heats of multistep transitions in ribonucleic acids are not very sensitive to the structural features, like the contents G · C and A · U base pairs. Therefore it is difficult to explain in structural terms the difference between the ΔQs observed for small compact RNA's and ribosomal 16S RNA. The complexity of the RNA melting curves gives, nevertheless, some opportunities for a more detailed analysis of thermodynamic data. These opportunities are based on the following properties of secondary structure unfolding. There is a considerable amount of evidence that the melting of short helices is nearly a two-state process [3, 23, 51]

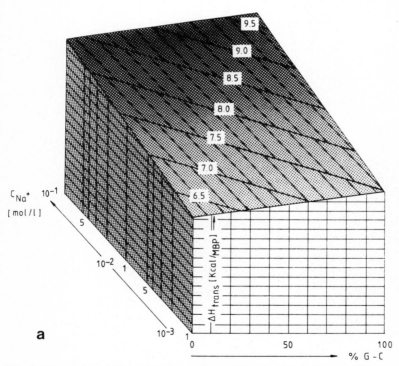

Fig. 4. a Dependence of the transition enthalpy ΔH of DNA sequences on GC-content and cation concentration (experimental results from Table 4 and unpublished data). **b** Dependence of the transition entropy ΔS of DNA sequences on GC-content and cation concentration (calculated from Gibbs-Helmholtz equation). **c** Dependence of the standard transition Gibbs energy ΔG^0 on GC-content and cation concentration

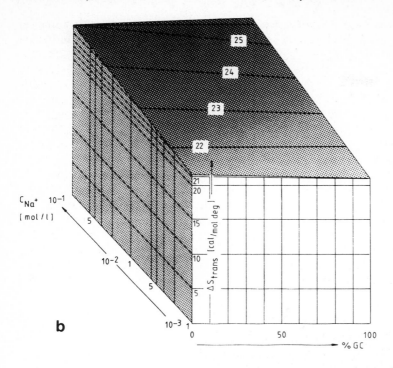

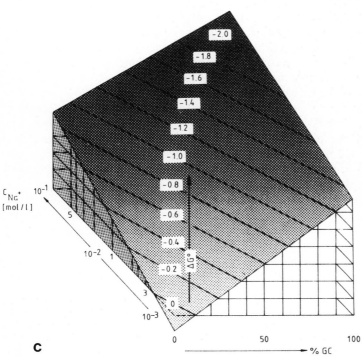

Table 4. Dependence of various thermodynamic parameters on the percentage of G-C base pairs present in DNA [71, 72]

DNA from	%-GC	$\langle m \rangle$ [a]	ΔH_{cal} kJ (mol base pair)$^{-1}$	T_m °C	b_c [b] Å	ΔG kJ (mol base pair)$^{-1}$	ΔS J K^{-1} (mol base pair)$^{-1}$
1 Salmon sperm	41.2	24	33.5	67.0	5.2	−4.4	83.7
2 Herring sperm	42.2	19	33.1	68.0	5.2	−5.0	81.0
3 Human placenta	41.2	20	33.5	68.0	5.1	−5.2	101.8
4 Chicken blood	41.7	20	33.5	67.0	5.0	−5.2	113.0
5 Calf thymus	41.9	16	33.5	66.0	5.2	−5.2	98.7
6 CL. perfringens	31.0	31	32.2	61.0	5.4	−3.4	96.2
7 E. coli	50.0	33	34.3	73.0	5.0	−5.3	99.5
8 M. lysodeikt	72.0	36	36.0	81.0	4.4	−6.1	101.6
9 Phage T7	48.0	43	34.7	70.0	5.0	−5.9	101.3
10 Phage PM2	43.0	33	33.1	72.0	5.0	−5.4	95.8
11 Plasmid pzme 134	50.0	33	33.1	72.0	5.0	−5.4	97.2
12 poly d(A-T)	–	75	31.4	45.0	6.0	−1.7	98.7
13 poly d(G-C)	100.0	60	37.7	97.0	3.9	−7.7	101.8

[a] Average cooperative length in base pairs.
[b] Linear phosphate distance as obtained from projection on the helix axis.

and that the disruption of nearest-neighbor contacts in single-stranded regions is noncooperative and could be considered as a background for the melting of secondary structure elements. It seems also that in Mg-free solutions the tertiary structure is very unstable and melts out during the first steps of unfolding [45, 51]. Therefore, under these conditions any RNA molecule could be considered as a set of cooperative blocks, whose thermodynamic parameters (particularly the enthalpy change) are determined mainly by their structure. Thus, to a first approximation one can consider the excess heat capacity curve as a superposition of several heat absorbance peaks, whose shapes, areas and positions on the temperatures scale are determined by the enthalpies and melting temperatures of each block. Under these assumptions, Privalov and Filimonov analyzed the excess heat capacity curves of six individual tRNA's and showed that the results of the analysis correlate with known physicochemical data and with the clover-leaf model of the tRNA secondary structure [10, 11, 22, 45].

Later Freire and Biltonen suggested a formal algorithm of deconvolution of the excess heat capacity curves based on the statistical-thermodynamical analysis of the general sequential scheme of unfolding [14]. They showed that scanning calorimetry data allow evaluation of the partition function of the system and a stepwise determination of the amount of macroscopic states populated during conformational transition as well as their thermodynamic parameters. It was proved that when $\delta C_p^{exc}(T)$ is known with infinite precision, its deconvolution gives unambiguous results. In practical cases when ΔC_p^{exc} is known with some uncertainty, the fidelity of deconvolution depends on many factors. The most important are the experimental error, the number of populated intermediate states, and the degree of overlapping of individual transitions, which in fact determines the concentrations of intermediates. The increase of these factors leads to an increase of the error propagation which is typical for all sequential algorithms. Therefore for practical purposes, Matveev et al. [40] combined sequential deconvolution with least-squares fitting of the melting curves, which has raised the precision of the analysis. A more detailed consideration of deconvolution techniques and results of analysis of experimental data would take too much space and for details the reader is referred to the original papers [14, 15, 40, 44, 45]. The general conclusion which has been drawn from the practical use of the deconvolution techniques for the analysis of calorimetric melting curves for small RNA's and multidomain proteins is that its application is promising. The analysis gives correct results for the polymers which undergo thermally induced unfolding through a relatively small number of intermediate states (less than 8–10) when individual transitions are separated by more than 3–5 K. This was proved by a parallel analysis of calorimetric data obtained for multidomain proteins and for individual domains prepared by limited proteolysis. In the case of small RNA's the results of deconvolution are supported by other physicochemical data. A further accumulation of calorimetric data for small RNA's will reveal more exact limits of the applicability of the deconvolution analysis to RNA melting curves.

5 Concluding Remarks

A large amount of thermodynamic data on conformational transitions in nucleic acids has been accumulated during the past decades. A general description of the thermodynamics of helix-coil transitions in polynucleotides has been developed both on theoretical and experimental grounds. Direct calorimetric techniques have played a very important role in this process.

However, most of the thermodynamic data available now are either fragmentary or insufficiently precise to allow their adequate structural interpretation. The solution of many important problems, such as elucidation of the nature and relative strength of forces stabilizing the structure of nucleic acids, as well as the development of methods of RNA structure prediction from its sequence, requires more complete and precise thermodynamic information. Further progress can be achieved only by thorough and systematic thermodynamic studies of natural and synthetic nucleic acids.

References

1. Applequist J, Damle V (1966) J Am Chem Soc 88:3895–3900
2. Baba Y, Kagemoto A (1974) Biopolymers 13:339–344
3. Blomfield VA, Crothers DM, Tinoco I Jr (1974) Physical chemistry of nucleic acids. Harper and Row, New York
4. Breslauer KJ, Sturtevant JM, Tinoco I Jr (1975) J Mol Biol 99:549–565
5. Brahms J, Michelson AM, Van Holde KE (1966) J Mol Biol 15:467–488
6. Cassani GR, Bollum FJ (1969) Biochemistry 8:3928–3936
7. De Murcia G, Wilhelm B, Wilhelm FX, Daune MP (1978) Biophys Chem 8:377–383
8. Edelhoch H, Osborne JC Jr (1976) Adv Protein Chem 30:183–251

9. Epand RM, Scheraga HA (1967) J Am Chem Soc 89:3888–3892
10. Filimonov VV, Privalov PL, Hinz H-J, von der Haar F, Cramer F (1976) Eur J Biochem 70:25–31
11. Filimonov VV, Privalov PL, Gangloff J, Dirheimer G (1978) Biochim Biophys Acta 521:209–216
12. Filimonov VV, Privalov PL (1978) J Mol Biol 122:465–470
13. Freier SM, Hill KO, Dewey TG, Marky LA, Breslauer KJ, Turner DH (1981) Biochemistry 20:1419–1426
14. Freire E, Biltonen RL (1978) Biopolymers 17:463–469
15. Freire E, Biltonen RL (1978) Biopolymers 17:498–510
16. Gruenwedel DW (1974) Biochim Biophys Acta 340:16–30
17. Gruenwedel DW (1975) Biochim Biophys Acta 395:246–257
18. Gukowski IY, Gukowskaya, AS, Sukhomudrenko AG, Suchorukov BI (1981) Nucleic Acids Res 9:4061–4079
19. Fasman GD (ed) (1976) Handbook of biochemistry and molecular biology. CRC, Cleveland
20. Hinz H-J (1979) In: Jones MN (ed) Biochemical thermodynamics. Elsevier, Amsterdam, pp 116–167
21. Hinz H-J, Haar W, Ackermann T (1970) Biopolymers 9:923–936
22. Hinz H-J, Filimonov VV, Privalov PL (1977) Eur J Biochem 72:79–86
23. Kallenbach NR, Berman HM (1977) Q Rev Biophys 10:138–147
24. Klump H (1976) Biophys Chem 5:359–361
25. Klump H (1977) Biochim Biophys Acta 475:605–610
26. Klump H, Ackermann T (1971) Biopolymers 10:513–529
27. Klump H, Burkart W (1977) Biochim Biophys Acta 475:601–604
28. Krakauer H, Sturtevant JM (1968) Biopolymers 6:491–512
29. Krakauer H (1972) Biopolymers 11:811–828
30. Lasurkin YS, Frank-Kamenetski MD, Trifonov EN (1970) Biopolymers 9:1252–1306
31. Leng M, Felsenfeld G (1966) J Mol Biol 15:455–466
32. Leng M, Michelson AM (1968) Biochim Biophys Acta 155:91–97
33. Lyubchenko YL, Vologodskii AV, Frank-Kamenetskii MD (1978) Nature 271:28–31
34. Mandel MI, Igambi L, Bergendahl J, Dodson ML, Scheltzen E (1970) J Bacteriol 101:333–338
35. Manning J (1972) Biopolymers 11:937–949
36. Marky LA, Blumenfeld KS, Breslauer KJ (1983) Nucleic Acid Res 11:2857–2870
37. Marky LA, Breslauer KJ (1982) Biopolymers 21:2185–2194
38. Marky LA, Patel D, Breslauer KJ (1981) Biochemistry 20:1427–1431
39. Matveev SV, Filimonov VV, Privalov PL (1982) Mol Biol (Mosc) 16:1234–1251
40. Matveev SV, Privalov PL (1979) Dokl Akad Nauk SSSR 247:985–989
41. Michelson AM (1968) In: Pullman B (ed) Molecular associations in biology. Academic Press, New York, pp 93–106
42. Michelson AM, Massoulie J, Guschlbauer W (1967) Prog Nucleic Acid Res Mol Biol 6:83–141
43. Neumann E, Ackermann T (1969) J Phys Chem 73:2170–2178
44. Privalov PL (1980) In: Beezer AE (ed) Biological microcalorimetry. Academic Press, New York, pp 413–451
45. Privalov PL, Filimonov VV (1978) J Mol Biol 122:447–464
46. Privalov PL, Ptitsyn OB, Birstein TM (1967) Biopolymers 8:559–571
47. Pörschke D, (1976) Biochemistry 15:1495–1499
48. Poland D, Vournakis JN, Scheraga HA (1966) Biopolymers 4:223–235
49. Rawitscher MA, Ross PD, Sturtevant JM (1963) J Am Chem Soc 85:1985–1918
50. Record MT, Anderson CF, Lohman TM (1978) Q Rev Biophys 11:103–178
51. Riesner D, Römer R (1973) In: Dudusne J (ed) Physico-chemical properties of nucleic acids, vol 2. Academic Press, London, pp 237–318
52. Riley M, Maling B, Chamberlin MJ (1966) J Mol Biol 20:359–389
53. Ross PD, Scruggs RL (1965) Biopolymers 3:491–496
54. Ross PD, Scruggs RL (1969) J Mol Biol 45:567–569

55. Ross PD, Scruggs RL (1974) Biopolymers 13:417
56. Ross PD, Scruggs RL, Howard FB, Miles HT (1971) J Mol Biol 61:727–733
57. Rüterjans H (1965) Doctoral Thesis, University of Münster, NRW, FRG
58. Scruggs PL, Ross PD (1970) J Mol Biol 47:29–40
59. Scheffler IE, Sturtevant JM (1969) J Mol Biol 42:577–580
60. Shiao DDF, Sturtevant JM (1973) Biopolymers 12:1829–1836
61. Stannard BS, Felsenfeld G (1975) Biopolymers 14:299–306
62. Steiner RF, Kitzinger (1962) Nature 194:1172–1173
63. Stulz J (1981) Doctoral Thesis, Freiburg University, Freiburg, FRG
64. Suurkuusk J, Alvarez J, Freire E, Biltonen RL (1977) Biopolymers 16:2641–2652
65. Tso POP (1968) In: Pullman B (ed) Molecular association in biology. Academic Press, New York, pp 39–92
66. Tong BY, Battersby SJ (1979) Biopolymers 18:1917–1936
67. Thrierr JC, Dourlent M, Leng M (1971) J Mol Biol 58:815–830
68. Zimm BH (1960) J Chem Phys 33:1349–1356
69. Breslauer KJ, Sturtevant JM (1977) Biophys Chem 7:205–209
70. Zimmerman SB (1975) J Mol Biol 101:563–565
71. Klump H (1985) Thermochim Acta 85:457–463
72. Klump H, Herzog K (1984) Ber Bunsen-Ges Phys Chem 88:20–24
73. Privalov PL (1980) Pure Appl Chem 52:479–497
74. Hinz H-J (1986) Methods Enzymol 130:59–79

Chapter 15 Methods for Obtaining Thermodynamic Data on Oligonucleotide Transitions

KENNETH J. BRESLAUER

Symbols:

K:	equilibrium constant
$K(T_m)$:	equilibrium constant at T_m
T_m:	melting temperature at $\alpha = 0.5$
ΔH_{vH}:	van't Hoff enthalpy as derived from temperature dependence of equilibrium constant [kJ (mol of cooperative unit)$^{-1}$]
$\Delta H_{vH}^{con.\,dep.}$:	van't Hoff enthalpy calculated using Eq. (9)
ΔH_{vH}^{shape}:	van't Hoff enthalpy calculated using Eq. (5) or (6)
ΔH_{vH}^{cal}:	van't Hoff enthalpy calculated using Eq. (8)
ΔH_{cal}:	calorimetrically obtained molar enthalpy. The numerical value depends on the unit of the mol (strands or base pairs)
ΔH_{cal}^0:	calorimetrically obtained standard molar enthalpy
ΔH:	transition enthalpy per mole of base pair stack
ΔH_{obs}^0:	observed standard transition enthalpy per mol of duplex
ΔH_{pred}^0:	predicted standard transition enthalpy per mol of duplex
ΔH^0:	standard enthalpy change
ΔG^0:	standard Gibbs energy change
ΔS^0:	standard entropy change
C_T:	total strand concentration
c_T^i:	total strand concentration of substance i
c_T^j:	total strand concentration of substance j
α:	fraction of oligonucleotide strands in the double-stranded state (degree of association)

1 Introduction

The goal of this chapter is to present a critical survey of spectroscopic and calorimetric methods for obtaining thermodynamic data on conformational transitions in oligonucleotides. A comprehensive compilation of the large body of published thermodynamic data on oligo- and polynucleotides will not be included here since these results have been tabulated and reviewed elsewhere [1–4]. Instead, selected data will be presented to help illustrate the relative merits of each experimental method described for extracting thermodynamic data from optical and calorimetric studies on oligonucleotides.

2 Why Are We Interested in Thermodynamic Data?

From a historical point of view, the early desire for thermodynamic data on nucleic acids reflected a fundamental interest in defining the nature of the inter- and intramolecular forces that stabilize single- and double-stranded DNA and RNA structures (Chap. 6 in reference [1], Chap. 6 in reference [2], and Chap. 3 in reference [3]). This interest intensified when it was shown that the thermal stabilities of nucleic acids depended not only on base composition, but also on nucleotide sequence [5–8]. Immediately, more studies focused on the thermodynamic characterization of nearest-neighbor base-stacking interactions [9–44]. Some laboratories needed such data to fit calculated with observed melting curves so that the validity of various helix-to-coil models could be assessed [24, 45–57]. Other laboratories used the thermodynamic data to resolve and interpret multiphasic transition curves or as thermodynamic "baselines" for drug-binding studies [48–60]. Thermodynamic data also were needed to meaningfully interpret the results of kinetic studies on DNA and RNA conformational changes. Still other laboratories desired such data as a basis for predicting sequence-specific secondary structural features along regions of the nucleic acid polymer chain [25, 26, 32, 60]. This predictive power probably represents the most intriguing and potentially powerful application of the thermodynamic data. It has been suggested that local, sequence-specific structural domains might provide unique binding sites and/or control switches for biological events such as protein-nucleic acid interactions, drug-DNA binding, gene expression, DNA packing, etc [61–64]. For example, bulge loops resulting from imperfect sequence complementarity have been proposed as intermediates in frameshift mutagenesis [63, 65]. Sequences favoring hairpin loops have been found near functional loci in DNA, thereby suggesting a structural basis for control mechanisms [66]. Consequently, in evaluating and proposing possible biological roles for specific sequences in naturally occurring nucleic acids, it would be extremely useful if one could predict the formation of particular secondary structural features along the polymer chain from the primary sequence. Such predictive powers are only possible if appropriate thermodynamic data are available.

The relatively recent discovery of new and unusual double helical forms (e.g., Z, alternating B, etc.) [67, 68] has stimulated interest in thermodynamic data on the stability and conformational flexibility of these double helices. Such thermodynamic data will allow an assessment of the relative stabilities of these new and potentially biologically significant structural forms [69, 70].

For all of the reasons noted above, several laboratories have dedicated much of their research effort to thermodynamic investigations of DNA and RNA molecules.

3 Why Study Oligonucleotides?

Much of the early thermodynamic work on DNA and RNA involved investigations of the solution behavior of natural polymers [5–22, 31, 71–79]. Although these studies produced a great deal of useful information, the vast base heterogeneity present in these systems made it difficult to correlate experimental observables with specific molecular events. This difficulty was largely circumvented by studies on synthetic DNA and RNA polymers of defined sequences. These studies yielded much new and important information on the relationship between base composition and nucleic acid stability. However, fundamental questions remained unanswered: Do particular base sequences favor specific structural forms? What are the relative stabilities of hairpin loops, internal loops, bulge loops, etc.? Do the synthetic polymer results provide meaningful models for the stability and conformational flexibility of short helical regions? These questions, and many others, only could be addressed by the investigation of specially designed and synthesized oligonucleotides containing specific base sequences. By systematically varying the sequences of such well-defined compounds, an experimental observable could be correlated umambiguously with a specific structural feature. Consequently, DNA and RNA oligomers became the subject of thermodynamic investigations in a number of laboratories [18, 26–30, 32, 35–46, 80–86]. The recent development of new and improved techniques for oligonucleotide sequences has further stimulated activity in this important area [87–90].

4 Methods for Obtaining Thermodynamic Data

4.1 Optical Techniques for Determining Transition Enthalpies

Various spectroscopic methods have been used to determine indirectly the enthalpy change accompanying a conformational transition in an oligonucleotide. The general approach usually involves monitoring the temperature dependence of some optical property. The temperature-dependent change in this optical property is then correlated with the corresponding change in the equilibrium constant by assuming a model for the thermally induced transition of the oligomer. Subsequent application of the van't Hoff relationship [Eq. (1)] permits the transition enthalpy (ΔH_{vH}) of the oligomer to be calculated using one of the protocols described below

$$\frac{d \log K}{dT} = \frac{\Delta H^0}{2.3\,RT^2}. \tag{1}$$

4.1.1 Shape Analysis of Absorbance Versus Temperature Profiles

A variety of observables have been used to monitor oligonucleotide transitions. However, for illustrative purposes we will focus on the change in UV absorbance at or near 260 nm. This optical window represents the most frequently monitored observable for following order-disorder transitions in oligonucleotides.

The transition of interest can be induced by a variety of methods which involve changes in solution conditions (e.g., temperature, pH, metal ions, added organic solvents, salt, etc.). In the discussions that follow, the transitions to be examined have been induced by increasing temperature. Consequently, we will describe methods for obtaining thermodynamic data from UV absorbance versus temperature profiles, so-called UV melting curves. However, the protocols outlined are general and with minor modifications can be used to thermodynamically analyze oligomer transitions which are induced by different structural perturbants and monitored using different optical and nonoptical techniques.

THERMALLY INDUCED HELIX — COIL TRANSITION

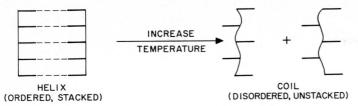

HELIX
(ORDERED, STACKED)

COIL
(DISORDERED, UNSTACKED)

Fig. 1. Schematic representation of a thermally induced transition from a fully base-paired and base-stacked double helix to two unpaired and unstacked single strands

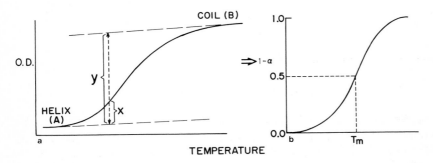

Note: $1-\alpha = \dfrac{x}{y}$ = fraction in coil (B) state

Fig. 2. a Schematic of an optical density (*O.D.*) vs. temperature profile (a UV melting curve) showing the hyperchromic effect associated with a transition from a base stacked (helix) to an unstacked (coil) state. **b** A derived melting curve showing the fraction in the coil state $(1-\alpha)$ as a function of temperature. The two figures together illustrate how the experimental absorbance vs. temperature profile can be converted into a $(1-\alpha)$ versus temperature melting curve

UV Melting Curves. At low temperatures in solution, two complementary single-stranded oligomer sequences will associate to form a base-paired and base-stacked duplex structure. As the temperature is increased, enough thermal energy (kT) becomes available to disrupt the helix-stabilizing forces and to melt the duplex structure. Such a thermally induced helix-to-coil transition can be represented schematically as shown in Fig. 1.

When the base pairs are unstacked in the final, single-stranded state(s), they absorb more 260 nm UV radiation than when they are stacked in the initial duplex state(s). This increase in absorbance is called the hyperchromic effect. Because of this stacking-dependent optical effect, the thermally induced helix-to-coil transition of an oligomer can be monitored by following the increase in 260 nm UV absorption with increasing temperature. Figure 2a shows a typical absorbance versus temperature profile in which this hyperchromic effect is illustrated for the thermally induced disruption of an oligomer from an initial helical state, A, to a final coil state, B. As described below and illustrated in Fig. 2, this curve can be analyzed to obtain a value for the transition enthalpy.

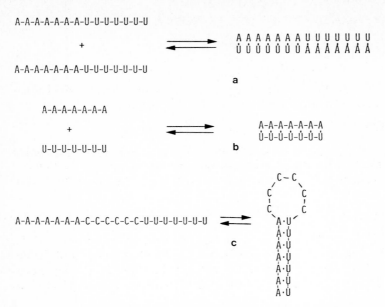

A-A-A-A-A-A-A-U-U-U-U-U-U-U

+

A-A-A-A-A-A-A-U-U-U-U-U-U-U

a

A-A-A-A-A-A-A

+

U-U-U-U-U-U-U

b

A-A-A-A-A-A-A-C-C-C-C-C-C-U-U-U-U-U-U-U

c

Fig. 3. a The molecular formation of a double helix from two self-complementary strands of rA_7U_7. **b** The molecular formation of a double helix from two separate non-self-complementary strands, rA_7 and rU_7. **c** The intramolecular formation of a duplex to form a hairpin structure

If one assumes that the fractional change in absorbance at any temperature monitors the extent of reaction, then any absorbance versus temperature curve can be converted into a 1-α versus temperature curve, where α equals the fraction of strands in the double-stranded state. This conversion is accomplished by taking the ratio at each temperature of the height between the lower baseline and the experimental curve and the height between the lower and upper baselines. As illustrated in Fig. 2, this graphical procedure converts an experimental absorbance versus temperature curve into a 1-α versus temperature profile. If one assumes a two-state (all-or-none) transition, we can write the equilibrium constant, K, for duplex *formation* in terms of alpha, α, and C_T, the total strand concentration. For the case of a duplex formed from a self-complementary oligomer sequence such as rA_7U_7 (Fig. 3a), where the equilibrium is $2\,A \rightleftarrows A_2$ one obtains

$$K = \frac{\alpha/2}{(1-\alpha)^2 C_T}. \tag{2}$$

For a duplex formed from two nonidentical, complementary oligomers, such as $A_7 + U_7$ (Fig. 3b), where the equilibrium is $A + B \rightleftarrows AB$, one obtains

$$K = \frac{2\alpha}{(1-\alpha)^2 C_T}. \tag{3}$$

For the case of a duplex formed intramolecularly from two complementary regions within the same strand such as $A_7 C_6 U_7$, (Fig. 3c), where the equilibrium is $A \rightleftarrows B$, one obtains

$$K = \frac{\alpha}{1-\alpha},$$ (4)

The appropriate expression for K is then substituted into the van't Hoff equation given earlier [Eq. (1)]. After enduring the algebra, one obtains

$$\Delta H_{vH} = 6\,RT^2(\partial\alpha/\partial T).$$ (5)

This expression can be used to determine the van't Hoff enthalpy for *intermolecular* duplex formation involving either complementary or self-complementary oligonucleotide sequences. For *intramolecular* duplex formation the corresponding equation is

$$\Delta H_{vH} = 4\,RT^2(\partial\alpha/\partial T).$$ (6)

A value for $\partial\alpha/\partial T$ can be obtained at the melting temperature, T_m (where $\alpha = 0.5$), by measuring the slope of a tangent line drawn at the inflection point of the derived 1-α versus T melting curve. Substituting this value and T_m into the appropriate van't Hoff expression allows the van't Hoff transition enthalpy to be calculated.

This approach for calculating ΔH_{vH} from experimental absorbance versus temperature profiles is quite convenient. The required UV melting curves can be readily measured on standard laboratory spectrophotometers. Furthermore, the high absorbtivity of the nucleic acid bases (extinction coefficients $> 10^3$), permits the measurement to be performed on very small quantities of material (less than 1 absorbance unit). These conveniences, however, at times can be overshadowed by the assumptions and approximations inherent in the analysis. To begin with, it has been demonstrated that the derived ΔH_{vH} can be quite sensitive to the choice of initial and final baselines used to analyze the experimental absorbance vs. temperature curves. The enthalpy data in Table 1 illustrate this point. Inspection of these data reveal that baseline selection can result in differences of some 20% in the calculated transition enthalpies. For sequences that are not self-complementary, the temperature-dependence of the final state (that is, the final baseline) can be determined experimentally by melting each of the separate strands [62]. However, the problem of defining the initial baseline remains.

Even when baseline selection is not a problem, the data analysis described above is usually limited by the assumption that the transition proceeds in a two-

Table 1. Dependence of van't Hoff transition enthalpies on lower baseline selection

Oligomer sequence	Method of analysis	ΔH_{vH} (kJ)	Reference
rA$_7$U$_7$	Eq. (5) using a flat lower baseline	−240	[37]
rA$_7$U$_7$	Eq. (5) using a sloping lower baseline	−294	[37]
d(GC)$_3$	Eq. (5) using a flat lower baseline	−173	[38]
d(GC)$_3$	Eq. (5) using a sloping lower baseline	−238	[38]
dCA$_5$G+dCT$_5$G	Eq. (5) using a flat lower baseline	−176	[62]
dCA$_5$G+dCT$_5$G	Eq. (5) using a sloping lower baseline	−213	[62]

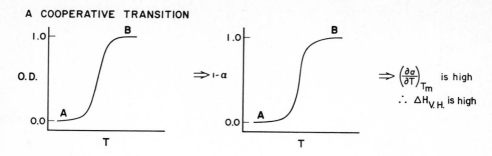

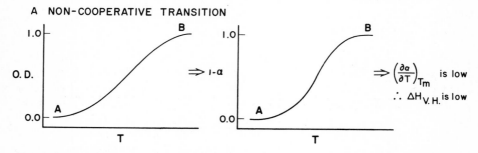

Fig. 4. The shape of the derived $(1-\alpha)$ vs. temperature melting curve depends on the shape of the experimental O.D. vs. temperature melting profile. Noncooperative transitions have broad melting profiles which yield $(1-\alpha)$ vs. temperature curves with reduced $(\partial\alpha/\partial T)$ at T_m slopes which results in a lower value for the transition enthalpy

state manner. This assumption may or may not be correct. When the transition does proceed in a two-state manner and the baselines are properly selected, this method of data analysis yields meaningful transition enthalpies that agree with those determined directly by calorimetry. However, if a transition proceeds with a significant population of intermediate states (noncooperatively), the experimental absorbance versus temperature profile and the derived $1-\alpha$ vs. T curve will be broadened as illustrated in Fig. 4. This broadening will result in a lower value for $(\partial\alpha/\partial T)_{T_m}$ which according to Eq. (5) or (6) will lead to a reduction in the calculated value for the transition enthalpy. Thus, for transitions that do not proceed in a two-state manner, the transition enthalpy calculated as described above will be lower than the value associated with disruption of the entire duplex. Although more sophisticated treatments are possible in which the optical data are fit to multi-state models, the results obtained frequently are not unique.

An additional complication to this method of data analysis derives from the fact that application of the van't Hoff equation as given above assumes a temperature-independent enthalpy. Although for some systems this assumption may be reasonable, a more direct measure of the heat capacity is desirable.

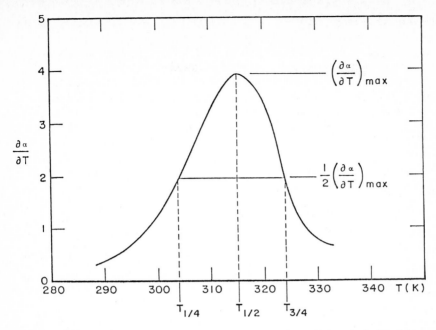

Fig. 5. A schematic of a differentiated melting curve. The half-width at the half-height can be used to calculate a van't Hoff transition enthalpy as described in the text

4.1.2 Analysis of the Shape of a Differentiated Absorbance versus Temperature Melting Curve

An approach very similar to the one just described can be used to obtain a van't Hoff transition enthalpy from a *differentiated* UV melting curve. Figure 5 shows a tracing of a typical differentiated melting curve. Such a curve can be derived from an experimental absorbance versus temperature profile or obtained directly from difference spectrometry or from temperature-jump measurements. The half-width of this curve at the half-height is inversely proportional to the van't Hoff transition enthalpy [91]. For mono- and bimolecular duplex formation, the relevant equations are (7) and (8), respectively.

$$\Delta H_{VH} = \frac{-3.2}{\dfrac{1}{T_{1/2}} - \dfrac{1}{T_{3/4}}} \tag{7}$$

$$\Delta H_{VH} = \frac{-4.37}{\dfrac{1}{T_{1/2}} - \dfrac{1}{T_{3/4}}}, \tag{8}$$

where $T_{1/2}$ and $T_{3/4}$ are the half and three quarters temperatures at the half height.

This analysis of temperature-dependent optical data still incorporates the assumption of a two-state transition and a temperature-independent enthalpy. However, because this analysis measures the half width at the half-height, it provides a means of circumventing the *lower* baseline problem which one frequently encounters in the analysis of the *overall shapes* of integral absorbance versus temperature curves. Specifically, for transitions with low T_m's where it is difficult to define the lower baseline (or to even obtain the lower half of the melting curve), this method of data analysis permits a transition enthalpy to be calculated from just the upper half of the melting profile (see Fig. 4).

When the differentiated melting curve is obtained directly from temperature-jump experiments, the problem of baseline selection may be reduced further. If the molecular events that give rise to the sloping baselines correspond to processes that are very fast, then the temperature-jump experiment provides a means of kinetically resolving the relatively slow, cooperative helix melting from these fast baseline effects [80, 91, 92]. In such a case, one obtains a differentiated melting curve that is kinetically filtered of the baseline problems encountered with integral melting curves.

4.1.3 Temperature Dependence of the Equilibrium Constant

As shown in Fig. 2, an experimental O.D. vs. T melting profile can be converted graphically into a $(1-\alpha)$ vs. T melting curve. From this curve, values of α can be calculated at different temperatures. These α values then can be used to calculate the K as a function of T by plugging them into the appropriate expression for the equilibrium constant [Eq. (2), (3), or (4)]. The resulting data are then plotted in the form $\ln K$ vs. $1/T$. Such a plot will produce a straight line if ΔH_{vH}^0 is constant over the temperature range. The slope of this line when multiplied by $-R$ is equal to ΔH_{vH}^0. As with the other methods just described, implicit in this approach for calculating the van't Hoff enthalpy is the assumption of a two-state transition.

4.1.4 Concentration Dependence of the Melting Temperature

The formation of a double helix from two separate self-complementary single strands is a bimolecular process. For short enough oligomers, this bimolecularity can be observed by monitoring the concentration dependence of the melting temperature. For a given oligomer, one measures UV melting curves over a range of oligomer concentrations. The resulting concentration-dependent melting temperature data then are interpreted using the equation

$$\frac{1}{T_m} = \frac{2.3\,R}{\Delta H}\,\log C_T + \frac{\Delta S}{\Delta H}, \tag{9}$$

where C_T equals the total oligomer strand concentration. From the slope of a plot of $1/T_m$ versus $\log C_T$ (Fig. 6), one can calculate the van't Hoff transition enthalpy.

As with the methods described above, this protocol usually incorporates the assumption of a two-state transition and a temperature-independent enthalpy.

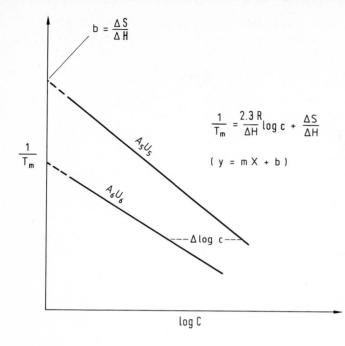

$$\frac{1}{T_m} = \frac{2.3\,R}{\Delta H}\log c + \frac{\Delta S}{\Delta H}$$

$$(\,y = m\,X + b\,)$$

Fig. 6. Typical plots of the concentration dependence of the melting temperature. When plotted as $1/T_m$ vs. $\log C$, the slope is equal to $R/\Delta H$ and the intercept is equal to $\Delta S/\Delta H$. At any temperature, the horizontal distance between the two lines ($\Delta \log C$) is proportional to the free energy difference between the two self-complementary oligomers as described in the text

However, this method of data analysis is much less sensitive to baseline problems. This lower sensitivity of the calculated ΔH_{vH} to the choice of baselines is associated with the fact that the slope of the $1/T_m$ versus $\log C_T$ line results from *differences* in the T_m values obtained from several UV melting curves. Consequently, as long as the baseline for all the melting curves is selected in a *consistent* manner, the slope of the line provides a good measure of the van't Hoff enthalpy for transitions that proceed in a two-state manner. Any deviations of the selected baselines relative to the "true" baselines simply produce a parallel line which has a different Y-axis intercept but the same slope.

This method of data analysis is uniquely applicable to short oligomer transitions where the concentration dependence of the melting temperature can be observed. For long oligomers and polymers, this approach cannot be applied, since the monomolecular helix growth steps dominate the bimolecular helix initiation step, thereby producing a pseudo first-order equilibrium for which a concentration-dependent melting temperature cannot be observed.

Calorimetric Determination of Transition Enthalpies. Differential scanning calorimetry (DSC) also can be used to detect and follow thermally-induced order-disorder transitions in oligonucleotides. However, with DSC the heat capacity rather than the absorbance of the solution is monitored. Figure 7 shows a schematic of the cell compartment of a DSC. In a typical experiment, the reaction and the reference cells each are filled with approximately 1 ml of solution and the temperature is scanned from about 5 °C to about 100 °C. For a thermally induced endothermic transition, the temperature of the reaction cell will lag behind that of the

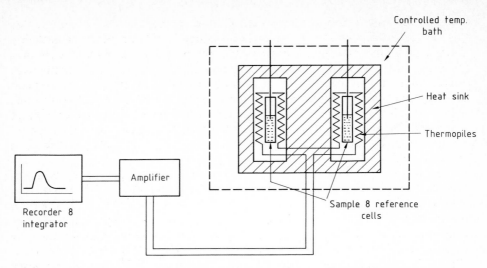

Fig. 7. Schematic of the components of a typical differential scanning calorimeter

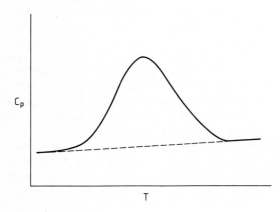

Fig. 8. A typical heat capacity (C_p) vs. temperature transition curve obtained by differential scanning calorimetry

solvent reference cell. The additional electrical energy feedback to the reaction cell to maintain it at the same temperature as the solvent reference cell is continuously measured. The instrument is calibrated by measuring the area produced by a controlled electrical pulse. These data, along with the known concentration of the oligomer, permit construction of a heat capacity versus temperature curve as shown in Fig. 8. The area under the curve is equal to the transition enthalpy of the oligomer.

In contrast to the model-dependent (usually two-state) ΔH_{vH} values indirectly derived from the optical data, the calorimetrically determined transition enthalpy does *not* depend on the nature of the transition. In calorimetry one simply measures the total energy required to go from the initial to the final state. The shape of the curve, which does depend on the nature of the transition, need not be analyzed as is done with the optical data. Thus, the calorimetric measurement provides a direct, model-independent determination of the transition enthalpy. For this reason, a comparison of the model-dependent van't Hoff transition enthalpy and the model-independent calorimetric transition enthalpy provides insight into the nature of the transition. If ΔH_{vH} is less than ΔH_{cal}, then the transition involves a significant population of intermediate states. However, if $\Delta H_{vH} = \Delta H_{cal}$, we can conclude that the transition proceeds in a two-state manner. In addition to this ability to define the nature of the transition, calorimetry also provides a direct measure of the heat capacity change accompanying the transition. Consequently, one need not assume that ΔC_p is zero, as usually, but not always [95], is done when analyzing optical data.

The features just described represent significant advantages that calorimetry provides over optical determinations of transition enthalpies. Table 2 lists the transition enthalpies for rA_7U_7 and $d(GC)_3$ that have been obtained using both optical and calorimetric techniques. For the rA_7U_7 data, the significant observation is that all the van't Hoff transition enthalpies are considerably lower than the calorimetrically-determined value. This observation reflects the fact that the transition of rA_7U_7 does *not* proceed in a two-state manner. For such multi-state transitions, calorimetry alone provides a meaningful measure of the transition enthalpy. Furthermore, the ΔH_{vH} value calculated from UV melting curves using Eq. (5) only agrees with the ΔH_{vH} derived from Eq. (9) and/or temperature-jump data using Eq. (8) when sloping baselines are employed.

For the $d(CG)_3$ data (Table 2), the significant observation is that the optically-derived van't Hoff and the calorimetrically-measured transition enthalpies agree when the optical data are analyzed using Eq. (5) (with sloping baselines) or Eq. (9). These observations permit two conclusions: (a) The transition of $d(GC)_3$ proceeds in a two-state manner, and (b) Eq. (5) (with properly chosen sloping baselines) or Eq. (9) provide good methods for obtaining transition enthalpies of oligomers that undergo order-disorder transitions in a two-state manner.

Calorimetry is not without its shortcomings. As with the optical methods, the calorimetric data analysis also has to deal with a baseline problem. Furthermore, since thermal measurements do not benefit from the absorptivity-induced signal enhancement of optical measurements, calorimetry requires significantly more

Table 2. Comparison of van't Hoff and calorimetric transition enthalpies for rA_7U_7 and $d(GC)_3$

Oligomer sequence	Method of analysis	ΔH_{cal} (kJ mol^{-1})	ΔH_{vH} (kJ)
rA_7U_7 [a]	Area under calorimetric heat capacity curve	415	
	Eq. (5) with sloping baselines		294
	Eq. (8) from temperature-jump data		328
	Eq. (9)		317
$d(GC)_3$ [b]	Area under calorimetric heat capacity curve	249	
	Eq. (5) with sloping baselines		238
	Eq. (9)		240

[a] [37].
[b] [38].

material to monitor the broad transitions exhibited by oligonucleotides. However, the recent explosion in and automation of solid-state oligonucleotide synthesis makes such quantities of oligomers readily available.

Aggregation of oligomers at high concentrations represents a potential source of difficulty with the calorimetric measurement. In fact, equilibrium sedimentation experiments at 4 °C for 4 days revealed oligomer aggregation even at *optical* concentrations [62]. However, the long-term spinning of the oligomers at low temperatures in these experiments may well have produced aggregates that do not exist under the conditions of either the optical or calorimetric measurements. In this connection, we find that after long-term freezer storage the first calorimetric transition curve of an oligomer often exhibits a heat capacity curve with an anomolous lower baseline [93]. However, after the first heating, all repeat runs are "well-behaved". Furthermore, the anomolous lower baseline behavior can be avoided by simply preheating the oligomer solution before the first run. If some oligomer aggregation occurs during freezer storage, these results suggest that after thermal disruption, their reformation is kinetically limited, thereby preventing aggregate formation from interfering with either the optical or calorimetric measurements. In fact, in the few cases where a direct comparison of optical and calorimetric results is possible, the enthalpy data agree, thereby suggesting that aggregation is not a problem [94]. Nevertheless, when interpreting both optical and calorimetric data, one should always be cognizant of the possible role of aggregates.

4.1.5 Calculating Complete Thermodynamic Transition Profiles from the Enthalpy Data

The Optical Approach. As described above, a van't Hoff enthalpy can be calculated from either integral or differentiated optical melting curves by application of Eq. (5), (6), (7), (8), or (9). The corresponding Gibbs energy and entropy changes can be calculated by either of the two methods described below.

Optical Method 1. For any conformational transition, we can write an expression for the equilibrium constant for helix formation in terms of α, the fraction of strands in the duplex state. Equations (2), (3), and (4) represent several examples. Using any of these expressions, the equilibrium constant, K, can be evaluated at any value of α. Usually a value of K is determined at the melting temperature ($\alpha = 0.5$) since it can be measured most accurately. Thus, by substituting $\alpha = 0.5$ into the appropriate equilibrium expression one obtains a value for K at the T_m. This K value, $K(T_m)$, is then extrapolated back to some reference temperature (usually 25 °C) using the experimentally measured melting temperature, the calculated transition enthalpy (assumed to be temperature-independent), and the integrated form of the van't Hoff equation shown below.

$$\frac{\log K(T_m)}{\log K_{298}} = \frac{\Delta H}{2.3 R}\left(\frac{1}{298} - \frac{1}{T_m}\right). \tag{10}$$

This value of the equilibrium constant at 298 K permits the free energy of the transition at 298 K to be calculated using the standard thermodynamic relationship $\Delta G^0 = -RT\ln Keq$. From values of ΔG^0 and ΔH^0, one can calculate ΔS^0 using the equation $\Delta G^0 = \Delta H^0 - T\Delta S^0$. However, if ΔC_p is not zero, the temperature extrapolation involved in this method can introduce serious errors.

Optical Method 2. If the van't Hoff transition enthalpy is determined from concentration-dependent melting studies using Eq. (9), a more direct procedure can be employed. Inspection of Eq. (9) and Fig. 6 reveals that while the slope of a $1/T_m$ versus log C plot yields ΔH_{vH}, the intercept permits calculation of ΔS^0. These two thermodynamic parameters then can be used to calculate ΔG^0. Alternatively, K and therefore ΔG^0 can be calculated at any reference temperature using the transition enthalpy (from log C versus $1/T_m$), the melting temperature, and the integrated form of the van't Hoff Eq. (10). Using these two thermodynamic parameters (ΔH^0 and ΔG^0), the transition entropy, ΔS^0, can be calculated by application of the standard thermodynamic relationship, $\Delta G^0 = \Delta H^0 - T\Delta S^0$. In principle, for a two-state transition these approaches should be equivalent to each other and to Optical Method 1 described above. However, since transition enthalpies calculated by Method 1 are more sensitive to baseline assignments, the thermodynamic profiles calculated by Method 2 probably are more reliable.

For an $A \rightleftarrows B$ equilibrium, the equilibrium constant, K, equals 1 at the melting temperature where $\alpha = 0.5$ [see Eq. (4)]. Consequently, at T_m, $\Delta G^0 = 0$ and we can write

$$\Delta S^0 = \frac{\Delta H^0}{T_m}. \tag{11}$$

Thus, using the transition enthalpy derived from either Eqs. (5) or (9) and the experimentally observed melting temperature, T_m, one can calculate a value for ΔS^0.

Substituting Eq. (11) into the standard thermodynamic relationship $\Delta G^0 = \Delta H^0 - T\Delta S^0$, one obtains

$$\Delta G^0 = \Delta H^0 (1 - T/T_m). \tag{12}$$

This expression can be used to calculate the transition Gibbs energy at any temperature of interest. For a two-state transition with a temperature-independent enthalpy, this method is equivalent to the other approaches described above.

Gralla and Crothers [82] have described an approach in which the free energy *difference* between two bimolecular duplexes can be determined from a comparison of the concentration dependence of the melting temperatures of the two duplexes. Figure 6 shows the relevant log c versus $1/T_m$ plots for the duplexes formed by A_5U_5 and A_6U_6. Gralla and Crothers showed that the concentration difference required to cause two duplexes to have identical T_m's is proportional to the free energy difference between the two duplexes as given by the equation

$$\Delta G_i^0 - \Delta G_j^0 = 2.3RT(\log C_T^i - \log C_T^j). \tag{13}$$

The value of $\log C_T^i - \log C_T^j$ can be read directly from Fig. 6 as $\Delta \log c$. This approach yields the *difference* in Gibbs energy between two duplex structures.

Significantly, all of the optical methods outlined above for calculating thermodynamic parameters from optical data are applicable only to transitions that proceed in a two-state manner. Unfortunately, such a characterization of the nature of a transition is not possible based solely on absorption data. Consequently, additional, independent information concerning the nature of a transition is required before these optical approaches can be applied to a given oligomer.

The Calorimetric Approach. As described above, a single calorimetric transition curve gives a direct, model-independent measure of both the heat capacity and enthalpy changes accompanying a thermally-induced conformational change. Frequently, these calorimetric data have been used in conjunction with independent equilibrium measurements of ΔG^0 to calculate complete thermodynamic profiles. Significantly, however, as described below, this dependence on noncalorimetric data is not necessary.

In a given DSC experiment, one obtains directly a heat capacity (C_p) versus temperature (T) curve. This C_p versus T curve can be differentiated to obtain a C_p/T versus T curve. Since $\Delta S = \int \dfrac{C_p dT}{T}$, the area under such a curve provides a "direct" measure of entropy change. Thus, from a single calorimetric transition curve one can obtain ΔC_p, ΔH^0, and ΔS^0. Using these data, the corresponding value of ΔG^0 can be calculated at any temperature. This calorimetric approach has two significant advantages relative to optical methods. First, one obtains a *model-independent* measure of the transition enthalpy and entropy rather than the "two-state values" usually obtained from analysis of optical data. Secondly, the calorimetric experiment provides a direct measure of ΔC_p so that the temperature dependence of conformational stability can be assessed. Table 3 compares enthalpy and entropy data obtained exclusively from a calorimetric heat capacity curve and the corresponding data derived from optical melting curves using Eq.(9). The significant observation is that for transitions that proceed through multiple states the optically derived values can seriously deviate from the calorimetrically-determined values. Only for two-state transitions do the optical methods of analysis described above yield meaningful thermodynamic data.

Table 3. Comparison of calorimetric and optical ΔH^0 and ΔS^0 values[a]

Oligomer sequences	Transition	Optical data[b]		Calorimetric data	
		ΔH^0_{vH}	ΔS^0_{vH}	ΔH^0_{cal}	ΔS^0_{cal}
(GC)$_3$	Two-state	240	659	249	649
GGAATTCC	Two-state	245	699	244	799
CGTACG	Two-state	191	544	203	640
rA$_7$U$_7$	Multi-state	317	931	415	1163
GCGAATTCGC	Multi-state	301	816	335	1151
CGCGAATTCGCG	Multi-state	273	711	427	1246
CGTGAATTCGCG	Multi-state	330	933	444	1277
CGCAGAATTCGCG	Multi-state	284	791	435	1364
TATGCGCGCATA	Multi-state	238	611	312	872

[a] ΔH^0_{vH} in kJ (mol cooperative unit)$^{-1}$, ΔH^0_{cal} in kJ (mol duplex)$^{-1}$, ΔS^0_{vH} in J (mol cooperative unit $\cdot$ K)$^{-1}$, ΔS^0_{cal} in J (mol duplex $\cdot$ K)$^{-1}$.
[b] From Eq. (9).

Table 4. Comparison of various van't Hoff transition enthalpies for selected DNA oligomers

Oligomer sequence	$\Delta H_{vH}^{con.\ dep.\ a}$	$\Delta H_{vH}^{shape\ b}$	$\Delta H_{vH}^{cal\ c}$
ATGCAT	163	164	146
GCGCGC	240	238	217
CGCGCG	180	209	192
CGTACG	191	183	197
GGAATTCC	245	225	248
GCGAATTCGC	301	318	362
ATATATATAT	344	144	171
TATGCGCGCATA	238	234	310
CGCGAATTCGCG	273	272	318

[a] Calculated using Eq. (9).
[b] Calculated using Eq. (5).
[c] Calculated using Eq. (8).
ΔH_{vH} in kJ (mol cooperative unit)$^{-1}$.

It is of interest to note that a van't Hoff transition enthalpy also can be calculated from the calorimetric data. In a manner paralleling the analysis of a differentiated optical melting curve, the shape of a calorimetric heat capacity curve can be analyzed using Eq. (8) to yield a van't Hoff transition enthalpy that can be designated as ΔH_{vH}^{cal}. Table 4 lists ΔH_{vH}^{cal} values obtained in this manner for some self-complementary deoxyoligomers and compares them with the corresponding optically derived ΔH_{vH} values. The significant observation is that ΔH_{vH}^{cal} does not always agree with the corresponding optical ΔH_{vH} derived from Eqs. (5) or (9). The largest differences are observed for the longer oligomers where we generally find that ΔH_{vH}^{cal} is larger in magnitude than ΔH_{vH} derived from the optical data. Since the longer oligomers generally undergo multistate transitions ($\Delta H_{vH} < \Delta H_{cal}$), this result suggests that the intermediate states manifest nonequivalent thermal and optical weighing factors.

Nature of the Transition. The methods described above for extracting thermodynamic data from *optical* measurements only can be applied rigorously to transitions that proceed in a two-state (all-or-none) manner. For transitions in which intermediate states are significantly populated, the optical melting curves will be broadened. According to Eqs. (5) and (6), this broadening will lead to a reduced ΔH_{vH} relative to the true calorimetric value. By contrast, as noted earlier, the calorimetrically determined transition enthalpy is derived directly from the area under rather than the shape of the heat capacity curve. Thus, ΔH_{cal} is independent of the nature of the transition as one would expect for a state function.

Comparison of the model-dependent ΔH_{vH} and the model-independent ΔH_{cal} allows one to conclude if the transition proceeds in an all-or-none fashion. As noted earlier, this behavior provides the basis for testing the applicability of the two-state model to a given transition. If $\Delta H_{vH} < \Delta H_{cal}$, then the transition involves a significant population of intermediate states. However, if $\Delta H_{vH} = \Delta H_{cal}$, then the transition proceeds in a two-state manner so that optical techniques can be used to obtain meaningful thermodynamic data.

Table 5. Size of the cooperative unit

Oligomer sequence	$\Delta H_{vH}^{con.\ dep.}/\Delta H_{cal}$
ATGCAT	1.17
GCGCGC	0.99
CGTACG	0.97
GGAATTCC	0.99
GCGAATTCGC	0.84
TATGCGCGCATA	0.77
CGCGAATCGCG	0.64
CGTAATTCGCG	0.74
rA_7U_7	0.76

A quantitative comparison of the van't Hoff and calorimetric transition enthalpies provides further insight into the nature of the transition. Specifically, the ratio $\Delta H_{vH}/\Delta H_{cal}$ provides a measure of the fraction of the molecule that melts cooperatively. This ability to define the size of the cooperative unit represents an important and unique advantage of the calorimetric measurement. Table 5 lists the $\Delta H_{vH}/\Delta H_{cal}$ ratio for a number of oligomers that we have studied. The significant qualitative observation is that the longer helices ($n=8$, 10, 12) melt less cooperatively than the shorter ones ($n=6$).

Selected Compilations of Some Recently Published Thermodynamic Data. As noted earlier, most of the pioneering spectroscopic and calorimetric studies on the thermodynamics of oligonucleotide transitions have been compiled and discussed in previously published review chapters and books [1–4]. Consequently, in the sections that follow I will restrict my attention to several recent investigations that add new information to these previous compilations.

The Effect of Dangling Ends on RNA Stability. The Turner group at the University of Rochester has systematically investigated the role of dangling ends on the stability of RNA duplex structures. Their approach involved spectroscopic studies on a series of pentanucleotides which associate to form duplexes containing common GGCC or CCGG tetrameric cores. As shown in Scheme 1, such pentameric sequences are partially self-complementary and in solution form duplex structures with dangling ends, X.

$2(G-G-C-C-X)$ ⇌
$$G-G-C-C-X$$
$$X-C-C-G-G$$

$2(C-C-G-G-X)$ ⇌
$$C-C-G-G-X$$
$$X-G-G-C-C$$ **Scheme 1.**

The three questions of interest are: (1) Do the dangling ends stabilize, destabilize, or have no effect on the tetrameric duplex core? (2) If an effect exists, what is its thermodynamic origin (enthalpy versus entropy)? (3) Does the influence of the dangling end(s) depend on the nature of the extra, unbase-paired nucleotide? The answers to these questions require thermodynamic data on each duplex structure both in the presence and absence of the dangling ends.

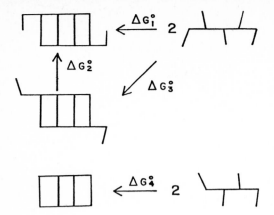

Fig. 9. A general representation of duplex formation from self-complementary pentameric sequences with dangling ends (ΔG_1^0, ΔG_2^0, ΔG_3^0) and self-complementary tetrameric sequences without dangling ends (ΔG_4^0). As described in the text, these data can be used to calculate the thermodynamic contribution of dangling ends to duplex stability

To obtain the relevant thermodynamic data, the Turner group set up the thermodynamic triangle shown in Fig. 9 [95]. Clearly, given values for ΔG_1^0, and ΔG_3^0 one can calculate a value for ΔG_2^0, the Gibbs energy contribution associated exclusively with the stacking of the two terminal, dangling bases. This indirect approach is required since one cannot experimentally measure ΔG_2^0, the quantity of interest. The Gibbs energy of duplex formation for the pentamer, ΔG_1^0, was determined indirectly using the spectroscopic methods described above. ΔG_3^0, which could not be measured directly, was assigned a value equal to ΔG_4^0, the Gibbs energy of duplex formation for the corresponding tetrameric core (GGCC or CCGG with or without a terminal phosphate). The quantity of interest ΔG_2^0 then was calculated from the difference between ΔG_3^0 and ΔG_1^0. To obtain the relevant ΔG^0 values, the required enthalpy and entropy data were derived from concentra-

Table 6. Thermodynamics of helix formation in 1 M NaCl, 0.01 M sodium cacodylate, 0.001 M EDTA, pH 7. [95]

Oligomer sequence	$-\Delta H^0$ (kJ mol)$^{-1}$	$-\Delta S^0$ J(mol·K)$^{-1}$	T_m [a] (°C)
GGCC	151	412	35.0
pGGCC	165	452	38.8
GGCCp	163	455	33.5
GGCCC	181	485	48.1
GGCCCp	174	467	47.5
GGCCUp	202	545	51.5
GGCCGp	219	585	57.5
GGCCAp	215	574	58.0
CCGG	144	403	27.4
pCCGG	152	421	32.9
CCGGCp	169	472	34.4
CCGGUp	176	490	38.0
CCGGGp	202	555	46.5
CCGGAp	201	556	44.7

[a] Calculated for 1×10^{-4} M.

Table 7. Excess stabilization added by 3′
dangling ends in 1 M NaCl. [95]

3′ terminal base	$-\Delta G_2^0\,(37\,^\circ C)$ [a] $(kJ\,mol^{-1})$ core helix:	
	CCGG	GGCCp
Cp	3.3	7.5
Up	5.0	10.9
Gp	10.5	15.5
Ap	9.6	15.5

[a] ΔG_2^0 is the difference between ΔG_1^0 for the
parameter and $\Delta G_4^0(=\Delta G_3^0)$ for the tetramer
core.

tion-dependent studies of the melting temperature using Eq. (9). The results obtained for each tetramer and pentamer sequence are summarized in Tables 6 and 7. The thermodynamic data for the stacking of a *single* dangling end actually is $\frac{1}{2}\Delta G_2^0$ since each pentamer duplex contains two dangling ends.

Inspection of the data in Table 7 reveals several significant features.

1. All the ΔG_2^0 data are negative. Thus 3′ dangling ends stabilize the duplex structure.

2. The favorable Gibbs energy of a 3′ dangling end stack is associated with a favorable enthalpy and an unfavorable entropy change. (Most, if not all, of the observed entropy change can be correlated with configurational entropy effects.) These data suggest that classical hydrophobic bonding does not make a major contribution to the stacking energy of 3′ dangling ends.

3. The order of stability for a given dangling end is $C < U < G, A$.

The Tinoco group of the University of California at Berkeley also have investigated the effects of dangling ends on duplex stability [96]. This group synthesized and spectroscopically investigated the melting of a series of DNA and RNA duplexes containing the sequences $rCA_mG + rCU_nG$, m, n = 5–7, and $dCA_mG + dCT_nG$, m, n = 5,6.

As shown in Scheme 2, these sequences in principle can form fully bonded, perfect duplexes, or imperfect, "perturbed" duplexes containing either bulge loops or dangling ends.

In practice, it was shown that for these sequences the dangling end structures are favored. The relevant thermodynamic data on the melting of these structures were obtained by analyzing the temperature- and concentration-dependent optical data using Eqs. (5) and (9). Tables 8 and 9 summarize the resulting thermodynamic data. The significant points to note are:

1. For the unperturbed fully bonded structures, the DNA duplex is more stable than the corresponding RNA duplex. This extra stabilization results from a more favorable enthalpy for the DNA duplex.

2. For the perturbed structures, a dangling end is more stable than a bulge loop.

3. For the RNA structures, an extra A or U destabilizes the double helix by 5.0 or 4.2 kJ mol^{-1}, respectively. For the DNA structures, an extra T destabilizes

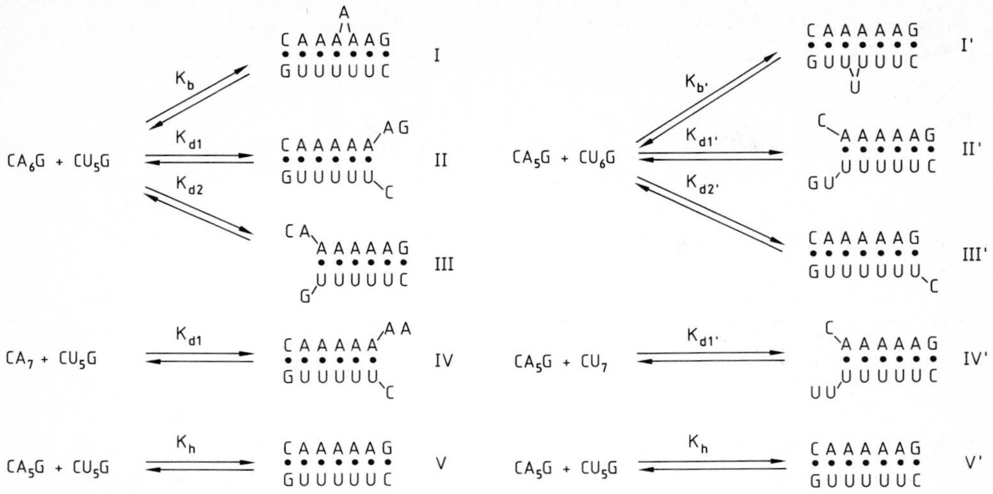

```
                       A
                       /\
                  C A A A A A G          I              C A A A A A G        I'
                  • • • • • •                           • • • • • •
                  G U U U U U C                         G U U U U U C
        Kb                                      Kb'           U
      ⟋                                       ⟋            U
   Kd1            C A A A A A  AG                  Kd1'         C
CA₆G + CU₅G  ⇌    • • • • • •        II       CA₅G + CU₆G  ⇌      ⟍A A A A A G    II'
   Kd2            G U U U U U                      Kd2'         • • • • • •
      ⟍                    ⟍C                                 G U ⟋U U U U U C
                  C A
                    ⟍A A A A A G   III                       C A A A A A G     III'
                    • • • • • •                              • • • • • •
                    ⟋U U U U U C                             G U U U U U U
                   G                                                    ⟍C

        Kd1            C A A A A A  AA                Kd1'           C
CA₇ + CU₅G  ⇌    • • • • • •        IV       CA₅G + CU₇  ⇌      ⟍A A A A A G    IV'
                  G U U U U U                                   • • • • • •
                          ⟍C                                  U U ⟋U U U U U C
                                                            U U

        Kh             C A A A A A G           Kh              C A A A A A G    V'
CA₅G + CU₅G  ⇌    • • • • • • •          V   CA₅G + CU₅G  ⇌     • • • • • • •
                  G U U U U U C                              G U U U U U C
```

Scheme 2.

Table 8. Thermodynamic parameters for $rCA_nG + rCU_nG$, 1.0 M NaCl. [96]

Oligomer	ΔH^0 (kJ mol^{-1})	ΔS^0 (J K^{-1} mol^{-1})	ΔG^0 (kJ mol^{-1})	T_m (200 μM, °C)
$dCA_5G + dCT_5G$	−205	−607	−25.5	28.9
$rCA_5G + rCU_5G$	−172	−502	−22.2	23.9
$rCA_6G + rCU_6G$	−209	−619	−25.9	29.2
$rCA_7G + rCU_7G$	−264	−782	−30.5	33.8

Table 9. Effect of mismatched bases on double-strand stability, 0.2 M NaCl. [96]

Oligomer	ΔH^0 (kJ mol^{-1})	ΔG^0 (kJ mol^{-1})	T_m (200 μM, °C)
$rCA_5G + rCU_5G$	−180	−27.2	19.2
$rCA_6G + rCU_5G$	−163	−22.2	10.8
$rCA_7 + rCU_5G$	−146	−20.5	7.7
$rCA_5G + rCU_6G$	−138	−23.0	12.9
$rCA_5G + rCU_7$	−155	−20.5	8.1
$dCA_5G + dCT_5G$	−209	−33.9	26.5
$dCA_5G + dCT_6G$	−188	−26.4	17.6

the duplex by about 7.5 kJ mol^{-1}. Thus, relative to the fully bonded duplexes, the perturbed structures are more destabilized for DNA than for RNA. More data however are required before such a trend can be considered general.

The results of the Turner and Tinoco labs described above add important new information to our understanding of the role of nonbonded bases on duplex stability.

Thermodynamics of DNA Oligomers. Most investigations of helix-to-coil transitions in oligonucleotides have used optical techniques and have focussed on RNA oligomers. We have begun a program in which calorimetric techniques are being used to systematically investigate helix-to-coil transitions in DNA oligomers. For each system studied, optical techniques also have been employed in an effort to critically evaluate which spectrophotometric method provides the best thermodynamic data. To date, our results indicate that evaluation of the slope at the T_m of a melting curve (using sloping baselines) and determination of the concentration dependence of the melting temperature [Eq. (9)] both represent good methods for determining thermodynamic data *on transitions that proceed in a two-state manner*. However, for multi-state transitions, calorimetric studies are required to obtain the relevant thermodynamic data.

As noted earlier, one of the most interesting and potentially powerful applications of the thermodynamic data lies in the ability to predict the structure and stability of DNA and RNA molecules (or segments) based purely on the primary sequence. Tinoco, Crothers, Uhlenbeck, and their collaborators have demonstrated the feasibility and utility of this approach for RNA molecules. We have begun a program in which calorimetric techniques are being used to obtain the corresponding thermodynamic data for DNA molecules.

We have used differential scanning calorimetry to determine directly the transition enthalpies accompanying the duplex-to-single strand transition of four synthetic DNA polymers containing different nearest-neighbor sequences. The calorimetric data allowed us to define the four average base stacking enthalpies listed in Table 10. If transition enthalpies mainly result from such pair-wise, nearest-neighbor base-stacking interactions, then these data can be used to predict transition enthalpies for new DNA duplex structures.

Using the data in Table 10, we have calculated transition enthalpies for a number of DNA duplexes by assuming that the whole is equal to the sum of the adjacent, pairwise parts. This calculation is illustrated in Fig. 10 for the hexamer d(CGTACG). The resulting predicted transition enthalpies (ΔH_{pred}) for some of the oligomers studied are listed in Table 11. Differential scanning calorimetry was used to obtain directly the transition enthalpies. These calorimetric values (ΔH_{cal}) are listed in Table 11 alongside the corresponding predicted values.

The following points should be noted:

1. In almost all cases, the predicted transition enthalpies are larger than the calorimetrically determined values. This result may be due to "end effects" in the

Table 10. Average nearest-neighbor DNA base stacking enthalpies

Compound	Interaction	ΔH (kJ/stack)[a]
poly d(AT)·poly d(AT)	AT/TA; TA/AT	−30.5
poly dA·poly dT	AA/TT	−36.8
poly d(AC)·poly d(TG)	AC/TG; CA/GT	−24.3
poly d(GC)·poly d(GC)	GC/CG; CG/GC	−53.6

[a] In all cases, the transitions occurred with negligible heat capacity changes. Thus, despite differences in the melting temperatures, a direct comparison of the enthalpy data is valid.

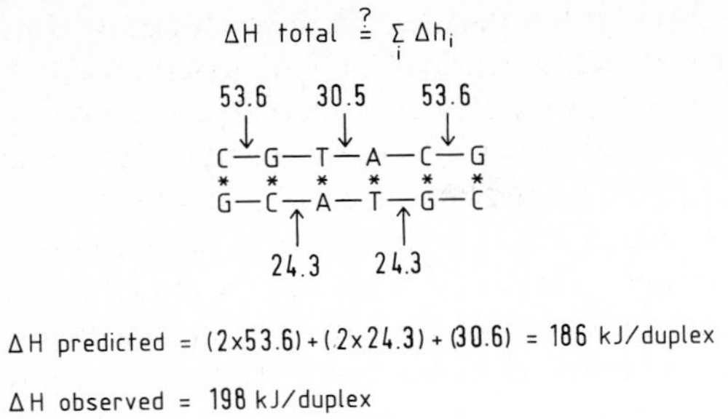

$$\Delta H \text{ total} \overset{?}{=} \sum_i \Delta h_i$$

ΔH predicted $= (2 \times 53.6) + (2 \times 24.3) + (30.6) = 186$ kJ/duplex

ΔH observed $= 198$ kJ/duplex

Fig. 10. An illustration of how a total transition enthalpy for an oligomer can be calculated as the sum of the pair-wise, nearest-neighbor base-stacking interactions

Table 11. Comparison of predicted and calorimetric transition enthalpies

Oligomer	ΔH^0 pred. (kJ/duplex)	ΔH^0 obs. (kJ/duplex)
ATGCAT	163	140
GCGCGC	268	249
CGCGCG	268	230
CGTACG	186	198
GGAATTCC	260	244
GCAAGCTTGC	332	255
GCGAATTCGC	367	357
ATATATATAT	275	161
TATGCGCGCATA	439	312
CGCGAATTCGCG	474	565

Table 12. Comparison of oligomer and polymer base stacking enthalpies

System	ΔH (kJ stack^{-1})
poly d(GC)	-53.6
d(GC)$_3$	-49.8
poly d(AT)	-29.7
d(AT)$_5$	-18.0

oligomers that are not manifest in the basis set of enthalpy data (Table 10) which was obtained from polymer systems where "end effects" are negligible. Support for this explanation comes from a comparison between the oligomer and corresponding polymer data listed in Table 12. For both (GC)$_3$ and (AT)$_5$, the reduced base stacking enthalpy of the oligomer compared with the corresponding polymer is consistent with "end effects"; in particular, for "AT" base pairs.

2. Substantial disagreement between the predicted and observed transition enthalpy occurs for those oligomers with "AT"-rich termini. This result may reflect "end effects" which are most pronounced for terminal "AT" base pairs.

3. For the self-complementary sequence GCAAGCTTGC, the two termini of the duplex are closed by "GC" base pairs. Thus, large "end effects" are not expected. Nevertheless, we find the predicted transition enthalpy to be considerably higher than the calorimetrically determined value. This observation suggests the equilibrium presence of competing "low-enthalpy" structures (e.g., hairpins) in addition to the duplex. In fact, for this sequence we have observed an equilibrium between duplex and hairpin structures which provides a reasonable explanation for the observed difference. This result suggests that such discrepancies between predicted and calorimetrically determined "duplex" transition enthalpies may provide a useful diagnostic indicator for the presence of multiple structural forms.

Comparison with RNA Base Stacking. Borer et al. [97] spectroscopically studied order-disorder transitions in a series of oligomeric RNA duplexes. From the temperature-dependence of the optical data, they calculated nearest-neighbor stacking enthalpies. To determine if any fundamental differences exist between DNA and RNA base-stacking energetics, we have compared our DNA enthalpy data with the corresponding RNA data of Borer et al. The relevant information is tabulated in Table 13.

The significant observation is the extraordinary similarity between corresponding DNA and RNA stacking enthalpies. All the interactions that can be compared agree within 5–10%, which is within the error limits of the data. This agreement suggests that the energetics of base stacking are not influenced significantly by whether the sugar is deoxyribo- or ribo-, but rather depend mainly on the identity of the bases participating in the stacking interaction. Consequently, stability differences between DNA and RNA duplexes containing identical sequences result primarily from either non-nearest-neighbor differential enthalpy effects or differential entropy effects.

Table 13. DNA vs. RNA base stacking enthalpies

Stacking interaction		$-$(Stacking enthalpies), kJ stack^{-1}	
DNA	RNA	DNA	RNA
AT/TA; TA/AT	AU/UA; UA/AU	30.5	27.2
AA/TT	AA/UU	36.8	34.3
AC/TG; TG/AC	AC/UG; UG/AC	24.3	24.7
GC/CG; CG/GC	GC/CG; CG/GC	53.6	57.7

[a] [97] Derived using a van't Hoff analysis of temperature-dependent optical data on RNA oligomers.

5 Concluding Remarks

Inspection of the stacking enthalpies listed in Table 10 clearly reveals the relatively small data base used for the predictive calculations just described. With the exception of AA/TT, we are employing *average* enthalpy values for 5' to 3' and 3' to 5'-directional isomers. Such a treatment assumes, for example, that the stacks AT/TA and TA/AT are energetically equivalent. In addition, for stacks between GC and AT base pairs, we are not energetically differentiating between isomers in which both purines are on the same or opposite strands. Thus we are assuming stacking isomers such as AG/TC and AC/TG to be energetically equivalent. These assumptions are not made out of a conviction that such energetic equivalencies are necessarily correct, but rather are born of necessity due to the current lack of available oligomeric and polymeric model systems that would allow us to calorimetrically differentiate between these stacking isomers.

As we investigate additional oligomer systems, we will expand our thermodynamic data base so as to more precisely define the effects of sequence on structural preferences and base-stacking thermodynamics. In this connection, recent studies on DNA oligomers have revealed that under appropriate salt conditions and oligomer concentrations, certain sequences form hairpin structures that are more stable than their duplex counterparts [98–100]. Thus, in polymer regions containing appropriate sequences, hairpin formation not only is reasonable, but can correspond to a stable secondary structure form. Clearly, more sequence-dependent thermodynamic data on the stability and conformational flexibility of all secondary structural features (e.g., internal loops, bulges, etc.) is required to assess the potential role of these structures in biological events such as frameshift mutagenesis, drug binding, and protein-nucleic acid interactions. Much more work needs to be done in this area, but the potential rewards are great.

Since the writing of the manuscript for this chapter, we have characterized the helix-to-coil transitions for a large number of additional polymeric and oligomeric DNA duplexes. These new data now have allowed us to resolve all ten nearest-neighbor interactions in DNA without the need for the assumed energy equivalences used above [see Breslauer and Marky (1980) Proc Natl Acad Sci USA 83].

Acknowledgments. The author wishes to acknowledge his research associates, in particular, Dr. Luis A. Marky, whose expert experimental work produced much of the DNA oligomer data reported here. This work was supported in part by grants from the National Institutes of Health (GM23509), the Charles and Johanna Busch Biomedical Research Fund, and the Research Corporation.

References

1. Cantor CR, Schimmel PR (1980) Biophysical chemistry, part I. The conformation of biological macromolecules, chapter 6. Freeman, San Francisco, CA
2. Bloomfield VA, Crothers DM, Tinoco I (1974) Physical chemistry of nucleic acids, chapter 6. Harper and Row, New York
3. Ts'o PO (1974) Basic principles in nucleic acid chemistry, vol II. Academic Press, New York
4. Hinz H (1974) In: Jones MN (ed) Biochemical thermodynamics. Elsevier Scientific, The Netherlands, pp 116–168

 5. Wells RD, Ohtzuka E, Khorana HG (1963) J Mol Biol 14:221
 6. Riley M, Maling B, Chamberlin MJ (1966) J Mol Biol 20:359
 7. Wells RD, Larson JE, Grant RC, Shortle BE, Cantor CR (1970) J Mol Biol 54:465
 8. Felsenfeld G, Hirschman SZ (1965) J Mol Biol 5:407
 9. Rawitscher MA, Ross PD, Sturtevant JM (1963) J Am Chem Soc 85:1915
10. Ross PD, Scruggs RL (1965) Biopolymers 3:491
11. Bunville LG, Geiduschek EP, Rawitscher MA, Sturtevant JM (1965) Biopolymers 3:213
12. Neumann E, Ackermann T (1967) J Phys Chem 71:2377
13. Krakauer H, Sturtevant JM (1968) Biopolymers 6:491
14. Scheffler I, Sturtevant JM (1969) J Mol Biol 42:577
15. Privalov PL, Ptitsyn OB, Birshtein TM (1969) Biopolymers 8:559
16. Klump H, Ackermann T, Neumann E (1969) Biopolymers 7:423
17. Scruggs RL, Ross PD (1970) J Mol Biol 47:29
18. Leng M, Felsenfeld G (1966) J Mol Biol 15:455
19. Stevens CL, Felsenfeld G (1964) Biopolymers 2:293
20. Brahms J, Michelson AM, Van Holde KE (1966) J Mol Biol 15:467
21. Fresco JR (1963) In: Informational macromolecules. Academic Press, New York, p 121
22. Scheffler IE, Elson E, Baldwin RL (1968) J Mol Biol 36:291
23. Scheffler IE, Elson E, Baldwin RL (1970) J Mol Biol 48:145
24. Elson E, Scheffler IE, Baldwin RL (1970) J Mol Biol 54:401
25. Tinoco I Jr, Uhlenbeck O, Levine MD (1971) Nature 230:362
26. Gralla J, Crothers DM (1973) J Mol Biol 73:497
27. Martin FH, Uhlenbeck OC, Doty P (1971) J Mol Biol 57:201
28. Uhlenbeck OC, Martin FH, Doty P (1971) J Mol Biol 57:217
29. Pörschke D, Eigen M (1971) J Mol Biol 62:361
30. Uhlenbeck OC, Borer PN, Dengler B, Tinoco I Jr (1973) J Mol Biol 73:483
31. Huang WM, Ts'o POP (1966) J Mol Biol 16:523
32. Tinoco I, Borer PN, Dengler B, Levine MD, Uhlenbeck OC, Crothers DM, Gralla J (1973)
    Nature New Biol 246:40
33. Davis RC, Tinoco I Jr (1968) Biopolymers 6:223
34. Suurkuusk J, Alvarez J, Freire E, Biltonen R (1977) Biopolymers 16:2641
35. Breslauer KJ, Charko G, Hrabar D, Oken C (1978) Biophys Chem 8:393
36. Breslauer KJ, Sturtevant JM (1977) Biophys Chem 7:205
37. Breslauer KJ, Sturtevant JM, Tinoco I Jr (1975) J Mol Biol 99:549
38. Albergo D, Marky LA, Breslauer KJ, Turner DH (1981) Biochemistry 20:1409
39. Freier SM, Hill KO, Dewey TG, Marky LA, Breslauer KJ, Turner DH (1981) Biochemistry
    20:1419
40. Patel DJ, Kozlowski SA, Marky LA, Broka C, Rice J, Itakura K, Breslauer KJ (1982) Bio-
    chemistry 21:428
41. Patel DJ, Kozlowski SA, Marky LA, Rice JA, Broka C, Itakura K, Breslauer KJ (1982)
    Biochemistry 21:437
42. Patel DJ, Kozlowski SA, Marky LA, Rice JA, Broka C, Itakura K, Breslauer KJ (1982)
    Biochemistry 21:445
43. Patel DJ, Kozlowski SA, Marky LA, Rice JA, Broka C, Itakura K, Breslauer KJ (1982)
    Biochemistry 21:451
44. Marky LA, Canuel L, Jones RA, Breslauer KJ (1981) Biophys Chem 13:141
45. Applequist J, Damle V (1966) J Am Chem Soc 88:3895
46. Applequist J, Damle V (1965) J Am Chem Soc 87:1450
47. Damle VN (1970) Biopolymers 9:353
48. Lyubchenko YL, Vologodskii AV, Frank-Kamenetskii MD (1978) Nature 271:28
49. Vologodskii AV, Frank-Kamenetskii MD (1978) Nucleic Acids Res 5:2547
50. Ueno S, Tachibana H, Husimi Y, Wada A (1978) J Biochem (Tokyo) 84:917
51. Vizard DL, White RA, Ansevin AT (1978) Nature 275:250
52. Tong BY, Battersby SJ (1978) Biopolymers 17:2933
53. Wada A, Ueno S, Tachibana H, Husimi Y (1979) J Biochem (Tokyo) 85:827
54. Tong BY, Battersby SJ (1979) Biopolymers 18:1917
55. Gabbarro-Appa J, Tougard P, Reiss C (1979) Nature 280:515

56. Azbel MY (1979) Proc Natl Acad Sci USA 76:101
57. Azbel MYA (1980) Biopolymers 19:61
58. Wada A, Yabuki S, Husimi Y (1980) CRC Crit Rev Biochem 9:87
59. Ornstein RL, Rein R, Breen DL, MacElroy RD (1978) Biopolymers 17:2341
60. Crescenzi V, Quadrifolglio F (1975) In: Rembaum A, Selegay E (eds) Polyelectrolytes and their applications. Reldel, Dardrecht-Holland
61. Gotah O, Tagashira Y (1981) Biopolymers 20:1033
62. Nelson JW, Martin FH, Tinoco I Jr (1981) Biopolymers 20:2509
63. Streisinger G, Okada Y, Emrich J, Newton J, Tsugita A, Terzagni E, Inouye M (1966) Cold Spring Harbor Symp Quant Biol 31:77
64. Huang C, Hearst JE (1980) Anal Biochem 103:127
65. Martin FH, Tinoco I Jr (1980) Nucleic Acids Res 8:2295
66. Grosschedl R, Hobom G (1979) Nature 277:621
67. Wang AHJ, Quigley GJ, Kolpak FJ, Crawford JL, van Boom JH, van der Marel G, Rich A (1979) Nature 282:680
68. Klug A, Jack A, Viswamitra MA, Kennard O, Shakked Z, Steitz TA (1979) J Mol Biol 131:669
69. Marky LA, Jones RA, Breslauer KJ (1982) Biophys J 37:3060
70. Nordheim A, Rich A, Wang JC (1982) Cell 31:309
71. Wells RD, Wartell RM (1974) In: Burton K (ed) Biochemistry of nucleic acids, vol 6. Butterworth Medical and Technical, London, pp 41–64
72. Arnott S, Chandrasekaran R, Selsing E (1975) In: Sundaralingam M, Rao ST (eds) Structure and conformation of nucleic acids and protein-nucleic acid interactions. University Park Press, Baltimore, pp 577–596
73. Wells RD, Blakesley RW, Hardies SC, Horn GT, Larson JE, Selsing E, Burd JF, Chan HW, Dodgson JB, Nes IF, Wartell RM (1977) CRC Crit Rev Biochem 4:305–340
74. Arnott S (1977) 1st Cleveland Symp Macromol. Elsevier, Amsterdam, pp 87–104
75. Leslie AGW, Arnott S, Chandrasekaran R, Ratliff RL (1980) J Mol Biol 143:49–72
76. Dodgson JB, Wells RD (1977) Biochemistry 16:2367–2374
77. Early TA, Kearns DR, Burd JF, Larson JE, Wells RD (1977) Biochemistry 16:541–551
78. Selsing E, Wells RD, Early TA, Kearns DR (1978) Nature 275:249–250
79. Kallenbach NR, Daniel Jr WE, Kaminker MA (1976) Biochemistry 15:1218–1224
80. Craig ME, Crothers DM, Doty PM (1971) J Mol Biol 62:383–401
81. Porschke D, Uhlenbeck OC, Martin FH (1973) Biopolymers 12:1313–1335
82. Gralla J, Crothers DM (1973) J Mol Biol 73:497–511
83. Cross AD, Crothers DM (1971) Biochemistry 10:4015
84. Cornelis AG, Haasnoot JHJ, den Hartog JF, de Rooij M, van Boom JH, Altona C (1979) Nature 281:235
85. Haasnoot CAG, den Hartog JHJ, de Rooij JF, van Boom JH, Altona C (1980) Nucleic Acids Res 8:169
86. Miller PS, Cheng DM, Dreon N, Jayaraman K, Kan LS, Leutzinger EE, Pulford SM, Ts'o POP (1980) Biochemistry 19:4688
87. Kössel H, Seliger H (1975) Fortschr Chem Org Naturstoffe 32:297
88. Amarnath V, Broom AD (1977) Chem Rev 77:183
89. Ikehara M, Ohtsuka E, Markham AF (1979) Adv Carbohydr Chem Biochem 36:135
90. Ohtsuka E, Ikehara M, Söll D (1982) Nucleic Acid Res 10:6553
91. Gralla J, Crothers DM (1973) J Mol Biol 78:301
92. Breslauer KJ, Bina-Stein M (1977) Biophys Chem 7:211
93. Marky LA, Breslauer KJ, unpublished results
94. Freier SM, Albergo DD, Turner DH (1983) Biopolymers 22:1107
95. Freier SM, Petersheim M, Hickey DR, Turner DH (1983) J Biomol Struct Dynam 1:1229
96. Nelson JW, Martin FH, Tinoco I Jr (1981) Biopolymers 20:2509
97. Borer PN, Dengler B, Tinoco I Jr, Uhlenbeck OC (1974) J Mol Biol 86:843
98. Marky LA, Breslauer KJ (1983) Biopolymers 22:1247
99. Haasnoot CAG, de Bruin SH, Berendsen RG, Janssen HGJM, Binnendijk TJJ, Hilbers CW, van der Marel GA, van Boom JH (1983) J Biomol Struct Dynam 1:115
100. Patel DJ, Ikuta S, Kozlowski S, Itakura K (1983) Proc Natl Acad Sci USA 80:2184

Section VI   **Enzyme Catalyzed Processes**

Chapter 16 Thermodynamics of Enzymatic Reactions

M. V. Rekharsky, G. L. Galchenko, A. M. Egorov, and I. V. Berenzin

Symbols

ΔG_r:	Gibbs energy change of the reaction at pH $=7$ and 298.15 K
ΔH_r:	reaction enthalpy at pH $=7$ and 298.15 K
ΔS_r:	reaction entropy at pH $=7$ and 298.15 K
ΔG_i, ΔH_i, ΔS_i:	Thermodynamic parameters referring to reaction i

1 Introduction

The application of thermodynamics to biochemical systems has given rise to the new discipline of biochemical thermodynamics [1–4], an important part of which is the thermodynamics of enzymatic reactions. The interest in this area extends beyond biochemistry, since enzymes appear to be promising catalysts for use in industrial organic sythesis [5–7]. The accumulated data in the literature on enzymatic reactions (equilibrium, enthalpy, entropy, Gibbs energy, heat capacity) need to be systematized and scrutinized. Armstrong et al. [8–12] have made an important contribution towards this goal by classifying the various sources of information.

The objective of this review is to make a further contribution to this goal by using data obtained predominantly in the last decade to choose "best fit" values in order to calculate thermodynamic parameters for a number of important enzymatic reactions.

2 Principle for the Systematization of Thermodynamic Data for Enzymatic Reactions

According to the IUB classification, all known enzymatic reactions fall into six large groups [13, 14]. The enzymes catalyzing reactions in these groups are termed: (1) oxidoreductases; (2) transferases; (3) hydrolases; (4) lyases; (5) isomerases, and (6) ligases.

Reliable calculations of thermodynamic parameters for various enzymatic reactions (bearing in mind their great diversity) require a special system or a bank of thermodynamic parameters. In our opinion, this system should correspond as closely as possible to the IUB classification [13, 14]. The system should also be based on key thermodynamic values, as are recent handbooks and data banks for the thermodynamics of inorganic and simple organic compounds [16–18].

In our earlier works [15, 19] we have applied the above approach to calculate the thermodynamic parameters of the first class of enzymatic reactions namely the redox reactions (catalyzed by oxidoreductases).

2.1 Thermodynamics of Enzymatic Reactions Catalyzed by Oxidoreductases

As key thermodynamic values in this case, we chose the Gibbs energy and enthalpy of the nicotinamide adenine dinucleotide (NAD^+) redox system (Table 1).

With reliable values of ΔH_1 and ΔG_1, it is possible to calculate the thermodynamic parameters for various redox reactions, because NAD^+ is involved in a large number of reactions (being a cofactor of more than 250 enzymes). The results of such calculations are given in Table 2.

It is to be emphasized that it is not always possible to choose reliable values of thermodynamic parameters for calculation from the available literature data. Consider the following two cases. First, Table 1 shows that all the enthalpy values of the NAD^+ redox reaction determined calorimetrically in various laboratories coincide within experimental error. Hence, the enthalpy value for this reaction is now entirely reliable and can be recommended as a key value. However, the situation is different in the second case, the enthalpy of the NAD^+-dependent oxidation of lactate (Table 2, reaction 1). This value was determined in four different laboratories, with the results given in Table 3.

Table 1. Enthalpy data for the reduction of NAD^+

$$NAD^+ + H_{2(gas)} = NADH + H^+ \quad (pH\ 7;\ 298.15\ K) \tag{1}$$

Hydrogen is in the gas form; ΔH_r values refer to a hypothetical buffer having zero ionization enthalpy

No.	Reactions	Method	ΔH_r kJ mol^{-1}	Reference
1.	Propane-2-ol + NAD^+ = acetone + $NADH + H^+$	Determination of d logK/dT	−29.2	[20]
2.	Unpublished results by Scott and Sturtevant cited from [20])	Calorimetry	−27.4±2.5	[20]
3.	Oxidation-reduction NAD/NADH on electrode	Determination of d E/dT	−14.9	[21]
4.	$NADH + 0.5\,O_{2(aq.\,soln)} + 2H^+$ = $2NAD^+ + 2H_2O$	Calorimetry	−28.0±5.4	[22]
5.	$HCOO^- + NAD^+ + H_2O$ = $HCO_3^- + NADH + H^+$	Calorimetry	−27.9±1.1	[23–25]
6.	$NAD^+ + H_{2(aq.\,soln)} = NADH + H^+$	Calorimetry	−27.2±1.5	[23–25]

The thermodynamic parameters suggested as key values for reaction (1) are:

$\Delta H_1(298.15\ K;\ pH\ 7) = −27.6±1.3\ kJ\ mol^{-1}$ [23–25];
$\Delta G_1(298.15\ K;\ pH\ 7) = −17.8\ kJ\ mol^{-1}$ [20].

Table 2. Thermodynamic parameters of some NAD reactions with organic compounds of biochemical interest (pH 7; 298.15 K; the ionization enthalpy of the buffer is equal to zero)

). Reaction	ΔH_r kJ mol^{-1}	ΔG_r kJ mol^{-1}	ΔS_r J mol$^{-1} \cdot$ K^{-1}	Reference
Experimental data				
$^-O_2C-CHOH-CH_3+NAD^+$	$+55.8\pm1.1$	$+25.9$	$+100.1$	[41, 33, 35]
(lactate)				
$=\,^-O_2C-CO-CH_3+NADH+H^+$				
(pyruvate)				
$^-O_2C-CH_2-CHOH-CO_2^-+NAD^+$	$+89.5\pm0.8$	$+29.7$	$+200.8$	[34]
(malate)				
$=\,^-O_2C-CH_2-CO-CO_2^-+NADH+H^+$				
(oxaloacetate)				
$^-O_2C-CH_2-CH_2-CO_2^-+NAD^+$	$+106.7\pm5.4$	$+73.6$	$+110.9$	[22]
(succinate)				
$=\,^-O_2C-CH=CH-CO_2^-+NADH+H^+$				
(fumarate)				
$^-O_2C-CH_2-CH_2-CHNH_3^+-CO_2^-$	$+60.2\pm2.1$	$+32.2$	$+94.1$	[36]
(glutamic acid)				
$+NAD^++H_2O$				
$=\,^-O_2C-CH_2-CH_2-CO-CO_2^-$				
(ketoglutarate)				
$+NADH+H^++NH_4^+$				
$HCOO^-+NAD^++H_2O$	-6.9 ± 1.0	-14.6	$+26.0$	[23–25]
(formate)				
$=HCO_3^-+NADH+H^+$				
$CH_3-CH_2OH+NAD^+$	$+50.6\pm1.7^a$	$+23.8$	$+90.0$	
(ethanol)				
$=CH_3-CHO+NADH+H^+$				
(acetaldehyde)				
Calculated data (using [16, 17, 37])				
$CH_3OH+2NAD^++H_2O$	$+50.8$	-15.1	$+221.0$	
(methanol)				
$=HCOO^-+NADH+3H^+$				
(formate)				
$CH_3CHOH-CH_3+NAD^+$	$+81.4$	$+6.7$	$+250.5$	
(propane-2-ol)				
$=CH_3-CO-CH_3+NADH+H^+$				
(acetone)				
$CH_3-CHO+NAD^++H_2O$	-19.4	-50.2	$+103.3$	
(acetaldehyde)				
$=CH_3-CO_2^-+NADH+2H^+$				
(acetate)				
$^-O_2C-CH=CH-CO_2^-+NAD^++2H_2O$	$+33.3$	-0.8	$+114.4$	
(fumarate)				
$=\,^-O_2C-CO-CH_3+HCO_3^-+NADH+H^+$				
(pyruvate)				
$^-O_2C-CO-CH_3+NAD^++2H_2O$	-37.9	-67.4	$+98.9$	
(pyruvate)				
$=\,^-O_2C-CH_3+HCO_3^-+NADH+2H^+$				
(acetate)				
$^-O_2C-CHNH_3^+-CH_3+NAD^++H_2O$	$+84.3$	$+38.5$	$+153.6$	
(alanine)				
$=\,^-O_2C-CO-CH_3+NADH+NH_4^++H^+$				

[a] Unpublished data of J. Jordan and W. Brattlie, Pennsylvania State University (cited from [19]).

Table 3. The results of the enthalpy value determinations for the NAD-dependent reaction of lactate oxidation

lactate^{1-} + NAD$^+$ = pyruvate^{1-} + NADH + H$^+$

ΔH_r (298.15 K) kJ mol^{-1}	Method	Reference
+61.9±1.3	Direct enthalpimetry [47]; calorimeter with an adiabatic sheath	[33]
+55.8±1.1	LKB-10700 system; batch microcalorimeter	[41]
+44.4	Calorimeter in the form of a Dewar vessel with isothermic shield	[42]
+49.0	Calorimetry (no experimental details)	Personal communication; cited from [43]

The ΔH-value suggested as ΔH_r (298.15 K; pH 7) = +55.8±1.1 kJ mol^{-1} [41].

The enthalpy values (Table 3) of the reaction determined by the different authors do not coincide (note that reference [43] does not contain experimental details). We used the enthalpy value of +55.8±1.1 kJ mol^{-1} [41] as a reliable value in our calculations because the instrument (LKB-10700; batch microcalorimeter) was calibrated and the enthalpy was calculated from concentrations of both NAD$^+$ and pyruvate, unlike reference [42].

In addition to NAD$^+$, other coenzymes participate in enzymatic redox reactions, these include nicotinamide adenine dinucleotide phosphate (NADP), flavin mononucleotide (FMN), flavin adenine dinucleotide (FAD), tetrahydrofolate (THF) and dihydrofolate (DHF). It should be noted that in some cases the natural cofactors can be replaced by synthetic redox mediators, e.g., methylviologen (MV), benzylviologen (BV), sodium dithionite, etc. The rate of some enzymatic reactions increases on substitution of a synthetic for the natural cofactor; this substitution could be promising for large-scale industrial enzymatic processes [26, 27]. We should also mention that some synthetic mediators react with natural cofactors in the absence of enzymes; this could be used in practice for the regeneration of cofactors [26, 27]. These considerations account for the interest in the thermodynamics of redox reactions of natural cofactors and synthetic mediators.

The thermodynamic parameters of the reactions of NAD$^+$ with natural cofactors and synthetic mediators are given in Table 4. The values of ΔH_r and ΔG_r in the table should be considered as secondary key thermodynamic values for the class of enzymatic reactions catalyzed.

The data in Tables 2 and 4 can be used to calculate thermodynamic parameters for the redox reactions of the organic compounds listed in Table 2 with the natural cofactors and synthetic mediators given in Table 4. Calculations can be performed for over 100 biologically important redox reactions [19].

With the aid of a relatively limited number of reliable key thermodynamic values, such as ΔH_r and ΔG_r, of reactions involving cofactors (Tables 1 and 4), it is possible to make thermodynamic calculations for a large number of en-

Table 4. Thermodynamic parameters of redox reactions of NAD with natural cofactors and synthetic mediators (pH 7; 298.15 K; the ionization enthalpy of the buffer is equal to zero)

No.	Reaction	ΔH_r kJ mol^{-1}	ΔG_r kJ mol^{-1}	ΔS_r J mol$^{-1}\cdot$K^{-1}	Reference
	$NAD^+ + H_{2(gas)} = NADH + H^+$	$-\ 27.6\pm1.3$	$-\ 17.8$	$-\ 32.9$	
1.	$FMNH_2 + NAD^+$ $= FMN + NADH + H^+$	$+\ 29.5\pm2.9$	$+\ 19.2$	$+\ 34.5$	[28, 29, 30]
2.	$2FMNH_2 + NAD^+$ $= (FMNH)_2 + NADH + H^+$	$+\ 1.9\pm4.2$	$-\ 6.3$	$+\ 27.5$	[30]
3.	$NADPH + NAD^+$ $+ NADP^+ + NADH$	$-\ 4.4\pm1.7$	$-\ 2.9$	$-\ 5.0$	[31]
4.	$S_2O_4^{2-} + NAD^+ + 2H_2O$ $= 2SO_3^{2-} + NADH + 3H^+$	$+\ 1.5\pm2.1$	$-\ 36.4$	$+127.1$	[30]
5.	$2/Fe(CN)_6/^{4-} + NAD^+ + H^+$ $= 2/Fe(CN)_6/^{3-} + NADH$	$+194.4\pm4.2$	$+128.9$	$+219.7$	[30, 32]
6.	$2MV + NAD^+ + H^+$ $= 2MV^+ + NADH$	$-\ 62.1\pm2.1$	$-\ 23.0$	-131.1	[30]
7.	$2BV + NAD^+ + H^+$ $= 2BV^+ + NADH$	$-\ 73.4\pm2.1$	$-\ 7.5$	-221.0	[30]
8.	$FADH_2 + NAD^+$ $= FAD + NADH + H^+$	$+\ 29.1\pm2.9$	$+\ 19.2$	$+\ 33.2$	[30]
9.	$THF + NAD^+$ $= DHF + NADH + H^+$	$+\ 48.3\pm1.8$	$+\ 24.1$	81.3	[38]

zymatic redox reactions. It is desirable, therefore, to decide on thermodynamic parameters for reactions of other cofactors as key values for other classes of enzymatic reactions.

2.2 Thermodynamics of Enzymatic Reactions Catalyzed by Transferases

The second class of enzymatic reactions is catalysed by transferases, enzymes transferring a group (e.g., a methyl or glycosyl group) from one compound (the donor) to another compound (the acceptor). In the transfer of various groups a number of compounds serve as cofactors, e.g., ATP, CoA, and UDP participate in the transfer of phosphoryl, acetyl and glycosyl groups, respectively.

In the reactions catalyzed by transferases, we suggest that the thermodynamic parameters of the hydrolysis reactions of ATP, CoA, and other cofactors should be chosen as key values, for the following reasons. First, knowing ΔH_r and ΔG_r for hydrolysis, it is easy to calculate the thermodynamic parameters for reactions catalyzed by transferases. For instance, knowing ΔH_r and ΔG_r for hydrolysis of ATP and of glucose-6-phosphate calculations can be made on the reaction in which ATP serves as the donor of a phosphoryl group to glucose (reaction 1 in Table 5). Second, the hydrolysis reactions have large negative free energy values and thus proceed essentially ($>99\%$) to completion, facilitating accurate determination of the enthalpies of reaction.

The key thermodynamic values for the phosphoryl group transfer reactions (catalyzed by phosphotransferases EC 2.7.X.Y) are ΔH_r and ΔG_r for hydrolysis of ATP:

$$ATP^{4-}+H_2O=ADP^{3-}+HPO_4^{2-}+H^+ \qquad (2)$$

with $\Delta H_r = 20 \pm 2$ kJ mol^{-1} [40, 44] and $\Delta G_r = -35$ kJ mol^{-1} [39, 45].

The ΔH value is slightly smaller than that determined for hydrolysis of the β- and γ-bonds of GTP which was found to be -23 ± 2 kJ mol^{-1} [46]. As the secondary key thermodynamic values, it is advisable to choose the thermodynamic parameters for the reactions in which a phosphoryl group is transferred from ATP^{4-} to UDP^{3-}, GDP^{3-}, CDP^{3-} etc. to form ADP^{3-} and the corresponding triphosphate nucleotides. According to [46], the enthalpies and free energies of the above reactions are close to zero (0 ± 2–3 kJ mol^{-1}).

The thermodynamic parameters of some biologically important phosphoryl transfer reactions, calculated by use of the key thermodynamic values for ATP hydrolysis [Eq. (2)] are listed in Table 5.

It is to be noted that ATP, ADP, and other cofactors participate in enzymatic reactions in the form of complexes with Mg^{2+} and Ca^{2+}. The Gibbs energy changes involved in complex formation have been studied, and their values can be found in reviews [54, 55, 74]. The differences in ΔG and ΔH for complex formation between Mg^{2+} and ADP^{3-} or ATP^{4-} are $+6$ kJ mol^{-1} and -5 kJ mol^{-1}, respectively. Therefore the thermodynamic parameters for the following reaction

$$[ATP\text{-}Mg]^{2-}+H_2O=[ADP\text{-}Mg]^-+HPO_4^{2-}+H^+ \qquad (2\,a)$$

are $\Delta G_r = 29$ kJ mol^{-1} and $\Delta H_r = -25$ kJ mol^{-1}.

The transfer of methyl groups is catalyzed by methyl transferases (EC 2.1.1.Y). Sturtevant studied three reactions of this subclass [56–58].

$$X\text{-}CH_3+\text{L-homocysteine}=X+\text{L-methionine.} \qquad (3)$$

Table 5. Thermodynamics of some enzymatic phosphorylation reactions (pH 7; 298.15 K, buffer with ionization enthalpy equal to zero)

No. Reaction	ΔH_r kJ mol^{-1}	ΔG_r kJ mol^{-1}	ΔS_r J mol$^{-1}\cdot$K^{-1}	Reference
1. Glucose$+ATP^{4-}=$glucose-6-phosphate$+ADP^{3-}+H^+$	-23.8	-31.9	$+27.2$	[50, 51]
2. Glycerol$+ATP^{4-}=$glycerol-3-phosphate$^{2-}+ADP^{3-}+H^+$	-26.6	-41.9	$+51.3$	[52]
3. Creatin$^{1-}+ATP^{4-}=$creatin-phosphate$^{3-}+ADP^{3-}+H^+$	$+21.5$	$+4.1$	$+58.4$	[52]
4. Pyruvate$^{-1}+ATP^{4-}=$phosphoenol-pyruvate$^{3-}+ADP^{3-}+H^+$	$+11.4$	$+14.1$	-9.1	[52]
5. Fructose-6-phosphate$^{2-}+ATP^{4-}=$fructose-1,6-diphosphate$^{4-}+ADP^{3-}+H^+$	-6.2	-34.9	$+96.3$	[52]
6. n-Nitrophenol$+ATP^{4-}=$n-nitrophenolphosphate$^{2-}+ADP^{3-}+H^+$	-8.5	$-$	$-$	[53]

The enthalpy (kJ mol^{-1}) of the above reaction equals respectively -36.8 ± 2.1 if X-CH$_3$ is betaine, -47.3 if X-CH$_3$ is dimethylacetothetin, and -32.2 if X-CH$_3$ is S-methylmethionine sulfonium bromide (in two last cases H$^+$ also is a product of reaction). The absence of experimental data precludes the determination of numerical values of the key thermodynamic parameters, i.e., those for the reaction

$$\text{L-methionine} + H_2O = \text{L-homocysteine} + CH_3OH. \tag{4}$$

The transfer of the acetyl group is effected with the aid of the coenzyme CoA and catalyzed by acetyl transferases (EC 2.3.X.Y). Wadsö has studied calorimetrically the hydrolysis of the compound CH$_3$CONHCH$_2$CH$_2$S-COCH$_3$ which is analogous to acetyl CoA. He obtained $\Delta H_5 = -3.8 \pm 0.3$ kJ mol^{-1} [59–61].

$$CH_3CONHCH_2CH_2S\text{-}COCH_3 + H_2O = CH_3CONHCH_2CH_2SH$$
$$+ CH_3COOH. \tag{5}$$

In the absence of a direct determination, the value of ΔH_5 may be used for the hydrolysis of acetyl CoA. The transfer of the acetyl group from CoA to an alcohol is associated, according to [60], with the following thermodynamic changes: $\Delta H_r = -5.5$ kJ mol^{-1} and $\Delta G_r = -10$ kJ mol^{-1}. Another set of enthalpy determinations have been generalized by Wadsö:

$$R - CH_2XCOCH_3 + R - CH_2NHCOCH_3$$
$$= R - CH_2XH + R - CH_2 - N \underset{\diagdown COCH_3}{\overset{\diagup COCH_3}{}}. \tag{6}$$

The enthalpies for these reactions equal $+55$ kJ mol^{-1}, $+46$ kJ mol^{-1} and $+96$ kJ mol^{-1} for X=O, S and NH respectively [60].

The coenzyme UDP-glucose is involved in enzymatic glucosyl group transfer reactions catalyzed by the glucosyl transferases (EC 2.6.X.Y). The thermodynamic parameters of the UDP-glucose hydrolysis reaction

$$\text{UDP-glucose}^{2-} + H_2O = UDP^{3-} + \text{glucose} + H^+ \tag{7}$$

are to be considered as key values. Unfortunately, no direct thermochemical studies have so far been reported. The knowledge of the thermodynamic parameters of Eq. (7), combined with ΔH_r and ΔG_r for the hydrolysis of sucrose, starch, cellulose, and other polysaccharides, would permit the calculation of the thermodynamic parameters of glucose transfer from UDP to relevant compounds. Some subclasses of enzymatic reactions, e.g., carboxyl and carbamoyl transferases, transketolases etc. are small, with fewer than 15–20 members.

The determination of key thermodynamic values for these small subgroups is nevertheless important. For instance, in the case of the subclass of enzymatic reactions which involve the transfer of hydroxymethyl, formyl, and related groups, the key thermodynamic values are the ΔH_r and ΔG_r for hydrolysis of hydroxymethyl and formyl derivatives of hydrofolic acid.

2.3 Thermodynamics of Enzymatic Reactions Catalyzed by Hydrolases

Hydrolases catalyze the hydrolysis of various types of bonds (peptide, ester etc.) generally without the need for cofactors. In this case, therefore, the principle of key thermodynamic values which we applied to oxidoreductase and transferase-catalyzed reaction is not operative. The only compound common to all reactions is water.

In order to facilitate thermodynamic calculations on hydrolase-catalyzed reactions, it is advisable to use the mean ΔH_r and ΔG_r values for hydrolysis of structurally similar compounds, e.g., ΔH_r and ΔG_r for hydrolysis of peptide, ester, and other bonds. When the experimental data have been accumulated, it is possible to divide the large subclasses of hydrolytic reactions (e.g., hydrolysis of the ester bond) into smaller groups in which the mean values of ΔH_r and ΔG_r will differ less from one another than they do in the large subclasses. Thus, there is no need to determine experimentally the thermodynamic parameters of hydrolytic reactions in all cases; in any particular case the approximate values can be assessed by identifying the group to which the reaction belongs.

This can be demonstrated by a few examples. Sturtevant [65–69] measured the enthalpies of hydrolysis for a number of synthetic peptides at 298.15 K (see Table 6).

Table 6 shows that the enthalpies of peptide bond hydrolysis range from -5.2 kJ mol^{-1} to -10.7 kJ mol^{-1}; the mean value is -7.5 kJ mol^{-1}. Analogous determination of the mean enthalpy values for hydrolysis of some other bonds in aqueous solution: ~ -25 kJ mol^{-1} for amide bonds [65–68]; $\sim +2$ kJ mol^{-1} for ester bonds [59–61, 72] and ~ -4 kJ mol^{-1} for thioester bonds [59–61].

Consider in greater detail the two reactions in which the amide bonds of asparagine and glutamine are hydrolyzed [65–68, 71]:

$$\begin{array}{c}
\overset{\displaystyle NH_3^+}{\underset{\displaystyle (\textit{L}\text{-asparagine})}{|}} \\
-O_2C-CH-CH_2-C\overset{\nearrow NH_2}{\underset{\searrow O}{}}+H_2O
\end{array}$$

$$\begin{array}{c}
\overset{\displaystyle NH_3^+}{|} \\
=-O_2C-CH-CH_2-CO_2^-+NH_4^+ \\
(\text{L-aspartate})
\end{array} \qquad (8)$$

$$\Delta H_r = -24.3 \text{ kJ mol}^{-1}; \Delta G_r = -12.1 \text{ kJ mol}^{-1}$$

$$\begin{array}{c}
\overset{\displaystyle NH_3^+}{\underset{\displaystyle (\text{L-glutamine})}{|}} \\
-O_2C-CH-CH_2-CH_2-C\overset{\nearrow NH_2}{\underset{\searrow O}{}}+H_2O
\end{array}$$

$$\begin{array}{c}
\overset{\displaystyle NH_3^+}{|} \\
=-O_2C-CH-CH_2-CH_2-CO_2^-+NH_4^+ \\
(\text{L-glutamate})
\end{array} \qquad (9)$$

$$\Delta H_r = -21.8 \text{ kJ mol}^{-1}; \Delta G_r = -11.7 \text{ kJ mol}^{-1}.$$

Table 6. Enthalpy of hydrolysis for the synthetic peptides at 25 °C

No.	Compound	Hydrolysed bond	ΔH_r kJ mol^{-1}	Reference
1.	Poly-L-lysine	Lysine-lysine	-5.2	[66]
2.	L-tyrosylglycinamide	Tyrosine-glycine	-5.4 ± 0.6	[65]
3.	Benzoyl-L-tyrosylglycine	Tyrosine-glycine	-5.6 ± 0.4	[65]
4.	Benzoyl-L-tyrosylglycinamide	Tyrosine-glycine	-6.5 ± 0.6	[68]
5.	Benzoyl-L-tyrosine	Benzoic acid-tyrosine	-8.3 ± 0.4	[65]
6.	Carbobenzoxyglycyl-L-lysine	Glycine-lysine	-8.8 ± 0.3	[67]

As expected, the structurally similar compounds aspargine and glutamine have similar thermodynamic parameters for the hydrolysis reactions. This is another example of a recommendation to use mean values ΔH_r and ΔG_r for various subclasses of hydrolase-catalyzed reactions in thermodynamic calculations.

2.4 Thermodynamics of Enzymatic Reactions Catalyzed by Lyases

The lyase-catalyzed reactions make up the fourth class of enzymatic reactions. Lyases catalyse the cleavage of C-C, C-O, C-N and other bonds in reactions other than hydrolytic and redox reactions, the overwhelming majority of enzymatic reactions of this class do not require cofactors. We assume that the principle of thermodynamic calculations for lyase-catalyzed reactions should be analogous to that for ΔH_r for ΔH_r and ΔG_r of hydrolytic reactions. Exceptions occur in the cases of the reactions catalyzed by oxoacid lyases (EC 4.1.3.7) in which CoA participate as a cofactor. In these cases, as with acyltransferase reactions, it is reasonable to use ΔH_r and ΔG_r for acetyl CoA hydrolysis as key thermodynamic values.

Thermodynamic parameters from the literature [22, 70, 73] and values calculated by us for some lyase-catalyzed reactions not requiring cofactors are summarized in Table 7.

As indicated in Table 7, data on the thermodynamic parameters of lyase-catalyzed reactions are scarce, and generalization analogous to those suggested for the various subclasses of hydrolase-catalyzed reactions cannot be made at present.

2.5 Thermodynamics of Enzymatic Reactions Catalyzed by Isomerases

The fifth group of enzymatic reactions involve conversions occurring within one molecule and are catalyzed by isomerases. As in the cases of hydrolases and lyases, isomerases do not generally require cofactors. Racemization of amino acids and conversions of L-lactate to D-lactate, L-acetoin to D-acetoin, etc. have enthalpies equal to zero and the free energy depends only on the entropy (Rln2). To calculate thermodynamic parameters of other reactions, it is expedient to use the approach we have discussed above for hydrolase-catalyzed reactions. The en-

Table 7. Thermodynamics of lyase-catalyzed reactions (pH 7; 298.15 K; buffers with the ionization enthalpy equal to 0; CO_2 is in gaseous form under standard conditions)

No. Reaction	ΔH_r kJ mol^{-1}	ΔG_r kJ mol^{-1}	ΔS_r J mol$^{-1} \cdot$ K^{-1}
1. $2\ ^-O_2C-CHOH-CH_2OPO_3^{2-}+2H^+$ (3-phosphoglycerate) $= {}^{-2}O_3POCH_2-CHOH-CHOH-CO$ (ribulose 1,5 diphosphate) $-CH_2OPO_3^{2-}+CO_2+H_2O$	$+32.0\pm2.5$	–	–
2. $CH_3-CO-CO_2^-+H^+$ (pyruvate) $=CH_3-CHO+CO_2$ (acetaldehyde)	$-\ 9.6$	-21.1	$+\ 38.6$
3. $HOOC-CO_2^-=HCO_2^-+CO_2$ (oxalate) (formate)	-18.2	-58.8	$+136.2$
4. $^-O_2C-CH_2-CO-CO_2^-+H^+$ (oxalacetate) $=CH_3-CO-CO_2^-+CO_2$ (pyruvate)	$+\ 5.2^a$	-33.5	–
5. $HOOC-CH_2-CH_2-COOH$ (succinate) $=CH_3-CH_2-COOH+CO_2$ (propionic acid)	$+36.6^a$	–	–
6. $^-O_2C-CH_2-COH-CH_2-CO_2^-$ $\qquad\qquad\quad\ \|$ $\qquad\qquad\quad CO_2^-$ (citrate) $=\ ^-O_2C-CH=C-CH_2-CO_2^-+H_2O$ $\qquad\qquad\quad\ \|$ $\qquad\qquad\quad CO_2^-$ (cis-aconitate)	$+33.5^a$	$+\ 8.4$	–
7. $^-O_2C-CH_2-CHOH-CO_2^-$ (malate) $=\ ^-O_2C-CH=CH-CO_2^-+H_2O$ (fumarate)	$+\ 6.7^a$	$+\ 3.3$	–
8. $^-O_2C-CH-CH_2CO_2^-$ $\qquad\ \|$ $\qquad NH_2^+$ (aspartate) $=\ ^-O_2C-CH=CH-CO_2^-+NH_4^+$ (fumarate)	$+33.5$	$+18.0$	$+\ 52.0$

^{a} Enthalpies under the conditions when initial substances and products are in standard pure state (solid, liquid, or gaseous) at 298.15 K.

Table 8. Isomerization enthalpies of carbohydrates (aqueous solution, 298.15 K) [62–64, 37]

No	Reaction	ΔH_r kJ mol^{-1}
1.	α-D-Glucose = β-D-Glucose	− 1.2
2.	α-D-Glucose = β-D-Mannose	+10.9
3.	eq-D-Glucose = eq-D-Mannose	+ 9.3
4.	eq-D-Glucose = eq-D-Fructose	+ 9.3
5.	α-D-Mannose = β-D-Mannose	+ 1.9
6.	α-D-Xylose = β-D-Xylose	− 2.2
7.	α-D-Galactose = β-D-Galactose	− 1.3
8.	α-D-Fructopyranose = β-D-Fructofuranose	+12.3
9.	α-Lactose = β-Lactose	− 1.1
10.	α-Maltose = β-Maltose	− 0.5
11.	α-Lactose = α-Maltose	− 5.9
12.	α-Lactose = Sucrose	+16.5

thalpies of some isomerase reactions involving carbohydrates are given in Table 8.

Using the data in Table 8, it is possible to calculate according to Hess' law the enthalpic parameters for other isomerase-catalyzed reactions involving carbohydrates not listed in the table. For instance, by combining reactions (3) and (4) the enthalpy for conversion of eq-D-mannose to eq-D-fructose can be calculated. Similarly by combining reactions (10) and (11) the enthalpy for conversion of α-lactose to β-maltose can be obtained.

2.6 Thermodynamics of Enzymatic Reactions Catalyzed by Ligases

The sixth class of enzymatic reactions are ligase-catalyzed reactions. Ligases catalyze the linkage of two molecules coupled with hydrolysis of a phosphate bond in ATP [Eq. (2)] or some other analogous triphosphate. Participation of ATP or other analogous triphosphates in this class of reactions determines the choice of key thermodynamic values; in this case the enthalpy and free energy of ATP hydrolysis [Eq. (2)]. As secondary key values, it is reasonable to accept the thermodynamic parameters of phosphoryl group transfer from ATP to such cofactors as GDP, UDP etc.

Earlier we gave the mean enthalpies for hydrolysis of various bonds. Using these data and the thermodynamic parameters of ATP hydrolysis we can calculate the enthalpies of various biosynthetic reactions. The results are shown in Table 9.

Table 9 shows that the formation of amide and C-N bonds in urea involves a greater consumption of enthalpy than is provided by hydrolysis of one ATP. In the remaining cases, the bond synthesis coupled with hydrolysis of an ATP proceeds with the liberation of enthalpy.

Table 9. The mean values of enthalpies for biosynthesis of some bonds in metabolic molecules in aqueous solution (pH $= 7.0$, 298.15 K)

No	Reaction	Synthetic bond	$\Delta H_{r;\,mean}$ kJ mol^{-1}
1.	$R_1OH + ATP^{4-} + R_2COOH = ADP^{3-}$ $+ R_1-C-CO-R_2 + HPO_4^{2-} + H^+$	Ester bond	-22
2.	$R_1SH + ATP^{4-} + R_2COOH = ADP^{3-}$ $+ R_1-S-CO-R_2 + HPO_4^{2-} + H^+$	Thioester bond	-16
3.	$R-CO_2^- + ATP^{4-} + NH_4^+ = R-C\begin{smallmatrix}\diagup O \\ \diagdown NH_2\end{smallmatrix}$ $+ ADP^{3-} + HPO_4^{2-} + H^+$	Amide bond	$+5$
4.	$R_1-\underset{\underset{NH_2}{\mid}}{CH}-CO_2^- + ATP^{4-} + R_3-\underset{\underset{NH_3^+}{\mid}}{CH}-CO-NHR_4$ $= R_1-\underset{\underset{NH_2}{\mid}}{CH}-CO-NH-CHR_3-CO-NHR_4 + ADP^{3-}$ $+ HPO_4^{2-} + H^+$	Peptide bond	-13
5.		$\alpha(1,4)$ Glucosidic bond	-16
6.		$(1,6)$ Glucosidic bond	-25
7.	$2NH_4^+ + ATP^{4-} + CO_2$ $= NH_2-CO-NH_2 + ADP^{3-} + HPO_4^{2-} + H^+$	Urea formation	$+35$

3 Conclusion

It is worthwhile noting that the literature data on the thermodynamic and especially thermochemical parameters of enzymatic reactions are often not in agreement. This paper is an attempt to elaborate the principles for ordering and systematizing experimental results. We hope that further studies will permit more precise determining of key thermodynamic values in enzymology which will, in turn, permit more reliable thermodynamic calculations for scientific and industrial purposes.

References

1. Krebs HA, Kornberg HL (1975) Transformations in living matter. Springer, Berlin Heidelberg New York
2. Mahler HR, Cordes EH (1968) Basic biological chemistry. Harper and Row, New York
3. Lehninger AL (1972) Biochemistry. Worth, New York
4. Jones MN (ed) (1979) Biochemical thermodynamics. Elsevier, Amsterdam
5. Shibata I, Tosa T, Sato T (1974) Appl Microbiol 27:878
6. Shibata I (1978) In: Enzyme engineering, vol 4. Plenum, New York, p 335
7. Beresin IV (1982) J Mendeleev All-Union Chem Soc (Russ) 27:62
8. Armstrong GT, Goldberg RN (1977) Proc 5th Biennial Int Codata Conf. Pergamon, Oxford, pp 339–348
9. Prosen EJ, Goldberg RN, Staples BR, Boyd RN, Armstrong GT (1974) In: Kambe H, Garn PD (eds) Thermal analysis: comparative studies on materials. Kadansha and Wiley, Tokyo, pp 253–289
10. Armstrong GT, Goldberg RN (1976) An annotated bibliography of compiled thermodynamic data sources for biochemical and aqueous systems (1930–1975). NBS Spec Publ 454
11. Armstrong GT (1980) Int Conf Chem Thermodynam. Abstracts of Poster. Merseburg, GDR
12. Armstrong GT (1981) Int Symp Biocalorimetry. Abstracts of Poster. Tbilisi, USSR
13. Enzyme Nomenclature (1972) Recommendations of IUB. Elsevier, New York
14. Enzyme nomenclature recommendations (1976) Biochim Biophys Acta 429:1–45
15. Rekharsky MV, Galchenko GL, Egorov AM (1982) 2nd Czechoslov Conf Calorimetry. Extended abstracts, Prague, pp 137–138
16. Glushko PV (ed) (1971) Thermal constants of substances, vol 4. Viniti, Moscow
17. Technical Note 270, NBS Washington
18. Westrum EF Jr (1982) 2nd Czechoslov Conf Calorimetry. Extended abstracts, Prague, p 159
19. Rekharsky MV, Egorov AM, Galchenko GL, Berezin IV (1981) Thermochim Acta 46:89
20. Burton K (1974) Biochem J 143:365
21. Rodkey FL (1959) J Biol Chem 234:188
22. Poe M, Gutfrend H, Estabrook RW (1967) Arch Biochem Biophys 122:204
23. Rekharsky MV (1981) Thesis for candidate of chemical sciences. Thermochemical investigation of some enzymatic reactions involving nicotinamide adenine dinucleotide. Moscow State University
24. Rekharsky MV, Egorov AM, Galchenko GL, Berezin IV (1979) Dokl Akad Nauk SSSR (Russ) 249:1156
25. Rekharsky MV, Egorov AM, Galchenko GL, Berezin IV (1980) Zh Obshch Khim (Russ) 50:2363
26. Fichter A (ed) Advances in biochemical engineering, (1979) vol 12. Immobilized Enzymes II. Springer, Berlin Heidelberg New York
27. Skinner KJ (1975) Chem Eng News 33:23
28. Beavdette MV, Langermann N (1974) Arch Biochem Biophys 161:125
29. Lowe HJ, Clark WM (1956) J Biol Chem 221:983
30. Watt GD, Burns A (1975) Biochem J 152:33
31. Engel OC, Dalziel K (1967) Biochem J 105:691
32. Hanania GIH, Irvine OH, Eaton WA, Geirge PJ (1967) J Physiol Chem 71:2022
33. Donnovan L, Barclay K, Otto K, Jespersen N (1975) Thermochim Acta 11:151
34. Jespersen N (1976) Thermochim Acta 17:23
35. Stinson RA, Holbrook JJ (1973) Biochem J 131:719
36. Subramanian S (1978) Biophys Chem 7:375
37. Brown HD (ed) (1969) Biochemical microcalorimetry. Academic Press, New York
38. Rothman RJ, Kisliuk N, Langermann K (1973) J Biol Chem 248:7845
39. Rosing J, Slater EC (1972) Biochim Biophys Acta 267:275
40. Podolsky R, Morales M (1956) J Biol Chem 218:945
41. Schmid FX, Hinz H-J (1979) Z Physiol Chem 360:1501
42. Katz S (1955) Biochim Biophys Acta 17:226

43. Watt G, Sturtevant JM (1976) Personal communication. In: Fasman GD (ed) Handbook of biochemistry and molecular biology. CRC, Cleveland Ohio, p 121
44. Podolsky R, Sturtevant JM (1955) J Biol Chem 217:603
45. Phillips RC, George SJ, Rutman RJ (1977) J Am Chem Soc 88:2631
46. Hinz H-J, Pollwein P, Schmidt R, Zimmermann F (1981) Arch Biochem Biophys 212:72
47. Jespersen ND, Jordan J (1970) Anal Lett 3:323
48. Monk P, Wadsö I (1968) Acta Chem Scand 22:1842
49. Wadsö I (1968) Acta Chem Scand 22:927
50. Goldberg RN (1976) Biophys Chem 4:215
51. Goldberg RN (1975) Biophys Chem 3:192
52. Redman-Furey NL (1982) Biochemical thermodynamics and analytical enthalpimetry. Thesis in Chemistry (Doctor of Philosophy), Pennsylvania State University
53. Sturtevant JM (1955) J Am Chem Soc 77:255
54. Christensen JJ, Izatt RM (1970) Handbook of metal ligand heats. Dekker, New York
55. Shikama K (1971) Arch Biochem Biophys 147:311
56. Durell J, Sturtevant JM (1957) Biochim Biophys Acta 26:282
57. Durrell J, Rawitscher M, Sturtevant JM (1962) Biochim Biophys Acta 56:552
58. Sturtevant JM (1962) In: Experimental thermochemistry. Academic Press, New York
59. Wadsö I (1962) Acta Chem Scand 16:487
60. Wadsö I (1965) Acta Chem Scand 19:1079
61. Wadsö I (1966) Acta Chem Scand 20:629
62. Takahashi K (1966) Agric Biol Chem 30:629
63. Takahashi K, Hiromi K, Ono S (1965) J Biochem (Tokyo) 58:255
64. Takahashi K, Yoshikawa Y, Hiromi K, Ono S (1965) J Biochem (Tokyo) 58:251
65. Rawitscher M, Wadsö I, Sturtevant JM (1961) J Am Chem Soc 83:3180
66. Sturtevant JM (1955) J Am Chem Soc 77:1495
67. Sturtevant JM (1953) J Am Chem Soc 75:2016
68. Dobry A, Fruton JS, Sturtevant JM (1952) J Biol Chem 195:149
69. Dobry A, Sturtevant JM (1952) J Biol Chem 195:141
70. Kitzinger C, Horecker BL, Weisbach A (1956) 20th Int Congr Physiol Sci, Brussels, p 502
71. Kitzinger C, Hems R (1959) Biochem J 71:395
72. Rekharsky MV, Rumsh LD, Antomov VK, Galchenko GL (1984) Thermochim Acta 81:167
73. Kitzinger C, Benzinger TH (1960) In: Glick D (ed) Heatburst microcalorimetry. Wiley Interscience, New York, p 309
74. Phillips RC, George S, Rutman RJ (1966) J Am Chem Soc 88:263
75. Belaich JP, Sari JC (1969) Proc Natl Acad Sci USA 64:763
76. Sari JC, Ragot M, Belaich JP (1973) Biochim Biophys Acta 305:1

Subject Index